Nanocomposites for Pollution Control

Nanocomposites for Pollution Control

edited by

Chaudhery Mustansar Hussain
Ajay Kumar Mishra

PAN STANFORD PUBLISHING

Published by

Pan Stanford Publishing Pte. Ltd.
Penthouse Level, Suntec Tower 3
8 Temasek Boulevard
Singapore 038988

Email: editorial@panstanford.com
Web: www.panstanford.com

British Library Cataloguing-in-Publication Data
A catalogue record for this book is available from the British Library.

Nanocomposites for Pollution Control

ISBN 978-981-4774-45-1 (Hardcover)
ISBN 978-1-315-14368-2 (eBook)

Contents

Part 2: Carbon Nanomaterial-Based Nanocomposites for Pollution Control

PART 4: BIO-NANOCOMPOSITES FOR POLLUTION CONTROL

PART 5: GREEN AND SUSTAINABLE FUTURE: NANOCOMPOSITES

Preface

The use of nanotechnology in composite technology has become increasingly important in addressing vital global needs in the 21st century for reliable, sustainable, and efficient access to clean energy, water, and environment. Environmental pollution has become a major problem nowadays for society and seriously threatens the existence of terrestrial life, as the natural environment cannot destroy pollutants on its own. Pollution must be taken seriously, as it has a negative effect on the natural elements that are crucial to the existence of life on the earth. Technology at the nanoscale has inspired the progress and use of novel and cost-effective techniques for catalytic degradation, adsorptive removal, and detection of pollutants in the environment. Nanocomposites, which integrate the benefits of both nanomaterials (NMs) and composite matrices, are the new avenues of modern scientific innovation and deep scientific reflection. They have been successfully utilized in pollution control devices and techniques at both research and industrial scales, show great promise toward the next generation of advanced materials, and have received increasing attention of researchers, scientists, and the industry. As a result, the pollution control arena in today's world is invariably linked with research and development initiatives in nanocomposites.

Nanocomposites possess outstanding mechanical properties and compatibility owing to their composite matrix, unique physical and chemical properties caused by unusually large surface-area-to-volume ratios, and high interfacial reactivity. In general, a nanocomposite is fabricated by combining composite materials with nanomaterials, where the incorporation of nanomaterials can provide a resultant material with several unique properties of nanomaterials and also possibly induce new characteristics and functions based on their synergetic effects. Nanocomposites of various compositions and morphologies can provide powerful tools for their pollution control applications. Moreover, freedom to functionalize NMs with various chemical groups can also increase their affinity toward target pollutants, which is very much desirable for selectively extracting target analytes in complex environmental matrices. However, the advanced comprehensive understanding and real-world applications of these nanocomposites in pollution control field are still away. This book summarizes the recent progress of nanocomposites and their properties, fabrication methods, and applications for pollution control, pollutant sensing, and detection at both experimental and theoretical model scales. Special attention has been paid to the approaches that tend to be green and eco-friendly. In the end, the research trends and future prospects are briefly discussed.

The book is divided into several parts. The first part discusses nanocomposites as a new dimension in the field of pollution control devices and techniques. The second part is focuses on carbon nanomaterial–based nanocomposites for pollution control. The third part is about recent developments in membrane separation techniques based on nanocomposites. The fourth part describes new trends such as the use of bionanocomposites in pollution control techniques. The last part discusses the future prospects of nanocomposites for the anticipated green and sustainable environment. The organization of these parts is based on the most recent research, teaching, and practical experience of the editors and the philosophy that pollution control techniques and devices are moving toward their next generation. Highly ranked researchers and scientists in academia and industry from across the world have contributed to the book. The diversity of authors and their disciplinary backgrounds reveals the interdisciplinary emphasis of the book.

This book is of significant interest to environmentalists, scientists, researchers, consultants, regulators, and engineers working on the issues surrounding real-time nanocomposite applications for pollution control, as well as those working in the industry on the commercial-scale exploration of nanocomposites. It is an up-to-date source of knowledge and guidelines for advanced undergraduate and graduate students. Overall, this book is aimed to be a reference for researchers and scientists who are searching for new and advanced materials, techniques, and devices for pollution control applications. We are grateful to all contributors for enriching the book with their distinctive and hard work.

Part 1

Nanocomposites: A New Dimension towards Pollution Control

Chapter 1

Nanocomposites: An Approach towards Pollution Control

Rekha Sharma[a] **and Dinesh Kumar**[b]

[a]*Department of Chemistry, Banasthali University, Rajasthan 304022, India*
[b]*School of Chemical Sciences, Central University of Gujarat, Gandhinagar, Gujarat 382030, India*

dsbchoudhary2002@gmail.com

1.1 Introduction

Drinking water is the basic and essential part of the living world for vital activities. Unfortunately, because of analytical development, industrialization, society, household and farming activities, and other biological and ecological developments, the quality of our water resources is falling continuously (Nemerow et al., 1991; Tchobanoglous et al., 1991; Ali et al., 2004). Thus, water contamination has turned into a grave issue in the current scenario, influencing every single living being, households, fishing, transportation, and other businesses (Franklin, 1991; Droste, 1997; John, 1990). Administrations, researchers, and people

Nanocomposites for Pollution Control
Edited by Chaudhery Mustansar Hussain and Ajay Kumar Mishra

ISBN 978-981-4774-45-1 (Hardcover), 978-1-315-14368-2 (eBook)
www.panstanford.com

are concerned about overcoming this problem. A large number of natural, non-living, and organic toxins are included among water effluents (Laws et al., 2000). Additionally, among these toxicants, a number of toxins are lethal and cause cancer (Ali et al., 2006, 2009; Vettorazzi, 1979). These toxins are hazardous for human beings, aquatic creatures, and the biological system of the earth. Several heavy metal ions are prominent as water toxins having immense contagious and cancer-causing nature (Duffus, 2002). Various species of arsenic such as arsenite, arsenate, monomethyl arsenic acid, and dimethyl arsenic acid (Jain et al., 2000; Yamamura et al., 1980; Kiping, 1977; WHO, 1981; Pershagen, 1983; Csanady et al., 1995; Matschullat, 2000; Ng et al., 2003; Dhillon et al., 2015) are toxic in nature. Additionally, some other metal particles, for example, Cd(II), Cr(III), Hg(II), Se(IV), and Pb(II), have grave toxicities (Marin et al., 2007; Zhang et al., 2005; Ruiz, 2006; Bayramoglu et al., 2007; Sigel et al., 1986). Various diseases, such as vomiting, nausea, asthma, and carcinoma, occur due to the presence of cobalt in drinking water at a higher concentration than permissible. In addition, cobalt also causes thyroid, gastrointestinal, and liver problems (Suhasini et al., 1999; Rengaraj et al., 2002; Smiciklas et al., 2006). Among all heavy metal ions, zinc is essential for the development of individuals; however, its excess causes poor development and mental fever (Vatistas et al., 1999; Weng et al., 2004; Oren et al., 2006). In addition, various anions present in drinking water, such as NO_3^-, SO_4^{2-}, PO_4^{3-}, F^-, Cl^-, and CO_4^{2-}, also have dangerous consequences. For instance, blue baby syndrome (methemoglobinemia) in children results due to the high concentration of nitrate (Kross et al., 1992). Additionally, a higher concentration of fluoride present in drinking water causes fluorosis (WHO, 1993). The taste of drinking water changes owing to the presence of various anions, for example, SO_4^{2-}, PO_4^{3-}, Cl^-, and $C_2O_4^{2-}$. In addition to inorganic toxicants, various groups of organic toxicants have been found, for example, pesticides, fertilizers, hydrocarbons, phenols, plasticizers, biphenyls, detergents, oils, greases, and pharmaceuticals (Meyers, 1999; Damia, 2005). In a variety of books and research articles, the consequences and lethal effects of the above-mentioned pollutants have been discussed (WHO, 2010; Kohler et al., 2006). Various kinds of microorganisms might be responsible for different illnesses. Water contaminants exist in dissolvable, colloidal, or suspended forms, which cause waterborne diseases (Gadd et al., 1978; Gledhill, 1987). The surface water and groundwater contaminated because of these types of pollutants are not suitable for drinking. By 2020, the world population is expected to reach 7.9 billion (Dyson, 1996), and the world might face the serious shortage of water because of the growing population. Thus, the removal of these toxins from polluted water is critical for ensuring disease-free human beings. For the treatment of water, various methods have been developed and used in the past few decades (Weber et al., 1991; Williams, 1991; Zinkus et al., 1998; Ali et al., 2011; Koros, 1995; Lin et al., 1997; Zor et al., 1998; Saleh et al., 2011a, 2011b, 2012; Gupta et al., 2011, 2011, 2012). Various methods have been used for the treatment of water since the ancient times. Adsorption is one of them, owing to its easy process and accessibility and is considered an appropriate water treatment method. It can also be utilized for the removal of dissolved and undissolved natural and organic toxicities. However, adsorption has some limitations. It cannot perform well at the commercial scale owing to the nonavailability of suitable adsorbents of high adsorption point of confinement. Additionally, a single adsorbent cannot be used for a wide range of toxins. For different toxins, different adsorbents are utilized. The cost viability arrangement of various water treatment methods is adsorption > evaporation > aerobic > anaerobic > ion exchange > electrodialysis > micro- and ultra-filtration > reverse osmosis > precipitation > distillation > oxidation > solvent extraction. It has been illustrated that adsorption is a suitable water treatment method for wastewater. Through adsorption batch modes, a lot of work has been carried on for the removal of different contaminants from water. Pollutants have been removed by activated carbon, which has been replaced by cost-effective adsorbents (Ali et al., 2005, 2006). Among all branches of science, nanotechnology has made the greatest impact with its applications in the past couple of decades. Nanosized adsorbents have been utilized for the removal of water contaminants. Various actions have been made by adsorption, to develop nanotechnological utilities using nanoadsorbents to look at varied parts of water treatment.

1.2 Adsorption

The procedure by which analytes are adsorbed on the surface of solid is called adsorption. Adsorption takes place by physical forces and it is a surface phenomenon. However, in this method, weak chemical bonding is also involved frequently (Faust et al., 1983). An adsorbate is a pollutant molecule adhering to the solid surface, and the solid surface is recognized as an adsorbent. Different parameters were used to control the adsorption process, for example, the effect of temperature, inherent character of the adsorbate and the adsorbent, and the existence of different contaminations besides the exploratory conditions (pH, pollutant concentration, contact time, the size of the molecule, and temperature). The equilibrium condition is set by the adsorption of the pollutant concentration on the solid surface. The adsorption isotherm is the isotherm that shows the correlation between adsorbed amounts of contaminants in water. Adsorption can be explained by Langmuir, Freundlich, Halsey, Henderson, Smith, Elovich liquid film diffusion, intraparticle diffusion, and Lagergren models. These models depend approximately on the comparability rule with little distinction in their methodologies (Masel, 1996; Condon, 2006; Barbosa-Caìnovas et al., 1996; Lewinsky, 2007). The values of ΔH, $\Delta_f G°$, ΔS, and ΔE are calculated to study the dynamics of adsorption. The adsorption innovation was created by batch process took after by the column studies about connected at the pilot and mechanical scales by using large size columns.

1.3 Preparation and Characterization of Nanocomposites

For the treatment of wastewater, the development of nanoparticles and their characterization are essential. Various methods have been used for the preparation of nanoparticles, including pulsed laser ablation, inert gas condensation, spark discharge generation, spray pyrolysis, ion sputtering, laser pyrolysis, thermal plasma synthesis, photothermal synthesis, flame synthesis, flame spray pyrolysis, and low temperature reactive synthesis (Brinker et al., 1990; Palkar, 1999; Bessekhouad et al., 2003; Sarkar et al., 2007; Azadmanjiri, 2007). These nanoparticles are characterized using various nanotechnology tools, such as X-ray diffraction pattern, scanning electron microscopy, and transmission electron microscopy. Fundamentally, the described techniques were used to prepare various kinds of nanoparticles for different applications, for instance, restorative, gadgets, visual, mechanical, natural science, and so on. The choice of the method relies on the bulk, assets, nature of the starting material and the use of nanoparticles. The methods that have been used for the preparation of nanoparticles have been summarized in Table 1.1.

1.4 Water Treatment

The particles that offer the potential for the treatment of contaminated water are 1–100 nm in size and are known as nanoparticles (Kiss et al., 1999; Buzea et al., 2007). Due to their unique properties, nanoparticles have been proved as excellent adsorbents and utilized for water treatment. The specific surface area (BET) of $MnFe_2O_4$, $MgFe_2O_4$, $ZnFe_2O_4$, $CuFe_2O_4$, $NiFe_2O_4$, and $CoFe_2O_4$ magnetic nanoparticles is 180.0, 70.3, 79.6, 93.8, 101.2, and 55.1 m^2/g, respectively, which is used in water treatment. For the treatment of wastewater, various nanoparticles were used, such as nanoparticles of Al_2O_3, anatase, akaganeite, CdS, $CoFe_2O_4$, CuO, Au, γ-Fe_2O_3, Fe, Fe_2O_3, $Fe(OH)_2$, NiO, SiO_2, SnO, TiO_2, ZnS, ZnO, and ZrO_2.

1.5 Removal of Inorganic Pollutants

Inorganic pollutants from water can be removed by the use of the adsorption technology. The generally used nanoparticles for metal ion removal are aluminum oxides, iron, and titanium. For effective removal of metal ions, nanoparticles of iron oxide have used because of their high surface area and simple preparation methods (Deliyanni et al., 2004; Huang et al., 2006). Different types of iron oxide, for example, goethite, indistinct ferric oxide, and crystalline ferric oxide have been utilized for the removal of metal particles from water (Huang et al., 2006; Johnson, 1990; Manceau et al., 1994; Venema et al., 1998; Zeng, 2003;

Table 1.1 Different methods for synthesizing nanoparticles used as adsorbents for water treatment

Nanomaterial	Method	Diameter (nm)	Starting Material	BET (m^2/g)	Refs.
Akaganeite	Precipitation	2.6	$FeCl_3$, $(NH_4)_2CO_3$	330	Deliyanni et al. (2005)
Alumina	Sol–gel Hydrolysis	6–30 80	$AlCl_3 . 6H_2O$, $C_9H_{21}O_3Al$, Bis(2-ethylhexyl) sulfosuccinate sodium salt	390.92	Zeng et al. 1998) Park et al. (2005)
Al_2O_3-SiO_2	Sol–gel	30	$SiC_8H_{20}O_4$, C_2H_5OH, NH_4OH, $(CH_3)_2CHOH$, $Al[OCH(CH_3)C_2H_5]_3$	—	Pacheco et al. (2006)
Anatase	Solvothermal	8–20	$C_8H_{24}O_4Ti$, C_2H_5OH	75–195	Gao et al. (2004)
$CoFe_2O_4$	Wet chemical reduction	15–48	$FeCl_3$, $CoCl_2$, NaOH	178	Hu et al. (2005)
Cr-ZnO	Chemical vapor Synthesis	18	$Zn(C_5H_7O_2)_2$, $C_{15}H_{21}CrO_6$	—	Jin et al. (2007)
$CoFe_2O_4$	Combustion wave	2.7–17	$Fe(NO_3)_3$, $Co(NO_3)_2$, $C_2H_5NO_2$	—	Biasi et al. (2007)
CeO_2	Flame electrospray Pyrolysis	2.4–6	• $Ce(NO_3)_3$ • $6H_2O$, C_2H_5OH, $CH_3(CH_2)_3OCH_2CH_2OCH_2CH_2OH$	—	Oh et al. (2007)
CuO	Reverse micelles	5–25	$CuCl_2$, NH_3, Triton-X-100, *n*-hexanol, *n*-pentanol, C_6H_{12}		Han et al. (2008)

Nanomaterial	Method	Diameter (nm)	Starting Material	BET (m^2/g)	Refs.
CdS	Sol–gel	1.66	$Si(OC_2H_5)_4$, C_2H_5OH, HCl, $Cd(CH_3CO_2)_2$, Na_2S	—	Hullavarad et al. (2007)
Au	Sonochemical Photochemical Intramolecular Photoreduction	22 7.5 6.7–50.9	$NaAuCl_4$, chitosan powder, C_3H_8O, NaBr, $NaCl_2H_{25}SO_4$, $AuCl_3$, macromolecular polymer, dendrimers, Au, $C_{10}H_{16}N_2O_8$, $Fe_2(SO_4)_3.6H_2O$	—	Okitsu et al. (2007); Dong et al. (2007); Dong et al. (2007)
Gum Arabic modified magnetic nanoadsorbent	Co-precipitation	13–67	$FeCl_3.6H_2O$, gum Arabic, NH_4OH	—	Banerjee et al. (2007)
Iron-nickel alloy	Reverse micelle	4–12	$FeCl_2$, $NiCl_2$, $NaBH_4$, C_8H1_8, *n*-butanol, $C_{19}H_{42}BrN$	160.21	Ban et al. (2006)
Fe_2O_3	Hydrothermal	14–25	$FeSO_4$, $C_{10}H_{20}O_2$ or $CH_3(CH_2)9NH_2$	—	Takami et al. (2007)
γ-Fe_2O_3	Sol–gel	10	$FeCl_3$, $FeCl_2$, NH_4OH	—	Hu et al. (2005)
TiO_2 NPs	Hydrolysis	6	$Ti_2(SO_4)_{24}$	138	Pena et al. (2005)
NiO_2 NPs	Sol–gel	4–22	$Ni(CH_3COO)_4 \cdot H_2O$, $(COOH)_2$, C_2H_5OH	—	Thota et al. (2007)
SiO_2	Sol–gel	15–700	$Si(OC_2H_5)O_4$	—	Chrusciel et al. (2003)

(Countinued)

Table 1.1 (*Continued*)

Nanomaterial	Method	Diameter (nm)	Starting Material	BET (m^2/g)	Refs.
SnO_2NPs	Precipitation	20–60	$SnCl_2 \cdot 2H_2O$, NH_4OH	19.5	Ibarguen et al. (2007)
TiO_2 NPs	Combustion synthesis	100–1000	titanium powder, $NaClO_4$,	3.05	Kitamura et al. (2007); Figgemeier et al (2007);
	Laser pyrolysis	14	$TiCl_4$		
	Solvothermal	10, 20	$Ti(OR)_4$, toluene		Kim et al. (2007);
	Sol–gel method	5	*n*-butyl-methylimidazolium hexafluorophosphates [BMIM][PF_6]		Zhai et al. (2007)
ZnO	Sonochemical	70–80	$ZnCl_2$, KOH		Kandjani et al. (2008)
ZnS	Sol–gel	2.3–4.5	$SiC_8H_{20}O_4$, C_3H_8O, CH_4N_2S, $Zn(NO_3)_2 \cdot 6H_2O$		Bhattacharjee et al. (2003)
ZnO	Laser ablation	14–20	C_3H_8O, C_3H_6O, Zn metal		Thareja et al. (2007)
ZnO	Thermolysis	18–45	$Zn(NO_3)_2$, $(C_2H_4O)_x$		Patil et al. (2007)
ZrO_2	Precipitation	15	$ZrOCl_2.8H_2O$, $Cl_2H_{18}O_9Zr$, NH_4OH		Guo et al. (2005)

Source: Reprinted from *Chemical Reviews*, vol. 112, Imran Ali, New generation adsorbents for water treatment, pp. 5073–5091, Copyright (2012), with permission from American Chemical Society.

Pokhrel et al., 2007; Fendrof et al., 1997). Oxides of iron have been utilized for the removal of various heavy metals, such as copper (Huang et al., 2006) and arsenic (Dixit et al., 2003). Alternatively, for the adsorption of heavy metals of hazardous nature such as As(III), Cd(II), Cr(II), Se(IV), Pb(II), and Zn(II), nanoparticles of zerovalent iron have additionally been utilized (Ponder et al., 2000; Kanel et al., 2005). Furthermore, alumina nanoparticles have various properties such as high surface area, and excellent thermal stability, which makes them suitable for the commercial scale (Ayuso et al., 2007; Valente et al., 2004). These alumina nanoparticles are used to remove various toxic metal ions such as cadmium, copper, chromium, lead, and mercury. The capacities and characteristics of nanoparticles with respect to the removal of metal particles are discussed further in the text.

Arsenic. Due to arsenic contamination of groundwater, around 150 million individuals all around the world are drinking arsenic-contaminated water (Brammer et al., 2009). Bangladesh is one of the countries that are most affected by arsenic contamination. The hazard of arsenic contamination is higher than ever seen before globally. In Bangladesh, the first instance of arsenic-caused skin injuries was reported in 1987; however, the disaster received worldwide attention in 1995. The groundwater in around 900 villages had a higher concentration of arsenic than the allowable limits (Chatterjee et al., 1995; Das et al., 1995). For the removal of arsenic from groundwater, various treatment plants have been utilized in Bangladesh, but owing to the high cost and handling issues, none is viable at the large scale. The highest removal of arsenic was recorded at 120.0 mg/g using akaganeite (β-FeOOH) nanoparticles (Deliyanni et al., 2003). The adsorption studies fitted to the Langmuir adsorption isotherm and were found to be endothermic in nature. Moreover, As(III) was removed using nanocrystalline hybrid akaganeite nanoparticles with a specific surface area of 231 m^2/g (Deliyanni et al., 2006) from the wastewater. The Freundlich isotherm used to fit the adsorption data and the adsorption took place after the pseudo-second-order reaction. Pena et al. (2005) reported arsenic(III) and arsenic(V) adsorption on TiO_2 nanocomposites. The Freundlich isotherm used to fit the adsorption data and the adsorption took place after the pseudo-second-order reaction. As(III) and As(V) were removed at pH 7.5–8.0. Kanel et al.

(2006) observed that nanoscale iron is a better adsorbent than zerovalent iron. The maximum adsorption was described by the study of various parameters, such as the pH effect, the effect of anions, kinetics of the adsorption process, and the adsorption mechanism. From the examination of the pseudo-first-order reaction (K_{obs} = 0.02–0.71 min^{-1}), it was observed that the adsorption of As(V) increased. The proposed mechanism was inner-sphere surface complexation described by the laser light scattering analysis. Carbonic acid, phosphoric acid, and silicic acid were the interfering anions in the adsorption process. The most important nanoparticles of metal oxides are TiO_2, Fe_2O_3, ZrO_2, and NiO NPs (Hristovski et al., 2007) for the removal of As(V). These nanoparticles have ~98% removal capacity excluding ZrO_2 NPs. To remove arsenic from wastewater, Zhu et al. (2009) synthesized NZVI onto activated carbon. The adsorption capacity of the synthesized nanoparticles was 1.997 mg/g at pH 6.5 in 2.0 mg/L of As(III) solution. The presence of PO_4^{3-} and SiO_4^{4-} ions reduces the removal of arsenic. However, no effect is observed in the case of sulfate, carbonate, and oxalate ions. The adsorbent was an excellent adsorbent for arsenic contaminated water. Additionally, for the removal of arsenate, Zhu et al. (2009a) prepared activated carbon impregnated with NZVI. The results were explained using the intraparticle diffusion model, which showed that the adsorption capacity of 15.4 mg/g was achieved. The arsenate removal decreased in the presence of phosphate and silicate but there was the insignificant effect of other anions and cations. Because of its good removal capacity, the authors declared this process to be appropriate for drinking water treatment. Repeatedly, for the removal of arsenic from drinking water, Zhu et al. (2009) reported activated carbon-supported nanosized zerovalent iron as an effective adsorbent. Langmuir adsorption isotherms were used to determine the adsorption capacity of arsenite and arsenate. As indicated by the author, the removal of both As(III) and As(V) decreased due to the presence of phosphate and silicate, whereas the effect of other anions and humic acid remained negligible. In addition, the adsorption process increased due to the presence of calcium and magnesium but decreased in presence of ferrous ion. Mostafa et al. (2010) described 100% As(V) removal in 5.0 min on a nanosized iron oxide coated with quartz (IOCQ). Chowdhury and Yanful (2010) described 96–99% arsenic removal on magnetite–maghemite nanoparticles under controlled pH conditions in the aqueous solution. The highest adsorption of As(III) and As(V) was 3.69 and 3.71 mg/g, respectively, at pH 2.0 with an initial concentration of 1.5 mg/L of both species. The limitation of arsenic uptake has also been described. Increasing the phosphate anion concentration decreases the uptake of arsenic. Jegadeesan et al. (2010) reported nanosized amorphous and crystalline TiO_2 NPs for the adsorption of arsenic(III) and arsenic(V) species. The solute having a pH range of 4–9 has considerable sorption of As(V) on amorphous TiO_2 nanoadsorbent. The sorption limits of the characteristic TiO_2 polymorphs were subject to the adsorption site thickness and specific surface area (size of the molecule), and the crystalline structure was demonstrated by arsenic(III) and arsenic(V) adsorption isotherms. The surface area of adsorbent remained approximately constant for the particles that have the size in the range of 5–20 nm. However, the surface area of As(V) increases with an increase in the degree of crystallinity.

Cadmium. The removal of cadmium from water using thiol acetic acid-modified TiO_2 NPs has been described (Skubul et al., 2002). Nanocrystalline akaganeite particles (Lazaridis et al., 2005) were also used for cadmium removal. The removal of cadmium from water on anatase nanoparticles (8–20 nm) has been described (Gao et al., 2004) and Langmuir isotherm followed the adsorption. Moreover, the cadmium adsorption from water on akaganeite nanocrystals was calculated (Deliyanni et al., 2005). The adsorption capacity of chemisorption was 17.1 mg/g. On raising the temperature from 25 to 65°C, the adsorption capacity rose from 30% to 90%, and the process was endothermic. Alumina-silica nanoparticles (Pacheco et al., 2006) were used to remove Cd(II) metal ions. The authors reported SiO_2–AlO_3 particles with 96.4% adsorption.

Chromium. Environmental contamination due to chromium metal ions is a serious issue globally. Ponder et al. (2000) reported that by using nanoscale zerovalent iron particles having a diameter of 10–30 nm, hexavalent chromium was removed. The adsorption follows pseudo-first-order kinetics. Shao-Feng et al. (2005) used starch-stabilized Fe NPs as a better adsorbent for the removal of Cr(VI) from the groundwater in contrast with fundamental particles. Moreover, the authors asserted that balanced out Fe NPs were superior to anything Fe powder and agitation. Lazaridis et al. (2005) reported Cr(VI) removal on the rod-like nanocrystalline akaganeite adsorbent of 3–6 nm in diameter at pH 5.5 with 80.0 mg/g removal efficiency. The Freundlich isotherm was followed by the calculated data. Hu et al. (2005) synthesized maghemite (γ-Fe_2O_3, 10 nm diameter) nanoparticles for the removal of Cr(VI). The highest removal of Cr(VI) was 19.2 mg/g at pH 2.5 within 15 min. The process was exothermic. Cr(VI) metal ions in the presence of Ni(II) and Cu(II) were removed from aqueous solutions using maghemite (γ-Fe_2O_3) nanoparticles (Hu et al., 2006). The sol–gel method was used to synthesize these nanoparticles with 198 m^2/g specific surface area and equilibrium time of 1.0 min. The adsorption of copper, chromium, and nickel on maghemite nanoparticles has also been described using some parameters, such as contact time. Hu et al. (2007) prepared various sizes of magnetic nanoparticles for the removal of chromium having 5–60 min equilibrium times. By the compound co-precipitation technique, different magnetic nanoparticles were prepared for the removal of heavy metals, such as manganese ferrite, cobalt ferrite, zinc ferrite, cuprous ferrite, nickel ferrite, and magnesium ferrite. Manganese ferrite is one of the nanoparticles that have the highest adsorption efficiency with 99.5% compared to the above-mentioned magnetic nanoparticles, at pH 2.0. The high adsorption efficiency is because of its highest surface area (180 m^2/g). Guan et al. (2007) adsorbed 0.0217 mM/g Cr(VI) at 3.0 pH on Fe_3O_4/*Sphaerotilus natans*. According to the authors, because of –$CONH_2$– and –NH–, functional groups of Fe_3O_4/*Sphaerotilus natans* Cr(VI) were trapped on the adsorbent by electrostatic attraction. The removal efficiency of Cr(VI) on akaganeite nanocrystals (Deliyanni et al., 2005), carbon nanotubes-supported ceria nanoparticles (Hu et al., 2006), and maghemite nanoparticles (Hu et al., 2007) was 80.00, 30.20, and 19.20 mg/g respectively, which distinctly demonstrates the most elevated adsorption limit of akaganeite nanocrystals.

Cobalt. Cobalt metal ions are responsible for various types of pollutions worldwide. Literature exists on the removal of cobalt using low-toxicity nanoparticles. Magnetic chitosan nanoparticles were reported (Chang et al., 2006) for the removal of cobalt from aqueous solutions. At pH 3.0–7.0, the maximum removal (27.5 mg/g) was accomplished in 1.0 min. Because of low electrostatic association among sorbents and nanoparticles, the procedure was found to be exothermic at a higher temperature. The Langmuir model fitted for the adsorption data. Removal of cobalt was also reported (Uheida et al., 2006) on nanoparticles of Fe_3O_4 and γ-Fe_2O_3 with adsorption capacity of 5.8×10^{-5} and 3.7×10^{-5} Mm^{-2}, respectively. The process is endothermic, and maximum adsorption was achieved within 5.0 min.

Copper. The most common industries are compost waters, electrical, metal completing, paint, electroplating, dyestuff, and wood fabricating industries, which is the main cause of copper pollution (Ajmal et al., 1998; Deng et al., 2003; Sun et al., 2006). Monodispersed Fe_3O_4 NPs incorporated with chitosan were fabricated (Chang et al., 2005b). These Fe_3O_4 NPs were moderately capable for Cu(II) ion removal at pH >2. The greatest adsorption capacity of monodispersed Fe_3O_4 NPs accomplished as 21.5 mg/g with 0.0165 mg/L equilibrium constant took place after the Langmuir model. On expanding pH from 2 to 5, the adsorption limit was expanded because of the impact of pH and temperature. The process of adsorption was exothermic. As a result of the nonexistence of internal diffusion resistance, the authors asserted quick removal within 1 min. Banerjee et al. (2007) synthesized magnetic nanoadsorbent using Fe_3O_4 NPs incorporated with gum Arabic to remove Cu(II) metal ion. Gum Arabic attached to Fe_3O_4 NPs due to the interaction between carboxylic group and the surface hydroxyl group of gum Arabic and Fe_3O_4 NPs, respectively. The nonexistence of internal diffusion resistance required to reach equilibrium within 2.0 min on rising pH of the solution. The maximum adsorption of Cu(II) was obtained due to complexation between the amine group and the surface hydroxyl group of gum Arabic and iron oxide, respectively. The adsorption data fitted to Langmuir model. The nanohydroxyapatite (n-HAp) nanoparticles were synthesized to remove copper(II) ions with chitin and chitosan (Gandhi et al., 2010). The adsorption data fitted to not only Freundlich isotherm but Langmuir. To explain the nature of adsorption, $\Delta G°$, $\Delta H°$, and $\Delta S°$ thermodynamic parameters were calculated. Hao et al. (2010) prepared magnetic nanoadsorbent (MNP-NH_2)

by covalent binding of 1,6-hexadiamine on the surface of Fe_3O_4 NPs for the removal of a copper ion from aqueous solutions. The rate limiting step of adsorption was the pseudo-second-order mechanism. The maximum adsorption capacity was 25.77 mg/g obtained with Langmuir isotherm at pH 6 and 298 K with 98% removal. The FeO(OH) (nanogoethite) and Fe_2O_3 (nanohematite) NPs were developed (Chen et al., 2010) to remove copper(II) ions. The author claimed good adsorption capacity with nanohematite and nanogoethite nanoparticles for copper ions. The maximum adsorption capacity of nanogoethite and nanohematite was 149.25 mg/g and 84.46 mg/g, respectively.

Selenium. The agricultural process, industrial development, for example, petrochemical industry are the main sources of selenium pollution. TiO_2 NPs were synthesized (Zhang et al., 2009) for the adsorption of Se(IV) ions from water. Contact time, pH, and temperature parameters were studied. The Dubinin–Radushkevich (D–R) adsorption isotherm was used to calculate the mean energy of adsorption at room temperature and was found to be 14.46 kJ mol^{-1}. Pan et al. (2010) removed selenite (SeO_3^{2-}) on a hybrid adsorbent (HFO-201). The adsorbent had good adsorption capacity for selenite. Removal of selenite on HFO-201 decreased with increasing pH and temperature. The Freundlich model was used to fit the adsorption data.

Miscellaneous Metal Ions. Generally, a specific nanoparticle is used to remove a specific metal ion. Polymeric hybrid nanoparticles sorbent (ZrPS-001) were developed (Zhang et al., 2008) for the removal of lead, cadmium, and zinc ions from water. The treated sewage or effluents reached the WHO drinking water standard. Afkhami et al. (2010) synthesized 2,4-dinitrophenylhydrazine (DNPH) immobilized on sodium dodecylsulfate-coated nanoalumina for wastewater treatment. From the various metal cations, Pb(II), Cr(III), and Cd(II) ions have the highest adsorption capacity. Various parameters were studied to explain the adsorption process, such as the effect of pH, adsorbent dosage, and contact time. The Langmuir and Freundlich models fitted for the adsorption data. The adsorption data of Mn(II), Pb(II), Cr(III), and Cd(II) metal cations best fitted to the Freundlich model and adsorption data of Ni(II) and Co(II) fitted to the Langmuir model. The adsorption capacity of zinc ion was reported on akaganeite NPs (Deliyanni et al., 2007). The maximum adsorption was found at pH 7.5 with 95% efficiency, and the process was exothermic.

Anions. Anions cause some disease symptoms and sicknesses in their higher concentration of metal particles. Some anions have serious side effects as already discussed. The significant wellsprings of water pollution by anions are residential and farming exercises. Different products are utilized as a part of household activities which discharge anions to our water assets. Furthermore, the waste products from industrial and municipal customers are also responsible for water pollution. In the following paragraphs, we discuss nanoparticles that have been used for the removal of anions from polluted water. Sairam et al. (2008) discussed fluoride removal on nanohydroxyapatite (n-HAp) particles with adsorption capacity of 1845.0 mg/kg. The mechanisms of removal were explained by ion exchange and adsorption processes. The adsorption process was characterized by various thermodynamic parameters, such as the effect of pH, contact time, adsorbent dose, etc. In addition, nanohydroxyapatite/chitin (n-HApCh) composites (Sairam et al., 2009) were synthesized for the removal of fluoride. For the adsorption process, effects of various parameters were studied, such as the pH effect, the effect of anions, and contact time. Compared with nanohydroxyapatite (n-HAp) nanocomposites, n-HApCh composite have a higher defluoridation capacity (DC) at 2840.0 mg/kg. Various nanomaterials have been utilized for the removal of fluoride from wastewater such as cubical ceria nanoadsorbents (Dhillon et al., 2016), Fe–Ca–Zr hybrid metal oxide nanomaterial (Dhillon et al., 2015a), crystalline and hybrid Fe–Ce–Ni nanoporous adsorbent (Dhillon et al., 2015), Fe–Ca–Ce oxide material (Dhillon et al., 2015). The author depicted adsorption isotherm, kinetics, and mechanism of adsorption. With increasing iron dosage, the removal efficiency increased, and the initial concentration and pH decreased. At an initial concentration of anions (less than 100 mg/L), the removal efficiency was 99.9%. The maximum adsorption was observed at 25°C of 19.17 mg/g. The Langmuir and Freundlich models fitted for the adsorption data. The pseudo-second order equation was followed by the sulfide adsorption. Magnetic nanoadsorbent prepared by covalent binding of poly (acrylic acid) (PAA) on the surface of Fe_3O_4 NPs (Huang et al., 2009) followed by amino functionalization for the removal of some toxic anions. The use of diethylenetriamine (DETA) using carbodiimide actuation describes the adsorption process. The prepared nanoadsorbent was used to remove various anions from water effectively.

1.6 Removal of Organic Pollutants

For the removal of normal and halogenated hydrocarbons, effective nanoparticles are aluminum, iron, titanium oxides, and native iron metal. The efficiency to remove these pollutants from wastewater is discussed below.

Dyes. Textile, paper, rubber, plastics, paints, printing, and leather industries are responsible for dye pollution. Here, the removal of dyes on nanoadsorbent is discussed. To remove reactive red 195 azodyes, TiO_2 NPs were prepared (Belessi et al., 2009). The pseudo-second-order rate kinetics was followed by the equilibrium data and fitted to Langmuir isotherms. The highest adsorption capacity at pH 3.0 and 30°C was 87.0 mg/g. The kinetic studies indicate that the dye uptake was fast for first 30 min with equilibrium at 1 h. MgO NPs of 38–44 nm size with surface area 153.7 m^2/g were synthesized (Moussavi et al., 2009) to remove azo and anthraquinone reactive dyes. The percentage removal of blue 19 and reactive red 198 was about 98% at 0.2 g dose, pH 8.0, 50–300 mg/L, and 5.0 min contact time. The maximum predicted adsorption capacities for RB 19 and RR 198 were 166.7 and 123.5 mg/g, respectively. The azo dyes have been removed using templated cross-linked chitosan, ECH-RB5, and ECH-3R NPs (Chen et al., 2011). The adsorption was controlled by controlling some parameters, such as initial pH, dye concentration, and temperature. The adsorption was spontaneous and exothermic in nature. The D-R model used to calculate the energy of activation indicated chemisorption. Two acid dyes orange G and acid green were removed using carboxymethylated chitosan conjugated Fe_3O_4 NPs (Chang et al., 2005a). The adsorption capacity simultaneously decreased on increasing the pH and ionic strength. The adsorption capacities for orange G and acid green dyes were 1883 and 1471 mg/g, respectively. Cheung et al. (2007) synthesized chitosan nanoparticle for the adsorption of acid dyes. The adsorption data fitted well to the Langmuir isotherms. The adsorption mechanism was mainly intraparticle dispersion. Chitosan/

montmorillonite (CTS/MMT) nanocomposites were synthesized for the removal of Congo red dye (Wang et al., 2007). Various effects were studied in the adsorption process, such as the initial content of adsorbent, pH, and temperature of CTS/MMT. The outcomes demonstrated that the adsorption limit of CTS/MMT nanocomposite was higher than the mean estimations of CTS and MMT. The adsorption data followed pseudo-second-order equation was well fitted by the Langmuir isotherms. Chitosan nanocomposites were prepared (Du et al., 2008) for the adsorption of eosin Y dye. The adsorption capacity of chitosan nanocomposites was 3.333 mg/g. The adsorption process was endothermic in nature, and the change in enthalpy was 16.7 kJ/mol. The adsorption of methylene blue dyes on the silica nanosheets was reported (Zhao et al., 2008), which resulted from vermiculite, through acid liberation. The adsorption was increased by increasing dye concentration and decreased by increasing pH and temperature. The activation energies were 3.42, 65.95, and 0.984 kJ/mol and −0.222 kJ/Kmol for ΔG, ΔH, and ΔS adsorption, respectively at 20°C. For methylene blue removal from aqueous solution, Sandoval et al. (2011) examined the effects of the in situ formation of ZrO_2 NPs on properties of granulated activated carbon (GAC). Iram et al. (2010) prepared Fe_3O_4 hollow nanospheres for the removal of dye contaminant. The Fe_3O_4 hollow nanospheres have maximum adsorption due to the function of initial dye concentration, pH, and contact time. The Langmuir and Freundlich adsorption isotherms fitted for the adsorption data. The magnetic hollow spheres (0.05 g) have the monolayer adsorption capacity about 105 mg/g. The adsorption process was endothermic in nature and spontaneous. The above-mentioned magnetic properties of Fe_3O_4 nanospheres are useful for dealing with environmental contamination.

Pesticides. The pesticide pollution has been attributed to industries, agriculture, and forestry through the air. The water bodies, sediments, and soil through rainwater have been contaminated due to the adsorption of the dust particles in the air containing pesticides, which are released from the pesticides spray. However, little work has been done for the removal of pesticides using nanoparticles. Li et al. (2008) reported the removal of pesticides through the adsorption of atrazine onto nano-SiO_2 and nanokaolin particles. Various parameters were studied for the adsorption process, such as effects of ionic strength, atrazine concentration, and pH. With increasing ionic strength, a decrease is observed in the adsorption of atrazine onto nano-SiO_2 and nano kaolin particles. The Freundlich model fits the adsorption data. The increment in dose from 5.0 to 20.0 g/L of nano kaolin the value of K_f was decreased from 71.55 to 37.22. The dominant factor for the adsorption phenomenon is the speciation change of atrazine.

Hydrocarbons. Cleaning arrangements, paint, spot remover, rubber cement, solvents, and different commercial ventures are the fundamental wellsprings of hydrocarbons with a remarkable contribution from the petrochemical business. Adsorption of phenanthrene was reported (Yang et al., 2009) on nano-TiO_2 and ZnO particles. Various experimental conditions were varied for the maximum adsorption. Moreover, adsorption of phenanthrene was also described on nano-Al_2O_3 particles coated with humic acid.

1.7 Removal of Biological Pollutants

Domestic activities are the key source of biological pollution. Consistently, water pollution by biological sources is also attributed to agriculture processes.

Viruses and Bacteria. For human beings, the most infamous pathogens are virus and bacteria, which are responsible for different types of diseases. Sometimes, these pathogens are deadly and hazardous. Generally, these pathogens are responsible for water pollution. The adsorption of the virus on α-Fe_2O_3 nanoparticles was reported (Shen et al., 2010) with 100% adsorption efficiency. When the initial concentration of virus decreased, the percentage of adsorption increased. The adsorption of virus decreased due to the presence of anions such as HPO_4^{2-} and HCO_3^-. Nanoscale zerovalent iron (NZVI) was adapted (Li et al., 2010) as a bactericidal for Escherichia coliform (*E. Coli*) groundwater. These nanoscale zerovalent nanoparticles were tested and exposed to their bactericidal properties for various molecules. The exposure of nanomaterials to bacterial activities a 2.2-log inactivation resulted after 10 min due to 100 mg/L of bare NZVI with 28% Fe NPs content and 5.2-log inactivation after 60 min. Less than 0.2-log inactivation and 1.8-log inactivation observed for NZVI with 7.0% Fe NPs due to adsorbed poly-

(styrene sulfonate) (PSS), poly(aspartate) (PAP), or NOM on NZVI with the same Fe NPs content significantly decreased its toxicity. Some significant pollutants were removed by the use of different adsorbents from aqueous solutions, as summarized in Table 1.2.

Applications of Nanotubes. Carbon nanotubes (CNTs) are an extraordinary kind of nanosized structures that have been utilized as adsorbents for water treatment. They show more efficiency for various pollutants. Thus, in this section, nanotubes have been discussed for the removal of metal ions such as single-walled carbon nanotubes (SWCNTs) and multiwalled carbon nanotubes (MWCNTs). Different metal ion removals from water are discussed here using nanotubes. Li et al. (2003) claimed CNTs oxidized with H_2O_2, $KMnO_4$, and HNO_3 to Cd(II). The ion exchange capacity of CNTs has increased due to the presence of these groups (Singh et al., 1998). The experimental results proposed that for three types of oxidized CNTs, Cd(II) adsorption limits expanded because of the functional groups presented by oxidation. CNTs showed the adsorption capacity at 1.1 mg/g, which reached 2.6, 5.1, and 11.0 mg/g for H_2O_2, $KMnO_4$, and HNO_3 oxidized CNTs, respectively. Accordingly, the adsorption was increased for $KMnO_4$ and HNO_3-treated CNTs strongly pH dependent. Li et al. (2005) also studied the adsorption of Pb(II) on CNTs. MWCNTs were prepared to remove Cd(II), Pb(II), and Cu(II) ions from aqueous solutions (Kochkar et al., 2009). The Pb(II), Cu(II) and Cd(II) metal ions showed the adsorption capacities at 97.08, 24.49, and 10.86 mg/g, respectively. Nickel adsorption on SWCNTs due to Stone–Wales imperfections was studied. On increasing, carbon-carbon bonds connected with Stone–Wales imperfections were more receptive than a perfect hexagon. Lu et al. (2006) used both types of CNTs for the removal of zinc metal ion from water. The removal of zinc increased when pH increases from 1.0 to 8.0 and equilibrium was obtained within 60 min. The SWCNTs and MWCNTs as adsorbent showed a good adsorption capacity for Zn(II). Aligned CNTs for the removal of chromium from drinking water were developed (Di et al., 2006). Activated carbon has 1.5 times lower adsorption capacity than that of the reported adsorption capacity at pH 7.0, which was 30.2 mg/g. CNTs and titanate nanotubes were used to adsorb copper metal ion, synthesized (Doong et al. (2008)). The adsorption capacity for copper and lead metal ions was 83–124 mg/g and

192–588 mg/g, respectively. The data fitted well to the Langmuir model. The CNTs of 5–10 nm size and 40–600 m^2/g surface area were synthesized (Stafiej et al., 2007). The acidified MWCNTs to adsorb Pb(II), Ag, Cu(II), and Co(II) ions were synthesized (Wang et al., 2007). The equilibrium was obtained at 5.0–20.0 min for the adsorption process. The HNO_3 and NaOCl-modified carbon nanotubes for the removal of copper(II) were described (Wu et al., 2007). Cu(II) metal ions showed an increment in adsorption capacities in the basic range. SWCNTs were used to remove dissolved organic matter (DOM) in raw water synthesized (Lou et al. (2011)). The controlling parameters for the adsorption were pH, ionic strength, and temperature. The adsorption capacity of SWCNTs increased when increased the concentration of DOM, and enhancement in ionic strength decreased the adsorption capacity. After a time interval of 120 min, equilibrium was obtained. The pore diffusion method was used to control the adsorption process, which was confirmed by intraparticle diffusion model. The maximum adsorption capacities of SWCNTs at 45°C were 26.1–20.8 mg/g. The MWCNTs used to remove reactive red M-2BE textile dye was detailed (Machado et al., 2011). The acidic pH (pH 2.0) was favorable for the adsorption of the dye. The equilibrium was obtained within 1 h. The adsorption data fitted well to the Avrami fractional-order kinetic model. Shim and co-workers (2010) reported t-MWCNTs for the removal of benzene, toluene, and m-xylene. The experimental adsorption isotherm data correlated with temperature-dependent to the isotherm model. The Van't Hoff equation was used to calculate the isosteric heat of adsorption. The adsorption capacity of some CNTs, such as oxidized H_2O_2, oxidized HNO_3, oxidized $KMnO_4$, oxidized NaOCl, obtained as 1.10, 2.60, 5.10, 11.0, and 47.40 mg/g having 82.20, 130.0, 84.30, 128.0, and 94.90 m^2/g surface area, respectively, used for the removal of metal ions which, demonstrated adequate adsorption capacities of carbon nanotubes for various pollutants. For the treatment of water by the adsorption method, CNTs have certain limitations. The coagulation of CNTs depends on the water quality and it occurs due to the presence of organic matter and algae. The coagulation of CNTs was also controlled by the doses of CNTs. Consequently, these are the limitations of CNTs in water treatment.

Table 1.2 Applications of nanoparticles for the removal of water pollutants by adsorption

Pollutant	Adsorbents (nanoparticles)	Removal capacities	BET (m^2/g)	pH	Refs
Arsenic(III)	ZVI	3.5 mg/g	24.4	7	Kanel et al. (2005)
Arsenic (III)	Akaganeite		231	7	Deliyanni et al. (2006)
Arsenic(III)	ZVI onto activated carbon	1.997 mg/g		6.5	Zhu et al. (2009)
Arsenic (III)	ZrO_2	1.85 mM/g		11.0	Zheng et al. (2012)
Arsenic(V) and Arsenic(III)	TiO_2	Arsenic(V) 95 μM/g and Arsenic(III) 50 μM/g	138	< 8 for As(V), > 7.5 for As(III)	Pena et al. (2005)
Arsenic(V) and Arsenic(III)	Fe_3O_4			4.8–8.0	Mayo et al. (2007)
Arsenic(V)	Akaganeite	120 mg/g	330	7.5	Deliyanni et al. (2003)
Arsenic(V)	ZVI		25	3–7	Kanel et al. (2006)
Arsenic(V)	alumina silicate treated with Fe(II)	22.5 mg/g for zeolite, <18 mg/g for clinoptilolite, and 10 mg/g for metakaoline	4.7 for zeolite, 19.5 for Clinoptilolite, 20.9 for metakaoline	5.5	Dousova et al. (2006)

(*Continued*)

Table 1.2 (*Continued*)

Pollutant	Adsorbents (nanoparticles)	Removal capacities	BET (m^2/g)	pH	Refs
Arsenic(V)	Metal-oxide				Hristovski et al. (2007)
Cadmium(II)	$C_3H_6O_2S$ modified TiO_2			pH < 5	Skubul et al. (2002)
Cadmium(II)	TiO_2	244.13 μM/g for RHT 47, and 497.90 μM/g for RHT 69	195.3	6.1	Gao et al. (2004)
Cadmium(II)	Akaganeite	17.1 mg/g	330	8	Deliyanni et al. (2005)
Cadmium(II)	Al_2O_3–SiO_2			6.5	Pacheco et al. (2006)
Cobalt(II)	Nanoparticles of Fe_3O_4 and γ-Fe_2O_3			5 for Fe_3O_4, 6 for γ- Fe_2O_3	Ajmal et al. (1998)
Cobalt(II)	magnetic chitosan	27.5 mg/g		5.5	Chang et al. (2006)
Chromium(VI)	Fe			7	Pacheco et al. (2006)
Chromium(VI)	γ-Fe_2O_3			2.5	Di et al. (2006)

Pollutant	Adsorbents (nanoparticles)	Removal capacities	BET (m^2/g)	pH	Refs
Chromium (VI)	Akaganeite	80 mg/g		5.5	Skubul et al. (2002)
Chromium(VI)	CNTs supported CeO_2	30.2 mg/g		3.0–7.4	Di et al. (2006)
Copper(II)	Gum Arabic modified magnetic	38.5 mg/g		2.0–5.1	Banerjee et al. (2007)
Copper (II)	chitosan-bound magnetite	21.5 mg/g		5	Chang et al. (2005b)
Chromium(III)	N_2-doped magnetic carbon			8	Shin et al. (2011)
Mercury(II)	Al_2O_3			6.5	Pacheco et al. (2006)
Mercury(II)	Fe_3O_4	125.0 mg/g	94.65	5.0	Song et al. (2011)
Mercury(II)	SH-Fe_3O_4-NMPs magnetic polymers (SH-Fe_3O_4-NMPs)			3.0	Fan et al. (2011)
Selenium(II)	Titanium dioxide			2–6	Zhang et al. (2009)
Zinc(II)	Akaganeite	27.61 mg/g	330	6.5	Deliyanni et al. (2007)
Zinc(II)	Fe_3O_4 modified chitosan	32.16 mg/g			Shengdong et al. (2012)

(*Continued*)

Table 1.2 (*Continued*)

Pollutant	Adsorbents (nanoparticles)	Removal capacities	BET (m^2/g)	pH	Refs
Chromium(VI), Copper(II) and Nickel (II)	γ-Fe_2O_3			2.5 for Cu(II), 6.5 for Cu(II), and 8.5 for Ni(II)	Hu et al. (2006)
Chromium(VI) and Lead(II)	ZVI		24.4	6–7	Ponder et al. (2000)
Red 195 azo	TiO_2	87.0 mg/g	155		Belessi et al. (2009)
Azo and anthraquinone reactive	MgO	98%	103.5	pH 8.0	Moussavi et al. (2009)
Orange G and acid green	Fe_3O_4	1883 and 1471 mg/g for orange G and acid green dyes			Chang et al. (2005a)
Acidic metanil yellow and reactive blue 15	Cross-linked chitosan	1334 mg/g for metanil yellow and 722 mg/g of dye reactive blue 15		4.0	Chiou et al. (2006)

Pollutant	Adsorbents (nanoparticles)	Removal capacities	BET (m^2/g)	pH	Refs
Eosin Y	Chitosan	3.333 mg/g			Du et al. (2008)
methylene blue	ZrO_2	0.5 mg/g	615 ± 2	7–9	Sandoval et al. (2011)
methylene blue	Fe_3O_4 coated with γ-glutamic acid	78.67 mg/g		1.0	Stephen et al. (2011)
Methylene blue, neutral red and methyl orange	Fe_3O_4	1–2 mg/g		6–7	Wu et al. (2011)
Methylene blue (MB) and cresol red (CR)	Fe_3O_4	6–35 mg/g		6–8	Zhang et al. (2011)
methylene blue and Congo red (CR)	Fe_3O_4	70.4 mg/g and 172.4 mg/g for MB and CR			Giri et al. (2011)
Methylene blue (MB)	$FeTiO_3$	71.9 mg/g			Chen (2011)
Congo red	Zinc-Ferrite	16.58 mg/g		6.0	Rahimi et al. (2011)

(*Continued*)

Table 1.2 (*Continued*)

Pollutant	Adsorbents (nanoparticles)	Removal capacities	BET (m^2/g)	pH	Refs
Rhodamine 6 G (Rh6G)	hexadecyl functionalized magnetic silica	35.6 mg/G		11.0	Yan-Ping et al. (2011)
Malachite green (MG)	halloysite nanotubes	99.6 mg/g		9.5	Kiani et al. (2011)
reactive red-120 and 4-(2-pyridylazo) Resorcinol	Fe_3O_4	166.67 and 49.26 mg/g for both dyes		2.5	Absalan et al. (2011)
N719 dye	TiO_2	65.2 (30°C), 68.2 (40°C), and 76.6 (50°C) mg/g	101		Fan et al. (2011)
acid orange 7(AO-7) and acid orange 10 (AO-10)	$C_2H_8N_2$-modified magnetic Chitosan	3.47 mM/g for AO7 and 2.25 mM/g for AO10		10.0	Zhou et al. (2011)
Phenol	Fe_3O_4 poly(methylmethacrylate-codivinylbenzene)		83.46	6–7	Tai et al. (2011)

Note: MB = Methylene Blue, CR = Congo Red.

Source: Reprinted from *Chemical Reviews*, vol. 112, Imran Ali, New generation adsorbents for water treatment, pp. 5073–5091, Copyright (2012), with permission from American Chemical Society.

1.8 Future Perspectives of Nanocomposites

In the near future, nanocomposites are a vital problem and have the novel individuality to the environment. Various nanocomposites are non-biodegradable in nature and enter the human body (Gatti et al., 2005; Hoet et al., 2004). Nanocomposites are sometimes toxic in nature and not safe for human beings. Water, soil, and air are the main sources by which nano-composites can enter the human body and other animals. These nanocomposites can accumulate in the edible parts of plants by absorption from water and soil. Thus, it is vital to develop safe and suitable strategies for the disposal of these nanocomposites material. The adverse effects of nanoparticles on the environment have been studied in a research project funded by the U.S. Environmental Protection Agency (Mnyusiwalla et al., 2003) in 2003. Eco-friendly nanoparticles were synthesized for the environmental sustainability and biodegradability of these nanoparticles. The improvement can be made by an alteration in the size and shape of nanoparticles. For the synthesis of suitable nanoparticles, the formulation steps, sterilization, and storage have to be optimized. These materials show many characteristic properties for the production of clean water, which can be helpful for scientists. Additionally, an important issue is the formation of more toxic products from these pollutants. A comparison with low-cost adsorbents to determine the future perspectives was carried out. These nanoparticles do not have paper portraying pilot and business scale uses. Despite this, the reuse of nanoparticles is still in the improvement stage. Furthermore, the management of the unusable nanoparticle is not totally created. Along these lines, it is unrealistic to compare the working capacities of nanoparticles and ease adsorbents. In any case, efforts were made to compare nanoparticles with conventional adsorbents. These inexpensive adsorbents were used to eliminate natural and inorganic contaminants from water. Due to varied experimental conditions for the elimination of the same contaminant, the exact comparison is not possible, used in the case of low-cost adsorbents and nanoparticles. Notwithstanding, endeavors were made to think about these two sorts of adsorbents for an assortment of contaminations. For the elimination of a wide range of contaminations, i.e., inorganic, natural, and organic, economical adsorbents have been utilized. On the other hand, for the removal of a few contaminants, nanoparticles have been used. Apparently, nanoparticles are less economical than adsorbents but by enhancing their arrangement, they can be modified for the modern scale. Due to the good working capacities of nanoparticles, they appeared to be better adsorbents. Nanoparticles show fast adsorption capacity compared with the ordinary adsorbent, which is confirmed by low contact time (1.0–15.0 min). Additionally, nanoparticles require a lower dosage regimen (in μg/L) than ordinary adsorbents, due to higher adsorption limits of nanoparticles. For the treatment of water, nanoparticles have been utilized under fluctuated states of pH (3–9). Normally, for the elimination of contaminants from water, low-cost adsorbents are feasible in their permissible limits (milligram levels). Besides, the permissible limit of nanoparticles is microgram level for the removal of contaminants. The permissible limit is 10.0 μg/L of arsenic by the WHO. Different contaminations, such as Cd(II), Hg(II), Lead(II), Se(IV), pesticides, and steroids can be eliminated by nanoparticles, due to their permitted ppb level limits. Concisely, nanoparticles may be considered as the need of the future in water treatment, in spite of some complications. These particles ought to be set up in an eco-accommodating way and utilized as a part of a controlled approach to maintaining a strategic distance from any ecological harm. By utilizing column operations, newly produced nanoparticles are needed to create, improve, and apply for water treatment at pilot and business levels. Because of poor water quality, nanoparticles are important for water treatment globally (Ali et al., 2006).

1.9 Conclusion

In this chapter, a basic assessment of nanomaterials as adsorbents demonstrates that nanoparticles have been utilized for the removal of metal particles, anions, and natural and organic species from water effectively. These contaminations can be removed by nanoparticles under fluctuated states of pH and temperature at low concentration. The very low dose of nanoparticles is responsible for the economical application of these nanoparticles. Besides, the removal time is 1.0 to 15.0 min, which is very quick. Due to these properties, nanoparticles are perfect for quick and cheap water treatment innovation. The adsorption of contaminants at the lab, pilot, and business scale columns is vitally required. For a complete treatment of water, other technologies also opt for nanoparticle formation. For the treatment of water on nanoparticles at a large scale, research should be promoted. The improvement of novel nanomaterials is still required with expanded affinity, limit, selectivity, and the capacity to work at column operations. Nanoadsorption innovation might be used for the welfare of individuals because the batch mode adsorption conditions ought to be exchanged to column operations. Additionally, the administration of the recuperated toxins and the depleted nanoparticles has not been tended to in the literature. To stay away from hazards of the recouped toxins and nanotoxicology of utilized nanoparticles, researchers ought to build up some eco-accommodating waste administration techniques. Nanoparticles play a wide role in the treatment of water. Hence, for the rapid, efficient, and achievable water treatment innovation it needs collaborative research at large scale.

Acknowledgment

We gratefully acknowledge support from the Ministry of Science and Technology and the Department of Science and Technology, Government of India, under the scheme of Establishment of Women Technology Park, for providing the necessary financial support to carry out this study vide letter No, F. No SEED/ WTP/063/2014.

References

Absalan G, Asadi M, Kamran S, Sheikhian L, Goltz DM (2011). Removal of reactive red-120 and 4-(2-pyridylazo) resorcinol from aqueous samples by Fe_3O_4 magnetic nanoparticles using ionic liquid as modifier, *J Hazard Mater*; **192**:476–484.

Afkhami A, Saber-Tehrani M, Bagheri H (2010). Simultaneous removal of heavy-metal ions in wastewater samples using nano-alumina modified with 2,4-dinitrophenylhydrazine, *J Hazard Mater*; **181**: 836–844.

Ajmal M, Khan AH, Ahmad S, Ahmad A (1998). Role of sawdust in the removal of copper(II) from industrial wastes, *Water Res*; **32**: 3085–3091.

Ali I, Aboul-Enein HY (2004). *Chiral Pollutants: Distribution, Toxicity and Analysis by Chromatography and Capillary Electrophoresis*, John Wiley & Sons: Chichester; UK.

Ali I, Aboul-Enein HY (2006). *Instrumental Methods in Metal Ions Speciation: Chromatography, Capillary Electrophoresis and Electrochemistry*, Taylor & Francis Ltd.: New York.

Ali I, Aboul-Enein HY, Gupta VK (2009). *Nano Chromatography and Capillary Electrophoresis: Pharmaceutical and Environmental Analyses*, John Wiley & Sons: Hoboken, NJ.

Ali I, Khan TA, Asim M (2011). Removal of arsenic from water by electrocoagulation and electrodialysis techniques, *Sep Purif Rev*; **40**: 25–42.

Ali I, Gupta VK (2006). *Adsorbents for Water Treatment: Development of Low-Cost Alternatives to Carbon. Encyclopedia of Surface and Colloid Science*, 2nd ed, Taylor & Francis: New York.

Ali I, Jain CK (1998). Groundwater contamination and health hazards by some of the most commonly used pesticides, *Curr Sci*; **75**: 1011–1014.

Ali I, Jain CK (2005). Wastewater treatment and recycling technologies. In *Water Encyclopedia: Domestic, Municipal, and Industrial Water Supply and Waste Disposal*, Lehr, J, ed, John Wiley & Sons: New York.

Ayuso EA, Sanchez AG, Querol X (2007). Adsorption of Cr(VI) from synthetic solutions and electroplating wastewaters on amorphous aluminium oxide, *J Hazard Mater*; **142**: 191–198.

Azadmanjiri JJ (2007). Preparation of Mn–Zn ferrite nanoparticles from chemical sol–gel combustion method and the magnetic properties after sintering, *Non-Cryst Solids*; **353**: 4170–4173.

Ban I, Drofenik M, Makovec D (2006). The synthesis of iron–nickel alloy nanoparticles using a reverse micelle technique, *J Magn Magn Mater*; **307**: 250–256.

Banerjee SS, Chen DH (2007). Fast removal of copper ions by gum arabic modified magnetic nano-adsorbent, *J Hazard Mater*; **147**: 792–799.

Barbosa-Caìnovas GV, Vega-Mercado H (1996). *Dehydration of Foods*, Chapman & Hall: New York.

Bayramoglu G, Arica MY (2007). Kinetics of mercury ions removal from synthetic aqueous solutions using by novel magnetic p(GMA-MMA-EGDMA) beads, *J Hazard Mater*; **144**: 449–457.

Belessi V, Romanos G, Boukos N, Lambropoulou D, Trapalis C (2009). Removal of Reactive Red 195 from aqueous solutions by adsorption on the surface of TiO_2 nanoparticles, *J Hazard Mater*; **170**: 836–844.

Bessekhouad Y, Robert D, Weber JV (2003). Preparation of TiO_2 nanoparticles by Sol-Gel route, *Int J Photoenergy*; **5**: 153–158.

Bhattacharjee B, Ganguli D, Chaudhuri S, Pal AK (2003). Synthesis and optical characterization of sol–gel derived zinc sulphide nanoparticles confined in amorphous silica thin films, *Mater Chem Phys*; **78**: 372–379.

Biasi RSD, Figueiredo ABS, Fernandes AAR, Larica C (2007). Synthesis of cobalt ferrite nanoparticles using combustion waves, *Solid State Commun*; **144**: 15–17.

Binns C (2010). *Introduction to Nanoscience and Nanotechnology*, Wiley & Sons: New York.

Brammer H, Ravenscroft P (2009). Arsenic in groundwater: A threat to sustainable agriculture in South and South-east Asia, *Environ Int*; **35**: 647–654.

Brinker CJ, Scherer WJ (1990). *Sol-Gel Science; the Physics and Chemistry of Sol-Gel Processing*, Academic Press: New York.

Buzea C, Pacheco I, Robbie K (2007). Nanomaterials and nanoparticles: Sources and toxicity, *Biointerphases*, **2**: MR17-71.

Chang YC, Chang SW, Chen DH (2006). Magnetic chitosan nanoparticles: Studies on chitosan binding and adsorption of Co(II) ions, *React Funct Polym*; **66**: 335–341.

Chang YC, Chen DH (2005a). Adsorption kinetics and thermodynamics of acid dyes on a carboxymethylated chitosan-conjugated magnetic nano-adsorbent, *Macromol Biosci*; **5**: 254–261.

Chang YC, Chen DH (2005b). Preparation and adsorption properties of monodisperse chitosan-bound Fe_3O_4 magnetic nanoparticles for removal of Cu(II) ions, *J Colloid Interface Sci*; **283**: 446–451.

Chatterjee A, Das D, Mandal BK, Chowdhury TR, Samanta G, Chakraborti D (1995). Arsenic in ground water in six districts of West Bengal, India: The biggest arsenic calamity in the world. Part I. Arsenic species in drinking water and urine of the affected people, *Analyst*; **120**: 643–650.

Chen YH (2011). Synthesis, characterization and dye adsorption of ilmenite nanoparticles, *J Non-Cryst Solids*; **357**:136–139.

Chen CY, Chang JC, Chen AH (2011). Competitive biosorption of azo dyes from aqueous solution on the templated crosslinked-chitosan nanoparticles, *J Hazard Mater*; **185**: 430–441.

Chen YH, Li FA (2010). Kinetic study on removal of copper(II) using goethite and hematite nano-photocatalysts, *J Colloid Interface Sci*; **347**: 277–281.

Cheung WH, Szeto YS, McKay G (2007). Intraparticle diffusion processes during acid dye adsorption onto chitosan, *Bioresour Technol*; **98**: 2897–2904.

Chi L (2010). Ed *Nanotechnology: Nanostructured Surfaces*, John Wiley & Sons: New York.

Chiou MS, Chuang GS (2006). Competitive adsorption of dye metanil yellow and RB15 in acid solutions on chemically cross-linked chitosan beads, *Chemosphere*; **62**: 731–740.

Chowdhury SR, Yanful EK (2010). Arsenic and chromium removal by mixed magnetite–maghemite nanoparticles and the effect of phosphate on removal, *J Environ Manage*; **91**: 2238–2247.

Chrusciel J, Slusarski L (2003). Synthesis of nanosilica by the sol-gel method and its activity toward polymers, *Mater Sci*; **21**: 461–469.

Condon JB (2006). *Surface Area and Porosity Determinations by Physisorption: Measurements and Theory*, Elsevier: New York.

Csanady M, Straub I (1995). Health damage due to pollution in Hungary. *Proceedings of the Rome Symposium*, September, 1994; International Association of Hydrological Sciences: Wallingford, Oxfordshire, UK; **233**: 147–152.

Damia B (2005). Ed *Emerging Organic Pollutants in Waste Waters and Sludge*, Springer: New York.

Das D, Chatterjee A, Mandal BK, Samanta C, Chakraborti D, Chanda B (1995). Arsenic in ground water in six districts of West Bengal, India: The biggest arsenic calamity in the world. Part 2. Arsenic concentration in drinking water, hair, nails, urine, skin-scale and liver tissue (biopsy) of the affected people, *Analyst*; **120**: 917–924.

Deliyanni EA, Bakoyannnnakis DN, Zouboulis AI, Matis KA (2003). Sorption of As(V) ions by akaganeite-type nanocrystals, *Chemosphere*; **50**, 155–163.

Deliyanni EA, Lazaridis NK, Peleka EN, Matis KA (2004). Metals removal from aqueous solution by iron-based bonding agents, *Environ Sci Pollut Res*; **11**: 18–21.

Deliyanni EA, Matis KA (2005). Sorption of Cd ions onto akaganeite-type nanocrystals, *Sep Purif Technol*; **45**: 96–102.

Deliyanni EA, Nalbandian LK, Matis A (2006). Adsorptive removal of arsenites by a nanocrystalline hybrid surfactant–akaganeite sorbent, *J. Colloid Interface Sci*; **302**: 458–466.

Deliyanni EA, Peleka EN, Matis KA (2007). Removal of zinc ion from water by sorption onto iron-based nanoadsorbent, *J Hazard Mater*; **141**: 176–184.

Deng S, Bai R, Chen JP (2003). Aminated polyacrylonitrile fibers for lead and copper removal, *Langmuir*; **19**: 5058–5064.

Dhillon A, Kumar D (2015a). Development of a nanoporous adsorbent for the removal of health-hazardous fluoride ions from aqueous systems, *J Mater Chem A*; **3**: 4215–4228.

Dhillon A, Kumar D (2015b). Nanocomposite for the detoxification of drinking water: Effective and efficient removal of fluoride and bactericidal activity, *New J Chem*; **39**, 9143–9154.

Dhillon A, Nair M, Bhargava SK, Kumar D (2015). Excellent fluoride decontamination and antibacterial efficacy of Fe–Ca–Zr hybrid metal oxide nanomaterial, *J Colloid Interface Sci*; **457**: 289–297.

Dhillon A, Nair M, Kumar D (2015). Analytical methods for sensing of health-hazardous arsenic from biotic and abiotic natural resources, *Anal Methods*; **7**: 10088–10108.

Dhillon A, Sharma TK, Soni SK, Kumar D (2016). Fluoride adsorption on a cubical ceria nanoadsorbent: Function of surface properties, *RSC Adv*; **6**: 89198–89209.

Di ZC, Ding J, Peng XJ, Li YH, Laun ZK, Liang J (2006). Chromium adsorption by aligned carbon nanotubes supported ceria nanoparticles, *Chemosphere*; **62**: 861–865.

Dixit S, Hering JG (2003). Comparison of arsenic(V) and arsenic(III) sorption onto iron oxide minerals: Implications for arsenic mobility, *Environ Sci Technol*; **37**: 4182–4189.

Dong S, Yang S, Tang C (2007). Rapid synthesis of size-controlled gold nanoparticles by complex intramolecular photoreduction, *Chem Res Chin Univ*; **23**: 500–504.

Dong S, Zhou S (2007). Photochemical synthesis of colloidal gold nanoparticles, *Mater Sci Eng B*; **140**: 153–159.

Doong RA, Chiang LF (2008). Coupled removal of organic compounds and heavy metals by titanate/carbon nanotube composites, *Water Sci Technol*; **58**: 1985–1992.

Dousova B, Grygar T, Martaus A, Fuitova L, Kolousek D, Machovi V (2006). Sorption of As(V) on aluminosilicates treated with Fe^{II} nanoparticles, *J Colloid Interface Sci*; **302**: 424–431.

Droste RL (1997). *Theory and Practice of Water and Wastewater Treatment*, John Wiley & Sons: New York.

Du WL, Xu ZR, Han XY, Xu YL, Miao ZG (2008). Preparation, characterization and adsorption properties of chitosan nanoparticles for eosin Y as a model anionic dye, *J Hazard Mater*; **153**: 152–156.

Duffus JH (2002). Heavy metals—a meaningless term, *Pure Appl Chem*; **74**: 793–807.

Dyson T (1996). *Population and Food: Global Trends and Future Prospects*, Routledge: London.

Environmental Health Criteria, 18: Arsenic, World Health Organization: Geneva; 1981.

Fan J, Cai W, Yu J (2011). Adsorption of N719 Dye on anatase TiO_2 nanoparticles and nanosheets with exposed (001) facets: Equilibrium, kinetic, and thermodynamic studies, *Chem Asian J*; **6**: 2481–2490.

Fan L, Luo C, Zhen LV, Lu F, Qiu H (2011). Preparation of magnetic modified chitosan and adsorption of Zn^{2+} from aqueous solutions, *Colloids Surf B*; **88**: 574–581.

Fahrner W (2005). *Nanotechnology and Nanoelectronics Materials, Devices, Measurement Techniques*, Springer: New York.

Faust SD, Aly OM (1983). *Chemistry of Water Treatment*, Butterworth: Stoneham, MA.

Fendrof S, Eick MJ, Grossl P, Sparks DL (1997). Arsenate and chromate retention mechanisms on goethite, *Environ Sci Technol*; **31**: 315–320.

Figgemeier E, Kylberg W, Constable E, Scarisoreanu M, Alexandrescu R, Morjan I, Soare I, Birjega R, Popovici E, Fleaca C, Gavrila-Florescu L, Prodan G (2007). Titanium dioxide nanoparticles prepared by laser pyrolysis: Synthesis and photocatalytic properties, *Appl Surf Sci*; **254**: 1037–1041.

Franklin LB (1991). *Wastewater Engineering: Treatment, Disposal and Reuse*, McGraw Hill Inc.: New York.

Gadd GM, Griffiths AJ (1978). Microorganisms and heavy metal toxicity, *Microb Ecol*; **4**: 303–317.

Gandhi MR, Kousalya GN, Meenakshi S (2010). Removal of copper(II) using chitin/chitosan nano-hydroxyapatite composite, *Int J Biol Macromol*; **48**: 119–124.

Gao Y, Wahi R, Kan AT, Falkner JC, Colvin VL, Tomson MB (2004). Adsorption of cadmium on anatase nanoparticles effect of crystal size and pH, *Langmuir*; **20**: 9585–9593.

Gatti AM (2005). *Nanotoxicity and Health Risk Related to Managing Nanoparticles: The European Experience of Nanopathology*; University of Modena and Reggio Emilia, Laboratory of Biomaterials: Modena, Italy.

Giri SK, Das NN, Pradha GC (2011). Synthesis and characterization of magnetite nanoparticles using waste iron ore tailings for adsorptive removal of dyes from aqueous solution, *Colloids Surf A*; **389**: 43–49.

Gledhill WE (1987). Microbial toxicity and degradation test methodology: An industrial perspective, *Toxicol Assess*; **2**: 89–96.

Guan XH, Qin YC, Qin YH, Yin R, Sun MJ (2007). Removing Cr(VI) by composite biosorbent of nano Fe_3O_4/Sphaerotilus natans, *Huan Jing Ke Xue*; **28**: 2096–3000.

WHO (1993). *Guidelines for Drinking-Water Quality*, 2nd ed; World Health Organization: Geneva; vol. 1.

Guo GY, Chen YL (2005). A nearly pure monoclinic nanocrystalline zirconia, *J Solid State Chem*; **178**: 1675–1682.

Gupta VK, Agarwal S, Saleh TA (2011a). Chromium removal by combining the magnetic properties of iron oxide with adsorption properties of carbon nanotubes, *Water Res*; **45**: 2207–2212.

Gupta VK, Agarwal S, Saleh TA (2011b). Synthesis and characterization of alumina-coated carbon nanotubes and their application for lead removal, *J Hazard Mater*; **185**: 17–23.

Gupta VK, Nayak A (2012). Cadmium removal and recovery from aqueous solutions by novel adsorbents prepared from orange peel and Fe_2O_3 nanoparticles, *Chem Eng J*; **180**: 81–90.

Hao YM, Man C, Hu ZB (2010). Effective removal of Cu (II) ions from aqueous solution by amino-functionalized magnetic nanoparticles, *J Hazard Mater*; **184**: 392–399.

Han D, Yang H, Zhu C, Wang F (2008). Controlled synthesis of CuO nanoparticles using Triton X-100-based water-in-oil reverse micelles, *Powder Technol*; **185**: 286–290.

Hu J, Chen G, Lo IMC (2005). Removal and recovery of Cr(VI) from wastewater by maghemite nanoparticles, *Water Res*; **39**: 4528–4536.

Hu J, Chen G, Lo ICM (2006). Selective removal of heavy metals from industrial wastewater using maghemite nanoparticle: Performance and mechanisms, *J Environ Eng*; **132**: 709–715.

Hu J, Lo IMC, Chen G (2007). Comparative study of various magnetic nanoparticles for Cr(VI) removal, *Sep Purif Technol*; **56**: 249–256.

Huang SH, Chen DH (2009). Rapid removal of heavy metal cations and anions from aqueous solutions by an amino-functionalized magnetic nano-adsorbent, *J Hazard Mater*; **163**: 174–179.

Huang YH, Hsueh CI, Cheng HP, Su LC, Chen CY (2006). Thermodynamics and kinetics of adsorption of Cu(II) onto waste iron oxide, *J Hazard Mater*; **144**: 406–411.

Hullavarad NV, Hullavarad SS (2007). Synthesis and characterization of monodispersed CdS nanoparticles in SiO_2 fibers by sol–gel method, *Photonics Nanostruct: Fundam Appl*; **5**: 156–163.

Hristovski K, Baumgardener A, Westerhoff P (2007). Selecting metal oxide nanomaterials for arsenic removal in fixed bed columns: From nanopowders to aggregated nanoparticle media, *J Hazard Mater*; **147**: 265–274.

Ibarguen C, Mosquera AA, Parra R, Castro MS, Rodríguez-Paez J (2007). Synthesis of SnO_2 nanoparticles through the controlled precipitation route, *E Mater Chem Phys*; **101**: 433–440.

Iram M, Guo C, Guan Y, Ishfaq A, Liu H (2010). Adsorption and magnetic removal of neutral red dye from aqueous solution using Fe_3O_4 hollow nanospheres, *J Hazard Mater*; **181**: 1039–1050.

Jain CK, Ali I (2000). Arsenic: Occurrence, toxicity and speciation techniques, *Water Res*; **34**: 4304.

Jegadeesan G, Al-Abed SR, Sundaram V, Choi H, Scheckel KG, Dionysiou DD (2010). Arsenic sorption on TiO_2 nanoparticles: Size and crystallinity effects, *Water Res*; **44**: 965–973.

John DZ (1990). *Handbook of Drinking Water Quality: Standards and Controls*, Van Nostrand Reinhold: New York. 4312.

Jin W, Lee IK, Kompch A, Dorfler U, Winterer M. (2007). Chemical vapor synthesis and characterization of chromium doped zinc oxide nanoparticles, *J Eur Ceram Soc*; **27**: 4333–4337.

Johnson BB (1990). Effect of pH, temperature, and concentration on the adsorption of cadmium on goethite, *Environ Sci Technol*; **24**: 112–118.

Kandjani AE, Tabriz MF, Pourabbas B (2008). Sonochemical synthesis of ZnO nanoparticles: The effect of temperature and sonication power, *Mater Res Bull*; **43**: 645–654.

Kanel SR, Charlet B, Choi L (2005). Removal of arsenic (III) from groundwater by nanoscale zero-valent iron, *Environ Sci Technol*; **39**: 1291–1298.

Kanel SR, Greneche JM, Choi H (2006). Arsenic(V) removal from groundwater using nano scale zero-valent iron as a colloidal reactive barrier material, *Environ Sci Technol*; **40**: 2040–2050.

Kiani G, Dostali M, Rostami A, Khataee AR (2011). Adsorption studies on the removal of Malachite Green from aqueous solutions onto halloysite nanotubes, *Appl Clay Sci*; **54**: 34–39.

Kim CS, Kwon IM, Moon BK, Jeong JH, Choi BC, Kim JH, Choi H, Yi S, Yoo DH, Hong KS, Park JH, Lee HS (2007). Synthesis and particle size effect on the phase transformation of nanocrystalline TiO_2, *Mater Sci Eng C*; **27**: 1343–1346.

Kiping MD. In *Arsenic, The Chemical Environment, Environment and Man*; Lenihan J, Fletcher WW (1977). Eds, World Health Organization: Glassgow, UK; **6**.

Kiss LB, Soderlund J, Niklasson GA, Granqvist CG (1999). New approach to the origin of lognormal size distributions of nanoparticles, *Nanotechnology*; **10**: 25–28.

Kitamura Y, Okinaka N, Shibayama T, Mahaney OOP, Kusano D, Ohtani B, Akiyama T (2007). Combustion synthesis of TiO_2 nanoparticles as photocatalyst, *Powder Technol*; **176**: 93–98.

Kochkar H, Turki A, Bergaoui L, Berhault G, Ghorbel A (2009). Study of Pd(II) adsorption over titanate nanotubes of different diameters, *J Colloid Interface Sci*; **331**: 27–31.

Kohler A, Hellweg S, Escher BI, Hungerbuhler K (2006). Organic pollutant removal versus toxicity reduction in industrial wastewater treatment: The example of wastewater from fluorescent whitening agent production, *Environ Sci Technol*; **40**: 3395–3401.

Koros WJ (1995). Membranes: Learning a lesson from nature, *Chem Eng Prog*; **91**: 68–81.

Kross BC, Ayebo AD, Fuortes L (1992). Methemoglobinemia: Nitrate toxicity in rural America, *J Am Fam Physician*; **46**: 183–188.

Laws EA (2000). *Aquatic Pollution: An Introductory Text*, 3rd ed, John Wiley & Sons: New York.

Lazaridis NK, Bakoyannakis DN, Deliyanni EA (2005). Chromium(VI) sorptive removal from aqueous solutions by nanocrystalline akaganeite, *Chemosphere*; **58**: 65–73.

Lewinsky AA (2007). *Hazardous Materials and Wastewater: Treatment, Removal and Analysis,* Nova Science Publishers: New York.

LI YH, Di Z, Ding J, Wu D, Luan Z, Zhu Y (2005). Adsorption thermodynamic, kinetic and desorption studies of Pb^{2+} on carbon nanotubes, *Water Res*; **39**: 605–609.

Li Z, Greden K, Alvarez PJ, Gregory KB, Lowry GV (2010). Adsorbed polymer and NOM limits adhesion and toxicity of nano scale zerovalent iron to E. coli, *Environ Sci Technol*; **44**: 3462–3467.

Li Y, Lu JJ, Shi BY, Wu YY (2008). Sorption of atrazine onto nano-SiO_2 and nano-kaolin particles, *Huan Jing Ke Xue*; **29**: 1687–1692.

Li YH, Wang S, Luan Z, Ding J, Xu C, Wu D (2003). Adsorption of cadmium(II) from aqueous solution by surface oxidized carbon nanotubes, *Carbon*; **41**: 1057–1062.

Lin SH, Chen ML (1997). Treatment of textile wastewater by chemical methods for reuse, *Water Res*; **31**: 868–876.

Lou JC, Jung MJ, Yang HW, Han JY, Huang WH (2011). Removal of dissolved organic matter (DOM) from raw water by single-walled carbon nanotubes (SWCNTs), *J Environ Sci Health Part A: Toxic/Hazard Subst Environ Eng*; **46**: 1357–1365.

Lu C, Chiu H (2006). Adsorption of zinc(II) from water with purified carbon nanotubes, *Chem Eng Sci*; **61**: 1138–1145.

Machado FM, Bergmann CP, Fernandes TH, Lima EC, Royer B, Calvete T, Fagan SB (2011). Adsorption of Reactive Red M-2BE dye from water solutions by multi-walled carbon nanotubes and activated carbon, *J Hazard Mater*; **192**: 1122–1131.

Manceau A, L Charlet (1994). The mechanism of selenate adsorption on goethite and hydrous ferric oxide, *J Colloid Interface Sci*; **168**: 87–93.

Marin ABP. Zapata VM, Orturao JF, Aguilar M, Saez J, Lloren MJ (2007). Removal of cadmium from aqueous solutions by adsorption onto orange waste, *Hazard Mater*; **139**: 122–131.

Masel RI (1996). *Principles of Adsorption and Reaction on Solid Surfaces*, John Wiley & Sons: New York.

Matschullat J (2000). Arsenic in the geosphere a review, *Sci Total Environ*; **249**: 297–312.

Mayo JT, Yavuz C, Yean S, Cong L, Shipley H, Yu W, Falkner J, Kan A, Tomson M, Colvin VL (2007). The effect of nanocrystalline magnetite size on arsenic removal, *Sci Technol Adv Mater*; **8**: 71–75.

Meyers RA (1999). Ed *Encyclopedia of Environmental Pollution and Clean-up*, John Wiley & Sons: New York.

Mnyusiwalla A, Daar AS, Singer PA (2003). 'Mind the gap': Science and ethics in Nanotechnology, *Nanotechnology*; **14**: R9–R13.

Mostafa MG, Chen YH, Jean JS, Liu CC, Teng H (2010). Adsorption and desorption properties of arsenate onto nano-sized iron-oxide-coated quartz, *Water Sci Technol*; **62**: 378–386.

Moussavi G, Mahmoudi M (2009). Removal of azo and anthraquinone reactive dyes from industrial wastewaters using MgO nanoparticles, *J Hazard Mater*; **168**: 806–812.

Nemerow N, Dasgupta A (1991). *Industrial and Hazardous Waste Treatment*, VNR: New York.

Ng JC, Wang J, Shrain A (2003). A global health problem caused by arsenic from natural sources, *Chemosphere*; **52**: 1353–1359.

Oh H, Kim S (2007). Synthesis of ceria nanoparticles by flame electrospray pyrolysis, *J Aerosol Sci*; **44**: 1185–1196.

Okitsu K, Mizukoshi Y, Yamamoto TA, Maeda Y, Nagata Y (2007). Sonochemical synthesis of gold nanoparticles on chitosan, *Mater Lett*; **61**: 3429–3431.

Oren AH, Kaya A (2006). Factors affecting adsorption characteristics of Zn^{2+} on two natural zeolites, *J Hazard Mater*; **131**: 59–65.

Pacheco S, Medina M, Valencia F, Tapia J (2006). Removal of inorganic mercury from polluted water using structured nanoparticles, *J Environ Eng*; **132**: 342–349.

Pacheco S, Tapia J, Medina M, Rodriguez R (2006). Cadmium ions adsorption in simulated wastewater using structured alumina–silica nanoparticles, *J Non-Cryst Solids*; **352**: 5475–5481.

Palkar VR (1999). Sol-gel derived nanostructured γ-alumina porous spheres as an adsorbent in liquid chromatography, *Nanostruct Mater*; **11**: 369–374.

Pan B, Xiao L, Nie G, Pan B, Wu J, Lv L, Zhang W, Zheng S (2010). Adsorptive selenite removal from water using a nano-hydrated ferric oxides (HFOs)/polymer hybrid adsorbent, *J Environ Monit*; **12**: 305–310.

Park YK, Tadd EH, Zubris M, Tannenbaum R (2005). Size-controlled synthesis of alumina nanoparticles from aluminum alkoxides, *Mater Res Bull*; **40**: 1506–1512.

Patil PR, Joshi SS (2007). Polymerized organic–inorganic synthesis of nanocrystalline zinc oxide, *Mater Chem Phys*; **105**: 354–361.

Pena ME, Koratis GP, Patel M, Lippincott L, Meng X (2005). Adsorption of As(V) and As(III) by nanocrystalline titanium dioxide, *Water Res*; **39**: 2327–2337.

Pershagen G (1983). *The Epidemiology of Human Arsenic Exposure*; Fowler BA, ed, Elsevier: Amsterdam, The Netherlands.

Persistent Organic Pollutants: Impact on Child Health; NLM classification: WA 671, WHO report; World Health Organization: Geneva; 2010.

Pokhrel D, Viraraghavan T (2007). Arsenic removal in an iron oxide-coated fungal biomass column: Analysis of breakthrough curves, *Bioresour Technol*; **99**: 2067–2071.

Ponder SM, Darab JG (2000). Remediation of Cr(VI) and Pb(II) aqueous solutions using supported, nanoscale zero-valent iron, *Environ Sci Technol*; **34**: 2564–2569.

Ramsden J (2011). *Nanotechnology: An Introduction*, Elsevier: The Netherlands.

Rengaraj S, Moon SH (2002). Kinetics of adsorption of Co(II) removal from water and wastewater by ion exchange resins, *Water Res*; **36**: 1783–1793.

Ruiz CG (2006). Mercury(II) removal from aqueous solutions by nonviable Bacillus sp. from a tropical estuary, *Bioresour Technol*; **97**: 1907–1911.

Saleh TA, Agarwal S, Gupta VK (2011a). Synthesis of MWCNT/MnO_2 and their application for simultaneous oxidation of arsenite and sorption of arsenate, *Appl Catal B*; **106**: 46–53.

Saleh TA, Gupta VK (2011b). Functionalization of tungsten oxide into MWCNT and its application for sunlight-induced degradation of rhodamine B, *J Colloids Interface Sci*; **362**: 337–344.

Saleh TA, Gupta VK (2012). Column with CNT/magnesium oxide composite for lead(II) removal from water, *Environ Sci Pollut Res*; **19**: 1224–1228.

Sairam SC, Viswanathan N, Meenakshi S (2008). Defluoridation chemistry of synthetic hydroxyapatite at nano scale: Equilibrium and kinetic studies, *J Hazard Mater*; **155**: 206–215.

Sairam SC, Viswanathan N, Meenakshi S (2009). Fluoride sorption by nano-hydroxyapatite/chitin composite, *J Hazard Mater*; **172**: 147–151.

Sandoval R, Cooper AM, Aymar K, Jain A, Hristovski K (2011). Removal of arsenic and methylene blue from water by granular activated carbon media impregnated with zirconium dioxide nanoparticles, *J Hazard Mater*; **193**: 296–303.

Sarkar D, Mohapatra D, Roy S, Bhattacharya S, Adak S, Mitra N (2007). Synthesis and characterization of sol–gel derived ZrO_2 doped Al_2O_3 nanopowder, *Ceram Int*; **33**: 1275–1282.

Shao-Feng N, Yong L, Xin-hua X, Zhang-hua L (2005). Removal of hexavalent chromium from aqueous solution by iron nanoparticles, *J Zhejiang Univ Sci B*; **6**: 1022–1027.

Shen LL, Zhao BZ, Zhang JB, Chen J, Zheng H (2010). Virus adsorption onto nano-sized iron oxides as affected by different background solutions, *Huan Jing Ke Xue*; **31**: 983–989.

Shengdong P, Haoyu S, Qihong X, Jian L, Meiqin H (2012). Surface mercapto engineered magnetic Fe_3O_4 nanoadsorbent for the removal of mercury from aqueous solutions, *J Colloid Interface Sci*; **365**: 204–212.

Shim WG, Balathanigaimani MS, Lee SG, Kim SC, Moon H (2010). Structural and energetic characterizations of thin multi-walled carbon nanotubes using adsorption isotherms, *J Nanosci Nanotechnol*; **10**: 3680–3685.

Shin KY, Hong JY, Jang J (2011). Heavy metal ion adsorption behavior in nitrogen-doped magnetic carbon nanoparticles: Isotherms and kinetic study, *J Hazard Mater*; **190**: 36–44.

Sigel H, Sigel A (1986). *Concepts on Metal Ion Toxicity*, Marcel Dekker, Inc.: New York.

Singh DB, Rupainwar DC, Prasad G, Jayaprakas KC (1998). Studies on the Cd(II) removal from water by adsorption, *J Hazard Mater*; **60**: 29–40.

Skubul LR, Meshkov NK, Rajh T, Thurnaur M (2002). Cadmium removal from water using thiolactic acid-modified titanium dioxide nanoparticles, *J Photochem Photobiol A*; **148**: 393–397.

Smiciklas I, Dimovic S, Plecas I, Mitrio M (2006). Removal of Co^{2+} from aqueous solutions by hydroxyapatite, *Water Res*; **40**: 2267–2274.

Song J, Kong H, Jang J (2011). Adsorption of heavy metal ions from aqueous solution by polyrhodanine-encapsulated magnetic nanoparticles, *J Colloid Interface Sci*; **359**: 505–511.

Stafiej A, Pyrzynska K (2007). Adsorption of heavy metal ions with carbon nanotubes, *Sep Purif Technol*; **58**: 49–52.

Stephen Inbaraj B, Chen BH (2011). Dye adsorption characteristics of magnetite nanoparticles coated with a biopolymer poly(c-glutamic acid), *Bioresour Technol*; **102**: 8868–8876.

Suhasini IP, Sriram G, Asolekar SR, Sureshkumar GK (1999). Biosorptive removal and recovery of cobalt from aqueous systems, *Process Biochem*; **34**: 239–247.

Sun C, Qu R, Ji C, Wang C, Sun Y, Yue Z, Cheng G (2006). Preparation and adsorption properties of crosslinked polystyrene-supported low-generation diethanolamine-typed dendrimer for metal ions, *Talanta*; **70**: 14–19.

Tai Y, Wang L, Gao J, Amer WA, Ding W, Yu H (2011). Synthesis of Fe_3O_4@ poly(methylmethacrylate-co-divinylbenzene) magnetic porous microspheres and their application in the separation of phenol from aqueous solutions, *J Colloid Interface Sci*; **360**: 731–738.

Takami S, Sato T, Mousavand T, Ohara SI, Umetsu M, Adschiri T (2007). Hydrothermal synthesis of surface-modified iron oxide nanoparticles, *Mater Lett*; **61**: 4769–4772.

Tchobanoglous G, Franklin LB (1991). *Wastewater Engineering: Treatment, Disposal and Reuse*, McGraw Hill, Inc.: New York.

Thareja RK, Shukla S (2007). Synthesis and characterization of zinc oxide nanoparticles by laser ablation of zinc in liquid, *Appl Surf Sci*; **253**: 8889–8895.

Thota S, Kumar J (2007). Sol–gel synthesis and anomalous magnetic behavior of NiO nanoparticles, *J Phys Chem Solids*; **68**: 1951–1964.

Uheida A, Alvarz GS, Bijorkman E, Yu Z, Muhammed M (2006). Fe_3O_4 and γ-Fe_2O_3 nanoparticles for the adsorption of Co^{2+} from aqueous solution, *J Colloid Interface Sci*; **298**: 501–507.

Valente S, Bokhimi X, Toledo JA (2004). Synthesis and catalytic properties of nanostructured aluminas obtained by sol–gel method, *Appl Catal A*; **264**: 175–181.

Vatistas N, Bartolozzi MJ (1999). Zinc contamination in the cathodic material of exhausted alkaline manganese dioxide batteries, *Power Sources*; **79**: 199–204.

Vettorazzi G (1979). *International Regulatory Aspects for Pesticide Chemicals*, CRC Press Inc.: Boca Raton, FL.; vol. 1: 14.

Venema P, Hiemstra T, Weidlu PG, Riemsdizk WHV (1998). Intrinsic proton affinity of reactive surface groups of metal hydroxides: Application to iron hydroxides, *J Colloid Interface Sci*; **198**: 282–295.

Wang L, Wang A (2007). Adsorption characteristics of Congo Red onto the chitosan/montmorillonite nanocomposite, *J Hazard Mater*; **147**: 979–985.

Wang H, Zhou A, Peng F, Yu H, Yang J (2007). Mechanism study on adsorption of acidified multiwalled carbon nanotubes to Pb(II), *J Colloid Interface Sci*; **316**: 277–283.

Weber B, Holz F (1991). Landfill leachates treatment by reverse osmosis. In *Effective Industrial Membrane Processes, Benefits and Opportunities*, Turner MK, ed; Elsevier Science Publishers Ltd.: Barking, Essex, UK.

Weng CH, Huang CP (2004). Adsorption characteristics of Zn(II) from dilute aqueous solution by fly ash, *Colloids Surf A*; **247**: 137–143.

Weng CH, Wang JH, Huang CP (1997). Adsorption of Cr(VI) onto TiO_2 from dilute aqueous solutions, *Water Sci Technol*; **35**: 55–62.

Williams AR (1991). The use of reverse osmosis for the purification of coal gasification liquors. In *Effective Industrial Membrane Processes, Benefits and Opportunities*, Turner MK, ed; Elsevier science Publishers Ltd.: Barking, Essex, UK.

Wu CH (2007). Studies of the equilibrium and thermodynamics of the adsorption of Cu^{2+} onto as-produced and modified carbon nanotubes, *J Colloid Interface Sci*; **311**: 338–346.

Wu D, Zheng P, Chang PR, Ma X (2011). Preparation and characterization of magnetic rectorite/iron oxide nanocomposites and its application for the removal of the dyes, *Chem Eng J*; **174**: 489–494.

Yamamura Y, Yamauchi H (1980). Arsenic metabolites in hair, blood and urine in workers exposed to arsenic trioxide, *Ind Health*; **18**: 203–210.

Yan-Ping C, Cui-Ling R, Yang Q, Zhen-Yang Z, Li-Jun D, Xing-Guo C (2011). Preparation and characterization of hexadecyl functionalized magnetic silica nanoparticles and its application in Rhodamine 6G removal, *Appl Surf Sci*; **257**: 8610–8616.

Yang K, Xing B (2009). Sorption of phenanthrene by humic acid-coated nanosized TiO_2 and ZnO, *Environ Sci Technol*; **43**: 1845–1851.

Zeng L (2003). A method for preparing silica-containing iron(III) oxide adsorbents for arsenic removal, *Water Res*; **379**: 4351–4358.

Zeng WM, Gao L, Guo JK (1998). A new sol-gel route using inorganic salt for synthesizing Al_2O_3 nanopowders, *Nanostruct Mater*; **10**: 543–550.

Zhang Z, Kong J (2011). Novel magnetic Fe_3O_4@C nanoparticles as adsorbents for removal of organic dyes from aqueous solution, *J Hazard Mater*; **193**: 325–329.

Zhang L, Liu N, Yang L, Lin Q (2009). Sorption behavior of nano-TiO_2 for the removal of selenium ions from aqueous solution, *J Hazard Mater*; **170**: 1197–1203.

Zhang FS, Nriagu JO, Itoh, H (2005). Mercury removal from water using activated carbons derived from organic sewage sludge, *Water Res*; **39**: 389–395.

Zhang Q, Pan B, Pan B, Zhang W, Jia K, Zhang Q (2008). Selective sorption of lead, cadmium and zinc ions by a polymeric cation exchanger containing nano-$Zr(HPO_3S)_2$, *Environ Sci Technol*; **42**: 4140–4145.

Zhai Y, Gao Y, Liu F, Zhang Q, Gao G (2007). Synthesis of nanostructured TiO_2 particles in room temperature ionic liquid and its photocatalytic performance, *Mater Lett*; **61**: 5056–5058.

Zhao M, Tang Z, Liu P (2008). Removal of methylene blue from aqueous solution with silica nano-sheets derived from vermiculite, *J Hazard Mater*; **158**: 43–51.

Zheng YM, Yu L, Wu D, Chen JP (2012). Removal of arsenite from aqueous solution by a zirconia nanoparticle, *Chem Eng J*; **188**: 15–22.

Zhou L, Jin J, Liu Z, Liang X, Shang C (2011). Adsorption of acid dyes from aqueous solutions by the ethylenediamine-modified magnetic chitosan nanoparticles, *J Hazard Mater*; **185**: 1045–1052.

Zhu HJ, Jia YF, Wu X, Wang H (2009a). Removal of arsenate from drinking water by activated carbon supported nano zero-valent iron, *Huan Jing Ke Xue*; **30**: 1644–1648.

Zhu H, Jia Y, Wu X, Wang H (2009b). Removal of arsenic from water by supported nano zero-valent iron on activated carbon, *J Hazard Mater*; **172**: 1591–1596.

Zhu HJ, Jia YF, Yao SH, Wu X, Wang SY (2009). Removal of arsenate from drinking water by activated carbon supported nano zero-valent iron, *Huan Jing Ke Xue*; **30**: 3562–3567.

Zinkus GA, Byers WD, Doerr WW (1998). Identify appropriate water reclamation technologies, *Chem Eng Prog*; **94**: 19–31.

Zor S, Yazici B, Erbil M, Galip H (1998). The electrochemical degradation of linear alkyl benzene sulfonate (LAS) on platinum electrode, *Water Res*; **32**: 579–586.

Chapter 2

Nanocomposites for Pollution Control

Ephraim Vunain,[a,b] Ajay Kumar Mishra,[b] and B. B. Mamba,[b]

[a]*Department of Chemistry, Chancellor College, University of Malawi, P.O. Box 280, Zomba, Malawi, Central Africa*
[b]*Nanotechnology and Water Sustainability Research Unit, College of Science, Engineering and Technology, University of South Africa, Florida Campus, Johannesburg, South Africa*

evunain@cc.ac.mw; evunain@gmail.com

2.1 Introduction

Rapid population, urbanization and industrialization in the past decades have led to emerging environmental problems. For example, air pollution due to fine particles (<2.5 nm) has turn out to be the major cause of cardiovascular and respiratory problems [1–4]. The presence of nitrogen dioxide, sulphur dioxide and ozone seriously affects people suffering from asthma [5–8]. Furthermore, arsenic in drinking water can be seriously affect human health as it can cause bladder, skin, and lung cancer as well as liver and kidney cancer [9–14]. Pollution can be defined as the introduction of undesirable substances to the environment. When toxic substances infiltrate into air, water and soil bodies, they dissolve or lie deposited on the bed, thus resulting in pollution. Several processes have emitted biodegradable and

Nanocomposites for Pollution Control
Edited by Chaudhery Mustansar Hussain and Ajay Kumar Mishra

ISBN 978-981-4774-45-1 (Hardcover), 978-1-315-14368-2 (eBook)
www.panstanford.com

non-biodegradable organic and inorganic pollutants into air, water and soil. Most often, some of the noxious waste has long resident times and remains in the environment for a long period producing imbalances in specific ecosystems and hence resulting in decrements of the biota. With the exception of the pollution of steams, anthropogenic pollution concentrates in a specific area and thus lasts for a longer period [15]. Anthropogenic pollution includes

(1) industrial wastewater that vary in terms of composition and flow rate;
(2) municipal wastewater, which contains, detergents, organic matter, inorganic compounds and urban runoff;
(3) wastewater from agricultural and livestock activities such as the use of fertilizers, pesticides, antibiotics, hormones and other synthetic products;
(4) toxic compounds such as CO, O_3, CO_2, SOx, NOx (leading to acid rain) and volatile organic compounds from vehicles and heavy-duty combustion engines [15].

It should be noted that all these noxious pollutants seriously harm animals, plants and human beings with serious environmental effects.

Undoubtedly, there is an urgent need for the development of better treatment technologies than the usual conventional technologies which are not sufficiently effective in removing these pollutants from our environment. Scientists and researchers are seeking to find some environmentally friendly methods to solve these environment problems. Nanotechnology appears to be viable option, where nanomaterials are employed for pollution remediation. Nanotechnology consists of the construction of functional materials, devices and systems with novel and valuable properties through the control, manipulation and organisation of mater at the nanometre length range (1–100 nm) [16]. Nanomaterials show different and/or enhanced physical and chemical behaviours with respect to bulk materials. They have a wide range of application in areas such as electronics, pharmaceuticals, medicine, agriculture, energy, transport, construction, cosmetic and personal care [17–25].

Indeed, the use of nanomaterials for environmental remediation has soared recently because of their enhanced properties over convention bulk materials [18, 26–31]. Today's technologies require a variety of materials with usual combinations of properties which are not met by conventional materials [32]. The development of such materials is currently a research area of great interest in most laboratories nowadays. Important among these materials are nanocomposites. Nanocomposites are the promising high-performance materials that exhibit excellent properties and unique design possibilities and as such are regarded as the materials of the 21st century [33]. Indeed, nanocomposites have emerged as suitable alternatives to overcome the shortcomings of conventional micro-composites, but some preparation challenges with respect to the control of elemental composition and stoichiometry in the nanocluster phase cause a concern [34]. Nanocomposites are the materials that are produced by introducing nanoparticles into a macroscopic sample material. They are the composites in which at least one of the phases shows dimensions in the nanometre range. A nanoscale dispersion of nanoparticles or sheet-like inorganic silicate particles into a based material (polymeric matrix, for example) leads to significant improvement in properties such as toughness, mechanical strength, and electrical or thermal conductivity over conventional composite materials. The effectiveness of the nanoparticles or filler materials added is normally only between 0.5% and 5%. Nanocomposites differ from conventional microscale composites in mechanical terms in that these materials possess very high surface area to volume ratio of the reinforcing phase and/or exceptionally high aspect ratio. Nanocomposites can be produced using simple and inexpensive techniques.

Generally, nanocomposites are a class of materials in which one or more phases with nanoscale dimensions (0-D, 1-D, 2-D) are embedded in a metal, ceramic or polymer matrix. These materials are made of distinctly dissimilar components and mixed at the nanoscale range. These solid materials thus have multiple phase domains with at least one of these domains in the nanoscale structure. Often, the materials show novel physical and chemical properties that depend on the morphology and interfacial characteristics of the component materials. The properties of nanocomposites depend on a range of factors such as the matrix material (which can exhibit nanoscale dimensions), size, shape, orientation of the nanoscale second phase, degree of dispersion, loading, processing methods and interaction between the matrix and the second phase [35]. Generally, nanocomposites show improved properties such as mechanical properties, including strength, stiffness, modulus, improved biodegradability, electrical conductivity, chemical resistance, thermal stability, flame retardancy, reduced permeability, surface appearance and optical clarity [32, 36, 37]. The potential applications of these materials range from packing to biomedical as well as in environmental remediation.

2.2 Classification of Nanocomposites

Nanocomposites can be classified in a number of ways and one of them is as follows:

(1) inorganic filler in an inorganic matrix (inorganic–inorganic nanocomposites)
(2) organic filler in an organic matrix (organic–organic nanocomposites)
(3) hybrid materials i.e., organic in inorganic or inorganic in organic matrix

Furthermore, hybrid nanocomposites are further divided into two categories: (a) The organic inorganic components could be weakly joined to each other by hydrogen bonding, van der Waals forces, π–π or weak interactions between them. (b) The second category of hybrid nanocomposites is seen where there is a very strong interaction by strong covalent or coordinative bonds between the organic and inorganic components [38]. Indeed, hybrid nanocomposites, particularly the inorganic/organic class of nanocomposites, have emerged as a new class of environmental materials where inorganic particles are dispersed into conventional host materials such as cellulose [39, 40], alginates [41], diatomite [42, 43], and porous polymeric resins [44, 45]. In all of these examples, the host materials of larger particle size greatly improve the permeability and separation ability of the resulting nanocomposites [46]. The organic materials dispersed into conventional host materials can be three-dimensional framework such as zeolites, two-dimensional layered materials such as metal oxides, metal phosphates, chalcogenides, clays, and even one-dimensional and zero-dimensional materials

[47–50]. Hybrid nanocomposites have received voluminous interest as a result of a combination of unique properties of organic and inorganic components in one material. It should be noted that a good number of nanocomposites have been fabricated and employed for the efficient remediation of contaminated environment. For example, hybrid sorbents based on surface charge polymeric supports (e.g. cation or anion exchanges) have proven to be effective sorbets toward the sorption of heavy-metal ions from the environment [46]. This has been attributed to the immobilized charged groups on a polymeric matrix for enhanced metal permeation and preconcentration prior to effective sorption by inorganic particles. This phenomenon is called the "Donnan membrane effect" [51, 52]. In addition, inorganic/organic nanocomposites are important for bio-ceramic and biomineralization in which *in situ* growth and polymerization of biopolymer and the inorganic matrix takes place.

2.3 Synthesis of Organic–Inorganic Nanocomposites

The synthesis of organic/inorganic nanocomposites can be carried out using four methods: (a) sol-gel process, (b) self-assembly process, (c) assembling or dispersing of nanobuilding blocks, and (d) interpenetrating networks. The synthesis of these nanocomposites has been well documented [53–60]. The sol-gel method stands as the most popular and common method for the preparation of organic/inorganic nanocomposites. This is a technique based on the hydrolysis of liquid precursors and the formation of colloidal *sols*.

2.4 Application of Nanocomposites in Environmental Remediation

The removal or degradation mechanism of several pollutants can be achieved through adsorption, catalytic degradation, sensing and detection of pollutants, redox reactions, although in some cases, the redox potential is not high enough to effectively reduce these pollutants. To achieve great success, nanoparticles, metal oxides, metals, and nanostructured bimetallic systems have been employed to remove or degrade contaminants such as heavy metals, halogenated compounds, dyes and microorganism from the environment.

Nanocomposites have been widely applied in the remediation of these pollutants from the environment. These materials are employed for the abatement of contaminants from various environmental media such as groundwater sources, soils, industrial effluents and gases. Adsorption and catalytic degradation remain the main mechanisms responsible for the environmental applicability of nanocomposites. Furthermore, nanocomposites are employed for the sensing and detection of pollutants, particularly at the trace level as well employed in green chemistry to minimize the discharge of pollutants into the environment [41]. In this chapter, we focus on the use of nanocomposites for the adsorption of pollutants, for catalytic and redox degradation of pollutants and sensing/detection of pollutants.

2.4.1 Use of Nanocomposites for Adsorption of Pollutants

Adsorption techniques are widely employed in gas and water/wastewater purification as one of the most effective and simplest approaches to removing toxic pollutants. Inorganic particles such as metal hydroxides (e.g. Fe(III) [61–64], Mn(IV) [62, 64–66] and $M(HPO_4)_2$ (M = Zr, Ti, Sn) [67–69] have been employed as environmentally benign adsorbents for the removal of some targeted pollutants. It is believed that at the nanoscale range, these particles appear to be more effective and efficient towards targeted pollutants because of their large surface areas and high reactivity. More often, these inorganic particles are incorporated into conventional polymers like cellulose, alginate, porous resins and ion-exchangers to avoid certain issues such as transition loss and excessive pressure drop caused by ultrafine particle size. Among these conventional polymers, adsorbents produced from porous polymeric resins and ion-exchangers have proven to be ideal hybrid adsorbents because of their excellent mechanical strength and adjustable surface chemistry of the polymeric supports [67, 69–71]. Here, enhanced permeation of inorganic pollutants of the counter charges is believed to occur as a result of the immobilized charged functional groups bound to the polymeric matrix, interpreted by the Donnan membrane principle [72, 73]. Many heavy metals such as arsenic (As^{3+}, As^{5+}), chromium (Cr^{6+}), mercury (Hg^{2+}), nickel (Ni^{2+}) and lead (Pb^{2+}) are commonly found in the environment. Water bodies contaminated with these pollutants seriously affect human beings, plants and animals. Their removal of these metal ions from contaminated water can be carried out through adsorption using polymer nanocomposites.

Magnetic nanoparticles offer a number of advantages over non-magnetic nanoparticles because they can be easily separated from the solution using a magnetic field. Recently, magnetic functional nanocomposites have become a feature of research owing to their magnetic separation features [74]. It should be noted that covalently immobilized polymers, molecules or even inorganic materials could be grafted onto the surfaces of magnetic nanoparticles for the removal of some heavy-metal ions from contaminated water. For example, some work has been done on the use of maghemite (Fe_2O_3), magnetite (Fe_3O_4) and jacobsite ($MnFe_2O_4$) loaded onto polymer matrix for removal of heavy-metal ions from contaminated water.

Su et al. [65] synthesized a hybrid polymer nanocomposite of supported nanosized hydrous manganese dioxide (HMO) for the removal of lead, cadmium and zinc from water. The sorbent HMO-001 was synthesized by impregnating nanosized hydrous manganese dioxide (HMO) onto a porous polystyrene cation exchanger resin (D-001). Lead adsorption onto HMO-001 adsorbent produced a maximum adsorption capacity of 395 mg/g. The HMO-001 adsorbent exhibited a high selectivity for lead retention from waters as compared to D-001 sorbent in the presence of competing ions such as Ca^{2+}, Mg^{2+} and N^{a+} at much greater level than the target heavy metal. Fixed-bed column adsorption of a simulated water showed lead retention on HMO-001 from 1 mg/L to below 0.01 mg/L (World Health Organization (WHO) drinking water standard). Furthermore, the exhausted sorbent was regenerated by binary NaAc-HAc solution for repeated use with any appreciable capacity loss.

Li et al. reported the preparation of polyaniline/grapheme oxide nanocomposites for the adsorption of mercury(II) ions from wastewater [75]. The composite of polyaniline and reduced grapheme oxide (PANI/RGO) was prepared through the polymerization of anile in the presence of grapheme oxide with a further reduction by hydrate hydrazine. Characterization of nanocomposites with EDX after synthesis, showed the preservation of most nitrogen-containing groups on PANI. Electron scanning microscopy (SEM) and BET analyses showed the PANI/RGO with porous nanostructure of a higher surface area than neat PANI. Results showed the 15 wt% GO loaded PANI as the most effective and efficient adsorbent with the highest adsorption for mercury(II) ions from aqueous solution. The adsorption of Hg(II) onto PANI involved chemical adsorption by coordinating/chelating with the N-containing groups of PANI. The adsorption process followed the pseudo-second order kinetic model with an increase in $Q_{\max}$ from 515.46 mg/g to 1000.00 mg/g at *p*H = 4.0. This excellent performance was attributed to the presence of the reduced grapheme oxide in the nanocomposite, which drastically enhanced the surface area and adsorption sites as revealed by SEM, TEM and BET characterization.

In a similar study, Najim et al. synthesized polyaniline (PANI) nanofibres and nanocomposites for the adsorptive removal of Cr(VI) and phosphate ions from aqueous solution [76]. In this study, polyaniline nanocomposites were prepared by in situ oxidative polymerization of aniline in acidic medium using ammonium persulphate as an initiator in the presence of natural silica (PANIS), acid treated natural silica (PANISA), fibre glass (PANIFG) and poly(ethylene terephthalate) powder from waste bottles (PANIPET). The results from this study showed PANISA as an effective and potential adsorbent for the removal of chromium and phosphate pollutant from aqueous solution due to its large surface area. This study showed that the mechanism of adsorption of chromium(VI) and phosphate ions onto the polymer nanocomposites occurred via electrostatic forces besides the ion-exchange process. Overall, the adsorbent (PANISA) showed lower efficiency for the adsorption of phosphate as compared to the adsorption of chromium(VI), which was attributed to the charge difference between the exchangeable anion Cl^- and divalent phosphate anion HPO_4^{-2}.

Sharma and Lee employed the use of acrylamide–titanium nanocomposites for the removal of cadmium(II) from aqueous solution [77]. Acrylamide was in situ doped into titanium during a sol-gel process to form the nanocomposites which were thereafter characterized and employed as adsorbent for Cd(II) ions removal. The adsorption behaviour of nanocomposites was examined through kinetics equilibrium batch experiments in a batch mode. The adsorption capacity Q_{max} (mg/g) value of the nanocomposites was found to be 322.58 at an optimum pH of 8.0, compared with 86.95 mg/g value for the nano-titanium. The kinetics model was well described by pseudo-second order-kinetics with a rate constant of 4.0×10^{-4} and 9.4×10^{-5} mg^{-1} min^{-1} at an initial cadmium concentration of 100 and 500 mg/L, respectively. An adsorption of 27% adsorption capacity of the nanocomposites was achieved at the fifth cycle indicating the effective regeneration of heavy-metal ions with 0.05 N acidic acid.

Furthermore, agricultural wastes affect our natural resources and cause ecological problems due to their low natural degradation rate and thus they can be eliminated through burning thereby constituting one of the major causes of atmospheric pollution. To utilize these agro-wastes in order to overcome the environmental problem, Shweta and Jha [78] fabricated rice-husk extracted lignin-TEOS (tretraethylorthosilicate) bionanocomposites for the removal of nickel from wastewater. In this study, lignosilicate nanocomposites were synthesized from pre-extracted (hot water, 80% ethanol, 0.3 N NaOH) lignin of rice husk using TEOS as matrix. This was followed by the surface modification of the extracted lignin by acetylation followed by in situ synthesis of lignosilicate nanocomposites via the sol-gel method. These nanocomposites materials were characterized and employed in the removal of nickel from wastewater. The findings from this study showed the alkali extracted adsorbent with a mean pore size distribution (PSD) of 14.89 nm with excellent thermal stability (T_m = 337°C) than the ethanol or the hot water extracted adsorbent. Furthermore, the ethanol extracted bionanocomposites showed a high adsorption capability of 38.74% for Ni^{2+}. This high adsorption exhibited by the alkali extracted adsorbent than the others was confirmed by scanning electron microscopy (SEM) and PSD (D_{50}) analyses through their rough surface and dispersive nature more suitable for biosorption [78]. Overall, more thermally stable adsorbent obtained from agro-wastes showed that the metal sorption capability was strongly influenced by the porous nature of the bio-sorbent rather than the composition and quantity of dispersed functional groups (–OH). Thus, the lignosilicate bionanocomposite materials from rice husk proved to be a potential biosorbent to chelate or adsorb environmental pollutants like heavy-metal ions.

Carbon dioxide has drawn significant attention as one of the main anthropogenic contributors to climate change. A wide range of approaches have been used for the capturing of CO_2 from the atmosphere, but the use of nanocomposites appears to be a viable option. Qi et al. [79], used highly active nano-composites sorbents for CO_2 capture based on amine-functionalized mesoporous capsules. The capsules offered increased amount of amine groups and reactive sites for the capture of CO_2. An excellent capturing performance of 7.93 mmol/g in simulated flue gas was achieved. Besides the excellent capturing performance of the nanocomposites, the sorbents were readily and fully regenerated at relative low temperatures (<100°C) and also exhibited good stability over repeated adsorption-desorption cycles.

Other researchers and scientists have also employed nanocomposites for capturing CO_2 from the atmosphere [80–86]. Besides the above-mentioned examples of the use of nano-composites for the adsorption of pollution, Table 2.1 illustrates more examples where nanocomposites could be employed for the adsorption of pollutants from the environment.

Table 2.1 Other examples of the possible application of nanocomposites for adsorption of pollutants from the environment

No.	Nanocomposite based-adsorbent	Targeted pollutant	Comments	Ref.
1	Conducting polymer nanocomposites	Cd(II) ions	Cd(II) removal performance using polypyrrole, polyaniline and polythiophene nanocomposites, are about 40.2%, 59% and 99.94%, respectively.	[87]

No.	Nanocomposite based-adsorbent	Targeted pollutant	Comments	Ref.
2	Graphene/Fe_3O_4@ polythiophene nanocomposite	Magnetic solid-phase extraction of polycyclic aromatic hydrocarbons	The optimum extraction conditions were 4 min for extraction time, 20 mg for sorbent amount, 100 mL for initial sample volume and toluene used as the desorption solvent. Good performance results were obtained at optimum conditions.	[88]
3	Cellulose derived magnetic mesoporous carbon nanocomposites.	Cr(VI)	The removal of Cr(VI) was highly *p*H dependent. The MC-O and MC-N nanocomposites adsorbents have the maximum adsorption capacities of 293.8 and 327.5 mg/g respectively. Adsorption behaviour was well fitted with pseudo-second-order-kinetics and Langmuir isotherm model.	[89]
4	Poly(γ-glutamic acid)-coated Fe_3O_4 magnetic nanocomposites	Cr^{3+}, Cu^{2+}, Pb^{2+}, Ni^{2+}	The γ-PGA/Fe_3O_4 MNPs composites were effective in the removal of over 99% of Cr^{3+}, Cu^{2+} and Pb^{2+} and over 77% of Ni^{2+} with an adsorption capacity of 24.60 mg/g from aqueous solution at pH higher than 6.	[90]

(Countinued)

Table 2.1 (*Countinued*)

No.	Nanocomposite based-adsorbent	Targeted pollutant	Comments	Ref.
5	Magnetic MOF@GO and MOF@CNT hybrid nanocomposites	Organic pollutants	Results showed that pollutant adsorption was significantly higher on the hybrid nanocomposites compared to that of the parent materials.	[91]
7	Magnetic zinc/ferrite-reduced graphene oxide nanocomposites.	Methylene blue in water	The nanocomposites exhibited extraordinary adsorption performance for MB in water and the adsorption activities increase with the increasing reduced-graphene oxide content in nanocomposites.	[92]
8	Oxidized MWCNTs nanocomposites	Cr(VI)	The removal of toxic Cr(VI) using the PPy/OMWCNTs NCs as adsorbent maximum sorption capacity of 294.18 mg/g. Results confirmed that the adsorption was dependent on initial solution pH with optimum adsorption achieved at pH 2.	[93]
9	Chitosan-coated iron oxide nanocomposites	Sequestration of As(III)	Maximum monolayer adsorption capacity of 267.2 mg/g was obtained at optimum conditions of pH 6.0,	

No.	Nanocomposite based-adsorbent	Targeted pollutant	Comments	Ref.
			temperature of 30°C and a contact time of 30 min. Adsorption of As(III) onto Chitosan coated iron-oxide nanocomposites was simple physisorption contributing to the reversibility of the process and regeneration of the adsorbent.	[94]
10	2-Ethylhexyl phosphonic acid mono-2-ethylhexyl ester-grafted magnetic silica nanocomposites	Lanthanum (III)	An optimal pH of 5.5 with a maximum adsorption capacity of 55.9 mg/g was obtained. The adsorption followed the second order kinetics equation and Langmuir isotherm model. After 10 times adsorption–desorption cycles, no obvious decrease in adsorption/desorption capacity was noticed.	[95]

2.4.2 Use of Nanocomposites for Catalytic and Redox Degradation of Dyes and Organic Pollutants

Nanoparticles possess great potential as catalysts and redox active substances owing to their high reactivity, large specific surface area and shape-dependent electronic, optical and catalytic properties. These excellent properties have attracted intense research from many researchers to design efficient/chem-catalytic nanocomposites materials for the purification of contaminated water and gases [41]. Common catalytic nanoparticles include among others nanosized semiconductor materials (TiO_2) [96, 97], CdS [98], WO_3 [99], ZnO [100], zero valent metals such as FeO [101, 102], and bimetallic nanoparticles such as Fe/Pd [103], Fe/Ni [104], Zn/Pd [105]. However, their use is limited because these nanoparticles tend to agglomerate, thereby decreasing their surface area and removal efficiency. In addition, inherent problems such as the separation of the fine particles from aqueous solution and recyclability of nanoparticles are always a concern. Therefore, to overcome these problems, researchers have tended to incorporate nanoparticles into polymer matrices, thus forming polymer nanocomposites which serve as better catalytic and redox sorbents for the degradation of environmental pollutants such as dyes [106–108], polychlorinated biphenyls (PCBs) [109–111], halogenated aliphatics [112–114], and nitro-aromatics [115, 116].

Dyes are useful materials in the paper, textile, food, paint and pigments industries. However, wastewater from these industries contains some of these dyes, which are regarded as serious pollutants. It should be noted that the remediation of coloured wastewater is very complex considering the fact that these dyes are stable organic molecules due to their conjugated electron systems. Most dyes used in industrial processes are made up of azoic compounds, which can only be effectively removed or degraded from the environment through adsorption–degradation mechanisms. For example, Li et al. [117] used sonocatalytic activity of Yb, B, Ga-codoped Er^{3+}:$Y_3Al_5O_{12}/TiO_2$ in the degradation of organic dyes using UV-Vis spectrometer under ultrasonic irradiation. In this study, sonocatalysts were prepared by the sol-gel process and characterized by X-ray diffraction (XRD), scanning electron microscopy (SEM), and energy dispersive X-ray spectroscopy (EDX). The (Yb,Er):(B, Ga)YAG/TiO_2 with a 1:3 mass ratio heat treated at 550°C for 60 min showed the highest sonocatalytic effect for the degradation of organic dyes. Furthermore, the optimized (Yb,Er):(B,Ga)YAG/TiO_2 adsorbent was very effective in the degradation of several organic dye pollutants under ultrasonic irradiation.

Rajendran and co-workers [118] prepared ZnO/CeO_2 nanocomposites via the thermal decomposition method with various proportions of ZnO and CeO_2, for photocatalytic degradation of methyl orange, methyl blue and phenol under visible light irradiation. After characterization of the nanocomposites, the XPS results confirmed the presence of Ce^{3+} ions in the ZnO/CeO_2 nanocomposites. The optimized and highly efficient sample comprised ZnO/CeO_2 in the ratio of 90:10 and showed excellent photocatalytic degradation performance for the degradation of methyl orange, methylene blue, phenol as well as industrial textile effluent compared to pure ZnO, CeO_2 or other nanocomposites. Furthermore, the same nanocomposite was used for the electrochemical detection of uric acid and the findings proved to be highly promising in terms of detection, sensitivity and performance compared to other investigated materials. The excellent performance was attributed to the presence of the Ce^{3+}, large surface area of the nanocomposite with less agglomeration of the nanoparticle. Overall, the (90:10) ZnO/CeO_2 nanocomposite exhibited an excellent performance and capability as a potential adsorbent for the photocatalytic degradation of pollutants from the environment [118].

Muthirulan et al. reported the efficiency of titania-graphene nanocomposites in the degradation of acid orange 7 dye (azoic compound) from aqueous solution [119]. In this study, it was found that graphene retarded charge recombination and electron transfer within the nanocomposites. The acid orange degradation into several aromatic intermediates was analysed using a GC-MS and degradation capacity of the nanocomposites only slightly decreased from 95.1% to 88.1% after four conservative cycles.

Panthi et al. [120] reviewed the use of electrospun ZnO hybrid nanofibres for photo-degradation of organic dyes from wastewater. In this review, various ideas were highlighted for a successful modification of electrospun ZnO nanofibres for improved photodegradation of organic dyes in the presence of UV light or visible light. Furthermore, the review also presented the fundamental principles and applications of electrospun nanofibres in terms of environmental and economic factors. Overall, it was concluded that the use of low-cost electrospun nanocomposite fibres in the photodegradation of organic dyes could accelerate in the coming years. Chowdhury and Balasubramanian [121] further reviewed recent advancement in the use graphene/semiconductor nanocomposites (GSNs) as a new catalyst for the heterogeneous photocatalytic treatment of industrial wastewater. Among others, this review presented the use of ternary nanocomposites such as Bi_2O_3/TiO_2/graphene, ZnO/Ag/graphene, Ti/Ce–GO, Au@Mo–GO, and Ag/AgBr/rGO for the degradation of rhodamine blue dyes from wastewater with a high level of success. Overall, the authors presented in detail the high potential capability of GSNs for the photodecolorization of textile effluents.

Both aromatic and aliphatic halogenated compounds are used as herbicides, pesticides and solvents and in wood preservation. These compounds are stable and non-biodegradable, some containing carcinogenic agents with long persistence in the environment. Several studies have been carried out for the removal or degradation of these pollutants through the use nanocomposites. For example, Tobajas et al. [122] employed the use of TiO_2–ZnO/clay nanoarchitectures for the degradation of emerging pollutants in water under solar irradiation. In this study, a novel TiO_2–ZnO/clay nanoarchitectures was developed via the sol-gel process where TiO_2–ZnO nanoparticles were obtained in a one-step pathway and supported on the surface of a delaminated layered clay. Structural characterization revealed a low ZnO loading (from 0.2 to 1%) was very possible to achieved crystallization of anatase without the formation of undesirable phases. The TiO_2–ZnO/cloisite with a 0.5% loading of ZnO gave the best performance for acetaminophen and antipyrine degradation under solar light. Good results were obtained after several operation cycles using TiO_2–ZnO/clay nanoarchitectures to promote the degradation of emerging pollutants under solar light. Overall, TiO_2–ZnO/clay nanoarchitectures showed the potentials for degradation of emerging water pollutants such as pharmaceuticals [122].

Phenolic compounds pose a serious threat to human health and environmental hazard. Therefore, Ganigar et al. [123] used polymer–clay nanocomposites for the removal of trichlorophenols and trinitrophenol from water. In this study, polycation–clay mineral nanocomposites were prepared and characterized and used as adsorbents for the removal of two pollutants, trinitrophenol (picric acid-PA) and trichlorophenols (TCP), from rivers, lakes and water reservoirs. The results from this study showed both the PA (anionic) and TCP (non-ionic) with high affinity to the less charged polycation PVPcoS (40% monomers charge) than the highly charged polycation (polydiallyl dimethyl ammonium chloride (PDADMAC)). Overall, both pollutants showed higher affinity towards PVPcoS-MMT driven by van der Waals interactions, thus corresponding to the linear curves obtained. However, PA exhibited a higher affinity to the nanocomposite relative to TCP, suggesting electrostatic interaction of forces.

Hernández-Hernández et al. [124] reviewed the used of polymer–clay nanocomposites and composites for the removal of organic compounds from the environment. The authors noted that organic pollutants such as phenol, 4-nitrophenol, pentachlorophenol, 2-nitrophenol, 2-chloro-4-dinitrophenol, and 2-chloro-4-dinitrophenol could be effectively from the environment using polymer–clay nanocomposites with much efficiency.

Furthermore, nanocomposites have also shown a great potential for the inhabitation of some microorganisms such as *Staphylococcus aureus*, *Pseudomonas aeruginosa*, *Escherichia coli*, *Enterococcus faecalis*, *Bacillus subtilis*, *Listonella anguillarum*, *Aspergillus niger* and *Bacillus cereus* [125]. For example, Song Jin et al. [125] removed bacteria and viruses from water using layered double hydroxide nanocomposites. In this study, the layered double hydroxide nanocomposites were synthesized and tested for microorganism removal. Results from this study showed the nanocomposites composed of magnesium–aluminium or zinc–aluminium with viral and bacterial adsorption efficiency of ≥99% at viral concentration between 5.9×10^6 and 9.1×10^6 plaque-forming units (pfu) and bacteria concentration between 1.6×10^{10} and 2.6×10^{10} colony-forming units (cfu) when exposed to the layered doubled hydroxide nanocomposites.

2.4.3 Use of Nanocomposites for Sensing and Detection of Pollutants

The use of rapid and precise sensors for the detection and sensing of pollutants at molecular levels from the environment is an effective means of understanding the types of pollutants present in our environment. A nano-sensor has the potential and capability to detect some pollutants such as metal ions and micro-pollutants from the environment without any preconcentration required. Furthermore, the use of these sensors is economical (cost effective) because the synthesis requires the use of conventional microelectronics manufacturing equipment using simple electrochemical techniques [126]. With this understanding, ecosystem monitoring, and environmental decision-making could be improved if more sensitive and cost-effective techniques are available for the detection of contaminants from the environment. Nanotechnology-based electrochemical biosensors for biomonitoring chemical exposures on humans who have been exposed to chemicals [126].

Air pollution has suddenly become a global environmental problem today. Therefore, the detection of gases, especially toxic gases such as SO_3 and NO_2 in our environment forms a strong basis for the control of air pollution. For example, NO_2 is a toxic compound produced by the combustion in power plants and combustion engines and very harmful to the environment because it is the major cause of acid rain, photochemical smog and pollution haze [127]. Actually, nanocomposites of late have been receiving a great deal of attention for pollutant detection owing to their large surface area, good compatibility, adjustable transport properties and chemical specificities, and the ease of processing coupled with scalable productions [41]. Studies have shown that NOx, SO_2 and CO_2 have been effectively removed by using carbon nanotubes (CNTs) on highly porous manganese oxide with gold nanoparticles [126].

Srivastava et al. [128] developed a graphene-based WO_3 nanocomposite for gas-sensing applications. The graphene–WO_3 nanocomposite thin-layer sensors were prepared by drop coating of dispersed solution onto alumina substrate and used for the detection of NO_2 gas. Transmission electron microscopy (TEM) characterization of these nanocomposites showed well-dispersed WO_3 nanoparticles within the graphene nanosheets. The addition of WO_3 nanoparticles to graphene increased the surface area for interaction. The results revealed that of the three different compositions (0.2, 0.5 and 1.0 wt%) of nanocomposites produced for the gas-sensing measurements, the graphene–

WO_3 nanocomposite with 0.5 wt% showed the highest response and recovery time. Results also showed the sensor response to NO_2 increasing to nearly three times with the 0.5 wt% nanocomposites sensor as compared to a pure WO_3 layer at room temperature.

Sakthinathan and Chen [129] used graphene-supported nanocomposites as sensor materials for the detection of environment pollutants such as hydrazine, phenolic compounds and arsenic. The authors used MnO_2/graphene oxide nanocomposites (MnO_2/GO), gold–graphene oxide nanocomposites (Au/GO) and silver–graphene oxide (Ag/GO) for the detection of hydrazine, phenolic and arsenic respective from the environment. In all the above-mentioned nanocomposites, excellent results were obtained. Overall, individual graphene oxide when combined with the nanoparticles offered excellent characteristics (such as enhanced physical and chemical characteristics) and displayed capability of being used as a sensor material for the detection and degradation of these pollutants.

Anbazhagan et al. [130] demonstrated biogenic silver nanoparticles (AgNPs) based on naturally occurring N-acylethanolamine (NAEA) as a capping agent and sodium borohydride as a reducing agent for the selective sensing of Hg^{2+} from biological fluid (blood plasma) and environmental samples such as tap water, pond water and sewage samples. In addition, the antimicrobial activity of NSEA-AgNPs against *S. aureus, B. thuringiensis, E. coli, S. typhi, Shigella, K. pneumoniae* was tested by a zone of inhibition and results were very promising with the NSEA-AgNPs being capable of inhibiting Gram-positive and Gram-negative bacteria species. The authors also demonstrated the use of NSEA-AgNPs as effective sorbents for the degradation of environmental organic dyes such as basic dye and methylene blue dye. Overall, the nanocomposite was recommended for commercial use for environmental applications.

Sadanand Pandey [131] reviewed extensively gas sensors based on polyaniline nanocomposites. The author summarizes the recent advantages in the use of PANI nanocomposites for the construction of sensors and sensing properties for the sensing of gas/vapours such as H_2, HCl, NO_2, NH_3, CO, CO_2, SO_2, LPG, chemical warfare agents (CWAs) and vapour of volatile organic compounds (VOCs) in the environment. The author highlighted the mechanisms involved in the sensing of various gases and concluded the effectiveness of the use of polyaniline nanocomposites for sensing of pollutants.

2.5 Conclusions and Perspectives

The traditional treatment methods for the remediation of environmental pollution are proving to be ineffective and sometimes expensive. Therefore, nanotechnology-based pollution treatment methods are key to supplement the already existing conventional methods. Undoubtedly, the unique properties of nanocomposites as mentioned in this chapter have necessitated the endless use of nanocomposites for many applications as well as in pollution control. Research on the use of nanocomposites for environmental remediation has attracted considerable investment from government and the business sectors worldwide. A vast number of publications regarding the potential use of nanocomposites for the removal of pollutants from air, water and soil have been published in the past decades. In summary, the findings presented in this work illustrate the high adsorptive capability and potentials of nanocomposites for environmental remediation compared to other nanomaterials.

However, researchers need to continue to investigate various strategies to optimize the synthesis of these nanocomposites in order to achieve better-quality materials with improved mechanical and transport properties. Furthermore, more research needs to be carried out to explore the use of low-cost nanomaterials in pollution control. This calls for more funding and more research to further assess the applicability of other hybrid nanomaterials in pollution control. Hopefully, when this is done, it will go a long way in addressing the few shortcomings of nanocomposites for pollution control.

Acknowledgments

Dr. Ephraim Vunain acknowledges the financial support from the College of Science, Engineering and Technology (CSET) of the University of South Africa (UNISA), South Africa. The authors acknowledge the University of South Africa Database and laboratory facility.

References

1. W. E. Cascio, Proposed pathophysiologic framework to explain some excess cardiovascular death associated with ambient air particle pollution: Insights for public health translation, *Biochim. Biophys. Acta: Gen. Subj.* 1860 (2016) 2869–2879. doi:10.1016/j.bbagen.2016.07.016.
2. B. A. Franklin, R. Brook, C. Arden Pope, Air pollution and cardiovascular disease, *Curr. Probl. Cardiol.* 40 (2015) 207–238. doi:10.1016/j.cpcardiol.2015.01.003.
3. L.-L. Cui, J. Zhang, J. Zhang, J.-W. Zhou, Y. Zhang, T.-T. Li, Acute respiratory and cardiovascular health effects of an air pollution event, January 2013, Jinan, China, Public Health. 1 (2015) 2013–2016. doi:10.1016/j.puhe.2015.11.003.
4. J. K. Vanos, C. Hebbern, S. Cakmak, Risk assessment for cardiovascular and respiratory mortality due to air pollution and synoptic meteorology in 10 Canadian cities, *Environ. Pollut.* 185 (2014) 322–332. doi:10.1016/j.envpol.2013.11.007.
5. A. Arnedo-Pena, L. García-Marcos, I. C. Urueña, R. B. Monge, M. M. Suárez-Varela, I. M. Canflanca, J. B. Garrido, A. B. Quirós, Á. López-Silvarrey Varela, G. G. Hernández, I. A. Ontoso, C. G. Díaz, Air pollution and recent symptoms of asthma, allergic rhinitis, and atopic eczema in schoolchildren aged between 6 and 7 years, *Arch. Bronconeumol.* (English ed. 45 (2009) 224–229. doi:10.1016/S1579-2129(09)72152-4).
6. K. H. Kim, S. A. Jahan, E. Kabir, A review on human health perspective of air pollution with respect to allergies and asthma, *Environ. Int.* 59 (2013) 41–52. doi:10.1016/j.envint.2013.05.007 Review.
7. H. Tong, Dietary and pharmacological intervention to mitigate the cardiopulmonary effects of air pollution toxicity, *Biochim. Biophys. Acta: Gen. Subj.* 1860 (2016) 2891–2898. doi:10.1016/j.bbagen.2016.05.014.
8. C. H. Chen, C. C. Chan, B. Y. Chen, T. J. Cheng, Y. Leon Guo, Effects of particulate air pollution and ozone on lung function in non-asthmatic children, *Environ. Res.* 137 (2015) 40–48. doi:10.1016/j.envres.2014.11.021.
9. S. Surdu, M. S. Bloom, I. A. Neamtiu, C. Pop, D. Anastasiu, E. F. Fitzgerald, E. S. Gurzau, Consumption of arsenic-contaminated drinking water and anemia among pregnant and non-pregnant

women in northwestern Romania, *Environ. Res.* 140 (2015) 657–660. doi:10.1016/j.envres.2015.05.020.

10. P. Mandal, S. R. Debbarma, A. Saha, B. Ruj, Disposal problem of arsenic sludge generated during arsenic removal from drinking water, *Procedia Environ. Sci.* 35 (2016) 943–949. doi:10.1016/j.proenv.2016.07.084.
11. T. G. Kazi, K. D. Brahman, H. I. Afridi, M. B. Arain, F. N. Talpur, A. Akhtar, The effects of arsenic contaminated drinking water of livestock on its total levels in milk samples of different cattle: Risk assessment in children, *Chemosphere.* 165 (2016) 427–433. doi:10.1016/j.chemosphere.2016.09.015.
12. A. Rasool, A. Farooqi, T. Xiao, S. Masood, M. A. Kamran, S. Bibi, Elevated levels of arsenic and trace metals in drinking water of Tehsil Mailsi, Punjab, Pakistan, *J. Geochemical Explor.* 169 (2016) 89–99. doi:10.1016/j.gexplo.2016.07.013.
13. Y.-Y. Cheng, N.-C. Huang, Y.-T. Chang, J.-M. Sung, K.-H. Shen, C.-C. Tsai, H.-R. Guo, Associations between arsenic in drinking water and the progression of chronic kidney disease: A nationwide study in Taiwan, *J. Hazard. Mater.* 321 (2016) 432–439. doi:10.1016/j.jhazmat.2016.09.032.
14. E. Vunain, A. K. Mishra, R. W. Krause, Fabrication, characterization and application of polymer nanocomposites for arsenic(III) removal from water, *J. Inorg. Organomet. Polym. Mater.* 23 (2013) 293–305. doi:10.1007/s10904-012-9775-8.
15. J. Trujillo-Reyes, J. R. Peralta-Videa, J. L. Gardea-Torresdey, Supported and unsupported nanomaterials for water and soil remediation: Are they a useful solution for worldwide pollution?, *J. Hazard. Mater.* 280 (2014) 487–503. doi:10.1016/j.jhazmat.2014.08.029.
16. A. Muñoz-Bonilla, M. Fernández-García, The roadmap of antimicrobial polymeric materials in macromolecular nanotechnology, *Eur. Polym. J.* 65 (2015) 46–62. doi:10.1016/j.eurpolymj.2015.01.030.
17. L. Zhang, L. Chen, J. Liu, X. Fang, Z. Zhang, Effect of morphology of carbon nanomaterials on thermo-physical characteristics, optical properties and photo-thermal conversion performance of nanofluids, *Renew. Energy.* 99 (2016) 888–897. doi:10.1016/j.renene.2016.07.073.
18. J. González-Sálamo, B. Socas-Rodríguez, J. Hernández-Borges, M. Á. Rodríguez-Delgado, Nanomaterials as sorbents for food sample analysis, *Trends Anal. Chem.* 85 (2016) 203–220. doi:10.1016/j.trac.2016.09.009.

19. Saloma, A. Nasution, I. Imran, M. Abdullah, Improvement of concrete durability by nanomaterials, *Procedia Eng.* 125 (2015) 608–612. doi:10.1016/j.proeng.2015.11.078.

20. I. Ganesh, Electrochemical conversion of carbon dioxide into renewable fuel chemicals—The role of nanomaterials and the commercialization, *Renew. Sustain. Energy Rev.* 59 (2016) 1269–1297. doi:10.1016/j.rser.2016.01.026.

21. R. J. B. Peters, H. Bouwmeester, S. Gottardo, V. Amenta, M. Arena, P. Brandhoff, H. J. P. Marvin, A. Mech, F. B. Moniz, L. Q. Pesudo, H. Rauscher, R. Schoonjans, A. K. Undas, M. V. Vettori, S. Weigel, K. Aschberger, Nanomaterials for products and application in agriculture, feed and food, *Trends Food Sci. Technol.* 54 (2016) 155–164. doi:10.1016/j.tifs.2016.06.008.

22. P. Bhanja, A. Bhaumik, Porous nanomaterials as green catalyst for the conversion of biomass to bioenergy, *Fuel.* 185 (2016) 432–441. doi:10.1016/j.fuel.2016.08.004.

23. S. Chaudhary, A. Umar, S. K. Mehta, Selenium nanomaterials: An overview of recent developments in synthesis, properties and potential applications, *Prog. Mater. Sci.* 83 (2016) 270–329. doi:10.1016/j.pmatsci.2016.07.001.

24. S. Ravi, S. Vadukumpully, Sustainable carbon nanomaterials: Recent advances and its applications in energy and environmental remediation, *J. Environ. Chem. Eng.* 4 (2015) 835–856. doi:10.1016/j.jece.2015.11.026.

25. L. Figueiredo, G. L. Erny, L. Santos, A. Alves, Applications of molecularly imprinted polymers to the analysis and removal of personal care products: A review, *Talanta.* 146 (2016) 754–765. doi:10.1016/j.talanta.2015.06.027.

26. Z. Wang, J. Yu, R. Gui, H. Jin, Y. Xia, Carbon nanomaterials-based electrochemical aptasensors, *Biosens. Bioelectron.* 79 (2016) 136–149. doi:10.1016/j.bios.2015.11.093.

27. P. Pashazadeh, A. Mokhtarzadeh, M. Hasanzadeh, M. Hejazi, Biosensors and bioelectronics nano-materials for use in sensing of salmonella infections: Recent advances, *Biosens. Bioelectron.* 87 (2017) 1050–1064. doi:10.1016/j.bios.2016.08.012.

28. N. Cuesta, I. Cameán, A. Ramos, S. De Llobet, A. B. García, Graphitic nanomaterials from biogas-derived carbon nanofibers, *Fuel Process. Technol.* 152 (2016) 1–6. doi:10.1016/j.fuproc.2016.05.043.

29. F. Varenne, J. Botton, C. Merlet, H. Hillaireau, F. Legrand, G. Barratt, Size of monodispersed nanomaterials evaluated by dynamic light

scattering: Protocol validated for measurements of 60 and 203 nm diameter nanomaterials is now extended to 100 and 400 nm, *Int. J. Pharm.* 515 (2016) 245–253. doi:10.1016/j.ijpharm.2016.10.016.

30. S. F. Ahmed, M. Khalid, W. Rashmi, A. Chan, K. Shahbaz, Recent progress in solar thermal energy storage using nanomaterials, *Renew. Sustain. Energy Rev.* 67 (2017) 450–460. doi:10.1016/j.rser.2016.09.034.
31. L. Wang, Q. Xiong, F. Xiao, H. Duan, 2D nanomaterials based electrochemical biosensors for cancer diagnosis, *Biosens. Bioelectron.* 89 (2016) 1–16. doi:10.1016/j.bios.2016.06.011.
32. N. B. Singh, S. Rai, S. Agarwal, Polymer nanocomposites and Cr(VI) removal from water, *Nanosci. Technol.* 1 (2014) 1–10.
33. W. P. Flauzino Neto, M. Mariano, I. S. V. da Silva, H. A. Silvério, J. L. Putaux, H. Otaguro, D. Pasquini, A. Dufresne, Mechanical properties of natural rubber nanocomposites reinforced with high aspect ratio cellulose nanocrystals isolated from soy hulls, *Carbohydr. Polym.* 153 (2016) 143–152. doi:10.1016/j.carbpol.2016.07.073.
34. P. Henrique, C. Camargo, K. G. Satyanarayana, F. Wypych, Nanocomposites: Synthesis, structure, properties and new application opportunities, *Mater. Res.* 12 (2009) 1–39. doi:10.1590/S1516-14392009000100002.
35. E. Vunain, A. K. Mishra, B. B. Mamba, Overview of polymer composites for heavy metal removal in *Heavy Metals: Sources, Toxicity and Remediation Techniques*, Nova Publishers (2016) pp. 35–58.
36. F. Z. Haque, P. Tripathi, M. Husain, K. S. Pandey, Polymer nanocomposites preparation characterization and application: An Overview, *Mat. Sci. Res. India* 4 (2007) 93–114.
37. C. C. Okpala, The benefits and applications of nanocomposites, *Int. J. Adv. Eng. Technol.* V (2014) 12–18.
38. B. Julia, M. Popall, Applications of hybrid organic–inorganic nanocomposites, 15 (2005) 3559–3592. doi:10.1039/b509097k.
39. Y. Takagai, A. Shibata, S. Kiyokawa, T. Takase, Synthesis and evaluation of different thio-modified cellulose resins for the removal of mercury (II) ion from highly acidic aqueous solutions, *J. Colloid Interface Sci.* 353 (2011) 593–597. doi:10.1016/j.jcis.2010.09.070.
40. D. W. O'Connell, C. Birkinshaw, T. F. O'Dwyer, Heavy metal adsorbents prepared from the modification of cellulose: A review, *Bioresour. Technol.* 99 (2008) 6709–6724. doi:10.1016/j.biortech.2008.01.036.

41. X. Zhao, L. Lv, B. Pan, W. Zhang, S. Zhang, Q. Zhang, Polymer-supported nanocomposites for environmental application: A review, *Chem. Eng. J.* 170 (2011) 381–394. doi:10.1016/j.cej.2011.02.071.

42. M. A. M. Khraisheh, M. F. Tutunji, Sorption of lead ions on diatomite and manganese oxides modified diatomite, *Water Res.* 35 (2001) 3724–3728.

43. N. Caliskan, A. R. Kul, S. Alkan, E. G. Sogut, Adsorption of Zinc(II) on diatomite and manganese-oxide-modified diatomite: A kinetic and equilibrium study, *J. Hazard. Mater.* 193 (2011) 27–36. doi:10.1016/j.jhazmat.2011.06.058.

44. Y. Chen, B. Pan, S. Zhang, H. Li, L. Lv, W. Zhang, Immobilization of polyethylenimine nanoclusters onto a cation exchange resin through self-crosslinking for selective Cu(II) removal, *J. Hazard. Mater.* 190 (2011) 1037–1044. doi:10.1016/j.jhazmat.2011.04.049.

45. G. S. Dhillon, S. Kaur, M. Verma, *Biopolymer-Based Nanomaterials: Potential Applications in Bioremediation of Contaminated Wastewaters and Soils*, 1st ed, Elsevier BV, 2012. doi:10.1016/B978-0-444-56328-6.00003-7.

46. Q. Zhang, T. Jiao, Polymer-supported organic-inorganic nanomaterials: Fabrication, characterization and environmental application in *Materials and Processes for Energy: Communicating Current Research and Technological Developments* (A. Méndez-Vila, Ed.), (2013) 903–912.

47. M. G. Kanatzidis, C.-G. Wu, Conductive polymer bronzes. Intercalated polyaniline in V_2O_5 xerogels, *J. Am. Chem. Soc.*, 111 (1989) 4139–4141.

48. L. Wang, C. Science, Lamellar polymer–Li MoO nanocomposites via encapsulative precipitation x, 7 (1997) 1277–1283.

49. Y. Liu, J. L. Schindler, D. C. Degroot, C. R. Kannewurf, W. Hirpo, M. G. Kanatzidis, Synthesis, Structure and reactions of poly(ethylene oxide)/V_2O_5 intercalative nanocomposites, 8 (1996) 525–534.

50. T. Section, A. M. Crystals, R. Bissessur, J. L. Schindler, C. R. Kannewurf, M. Kanatzidis, Nanoscale composites formed by encapsulation of polymers in MoS_2 from conjugated polymers to plastics. Detection of metal to insulator transition, (2006). doi:10.1080/10587259408051697.

51. A. K. Sengupta, The donnan membrane principle: Opportunities for sustainable engineered processes and materials, *Environ. Sci. Technol.* 44 (2010) 1161–1166.

52. B. Pan, H. Qiu, B. Pan, G. Nie, L. Xiao, L. Lv, W. Zhang, Q. Zhang, S. Zheng, Highly efficient removal of heavy metals by polymer-supported nanosized hydrated Fe(III) oxides: Behavior and XPS study, *Water Res.* 44 (2010) 815–824. doi:10.1016/j.watres.2009.10.027.

53. E Vunain, AK Mishra, BB Mamba, Dendrimers, mesoporous silicas and chitosan-based nanosorbents for the removal of heavy-metal ions: A review, *Int. J. Biol. Macromol.* 86 (2016) 570–586.

54. F. Hussain, M. Hojjati, M. Okamoto, R. E. Gorga, Review article: Polymer-matrix nanocomposites, processing, manufacturing, and application: An overview, *J. Composite Mater.* 40 (2006) 1511–1575.

55. X. Tan, Y. Liu, Y. Gu, Y. Xu, G. Zeng, X. Hu, S. Liu, X. Wang, S. Liu, J. Li, Biochar-based nano-composites for the decontamination of wastewater: A review, *Bioresour. Technol.* 212 (2016) 318–333. doi:10.1016/j.biortech.2016.04.093.

56. M. Othmani, A. Aissa, A. Grelard, R. K. Das, R. Oda, M. Debbabi, Synthesis and characterization of hydroxyapatite-based nanocomposites by the functionalization of hydroxyapatite nanoparticles with phosphonic acids, *Colloids Surf. A Physicochem. Eng. Asp.* 508 (2016) 336–344. doi:10.1016/j.colsurfa.2016.08.078.

57. X. Zhang, C. Gao, M. Liu, Y. Huang, X. Yu, E. Ding, Applied surface science synthesis and characterization of asymmetric polymer/inorganic nanocomposites with pH/temperature sensitivity, *Appl. Surf. Sci.* 264 (2013) 636–643. doi:10.1016/j.apsusc.2012.10.084.

58. Y. Haldorai, J. J. Shim, K. T. Lim, Synthesis of polymer-inorganic filler nanocomposites in supercritical CO_2, *J. Supercrit. Fluids.* 71 (2012) 45–63. doi:10.1016/j.supflu.2012.07.007.

59. M. Mansournia, L. Ghaderi, CuO@ZnO core-shell nanocomposites: Novel hydrothermal synthesis and enhancement in photocatalytic property, *J. Alloys Compd.* 691 (2017) 171–177. doi:10.1016/j.jallcom.2016.08.267.

60. N. Kannapiran, A. Muthusamy, P. Chitra, S. Anand, R. Jayaprakash, Poly(o-phenylenediamine)/$NiCoFe_2O_4$ nanocomposites: Synthesis, characterization, magnetic and dielectric properties, *J. Magn. Magn. Mater.* 423 (2017) 208–216. doi:10.1016/j.jmmm.2016.09.095.

61. K. C. Swallow, D. N. Hume, F. M. M. Morel, Sorption of copper and lead by hydrous ferric oxide, *Environ. Sci. Technol.* 14 (1980) 1326–1331. doi:10.1021/es60171a003.

62. M. Fan, T. Boonfueng, Y. Xu, L. Axe, T. A. Tyson, Modeling Pb sorption to microporous amorphous oxides as discrete particles and

coatings, *J. Colloid Interface Sci.* 281 (2005) 39–48. doi:10.1016/j.jcis.2004.08.050.

63. J. Hu, G. Chen, I. M. C. Lo, Removal and recovery of Cr(VI) from wastewater by maghemite nanoparticles, *Water Res.* 39 (2005) 4528–4536. doi:10.1016/j.watres.2005.05.051.

64. Y. Xu, T. Boonfueng, L. Axe, S. Maeng, T. Tyson, Surface complexation of Pb(II) on amorphous iron oxide and manganese oxide: Spectroscopic and time studies, *J. Colloid Interface Sci.* 299 (2006) 28–40. doi:10.1016/j.jcis.2006.01.041.

65. Q. Su, B. Pan, S. Wan, W. Zhang, L. Lv, Use of hydrous manganese dioxide as a potential sorbent for selective removal of lead, cadmium, and zinc ions from water, *J. Colloid Interface Sci.* 349 (2010) 607–612. doi:10.1016/j.jcis.2010.05.052.

66. M. Kawashima, Y. Tainaka, T. Hori, M. Koyama, T. Takamatsu, Phosphate adsorption onto hydrous manganese (IV) oxide in the presence of divalent cations, *Water Res.* 20 (1986) 471–475. doi:10.1016/0043-1354(86)90195-8.

67. B. Pan, Q. Zhang, W. Du, W. Zhang, B. Pan, Q. Zhang, Z. Xu, Q. Zhang, Selective heavy metals removal from waters by amorphous zirconium phosphate: Behavior and mechanism, *Water Res.* 41 (2007) 3103–3111. doi:10.1016/j.watres.2007.03.004.

68. K. Jia, B. Pan, Q. Zhang, W. Zhang, P. Jiang, C. Hong, B. Pan, Q. Zhang, Adsorption of Pb^{2+}, Zn^{2+}, and Cd^{2+} from waters by amorphous titanium phosphate, *J. Colloid Interface Sci.* 318 (2008) 160–166. doi:10.1016/j.jcis.2007.10.043.

69. B. C. Pan, Q. R. Zhang, W. M. Zhang, B. J. Pan, W. Du, L. Lv, Q. J. Zhang, Z. W. Xu, Q. X. Zhang, Highly effective removal of heavy metals by polymer-based zirconium phosphate: A case study of lead ion, *J. Colloid Interface Sci.* 310 (2007) 99–105. doi:10.1016/j.jcis.2007.01.064.

70. L. M. Blaney, S. Cinar, A. K. SenGupta, Hybrid anion exchanger for trace phosphate removal from water and wastewater, *Water Res.* 41 (2007) 1603–1613. doi:10.1016/j.watres.2007.01.008.

71. M. J. DeMarco, A. K. SenGupta, J. E. Greenleaf, Arsenic removal using a polymeric/inorganic hybrid sorbent, *Water Res.* 37 (2003) 164–176. doi:10.1016/S0043-1354(02)00238-5.

72. L. Cumbal, A. K. Sengupta, Arsenic removal using polymer-supported hydrated iron(III) oxide nanoparticles: Role of Donnan membrane effect, *Environ. Sci. Technol.* 39 (2005) 6508–6515. doi:10.1021/es050175e.

73. Q. Zhang, B. Pan, X. Chen, W. Zhang, B. Pan, Q. Zhang, L. Lv, X. S. Zhao, Preparation of polymer-supported hydrated ferric oxide based on Donnan membrane effect and its application for arsenic removal, *Sci. China Ser. B Chem.* 51 (2008) 379–385. doi:10.1007/s11426-007-0117-6.

74. C. Jing, Z. Zhaoxiang, X. U. Hong, Y. A. O. Zhong, Fabrication of poly (γ-glutamic acid)-coated Fe_3O_4 magnetic nanoparticles and their application in heavy metal removal, *Chinese J. Chem. Eng.* 21 (2013) 1244–1250. doi:10.1016/S1004-9541(13)60629-1.

75. R. Li, L. Liu, F. Yang, Preparation of polyaniline/reduced graphene oxide nanocomposite and its application in adsorption of aqueous Hg (II), *Chem. Eng. J.* 229 (2013) 460–468. doi:10.1016/j.cej.2013.05.089.

76. T. S. Najim, A. J. Salim, Polyaniline nanofibers and nanocomposites: Preparation, characterization, and application for Cr(VI) and phosphate ions removal from aqueous solution, *Arab. J. Chem.* (2014). doi:10.1016/j.arabjc.2014.02.008.

77. A. Sharma, B. Lee, Cd (II) removal and recovery enhancement by using acrylamide–titanium nanocomposite as an adsorbent, *Appl. Surf. Sci.* 313 (2014) 624–632. doi:10.1016/j.apsusc.2014.06.034.

78. K. Shweta, H. Jha, Rice husk extracted lignin–TEOS biocomposites: Effects of acetylation and silane surface treatments for application in nickel removal, *Biotechnol. Reports.* 7 (2015) 95–106. doi:10.1016/j.btre.2015.05.003.

79. G. Qi, Y. Wang, L. Estevez, X. Duan, N. Anako, A.-H. A. Park, W. Li, C. W. Jones, E. P. Giannelis, High efficiency nanocomposite sorbents for CO_2 capture based on amine-functionalized mesoporous capsules, *Energy Environ. Sci.* 4 (2011) 444–452. doi:10.1039/c0ee00213e.

80. J. Wang, X. Mei, L. Huang, Q. Zheng, Y. Qiao, K. Zang, S. Mao, R. Yang, Z. Zhang, Y. Gao, Z. Guo, Z. Huang, Q. Wang, Synthesis of layered double hydroxides/graphene oxide nanocomposite as a novel high-temperature CO<inf>2</inf> adsorbent, *J. Energy Chem.* 24 (2015) 127–137. doi:10.1016/S2095-4956(15)60293-5.

81. M. M. Gui, W. M. P. Wong, S.-P. Chai, A. R. Mohamed, One-pot synthesis of Ag-MWCNT@TiO_2 core–shell nanocomposites for photocatalytic reduction of CO_2 with water under visible light irradiation, *Chem. Eng. J.* 278 (2014) 272–278. doi:10.1016/j.cej.2014.09.022.

82. M. R. Awual, Novel nanocomposite materials for efficient and selective mercury ions capturing from wastewater, *Chem. Eng. J.* 307 (2017) 456–465. doi:10.1016/j.cej.2016.08.108.

83. N. H. Khdary, M. A. Ghanem, M. G. Merajuddine, F. M. Bin Manie, Incorporation of Cu, Fe, Ag, and Au nanoparticles in mercapto-silica (MOS) and their CO_2 adsorption capacities, *J. CO_2 Util.* 5 (2014) 17–23. doi:10.1016/j.jcou.2013.11.003.

84. A. Biswas, T. Tokoly, T. Wang, P. Ramidi, A. Ghosh, E. Dervishi, F. Watanabe, A. S. Biris, I. S. Bayer, M. Grant Norton, Design and synthesis of sprayable nanocomposite coatings for carbon capture and direct conversion into environmentally safe stable carbonates, *Chem. Phys. Lett.* 508 (2011) 276–280. doi:10.1016/j.cplett.2011.04.061.

85. G. Jing, F. Pan, B. Lv, Z. Zhou, Immobilization of carbonic anhydrase on epoxy-functionalized magnetic polymer microspheres for CO_2 capture, *Process Biochem.* 50 (2015) 2234–2241. doi:10.1016/j.procbio.2015.09.015.

86. W. Zhang, C. Xie, G. Zhang, J. Zhang, S. Zhang, D. Zeng, Porous $LaFeO_3$/SnO_2 nanocomposite film for CO_2 detection with high sensitivity, *Mater. Chem. Phys.* 186 (2017) 228–236.

87. L. Zoleikani, H. Issazadeh, B. Z. Nezhad, Preparation of new conductive polymer nanocomposites for cadmium removal from industrial wastewater, *J. Chem. Technol. Metall.* 50 (2015) 71–80.

88. A. Mehdinia, N. Khodaee, A. Jabbari, Fabrication of graphene/Fe_3O_4@polythiophene nanocomposite and its application in the magnetic solid-phase extraction of polycyclic aromatic hydrocarbons from environmental water samples, *Anal. Chim. Acta.* 868 (2015) 1–9. doi:10.1016/j.aca.2014.12.022.

89. B. Qiu, H. Gu, X. Yan, J. Guo, Y. Wang, D. Sun, Cellulose derived magnetic mesoporous carbon nanocomposites with enhanced hexavalent chromium removal, *J. Mater. Chem. A Mater. Energy Sustain.* 2 (2014) 17454–17462. doi:10.1039/C4TA04040F.

90. J. Chang, Z. Zhong, H. Xu, Z. Yao, R. Chen, Fabrication of Poly(γ-glutamic acid)-coated Fe_3O_4 magnetic nanoparticles and their application in heavy metal removal, *Chinese J. Chem. Eng.* 21 (2013) 1244–1250. doi:10.1016/S1004-9541(13)60629-1.

91. V. Jabbari, J. M. Veleta, M. Zarei-Chaleshtori, J. Gardea-Torresdey, D. Villagrán, Green synthesis of magnetic MOF@GO and MOF@CNT hybrid nanocomposites with high adsorption capacity towards organic pollutants, *Chem. Eng. J.* 304 (2016) 774–783. doi:10.1016/j.cej.2016.06.034.

92. P. Fei, Q. Wang, M. Zhong, B. Su, Preparation and adsorption properties of enhanced magnetic zinc ferrite-reduced graphene oxide

nanocomposites via a facile one-pot solvothermal method, *J. Alloys Compd.* 685 (2016) 411–417. doi:10.1016/j.jallcom.2016.05.279.

93. M. Bhaumik, S. Agarwal, V. Kumar, A. Maity, Enhanced removal of Cr(VI) from aqueous solutions using polypyrrole wrapped oxidized MWCNTs nanocomposites adsorbent, *J. Colloid Interface Sci.* 470 (2016) 257–267. doi:10.1016/j.jcis.2016.02.054.
94. N. Gerard, R. Santhana Krishnan, S. K. Ponnusamy, H. Cabana, V. K. Vaidyanathan, Adsorptive potential of dispersible chitosan coated iron-oxide nanocomposites toward the elimination of arsenic from aqueous solution, *Process Saf. Environ. Prot.* 104 (2016) 185–195. doi:10.1016/j.psep.2016.09.006.
95. D. Wu, Y. Sun, Q. Wang, Adsorption of lanthanum (III) from aqueous solution using 2-ethylhexyl phosphonic acid mono-2-ethylhexyl ester-grafted magnetic silica nanocomposites, *J. Hazard. Mater.* 260 (2013) 409–419. doi:10.1016/j.jhazmat.2013.05.042.
96. R. Verma, A. Awasthi, P. Singh, R. Srivastava, H. Sheng, J. Wen, D. J. Miller, A. K. Srivastava, Interactions of titania based nanoparticles with silica and green-tea: Photo-degradation and -luminescence, *J. Colloid Interface Sci.* 475 (2016) 82–95. doi:10.1016/j.jcis.2016.04.038.
97. P. Sagitha, K. Sarada, K. Muraleedharan, One-pot synthesis of poly vinyl alcohol (PVA) supported silver nanoparticles and its efficiency in catalytic reduction of methylene blue, *Trans. Nonferrous Met. Soc. China.* 26 (2016) 2693–2700. doi:10.1016/S1003-6326(16)64397-2.
98. M. Darwish, A. Mohammadi, N. Assi, Integration of nickel doping with loading on graphene for enhanced adsorptive and catalytic properties of CdS nanoparticles towards visible light degradation of some antibiotics, *J. Hazard. Mater.* 320 (2016) 304–314. doi:10.1016/j.jhazmat.2016.08.043.
99. N. Dirany, M. Arab, C. Leroux, S. Villain, V. Madigou, J. R. Gavarri, Effect of WO_3 nanoparticles morphology on the catalytic properties, *Mater. Today Proc.* 3 (2016) 230–234. doi:10.1016/j.matpr.2016.01.062.
100. Y. Lu, J. Zhang, L. Ge, C. Han, P. Qiu, S. Fang, Synthesis of novel AuPd nanoparticles decorated one-dimensional ZnO nanorod arrays with enhanced photoelectrochemical water splitting activity, *J. Colloid Interface Sci.* 483 (2016) 146–153. doi:10.1016/j.jcis.2016.08.022.
101. S. Groiss, R. Selvaraj, T. Varadavenkatesan, R. Vinayagam, Structural characterization, antibacterial and catalytic effect of iron oxide

nanoparticles synthesised using the leaf extract of Cynometra ramiflora, *J. Mol. Struct.* 1128 (2017) 572–578. doi:10.1016/j.molstruc.2016.09.031.

102. A. Nezamzadeh-Ejhieh, Z. Ghanbari-Mobarakeh, Heterogeneous photodegradation of 2,4-dichlorophenol using FeO doped onto nano-particles of zeolite P, *J. Ind. Eng. Chem.* 21 (2015) 668–676. doi:10.1016/j.jiec.2014.03.035.

103. S. Samiee, E. K. Goharshadi, P. Nancarrow, Successful degradation of Reactive Black 5 by engineered Fe/Pd nanoparticles: Mechanism and kinetics aspects, *J. Taiwan Inst. Chem. Eng.* 67 (2016) 406–417. doi:10.1016/j.jtice.2016.07.012.

104. X. Weng, M. Guo, F. Luo, Z. Chen, One-step green synthesis of bimetallic Fe/Ni nanoparticles by eucalyptus leaf extract: Biomolecules identification, characterization and catalytic activity, *Chem. Eng. J.* 308 (2017) 904–911. doi:10.1016/j.cej.2016.09.134.

105. E. Tavakolian, J. Tashkhourian, Z. Razmi, H. Kazemi, M. Hosseini-Sarvari, Ethanol electrooxidation at carbon paste electrode modified with Pd-ZnO nanoparticles, *Sensors Actuators, B Chem.* 230 (2016) 87–93. doi:10.1016/j.snb.2016.02.006.

106. R. Istratie, M. Stoia, C. Păcurariu, C. Locovei, Single and simultaneous adsorption of methyl orange and phenol onto magnetic iron oxide/carbon nanocomposites, *Arab. J. Chem.* (2015). doi:10.1016/j.arabjc.2015.12.012.

107. C. J. Pandian, R. Palanivel, S. Dhananasekaran, Green synthesis of nickel nanoparticles using Ocimum sanctum and their application in dye and pollutant adsorption, *Chinese J. Chem. Eng.* 23 (2015) 1307–1315. doi:10.1016/j.cjche.2015.05.012.

108. Y. Wang, M. Yao, Y. Chen, Y. Zuo, X. Zhang, L. Cui, General synthesis of magnetic mesoporous FeNi/graphitic carbon nanocomposites and their application for dye adsorption, *J. Alloys Compd.* 627 (2015) 7–12. doi:10.1016/j.jallcom.2014.12.017.

109. L. Wang, X. Wang, J. Bin Zhou, R. S. Zhao, Carbon nanotube sponges as a solid-phase extraction adsorbent for the enrichment and determination of polychlorinated biphenyls at trace levels in environmental water samples, *Talanta.* 160 (2016) 79–85. doi:10.1016/j.talanta.2016.07.005.

110. Q. L. Li, L. L. Wang, X. Wang, M. L. Wang, R. S. Zhao, Magnetic metal-organic nanotubes: An adsorbent for magnetic solid-phase extraction of polychlorinated biphenyls from environmental and

biological samples, *J. Chromatogr. A.* 1449 (2016) 39–47. doi:10.1016/j.chroma.2016.04.060.

111. Y. A. Shaban, M. A. El Sayed, A. A. El Maradny, R. K. Al Farawati, M. I. Al Zobidi, S. U. M. Khan, Photocatalytic removal of polychlorinated biphenyls (PCBs) using carbon-modified titanium oxide nanoparticles, *Appl. Surf. Sci.* 365 (2016) 108–113. doi:10.1016/j.apsusc.2016.01.001.

112. Y. hsin Shih, M. Y. Chen, Y. F. Su, Pentachlorophenol reduction by Pd/Fe bimetallic nanoparticles: Effects of copper, nickel, and ferric cations, *Appl. Catal. B Environ.* 105 (2011) 24–29. doi:10.1016/j.apcatb.2011.03.024.

113. N. min Zhu, Yi-Li, F. S. Zhang, Catalytic dechlorination of polychlorinated biphenyls in subcritical water by Ni/Fe nanoparticles, *Chem. Eng. J.* 171 (2011) 919–925. doi:10.1016/j.cej.2011.04.041.

114. E. J. García-Suárez, M. Tristany, A. B. García, V. Collire, K. Philippot, Carbon-supported Ru and Pd nanoparticles: Efficient and recyclable catalysts for the aerobic oxidation of benzyl alcohol in water, *Microporous Mesoporous Mater.* 153 (2012) 155–162. doi:10.1016/j.micromeso.2011.12.023.

115. B. Huang, W. Qian, C. Yu, T. Wang, G. Zeng, C. Lei, Effective catalytic hydrodechlorination of o-, p- and m-chloronitrobenzene over Ni/Fe nanoparticles: Effects of experimental parameter and molecule structure on the reduction kinetics and mechanisms, *Chem. Eng. J.* 306 (2016) 607–618. doi:10.1016/j.cej.2016.07.109.

116. S. Jian, Y. Li, Ni@Pd core-shell nanoparticles supported on a metal-organic framework as highly efficient catalysts for nitroarenes reduction, *Cuihua Xuebao/Chinese J. Catal.* 37 (2016) 91–97. doi:10.1016/S1872-2067(15)60940-8.

117. S. Li, C. Wei, J. Wang, L. Zhang, Y. Li, Y. Li, B. Wang, Sonocatalytic activity of Yb, B, Ga-codoped Er^{3+}:$Y_3Al_5O_{12}/TiO_2$ in degradation of organic dyes, *Mater. Sci. Semicond. Process.* 26 (2014) 438–447. doi:10.1016/j.mssp.2014.05.041.

118. S. Rajendran, M. M. Khan, F. Gracia, J. Qin, photocatalytic degradation and electrochemical activity of ZnO/CeO_2 nanocomposite, *Nat. Publ. Gr.* (2016) 1–11. doi:10.1038/srep31641.

119. P. Muthirulan, C. K. N. Devi, M. M. Sundaram, Fabrication and characterization of efficient hybrid photocatalysts based on titania and graphene for acid orange seven dye degradation under UV irradiation, *Adv. Mater. Lett.* 5 (2014) 163–171. doi:10.5185/amlett.2013.7507.

120. G. Panthi, M. Park, H. Kim, S. Lee, S. Park, Electrospun ZnO hybrid nanofibers for photodegradation of wastewater containing organic dyes: A review, *J. Ind. Eng. Chem.* 21 (2015) 26–35. doi:10.1016/j.jiec.2014.03.044.

121. S. Chowdhury, R. Balasubramanian, Graphene/semiconductor nanocomposites (GSNs) for heterogeneous photocatalytic decolorization of wastewaters contaminated with synthetic dyes: A review, *Applied Catal. B Environ.* 160–161 (2014) 307–324. doi:10.1016/j.apcatb.2014.05.035.

122. M. Tobajas, C. Belver, J. J. Rodriguez, Degradation of emerging pollutants in water under solar irradiation using novel TiO_2-ZnO/clay nanoarchitectures, *Chem. Eng. J.* 309 (2017) 596–606.

123. R. Ganigar, G. Rytwo, Y. Gonen, A. Radian, Y. G. Mishael, Polymer–clay nanocomposites for the removal of trichlorophenol and trinitrophenol from water, *Appl. Clay Sci.* 49 (2010) 311–316. doi:10.1016/j.clay.2010.06.015.

124. K. A. Hernández-Hernández, J. Illescas, M. del Carmen Díaz-Nava, C. R. Muro-Urista, S. Martínez-Gallegos and R. E. Ortega-Aguilar, Polymer–clay nanocomposites and composites: Structures, characteristics, and their applications in the removal of organic compounds of environmental interest, *Med. Chem. (Los. Angeles)*. 6 (2016) 201–210. doi:10.4172/2161-0444.1000347.

125. S. Jin, P. H. Fallgren, J. M. Morris, Q. Chen, S. Jin, P. H. Fallgren, J. M. Morris, Q. Chen, S. Jin, P. H. Fallgren, J. M. Morris, Q. Chen, Removal of bacteria and viruses from waters using layered double hydroxide nanocomposites removal of bacteria and viruses from waters using layered double hydroxide nanocomposites, *Sci. Technol. Adv. Mater.* 8 (2007) 67–70. doi:10.1016/j.stam.2006.09.003.

126. G. V. Padmaja, Biocompatible nanomaterials for pollutant treatment technologies, *Int. J. Sci. Res.* 4 (2015) 394–399.

127. H. Zhang, Q. Li, J. Huang, Y. Du, S. Ruan, Reduced graphene oxide/Au nanocomposite for NO_2 sensing at low operating temperature, *Sensors*. 16 (2016) 1152. doi:10.3390/s16071152.

128. S. Srivastava, K. Jain, V. N. Singh, S. Singh, N. Vijayan, N. Dilawar, G. Gupta, T. D. Senguttuvan, Faster response of NO_2 sensing in graphene–WO_3 nanocomposites, *Nanotechnology*. 23 (2012) 205501. doi:10.1088/0957-4484/23/20/205501.

129. S. Sakthinathan, S. M. Chen, Graphene supported nanocomposite for electrochemical detection of pollutant materials: A short review, *Int. J. Electrochem. Sci.* 10 (2015) 6527–6536.

130. V. Anbazhagan, K. Behlol, A. Ahmed, S. Janani, Synthesis of catalytically active silver nanoparticles using lipid derived signaling molecule, N-steroylethanolamine: Promising antibacterial agent and selective colorimetric sensor for mercury ion, *Sensors Actuators B. Chem.* 200 (2014) 92–100. doi:10.1016/j.snb.2014.04.033.

131. S. Pandey, Highly sensitive and selective chemiresistor gas/vapor sensors based on polyaniline nanocomposite: A comprehensive review, *Sci. Adv. Mater. Devices.* (2016). doi:10.1016/j.jsamd.2016.10.005.

Chapter 3

Nanocomposites for Abatement of Water Pollution

Mona Abdel Rehim[a] and Abdelrahman Badawy[b]

[a]*Packing and Packaging Materials Department,*
Division of Chemical Industries Research,
National Research Centre, Giza, Egypt
[b]*Physical Chemistry Department,*
Inorganic Chemical Industries and Mineral Resources Division,
National Research Centre, Giza, Egypt

monaabdelrehim23@gmail.com

3.1 Introduction

Water is a vital component of life. Nevertheless, despite that two thirds of the earth is covered with water, about 800 million people do not have access to fresh water and one fifth of the world's population faces water shortage [1]. Water scarcity is a problem facing many societies, and although it is not a global problem, many countries face water shortage. An increase in industrial and agricultural activities has aggravated the problem. Of the accessible amount of fresh water, only 10% is used for domestic purposes,

Nanocomposites for Pollution Control
Edited by Chaudhery Mustansar Hussain and Ajay Kumar Mishra

ISBN 978-981-4774-45-1 (Hardcover), 978-1-315-14368 2 (eBook)
www.panstanford.com

and the rest is used for irrigation (70%) and industrial processes (20%) [2]. Furthermore, the world's population is expected to grow by 3 billion people by 2050, of which roughly 2.7 billion will be in developing countries, where situation is already disastrous. As agriculture consumes 70% of all human fresh water withdrawals, water shortage will have an impact not only on health issues but also on food production (irrigation and fertilizer production) for the growing population [3]. However, increasing industrial activities led production of byproduct for industrial process known as industrial wastewater. This type of water needs special treatment before releasing in the sanitary or surface water. The composition of the industrial effluent varies according to the industrial process. For instance, agricultural wastewater contains large amounts of organic compounds from animal and plants sources. On the other hand, effluents of various industries, such as pharmaceutical, cosmetic, textile, and rubber, contain aliphatic organic solvents which are often flammable, malodorous and potentially toxic to aquatic organisms [4]. In order to improve the quality of water before disposal in water streams special treatment procedures are essential. However, many wastewater treatment facilities rely on biological systems that tend to be cheap and easy to operate [5]. For instance, the dairy industry combines different steps of decantation in addition to aerobic and anaerobic biological treatments [6]. Even, the pharmaceutical industry has developed biocolumn containing activated sludge that is continuously aired to treat its wastewaters [7]. Another type of "natural" wastewater treatment takes advantage of chemical, physical and biological processes to eliminate water contaminants. Biological processes such as the metabolic activity of bacteria and algae are key factors in these systems. Water macrophytes (helophytes) also play an important role in natural wastewater treatments since they consume nutrients released by the microbial degradation of organic matter and they are able to inject dissolved O_2 through their roots [8, 9]. Ozonation is becoming more widely used for industrial wastewater treatment. Ozone can rapidly react with organic and inorganic contaminants [10, 11] and shows a strong sterilizing and disinfecting effect. Also, it has proved its usefulness in improving iron and manganese precipitation [12] and in eliminating organic traces in water [13]. Lately, fabrication of nanomaterials or polymeric based nanocomposites is researched in depth for wastewater treatment by separation or adsorption techniques. Moreover, the use of nanomaterials to disinfect wastewater is described [14, 15]. Metal oxides such as TiO_2 and ZnO are among the photocatalysts with antimicrobial properties that gained interest for wastewater disinfection. On the other hand, polymeric materials are widely used for water purification in the form of separating membranes, flocculants or nanocomposites for the adsorption of pollutants and heavy metals from wastewater.

Hyperbranched polymers (HBP) are a major subclass of dendritic architecture characterized by their irregular structures with random branching topology [16, 17]. This type of macro-molecules have found applications in coatings and formulations, e.g., powder coatings, high solid coatings, flame retardant coatings, and barrier coatings for flexible packaging [18–26]. Hybrid materials are also described through blend HBP with linear polymers, to form semi-crystalline hybrid polymers. Moreover, they can be further functionalized to become cross-linkable [27–29]. Synthesis of HBP/clay nanocomposites is also studied, taking advantage of the globular structure of HBP besides the presence of a large number of end groups, which stabilize the exfoliated structure of the clay after drying [30], thus, preventing the molecule to collapse onto the clay layer [31, 32]. Investigation and use of HBP for industrial wastewater treatment and adsorption of heavy metals were demonstrated by many researchers as either polymeric material or nanocomposite [33–36].

This chapter highlights industrial wastewater purification using nanocomposites based on metal oxide photocatalysts such as TiO_2 and ZnO. It demonstrates the role of hyperbranched polymers in the adsorption of heavy metals either alone or as a matrix for photocatalyst.

3.2 Hyperbranched Polymers for Wastewater Purification

Hyperbranched polymers have unique structural features that give them versatile properties, including non-symmetrical polydisperse structure, a large number of end groups, and above all easy way of synthesis. Moreover, due to their globular structure, HBPs contain nano-cavities that are able to encapsulate different molecules. The presence of a large number of terminal groups in the HBP enables the modification of the macromolecule with various functionalities and changing its properties. Moreover, the terminal groups show the so-called polyvalency effect [37, 38], which facilitates their binding to different molecules. Investigation of encapsulation properties of alkylated HBP such as polyglycerol, poly(ethylene imine), and hyperbranched polyester amide was applied for the removal of polycyclic aromatic hydrocarbons [39]. Faster absorption rates were observed for HBP derivatives with flexible branching chains. This was attributed to the favorable formation of cavities through flexible chains. The size of these cavities can be adapted according to the size of the encapsulated molecules [40]. Moreover, the presence of large number of end groups enhances the adsorption capacity. Abdel Rehim et al. reported the removal of different dye staffs using hyperbranched poly esteramide [34]. The removal reaction followed a first-order kinetics and the extent of removal was reactive dye > acid dye > direct dye. The removal of the dyes was more pronounced at lower pH values.

Nanocapsules of hyperbranched polyglycerols have been utilized in the transportation of dye molecules [41]. Modification of HBP with fatty acid chlorides yielded the corresponding amphiphilic polyglycerols. Congo red, as a water-soluble dye, was dissolved in water and extraction of the aqueous phase was performed with an unpolar solvent containing amphiphilic polyglycrols. Geitner et al. investigated the hyperbranched poly(ethyleneimine) as oil dispersant. The polymer showed high dispersing capacity for polyaromatic as well as linear hydrocarbon [42].

3.3 Hybrid Materials Based on Hyperbranched Polymers for Wastewater Treatment

Several pollutants of industrial wastewater such as dyes and heavy metals are toxic and must be removed since they cause environmental damage. Removal of organic pollutants such as dyes can be carried out through adsorption or degradation. While heavy metals cannot undergo physical or chemical degradation, several approaches have been developed to get rid of inorganic pollutants. Adsorption is the most attractive method for the removal of heavy metals due to low cost and high efficiency [43]. Mesoporous silica is a promising material for the adsorption of heavy metals because of its large surface area, high pore volume, and high thermal stability [44]. Functionalization of MS would greatly increase its adsorption efficiency [45]. MS surface grafted with different functional groups through attachment and silanol groups can be seen on its surface [46–48]. Grafting of hyperbranched polymers on the MS groups has been reported taking advantage of the large numbers of terminal groups of HBP in order to increase adsorption efficiency of the final hybrid adsorbent [49]. Recently, Tao et al. described the synthesis of highly efficient adsorbent from hybrid mesoporous silica and amino-terminated hyperbranched polymer [50]. The obtained nanonetwork showed high adsorption efficiency and rate for both dyes and heavy metal ions (Cu^{2+} and Fe^{2+}) from their aqueous solutions. Water-soluble nanocomposite based on nanosilica as core and hyperbranched polyamide-amine as shell was synthesized for removal of oil from water [51]. Rheological and degradation stability studies have been carried out for the prepared 3D material. The results confirmed the robust structure, high stability, and temperature resistance of the nanocomposite. Moreover, core-flooding experiments confirmed the positive effect of HBP on tuning flooding phase mobility and sweep efficiency and the robust applicability for oil recovery from water. Dvonic et al. reported the synthesis of cross-linked silylated HBP with surface hydroxyl groups [52]. The nanostructured networks were prepared by consecutive hydrolysis and polycondensation. The obtained materials were covalently bounded to titanium oxide ceramic filters to prepare ultra-pure water [53]. Inclusion of kinetic experiments offered the ability to evaluate the encapsulation efficiency of these dendritic networks. The results confirmed that the nanostructured films have the ability to absorb polycyclic hydrocarbon such as pyrene, phenanthrene and β-naphthol from water. However, absorption of β-naphthol is slower than in case of as pyrene, and phenanthrene. This lower solubility is due to its higher water solubility compared to the other two hydrocarbons. This solubility of β-naphthol reduces its absorption inside the nanocavities of the dendritic network. Impregnation of ceramic filters with hyperbranched derivatives and then cross-linking by curing was carried out and these hybrid materials were used for water purification. Comparing the results with those obtained for the non-impregnated filters revealed the presence of HBP-enhanced absorption of polycyclic aromatic materials. It is suggested that curing led to the bridging of the siloxane groups. Hence, the structure of the polymer becomes more rigid compared to that originating from the HBP. Similarly, dendritic polymer/SiO_2 hybrid material [54] and attapulgite-grafted HBP aliphatic polyester [55] were synthesized and their complexing capacity toward Cu^{+2}, Hg^{+2}, Cd^{+2} and Zn^{+2} was studied. It should be pointed out that silica by itself has been studied as a heavy metal sorbent from water [56, 57]. The sorption process was explained as electrostatic interactions between the negatively charged silica ions and the positively charged metal ions. On the other hand, hybrids of HBP and silica showed higher sorption efficiency compared to polymer-free silica. It was attributed to the combination of chelation and electrostatic interactions, which allowed the adsorbed ions come to the surface of the sorbing material. Grafting of cellulosic fibers with hyperbranched polyester for removal of Zn (II), Cu (II), Hg (II) and Cd (II) from their mixture solution has been reported [33]. In this context, cotton fibers were soaked in 2,2-bis(hydroxymethyl)propion ic acid monomer solution in DMF and the polycondensation reaction proceeded in presence of p-toluenesulfonic acid as a catalyst. The results confirmed that cellulosic fibers grafted with HBP have high loading capacity and good adsorption properties beside***s*** another advantage coming from being biodegradable unlike most of used adsorbing composites based on synthetic polymers.

Magnetic nanoparticles have been investigated for heavy metals and dye removal due to their high surface area, high number of active sites, and high adsorption capacities [58, 59]. However, modification of the magnetic nanoparticles with different surface modifiers has been reported with the aim of increasing the surface area and thus increasing the adsorption capacity. Among these modifiers, carbon nanotubes [60], porous carbon [61] or organic compounds containing thiol or amino groups were used [62–65]. It was reported that the surface-anchored polymer chains offer larger surface functional group density, which subsequently enhances the heavy metals removal [66–68]. Liu et al. [69] have prepared nanocomposites of amino-terminated hyperbranched polymer and amino-functionalized mesoporous Fe_3O_4. The advantage of the modification of magnetic nanoparticles with amino-terminated HBP is the biocompatibility of the used HBP, which opens the way for heavy metals removal with no added toxicity [70]. Moreover, the three-dimensional structure of the HBP with the large number of end groups can enhance the adsorption of the contaminating heavy metals [71]. It was found that the amino-functionalized mesoporous Fe_3O_4 modified with HBP showed high adsorptive removal of heavy metal as well as adsorption of dyes methyl orange (MO), malachite green oxalate (MGO) methylene blue hydrate (MBH) rhodamine B (RB), and crystal violet (CV). The results also revealed that the nanocomposite generally has higher removal efficiency than amino-terminated Fe_3O_4 alone. On the other hand, adsorption of Pb^{2+} and MO was more pronounced over the other pollutants. Moreover, the adsorption rate was much faster and adsorption equilibrium was reached after 20 min for Pb^{2+} and 10 min for MO. Investigation of adsorption kinetics has revealed that the adsorption reaction is a pseudo-second order process.

3.4 Metal Oxide Photocatalysts

Photocatalysis is considered an effective system for the mineralization of many organics through the generation of radicals such as •OH and O_2-•. Metal oxides such as TiO_2, ZnO, CeO_2, CdS, and ZnS can be used as photocatalysts. TiO_2 is the most widely used semiconductor for photocatalytic studies and applications since it has the highest quantum yields. Some studies have confirmed that ZnO exhibits a better efficiency than TiO_2 in photocatalytic degradation of some dyes [72, 73]. The photochemical properties and photodegradation mechanism of ZnO is similar to that of TiO_2 [74, 75]. Light assimilation is the initial phase in the photocatalytic pathway. However, absorbance measurements with colloidal TiO_2 are complicated by strong scattering. Sun and Bolton utilized integrating sphere to quantify the true absorbance of anatase (100–200 nm) scattered in water [76]. The UV absorbance is around 0.007 a.u. from 300 to 330 nm, after which it decreases straightly with increasing wavelength to zero at 400 nm. Conventional UV spectroscopy overestimates the absorbance by around a variable of 2. Finally, the penetration depth of absorbed UV light into TiO_2 has been evaluated to be 25 nm [77]. When TiO_2 absorbs light of wavelength less than 380 nm, which has an energy greater than the TiO_2 band gap energy (3.2 eV), photoelectrons (e−) are generated in the conduction band and holes (h+) in the valence band [78, 79]. Hirakawa and Nosaka have described two types of holes: (1) those trapped on the particle surface and available for surface reaction and (2) interior holes, which cannot catalyze reactions. Hirakawa et al. distinguished between shallow traps, which are reversible, and irreversible deep traps [80]. The recombination of electrons with holes competes with photocatalyzed reactions [81]. Blake et al. summarized the properties of light sources most commonly used in photocatalysis studies [82]. UV wavelengths are often subdivided into UVA (380–315 nm), UVB (315–280 nm), and UVC (b280 nm), and 99% of the ultraviolet radiation that reaches the Earth's surface is UVA. Because natural UVA light flux is low and variable, most studies employ light sources. For example, the maximum output of the common fluorescent black light is about 365 nm, which is ideal for TiO_2 photo-activation. Medium-pressure mercury or xenon arc lights are also commonly used, although for photocatalytic disinfection studies, care must be taken to remove sub–300 nm light, which can directly kill bacteria [83].

3.5 Influence of Light on TiO_2 Properties

By definition, a catalyst does not change or regenerated over the course of a reaction, and TiO_2 obeys these criteria [84]. Nevertheless, TiO_2 does display temporary, measurable changes upon irradiation. Hashimoto et al. postulate that irradiation induces the conversion of surface Ti–O–Ti bonds to more hydrophilic TiOH (titanol) groups in the presence of water [84]. Photochemical changes are also slightly reflected in the TiO_2 hardness. Performance of TiO_2 for the removal of Cu(II) upon UV illumination and without UV has been also studied [85].

The photodegradation process of the suspension contained cyanide and copper ions and was enhanced upon using TiO_2 as photocatalyst. Moreover, the removal efficiency increased by increasing molar ratio of CN-: Cu(II) and was found maximum for 10:1 ratio which showed removal of both ions. This result suggests that the co-existence of CN-: Cu(II) enhanced the photodegradation reaction [86, 87].

3.6 Reactions of Water on Irradiated TiO_2

In many applications, photo-activated TiO_2 first reacts with air-saturated water. The reactions of holes and electrons are summarized in Table 3.1, which shows the most important initial products for organic mineralization (i.e., conversion of organics to carbon).

Table 3.1 Reaction products of photo-activated TiO_2 and water

$TiO_2 + h\nu \rightarrow TiO_2\,(e-; h+)$	Eq. (3.1)
$h^+ + H_2O \rightarrow H^+ + \bullet OH$	Eq. (3.2)
$h^+ + OH^- \rightarrow \bullet OH$	Eq. (3.3)
$e^- + O_2 \rightarrow O^{2-}$	Eq. (3.4)
$O^{2-} + H^+ \rightarrow HO_2$	Eq. (3.5)
$\bullet OH + \bullet OH \rightarrow H_2O_2$	Eq. (3.6)
$e- + H_2O_2 \rightarrow \bullet OH + OH^-$ (photo-Fenton reaction)	Eq. (3.7)
$H_2O_2 + M^{2+} \rightarrow HO^\bullet + OH^- + M^{3+}$ (Fenton reaction)	Eq. (3.8)

3.7 TiO_2 and ZnO as Photocatalysts

Generally speaking, photocatalytic efficiency is governed by several factors such as the photocatalyst bezel for harvesting light, number of active sites on the surface of the photocatalyst, and life time of hole and electrons generated by light. Elimination of toxic heavy metals by reducing the toxic form to the non-toxic form by photocatalysis was studied. Dasa et al. thoroughly investigated the photocatalytic reduction of Cr^{6+} due to its toxicity and harmful effect. A thin film of TiO_2 prepared by the sol-gel method immobilized over glass substrate has been used to reduce

Cr^{6+} in presence of UV light [88]. The results showed that Cr^{6+} was completely eliminated and the TiO_2 films have comparable efficiency with that of TiO_2 nanoparticles. Photocatalytic activity of modified TiO_2 with sulfate or TiO_2 loaded on other materials such as zirconium phosphate has been reported [89, 90]. Investigation of pH effect showed that samples synthesized at lower pH had higher surface area and better phocatalytic reactivity. Moreover, the photocatalytic activity of TiO_2 in the presence of organic compound, as an electron donor, has been investigated [91]. A marked enhancement in the reduction of Cr^{6+} to Cr^{3+} was observed for large-surface-area photocatalysts. These results showed that the reduction of Cr^{6+} was dependent on the crystalline form of TiO_2 and its surface area in the absence of the organic material. On the other hand, in the presence of organic compound, the surface area of the photocatalyst is the dominant factor due to the synergistic effect between the photocatalytic reduction of Cr^{6+} and oxidation of the organic compound photocatalytically. Doping of TiO_2 with neodymium (Nd) was synthesized by sol-gel method and used photocatalytic reduction of Cr^{6+} by Rengaraj et al. [92]. The presence of Nd on the surface of TiO_2 enhanced its photocatalyic reactivity since Nd acted as active sites for electrons accumulation on the surface of TiO_2. However, the good separation between electrons and holes on the photocatalyst surface led to improved reduction and oxidation processes rather than recombination reactions. Due to limitations of TiO_2 powder, immobilized TiO_2 electrode was used for the photocatalytic reduction of Cr^{6+} in its aqueous solution. The photocatalytic reaction was performed in acidic conditions with conversion of 98% to non-toxic Cr^{3+} [93]. Loading of iron oxide and TiO_2 on solid wasted melted slag was demonstrated by Zhang and Itoh as a heterogeneous photocatalyst. Oxidation of arsenite to arsenate using the heterogeneous adsorbent has been performed with complete transformation under UV illumination within 3 h [94].

Another type of metal oxides photocatalyst, ZnO, has been extensively investigated due to its unique properties such as oxygen vacancies, polar planes, surface area, non-toxicity, and environmental sustainability [95, 96]. Recently, a condensed review demonstrating the recent developments of ZnO as photocatalyst for water treatment has been published [97].

The article presented different factors affecting photocatalytic reaction such as light intensity and wavelength, pH, ZnO loading and structure, concentration of the organic material. Effect of ZnO morphology on the photocatalytic efficiency was studied among these morphologies, nanoparticles, nanorods, nanoflowers and nanotubes. It was reported that nanorods exhibited high photocatalytic activity compared to other morphologies [98]. However, the use of 1D morphology of the photocatalyst such as nanorods and nanotubes or doping the photocatalyst with another oxide led to improved light harvesting by the semiconductor [99–109]. Xu et al. prepared several morphologies of ZnO by the solvothermal method from Zinc acetylacetonate as precursor [110]. Different solvents were used for the synthesis, such as THF, decane, toluene, water, and ethanol. The change of the solvent led to obtaining different morphologies and consequently change of the photocatalytic activity was observed. Moreover, the results showed that ZnO prepared in the presence of THF as solvent has kinetic rate constant nine times higher than that of commercial ZnO. Nanowires (NWs) of hydrogen titanate were prepared by the hydrothermal method in the presence of KOH [111]. The obtained NWs were calcinated at temperature between 400 and 600°C and the photocatalytic activity of the material was investigated. The results showed that the calcinated NWs have higher photocatalytic activity than TiO_2 (P25). This was attributed to the large surface area of the formed NWs and larger pore volume. Moreover, the increase of the calcination temperature above 700°C led to the decrease in the photocatalytic activity due to the increase of TiO_2 crystallite. Nanofibers of TiO_2 were prepared by direct growth on the Ti-substrate. The obtained nanofibers had a diameter of 10–20 nm and length of 100 μm. A freestanding membrane with the thickness of a few millimeters cleaved from the Ti-foil showed photocatalytic degradation of pharmaceuticals under UV irradiation [112]. Degradation of phenol to evaluate the photocatalytic activity of different morphologies of ZnO has been demonstrated [113]. The effect of the surface area and the morphology of nanoparticles, tetra-needles, and micro-sized ZnO besides the effect of initial phenol concentration were investigated. The results showed that the tetra-needle morphology of ZnO has the highest photocatalytic activity and the increase in phenol concentration reduces the degradation rate.

Doping of semiconductor with another metal oxide in order to obtain binary photocatalyst system has gained much interest recently. The goal is to decrease electron loss and increase photon absorption while enhancing photocatalytic efficiency. It was observed that the nature of the dopant and the method of preparation greatly affect the properties of the binary oxide photocatalyst system. Moreover, high doping concentration greatly reduces the photocatalytic efficiency due to physical defects. It was found that the excess number of cations act as traps for holes and electrons and hence increase recombination reactions. Consequently, the generation of •OH and O_2-• radicals is reduced and the final decrease in the photodegradation reaction can be observed. Doping of ZnO with rare earth metals has recently received much attention since the redox properties and unique photocatalytic properties of rare earth metals are important to enhance the surface properties of ZnO [114–118]. Many researchers have also demonstrated the doping of ZnO with other co-dopants to reduce the recombination of holes and electrons [119, 120]. This technique was used in the removal of dyes such as methylene blue, methyl orange, and acid black [121–123].

3.8 Photocatalytic Activity of Hyperbranched/Metal Oxide Hybrids

Photocatalytic degradation of polyvinyl chloride (PVC) has been achieved through the incorporation of TiO_2 modified with hyperbranched polycaprolactone in the PVC matrix [124]. The results confirmed that this method can be considered an eco-friendly technique to get rid of toxic byproduct-emitting incineration of PVC. However, the utilization of HBP/metal oxide hybrid for wastewater is not found much in the literature, while the combination of HBP and with either TiO_2 or ZnO in order to obtain nanocomposite photocatalysts gives the opportunity to gather many advantages from both sides and improving the life time of the photocatalyst. In a previous work, TiO_2 nanowires were modified with hyperbranched polyester by in situ and ex situ techniques [125]. Mineralization of water using the as-prepared nanocomposites has been realized by the chemical oxygen demand (COD) analysis. The results confirmed that the photocatalytic activity of the TiO_2/HBP nanocomposite and the rate of reaction are higher than that of pure TiO_2 nanowires (Fig. 3.1). It is assumed that the presence of a large number of OH groups as terminal in the HBP structure enhanced the photocatalytic reaction. On the other hand, preparation of the nanocomposite by the in situ method led to the formation of TiO_2 of smaller crystallite size, as has been confirmed by XRD investigation [125]. Moreover, the durability study revealed that the nanocomposite showed good photocatalytic activity in the second run.

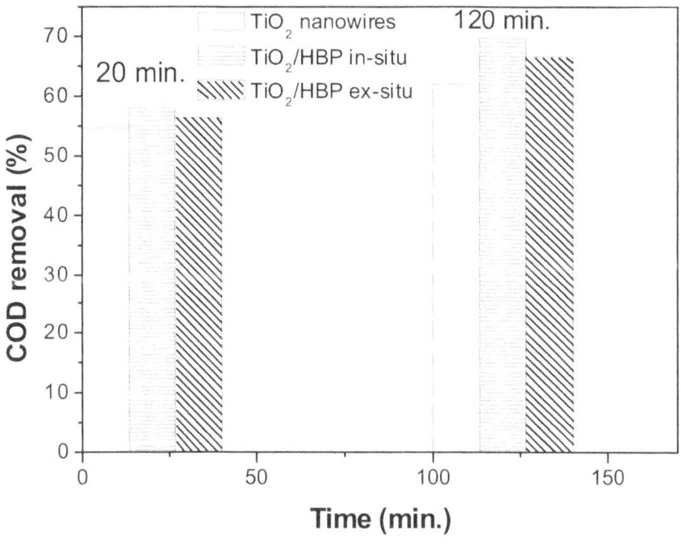

Figure 3.1 Comparison of the percentage COD removal using three photocatalytic systems after 20 and 120 min, respectively.

Comparison of photocatalytic efficiency of TiO_2 and ZnO nanowires toward the degradation of organic water pollutants (phenol, sulfanilamide and *N*-(1-naphthyl)-ethylenediamine dichloride) was carried out. The results shown in Fig. 3.2 reveal the following (i) No effect was observed by using UV only. (ii) COD removal was 67% and 62.2% on using ZnO and TiO_2, respectively, for 120 min, which might be attributed to the larger surface area of ZnO nanowires (90.3 m^2/g) than that of TiO_2 (81.7 m^2/g) as confirmed by BET measurements. Moreover, it was found that

80 min reaction time is enough for ZnO nanowires to reach the maximum efficiency and then its photocatalytic activity decreases. On the other hand, the increase of photocatalytic activity of TiO_2 nanowires is steady and reaches the maximum value after 120 min reaction time.

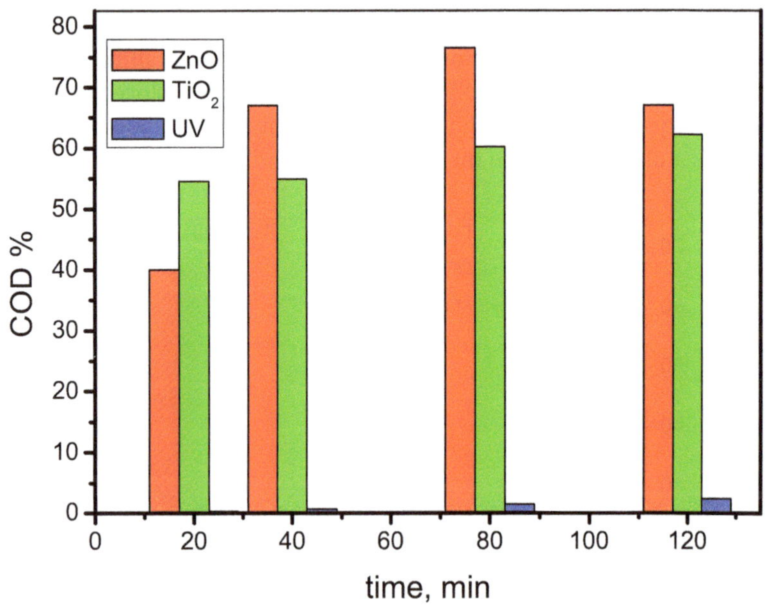

Figure 3.2 Percentage COD removal in case of TiO_2 and ZnO nanowires as photocatalysts.

Nanocomposites of ZnO nanowires/hyperbranched polyester (HPES) have been tested as photocatalysts and the obtained results were compared with the previously investigated system TiO_2/HPES (Fig. 3.3). It was found that HPES enhanced the catalytic activity of both ZnO and TiO_2, which confirms the effect of organic material on the enhancement of the photocatalytic efficiency. However, the increase in photocatalytic activity was more pronounced for ZnO (15.7%) than that for TiO_2 (7.2%). This effect might be attributed to the difference in the band gap between ZnO/HPES and that for TiO_2. More investigations should be focused on the effect of hyperbranched polymers on both the band gap and the crystallites size of the metal oxide semiconductor.

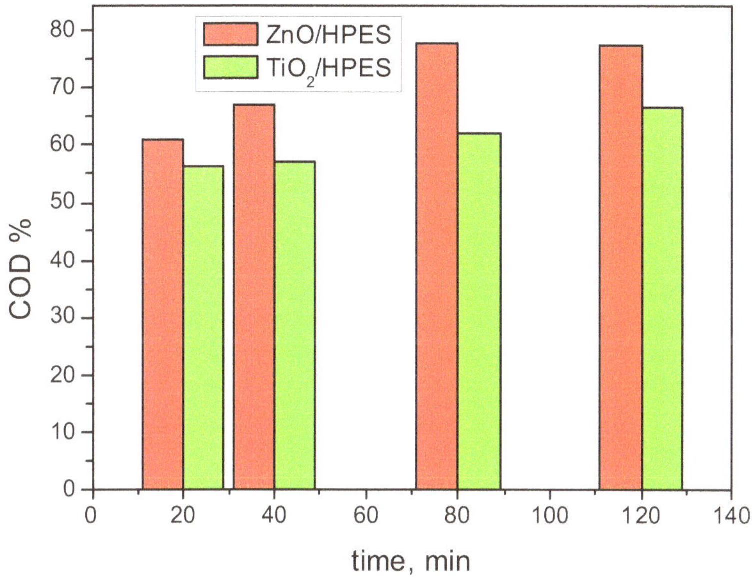

Figure 3.3 Percentage COD removal in case of ZnO/PES and TiO_2/HPES nanocomposites as photocatalysts.

The diversity of nanocomposites based on hyperbranched polymers and ZnO or TiO_2 as photocatalysts makes them the best solution for the removal of organic water pollutants. The use of different morphologies of metal oxides such as nanowires or nanorods enhances the photocatalytic activity due to the increased surface area. Combination of a photocatalyst with a hyperbranched polymer not only enhances photocatalytic efficiency but also increases durability as well. Therefore, the development of novel photocatalytic nanocomposites from hyperbranched polymers with different functionalities and metal oxide semiconductor can lead to materials of superior photocatalytic properties.

References

1. http://www.un.org/waterforlifedecade/scarcity.shtml.
2. Shannon, M. A., Bohn, W. P., Elimelech, M., Georgiadis, J. G., Mayes, A. M. (2008) Science and technology for water purification in the coming decades, *Nature*, **452**, pp. 301–310.

3. www.who.int/water_sanitation_health/publications/facts2004/en.
4. Henry, M. P., Donlon, B. A., Lens, P. N., Colleran, E. M. (1996) Use of anaerobic hybrid reactors for treatment of synthetic pharmaceutical wastewaters containing organic solvents, *J. Chem. Tech. Biotechnol.*, **66**, pp. 251–264.
5. Lagrega, M., Buckingham, P., Evans, J. (1996) *Gestion de Residuos Tóxicos-tratamiento, eliminación y recuperación de suelos*, McGraw-Hill, New York.
6. Zeeman, G., Sanders, W. T. M., Wang, K. Y., Gatze Lettinga, G. (1997) Anaerobic treatment of complex wastewater and waste activated sludge—Application of an Upflow Anaerobic Solid Removal (UASR) reactor for the removal and pre-hydrolysis of suspended COD, *Water Sci. Tech.*, **35**(10), pp. 121–128.
7. Jelić, A, Gros, M., Petrović, M., Ginebreda, A. (2012) Occurrence and elimination of pharmaceuticals during conventional wastewater treatment, emerging and priority pollutants in rivers, *Hdb. Env. Chem.*, **19**, pp. 1–24.
8. Brix, H. (1994) Use of constructed wetlands in water pollution control: Historical development, present status, and future perspectives. *Water Sci. Technol.*, **30**(8), pp. 209–223.
9. Gersberg, R. M., Elkins, B. V., Goldman, C. R. (1986) Role of aquatic plants in wastewater treatment by artificial wetlands, *Water Res.*, **20**, pp. 363–386.
10. Hoigne, J., Bader, H. (1983) Rate constants of reactions of ozone with organic and inorganic compounds in water—I: Non-dissociating organic compounds, *Water Res.*, **17**, pp. 173–183.
11. Hoigne, J., Bader, H. (1985) Rate constants of reactions of ozone with organic and inorganic compounds in water—II: Dissociating organic compounds, *Water Res.*, **17**, pp. 185–194.
12. Pillard, H., Legube, B., Boubigot, M. M., Lefevre, E. (1987) Iron and manganese removal with ozonation in the presence of humic substances, *Ozone Sci. Eng.*, **11**, pp. 93–113.
13. Staehelin, J., Hoigne, J., Environ, J. (1985) Decomposition of ozone in water in the presence of organic solutes acting as promoters and inhibitors of radical chain reactions, *Environ. Sci. Technol.*, **19**, pp. 1206–1213.
14. Hu, J., Chen, G., LO, I. M. C. (2005) Removal and recovery of Cr(VI) from wastewater by maghemite nanoparticles, *Water Res.*, **39**, pp. 4528–4536.

15. Mohan, D., Pittman Jr., C. U. (2007) Arsenic removal from water/wastewater using adsorbents—A critical review, *J. Haz. Mater.*, **142**, pp. 1–53.

16. Flory, P. J. (1952) Molecular size distribution in three dimensional polymers. VI. Branched polymers containing A–R–Bf-1 type units, *J. Am. Chem. Soc.*, **74**, pp. 2718–2723.

17. Kim, Y. H., Webster, O. W. (1990) Water soluble hyperbranched polyphenylene: "A unimolecular micelle" *J. Am. Chem. Soc.*, **112**, 4592–4593.

18. Jikei, M., Kakimoto, M. (2001) Hyperbranched polymers: A promising new class of materials, *Prog. Polym. Sci.*, **26**, pp. 1233–1285.

19. Voit, B. I. (2000) New Developments in hyperbranched polymers, *J. Polym. Sci. Part A: Polym. Chem.*, **38**, pp. 2505–2525.

20. Frings, R. B., Wend, M. (2003) New Hyperbranched polyesters for UV-curing, *DIC Tech. Rev.*, **9**, pp. 43–51.

21. Hult, A., Johansson, M., Malmstroem, E. (1999) Hyperbranched polymers, eds., *Advances in Polymer Science,* Springer-Verlag: Berlin, vol. **143**, pp. 1–34.

22. Diao, P., Guo, M., Hou, Q., Xiang, M., Zhang, Q. (2006) Electrochemically partitioned assembly of organosulfur monolayers and nanoparticles, *J. Phys. Chem. B,* **110**, pp. 20386–20391.

23. Whiting, G. L., Snaith, H. J., Khodabakhsh, S., Andreasen, J. W., Breiby, D. W., Nielsen, M. M., Greenham, N. C., Friend, R. H., Huck, W. T. S. (2006) Enhancement of charge-transport characteristics in polymeric films using polymer brushes, *Nano. Lett.,* **6**, pp. 573–587.

24. Xu, J., Qiu, F., Zhang, H., Yang, Y. (2006) Morphology and interactions of polymer brush-coated spheres in a polymer matrix, *J. Polym. Sci. Part B: Polym. Phys.*, **44**, pp. 2811–2820.

25. Di Gianni, A., Trabelsi, S., Rizza, G., Sangermano, M., Althues, H., Kaskel, S., Voit, B. (2007) Hyperbranched polymer/TiO_2 hybrid nanoparticles synthesized via an in situ sol-gel process, *Macromol. Chem. Phys.*, **208**, pp. 76–86.

26. Däbritz, F., Voit, B., Naguib, M., Sangermano, M. (2011) Hyperstar poly(ester-methacrylate)s as toughening agents in thermally and photocured epoxy resins, *Polymer*, **52**, pp. 5723–5731.

27. Trollsås, M., Hawker, C. J., Remenar, J. F., Hedrick, J. L., Johansson, M., Ihre, Hult, H. (1998) Highly branched radial block copolymers via dendritic initiation of aliphatic polyesters, *J. Polym. Sci. Part A: Polym. Chem.*, **36**, pp. 2793–2798.

28. Trollsås, M., Claesson, H., Attenhoff, B., Hedrick, J. L. (1998) Layered dendritic block copolymers, *Angew. Chem. Int. Ed,* **37**, pp. 3132–3136.
29. Johansson, M., Malmstroem, E., Jansson, A., Hult, A. (2000) Novel concept for low temperature curing powder coatings based on hyperbranched polyesters, *J. Coat. Technol.*, **72**, pp. 49–54.
30. Rodlert, M., Plummer, C. J. G., Garamszegi, L., Leterrier, Y., Grunbauer, H. J. M., Månson, J. A. E. (2004) Hyperbranched polymer/montmorillonite clay nanocomposites, *Polymer*, **45**, pp. 949–960.
31. Rodlert, M., Plummer, C. J. G., Leterrier, Y., Månson, J. A. E., Grunbauer, H. J. M. (2004) Rheological behavior of hyperbranched polymer/montmorillonite clay nanocomposites, *J. Rheol.*, **48**, pp. 1049–1056.
32. Abdel Rehim, M. H., Youssef, A. M., Essawy, H. A. (2010) Hybridization of kaolinite by consecutive intercalation: Preparation and characterization of hyperbranched poly (amidoamine)-kaolinite nanocomposites, *Mater. Chem. Phys.*, **119**, pp. 546–552.
33. Peng, L. (2007) A novel degradable adsorbent of the hyperbranched aliphatic polyester grafted cellulose for heavy metal ions, *Turk. J. Chem.*, **31**, pp. 457–462.
34. Abdel Rehim, M., Fahmy, H. M., Mohamed, Z. E., Abo-Shosha, M. H., Ibrahim, N. A. (2010) Synthesis, characterization and utilization of hyperbranched poly(ester-amide) for the removal of some anionic dyestuffs from their aqueous solutions, *Resins Pigment Technol.*, **39**, pp. 149–155.
35. Liu, P., Xue, Q. J., Tian, J., Liu, W. M. (2003) Self-assembly of functional silanes onto silica nanoparticles, *Chin. J. Chem. Phys.*, **16**, pp. 481–486.
36. Diallo, M. S., Arasho, W., Johnson, Jr., J. H., Goddard II, W. A. (2008) Dendritic chelating agents. 2. U(VI) binding to poly(amidoamine) and poly(propyleneimine) dendrimers in aqueous solutions, *Environ. Sci. Technol.*, **42**, pp. 1572–1579.
37. Kitov, P. I., Bundle, D. R. (2003) On the nature of the multivalency effect: A thermodynamic model, *J. Am. Chem. Soc.*, **125**, pp. 16271–16284.
38. Badjic, J. D., Nelson, A., Cantrill, S. J., Turnbull, W. B., Stoddart, J. F. (2005) Multivalency and cooperativity in supramolecular chemistry. *Acc. Chem. Res.*, **38**, pp. 723–732.
39. Arkas, M., Eleades, L., Paleos, C. M., Tsiourvas, D. J. (2005) Alkylated hyperbranched polymers as molecular nanosponges for the purification of water from polycyclic aromatic hydrocarbons, *Appl. Polym. Sci.*, **97**, pp. 2299–2305.

40. Arkas, M., Tsiourvas, D., Paleos, C. M. (2010) Functional dendritic polymers for the development of hybrid materials for water purification, *Macromol. Mater. Eng.*, **295**, pp. 883–898.

41. Stiriba, S. E., Kautz, H., Frey, H. (2002) Hyperbranched molecular nanocapsules: Comparison of the hyperbranched architecture with the perfect linear analogue, *J. Am. Chem. Soc.*, **124**, pp. 9698–9699.

42. Geitner, N. K., Bhattacharya, P., Steele, M., Chen, R., Ladner, D. A., Ke. P. C. (2012) Understanding dendritic polymer–hydrocarbon interactions for oil dispersion. *RSC Adv.*, **2**, pp. 9371–9375.

43. Yuan, M., Wang, S., Wang, X., Zhao, L., Hao, T. (2011) Removal of organic dye by air and macroporous $ZnO/MoO_3/SiO_2$ hybrid under room conditions, *Appl. Surf. Sci.*, **257**, pp. 7913–7919.

44. Walcarius, A., Mercier, L. (2010) Mesoporous organosilica adsorbents nanoengineered materials for removal of organic and inorganic pollutants, *J. Mater. Chem.*, **20**, pp. 4478–4511.

45. Rosenholm, J. M., Penninkangas, A., Linden, M. (2006) Amino-functionalization of large-pore mesoscopically ordered silica by a one-step hyperbranching polymerization of a surface-grown polyethyleneimine, *Chem. Commun. (Cambridge, U. K.)*, **37**, pp. 3909–3911.

46. Benhamou, A., Baudu, M., Derriche, Z., Basly, J. P. (2009) Aqueous heavy metals removal on amine-functionalized Si-MCM-41 and Si-MCM-48, *J. Hazard. Mater.*, **171**, pp. 1001–1008.

47. Ezzeddine, Z., Batonneau-Gener, I., Pouilloux, Y., Hamad, H., Saad, Z., Kazpard, V. (2015) Divalent heavy metals adsorption onto different types of EDTA-modified mesoporous materials: Effectiveness and complexation rate, *Microporous Mesoporous Mater.*, **212**, pp. 125–136.

48. Jeong, E.-Y., Ansari, M. B., Mo, Y.-H., Park, S.-E. (2011) Removal of Cu(II) from water by tetrakis(4-carboxyphenyl) porphyrin-functionalized mesoporous silica, *J. Hazard. Mater.*, **185**, pp. 1311–1317.

49. Zhang, D. S., Liu, X. Y., Li, J. L., Xu, H. Y., Lin, H., Chen, Y. Y. (2013) Design and fabrication of a new class of nano hybrid materials based on reactive polymeric molecular cages, *Langmuir*, **29**, pp. 11498–11505.

50. Tao, J., Xiong, J., Jiao, C., Zhang, D., Lin, H., Chen, Y. (2016) Hybrid mesoporous silica based on hyperbranch-substrate nanonetwork as highly efficient adsorbent for water treatment, *ACS Sust. Chem. Eng.*, **4**, pp. 60–68.

51. Pu, W.-F., Liu, R., Wang, K.-Y., Li, K.-X., Yan, Z.-P., Li, B., Zhao, L. (2015) Water soluble core-shell hyperbranched polymers for enhanced oil recovery, *Ind. Eng. Chem. Res.*, **54**, pp. 798–807.

52. Dvornic, P. R. (2006) PAMAMOS: The first commercial silicon-containing dendrimers and their applications, *J. Polym. Sci. Part A: Polym. Chem.*, **44**, pp. 2755–2773.

53. Arkas, M., Tsiourvas, D., Paleos, C. M. (2005) Organosilicon dendritic networks in porous ceramics for water purification, *Chem. Mater.*, **17**, pp. 3439–3444.

54. Ruckenstein, E., Yin, W. (2000) SiO_2–poly(amidoamine) dendrimer inorganic/organic hybrids, *J. Polym. Sci. Part A: Polym. Chem.*, **38**, pp. 1443–1449.

55. Liu, P., Wang, T. (2007) Adsorption properties of hyperbranched aliphatic polyester grafted attapulgite towards heavy metal ions, *J. Hazard. Mater.*, **149**, pp. 75–79.

56. Lantenois, S., Prélot, B., Douillard, J.-M., Szczodrowski, K., Charbonnel, M.-C. (2007) Flow microcalorimetry: Experimental development and application to adsorption of heavy metal cations on silica, *Appl. Surf. Sci.*, **253**, pp. 5807–5813.

57. Walcarius, A., Devoy, J., Bessiere, J. (1999) Electrochemical recognition of selective mercury adsorption on minerals, *Environ. Sci. Technol.*, **33**, pp. 4278–4284.

58. Nassar, N. N. (2010) Kinetics, mechanistic, equilibrium, and thermodynamic studies on the adsorption of acid red dye from wastewater by γ-Fe_2O_3 nanoadsorbents, *Sep. Sci. Technol.*, **45**, pp. 1092–1103.

59. Tan, Y., Chen, M., Hao, Y. (2012) High efficient removal of Pb(II) by amino-functionalized Fe_3O_4 magnetic nano-particles, *Chem. Eng. J.*, **191**, pp. 104–111.

60. Wang, H., Yan, N., Li, Y., Zhou, X., Chen, J., Yu, B., Gong, M., Chen, Q. (2012) Fe nanoparticle-functionalized multi-walled carbon nanotubes: One-pot synthesis and their applications in magnetic removal of heavy metal ions, *J. Mater. Chem.*, **22**, pp. 9230–9236.

61. Wang, H., Yu, Y.-F., Chen, Q.-W., Cheng, K. (2010) Carboxyl-functionalized nanoparticles with magnetic core and mesopore carbon shell as adsorbents for the removal of heavy metal ions from aqueous solution, *Dalton Trans.*, **40**, pp. 559–563.

62. Zargoosh, K., Abedini, H., Abdolmaleki, A., Molavian, M. R. (2013) Effective removal of heavy metal ions from industrial wastes using

thiosalicylhydrazide-modified magnetic nanoparticles, *Ind. Eng. Chem. Res.*, **52**, pp. 14944–14954.

63. Singh, S., Barick, K., Bahadur, D. (2011) Surface engineered magnetic nanoparticles for removal of toxic metal ions and bacterial pathogens, *J. Hazard. Mater.*, **192**, pp. 1539–1547.

64. Tao, S., Wang, C., Ma, W., Wu, S., Meng, C. (2012) Designed multifunctionalized magnetic mesoporous microsphere for sequential sorption of organic and inorganic pollutants, *Microporous Mesoporous Mater.*, **147**, pp. 295–301.

65. Badruddoza, A. Z. M., Shawon, Z. B. Z., Tay, W. J. D., Hidajat, K., Uddin, M. S. (2013) Fe_3O_4/cyclodextrin polymer nanocomposites for selective heavy metals removal from industrial wastewater, *Carbohydr. Polym.*, **91**, pp. 322–332.

66. Milner, S. (1991) Polymer brushes, *Science*, **251**, pp. 905–914.

67. Farrukh, A., Akram, A., Ghaffar, A., Hanif, S., Hamid, A., Duran, H., Yameen, B. (2013) Design of polymer-brush-grafted magnetic nanoparticles for highly efficient water remediation, *ACS Appl. Mater. Interf.*, **5**, pp. 3784–3793.

68. Nayab, S., Farrukh, A., Oluz, Z., Tuncel, E. I., Tariq, S. R., Rahman, H. U., Kirchhoff, K., Duran, H., Yameen, B. (2014) Design and fabrication of branched polyamine functionalized mesoporous silica: An efficient absorbent for water remediation, *ACS Appl. Mater. Interf.*, **6**, pp. 4408–4417.

69. Liu, M., Zhang, B., Wang, H., Zhao, F., Chen, Y., Sun, Q. (2016) Facile crosslinking synthesis of hyperbranch-substrate nanonetwork magnetite nanocomposite for the fast and highly efficient removal of lead ions and anionic dyes from aqueous solutions, *RSC Adv.*, **6**, pp. 67057–67071.

70. Tao, J., Xiong, J., Jiao, C., Zhang, D., Lin, H., Chen, Y. (2016) Hybrid mesoporous silica based on hyperbranch-substrate nanonetwork as highly efficient adsorbent for water treatment, *ACS Sustainable Chem. Eng.*, **4**, pp. 60–68.

71. Niu, Y., Qu, R., Chen, H., Mu, L., Liu, X., Wang, T., Zhang, Y., Sun, C. (2014) Synthesis of silica gel supported salicylaldehyde modified PAMAM dendrimers for the effective removal of Hg(II) from aqueous solution, *J. Hazard. Mater.*, **278**, pp. 267–278.

72. Amina, A. K., Tahar, S., Jean-Francois, P., Pierre, B. (2001) Photocatalytic degradation of 2-phenylphenol on TiO_2 and ZnO in aqueous suspensions, *J. Photoch. Photobio. A*, **141**, pp. 231–239.

73. Yaber, M. C., Roderiguez, J., Freer, J., Baeza, J., Dura, N., Mansilla, H. D. (1999) Advanced oxidation of a pulp mill bleaching wastewater, *Chemosphere*, **39**, pp. 1679–1688.

74. Dindar, B., Icli, S. (2001) Unusual photoreactivity of zinc oxide irradiated by concentrated sunlight, *J. Photoch. Photobio. A*, **140**, pp. 263–268.

75. Gautam, M., Verma, M., Misra, G. (2011) Structural and optical properties of ZnO nanocrystals, *J. Biomed. Nanotechnol.*, **7**, pp. 161–162.

76. Sun, L., Bolton, J. R. (1996) determination of the quantum yield for the photochemical generation of hydroxyl radicals in tio_2 suspensions, *J. Phys. Chem.*, **100**, pp. 4127–4134.

77. Nosaka, Y., Kishimoto, M., Nishino, J. (1998) Factors governing the initial process of TiO_2 photocatalysis studied by means of in-situ electron spin resonance measurements, *J. Phys. Chem. B*, **102**, pp. 10279–10283.

78. Diebold, U. (2003) The surface science of titanium dioxide, *Surf. Sci. Rep.*, **48**, pp. 53–229.

79. Daoud, W. A., Xin, J. H., Zhang, Y-H. (2005) Surface functionalization of cellulose fibers with titanium dioxide nanoparticles and their combined bactericidal activities, *Surf Sci.*, **599**, pp. 69–75.

80. Hirakawa, T., Nosaka, Y. (2002) Properties of O_2•– and OH• formed in TiO_2 aqueous suspensions by photocatalytic reaction and the influence of H_2O_2 and some ions, *Langmuir*, **18**, pp. 3247–3254.

81. Wold, A. (1993) Photocatalytic properties of titanium dioxide (TiO_2), *Chem. Mater.*, **5**, pp. 280–283.

82. Blake, D. M., Maness, P.-C., Huang, Z., Wolfrum, E. J., Huang, J., Jacoby, W. A. (1999) Application of the photocatalytic chemistry of titanium dioxide to disinfection and the killing of cancer cells, *Sep. Purif Methods*, **28**, pp. 1–50.

83. Roy, C. R., Gies, H. P., Lugg, D. J., Toomey, S., Tomlinson, D. W. (1998) The measurement of solar ultraviolet radiation, *Mutat. Res.*, **422**, pp. 7–14.

84. Hashimoto, K., Irie, H., Fujishima, A. (2005) TiO_2 Photocatalysis: A historical overview and future prospects, *Jpn. J. Appl. Phys.*, **44**, pp. 8269–8285.

85. Barakat, M. A., Chen, Y. T., Huang, C. P. (2004) Removal of toxic cyanide and Cu(II) ions from water by illuminated TiO_2 catalyst, *J. Appl. Catal. B: Environ.*, **53**, pp. 13–20.

86. Kajitvichyanukula, P., Ananpattarachaia, J., Pongpom, S. (2005) Sol–gel preparation and properties study of TiO_2 thin film for photocatalytic reduction of chromium(VI) in photocatalysis process, *Sci. Technol. Adv. Mater.*, **6**, pp. 352–358.

87. Mohapatra, P., Samantaray, S. K., Parida, K. (2005) Photocatalytic reduction of hexavalent chromium in aqueous solution over sulphate modified titania, *J. Photochem. Photobiol. A: Chem.*, **170**, pp. 189–194.

88. Dasa, D. P., Parida, K., Ranjan, B. D. (2006) Photocatalytic reduction of hexavalent chromium in aqueous solution over titania pillared zirconium phosphate and titanium phosphate under solar radiation, *J. Mol. Catal. A: Chem.*, **245**, pp. 217–224.

89. Gkika, E., Troupis, A., Hiskia, A., Papaconstantinou, E. (2006) Photocatalytic reduction of chromium and oxidation of organics by polyoxometalates, *Appl. Catal. B: Environ.*, **62**, pp. 28–34.

90. Wang, L., Wang, N., Zhu, L., Yu, H., Tang, H. (2008) Photocatalytic reduction of Cr(VI) over different TiO_2 photocatalysts and the effects of dissolved organic species, *J. Hazard. Mater.*, **152**, pp. 93–99.

91. Papadama, T., Xekoukoulotakis, N. P., Poulios, I., Mantzavinos, D. (2007) Photocatalytic transformation of acid orange 20 and Cr(VI) in aqueous TiO_2 suspensions, *J. Photochem. Photobiol. A: Chem.*, **186**, pp. 308–315.

92. Rengaraj, S., Venkataraj, S., Yeon, J. W., Kim, Y., Li, X. Z., Pang, G. K. H. (2007) Preparation, characterization and application of Nd–TiO_2 photocatalyst for the reduction of Cr(VI) under UV light illumination, *Appl. Catal. B: Environ.*, **77**, pp. 157–165.

93. Yoona, J., Shimb, E., Baec, S., Jooa, H. (2009) Application of immobilized nanotubular TiO_2 electrode for photocatalytic hydrogen evolution: Reduction of hexavalent chromium (Cr(VI)) in water, *J. Hazard. Mater.*, **161**, pp. 1069–1074.

94. Zhang, F. S., Itoh, H. (2006) Photocatalytic oxidation and removal of arsenite from water using slag–iron oxide–TiO_2 adsorbent, *Chemosphere*, **65**(1), 125–131.

95. Hosono, E., Fujihara, S., Honma, I., Zhou, H. (2005) The fabrication of an upright-standing zinc oxide nanosheet for use in dye-sensitized solar cells, *Adv. Mater.*, **17**(17), pp. 2091–2094.

96. Johar, M. A., Afzal, R. A., Ali Alazba, A. A., Manzoor, U. (2015) Photocatalysis and bandgap engineering using ZnO nanocomposites, *Adv. Mater. Sci. Eng.*, **2015**, Article ID 934587, pp. 1–22.

97. Lee, K. M., Lai, C. W., Ngai, K. S., Juan, J. C. (2016) Recent developments of zinc oxide based photocatalyst in water treatment technology: A review, *Water Res.*, **88**, pp. 428–448.

98. Xu, L., Hu, Y. L., Pelligra, C., Chen, C. H., Jin, L., Huang, H., Sithambaram, S., Aindow, M., Joesten, R., Suib, S. L. (2009) ZnO with different morphologies synthesized by solvothermal methods for enhanced photocatalytic activity, *Chem Mater.*, **21**, pp. 2875–2885.

99. Yu, H., Yu, J., Cheng, B. (2007) Photocatalytic activity of the calcined H-titanate nanowiresfor photocatalytic oxidation of acetone in air, *Chemosphere*, **66**, pp. 2050–2057.

100. Borgohain, B., Senapati, K. K., Sarma, K. C., Phukan, P. (2012) A facile synthesis of nanocrystalline $CoFe_2O_4$ embedded one-dimensional ZnO hetero-structure and its use in photocatalysis," *J. Mol. Cat. A: Chem.*, **363–364**, pp. 495–500.

101. Xue, X., Wang, T., Jiang, X., Jiang, J., Pan, C., Wu, Y. (2014) Interaction of hydrogen with defects in ZnO nanoparticles-studied by positron annihilation, Raman and photoluminescence spectroscopy, *Cryst. Eng. Commun.*, **16**, pp. 1207–1216.

102. Liu, S., Sun, H., Suvorova, A., Wang, S. (2013) One-pot hydrothermal synthesis of ZnO-reduced graphene oxide composites using Zn powders for enhanced photocatalysis, *Chem. Eng. J.*, **229**, pp. 533–539.

103. Hong, R. Y., Li, J. H., Chen, L. L., et al. (2009) Synthesis, surface modification and photocatalytic property of ZnO nanoparticles, *Powder Tech.*, **189**(3), pp. 426–432.

104. Georgekutty, R., Seery, M. K., Pillai, S. C. (2008) A highly efficient Ag-ZnO photocatalyst: Synthesis, properties, and mechanism, *J. Phys. Chem. C*, **112**(35), pp. 13563–13570.

105. Anandan, S., Vinu, A., Sheeja, K. L. P., et al. (2007) Photocatalytic activity of La-doped ZnO for the degradation of monocrotophos in aqueous suspension, *J. Mol. Cat. A: Chem.*, **266**(1–2), pp. 149–157.

106. Zhou, Y., Lu, S. X., Xu, W. G. (2009) Photocatalytic activity of Nd-doped ZnO for the degradation of C. I. Reactive Blue 4 in aqueous suspension, *Env. Prog. Sustain. Energy.*, **28**(2), pp. 226–233.

107. Karunakaran, C., Gomathisankar, P., Manikandan, G. (2010) Preparation and characterization of antimicrobial Ce-doped ZnO nanoparticles for photocatalytic detoxification of cyanide, *Mater. Chem. Phys.*, **123**, pp. 585–594.

108. Wu, C., Shen, A., Zhang, Y.-C., Huang, Q. (2011) Solvothermal synthesis of Cr-doped ZnO nanowires with visible light-driven photocatalytic activity, *Mater. Lett.*, **65**(12), pp. 1794–1796.

109. Wu, C., Shen, L., Yu, H., Zhang, Y.-C., Huang, Q. (2012) Solvothermal synthesis of Cu-doped ZnO nanowires with visible light-driven photocatalytic activity, *Mater. Lett.*, **74**, pp. 236–238.

110. Xu, L., Pelligra, C., Chen, C.-H., Jin, L., Huang, H., Sithambaram, S., Joesten, R., Suib, S. L. (2209) ZNO with different morphologies synthesized by solvothermal methods for enhanced photocatalytic activity, *Chem. Mater.*, **21**(13), pp. 2875–2885.

111. Yu, H., Yu, J., Cheng, B. (2007). Photocatalytic activity of the calcined H-titanate nanowires for photocatalytic oxidation of acetone in air, *Chemosphere*, **66**, pp. 2050–2057.

112. Hu, A., Zhang, X., Oakes, K. D., Peng, P., Zhou, Y. Z., Servos, M. R. (2011) Hydrothermal growth of free standing TiO_2 nanowire membranes for photocatalytic degradation of pharmaceuticals, *J. Hazard. Mater.*, **189**, pp. 278–285.

113. Yi, Z., Xu, X., Duan, X., Zhu, W., Zhou, Z., Fan, X. (2011) Photocatalytic activity and stability of ZnO particles with different morphologies, *Rare Metals*, **30** pp. 183–187.

114. Sin, J.-C., Lam, S.-M., Lee, K.-T., Mohamed, A. R. (2014) Preparation of rare earth doped ZnO hierarchical micro/nanospheres and their enhanced photocatalytic activity under visible light irradiation, *Ceram. Int.*, **40**(4), pp. 5431–5440.

115. Sin, J.-C., Lam, S.-M., Satoshi, I., Lee, K.-T., Mohamed, A. R. (2014) Sunlight photocatalytic activity enhancement and mechanism of novel europium-doped ZnO hierarchical micro/nanospheres for degradation of phenol, *Appl. Catal. B Environ.*, **148–149**, pp. 258–268.

116. Zhao, Z., Song, J.-L., Zheng, J.-H., Lian, J.-S. (2014) Optical properties and photocatalytic activity of Nd-doped ZnO powders, *Trans. Nonferrous Metals Soc. China*, **24**(5), pp. 1434–1439.

117. Phuruangrat, A., Yayapao, O., Thongtem, T., Thongtem, S. (2014) Synthesis and characterization of europium-doped zinc oxide photocatalyst, *J. Nanomater.*, **9**. Article ID 367529.

118. Phuruangrat, A., Yayapao, O., Thongtem, T., Thongtem, S. (2014) Preparation, characterization and photocatalytic properties of Ho doped ZnO nanostructures synthesized by sonochemical method, *Superlattices Microstruct.*, **67**, pp. 118–126.

119. Zhang, X., Dong, S., Zhou, X., Yan, L., Chen, G., Dong, S., Zhou, D. (2015) A facile onepot synthesis of Er-Al co-doped ZnO nanoparticles with enhanced photocatalytic performance under visible light, *Mater. Lett.*, **143**, pp. 312–314.

120. Zhou, X., Shi, T., Zhou, H. (2012) Hydrothermal preparation of ZnO-reduced graphene oxide hybrid with high performance in photocatalytic degradation, *Appl. Surf. Sci.*, **258**(17), pp. 6204–6211.

121. Senthilraja, A., Subash, B., Krishnakumar, B., Rajamanickam, D., Swaminathan, M., Shanthi, M. (2014) Synthesis, characterization and catalytic activity of co-doped Ag-Au-ZnO for MB dye degradation under UV-A light, *Mater. Sci. Semicond. Process.*, **22**, pp. 83–91.

122. Subash, B., Krishnakumar, B., Swaminathan, M., Shanthi, M. (2013) Highly active Zr co-doped Ag-ZnO photocatalyst for the mineralization of Acid Black 1 under UV-A light illumination, *Mater. Chem. Phys.*, **141**(1), 11–120.

123. Zhang, D., Zeng, F. (2012) Visible light-activated cadmium-doped ZnO nanostructured photocatalyst for the treatment of Methylene Blue dye, *J. Mater. Sci.*, **47**(5), 2155–2161.

124. Kim, S. H., Kwak, S. Y., Kim, S. H., Kwak, S.-Y., Suzuki, T. (2006) Photocatalytic degradation of flexible PVC/TiO_2 nanohybrid as an eco-friendly alternative to the current waste landfill and dioxin-emitting incineration of post-use PVC, *Polymer*, **47**, pp. 3005–3016.

125. Ghanem, A. F., Badawy, A. A., Ismail, N., Rayn Tian, Z., Abdel Rehim, M. H., Rabia, A. (2014) Photocatalytic activity of hyperbranched polyester/TiO_2 nanocomposites, *App. Cat. A: General*, **472**, pp. 191–197.

Chapter 4

Dimensions of Nanocomposites in Pollution Control

Pradeep Pratap Singh[a] **and Ambika**[b]

[a]*Department of Chemistry, Swami Shraddhanand College, University of Delhi, Delhi 110036, India*
[b]*Department of Chemistry, Hans Raj College, University of Delhi, Delhi 110007, India*

ambika@hrc.du.ac.in

Environmental pollution is a serious day-to-day problem faced by developing and the developed nations. Pollution due to the anthropogenic sources contributes a major share to the overall imbalance of the ecosystem, which interferes with human health, the quality of life, natural functioning of living things and their surroundings. Thus, there is an urgent need of technology that is able to monitor, sense and purify the contaminants from the environment. Nanotechnology can be utilized to obtain desirable inexpensive materials with low toxicity and high degradation activity for the protection of the environment. Nanocomposites (NCs) can remove a broad range of pollutants, such as bacteria, heavy metals, and organic pollutants from the surroundings. In addition to remediating pollution, NCs can be used as sensors

Nanocomposites for Pollution Control
Edited by Chaudhery Mustansar Hussain and Ajay Kumar Mishra

ISBN 978-981-4774-45-1 (Hardcover), 978-1-315-14368-2 (eBook)
www.panstanford.com

to monitor toxins, which are more sensitive and selective than conventional sensors. This chapter deals with the application of nanotechnology in the environmental field to prevent pollution.

4.1 Introduction

Owing to the increase in population around the world, the rate of consumption of resources is also increasing, which is a major contributor to pollution in today's World. Contaminants and deadly pollutants are constantly produced and introduced into the environment due to both natural (volcanic eruption, etc.), and anthropogenic activities resulting in the overall imbalance of the ecosystem. One of the great challenges is maintaining and re-improving the quality of water, air, and soil for sustainable life. Thus, there is an urgent need of technology that can monitor, sense, and, if possible, purify the contaminants from the air, water, and soil.

Recently, nanocomposite (NC) materials have attracted scientists' and researchers' attention owing to their exceptional characteristics and significant applications in sciences [1]. The protection of the environment and health is a great concern of study for the determination and removal of poisonous materials through a well-recognized method. NCs are multiphase solid materials that can be obtained by the combination of one or more separate components of nanomaterial in order to obtain the best properties of each component (composite). In NCs, nanoparticles (NPs) act as fillers in a matrix, usually polymer matrix. The properties of NCs differ markedly from that of the component materials [2]. The possibilities of producing materials with tailored physical and electronic properties at low cost could result in interesting applications ranging from drug delivery to corrosion prevention to electronic/automotive parts to industrial equipment and several others. NCs can be used to sense, detect, separate, and remove pollutants from the environment and allow facile recycling without major loss of sensor efficiency and potential. In treatment and remediation of pollutants of the soil and groundwater, minute nanocomposites are being extensively used to convert (respective break down of) pollutants on site [3]. The process can also be employed with contaminants that are difficult to treat and remove [4]. Thus, NCs can play an important role in shaping current environmental engineering and science. Hopefully, nanotechnological applications and products will lead to a cleaner and healthier environment.

4.2 Classification of Nanocomposites

Nanocomposites can be classified on the basis of dimensional morphology [5–7], microstructure [8], and matrices.

4.2.1 Classification Based on the Nanomaterial's Dimensional Morphology

(a) Zero-dimensional nanomaterial, e.g., nanoparticle
(b) One-dimensional nanomaterial, e.g., nanowire and nanotubes
(c) Two-dimensional nanomaterial, e.g., silicate layers
(d) Three-dimensional nanomaterial, e.g., zeolites

4.2.2 Classification Based on Microstructure

(a) Nanolayered composites made up of alternating layers of nanoscale dimension
(b) Nanofilamentary composites composed of a matrix with embedded (and generally aligned) nanoscale diameter filaments
(c) Nanoparticulate composites composed of a matrix with embedded nanoscale particles

4.2.3 Classification Based on Matrices

In most of the nanocomposite materials, generally one phase is continuous (matrix), while the other phase is the dispersed phase. On the basis of the matrix, nanocomposites can be classified into the following categories (Scheme 4.1).

4.2.3.1 Metal oxides nanocomposites

Metal oxide–based nanocomposites are those in which the matrix is of a metal oxide and the fillers could be metal oxide NPs, nanowires, etc. These NCs can be prepared by various mechanical and chemical methods. In the mechanical method, the metal is ground resulting in very small grain size and homogeneous mixture [9]. Repeated breaking up and joining of the component particles leads to alloying, which can be utilized for the preparation of highly metastable structures such as amorphous alloys and nanocomposite structures with high flexibility. They can also be prepared by sol-gel processes and the wet chemical synthesis, in which metal salts are reduced chemically [10] and the precursors are decomposed using thermal [11], photolytic, or sonochemical treatment [12]. Metal oxide NCs offer several advantages such as stability in air, cost-effectiveness, ease of preparing with tunable porosity, low-temperature encapsulation, negligible swelling, mechanical and biodegradable stability, and high sensitivity at lower operating temperatures for detection. Thus, they can be used as sensing materials in chemical sensors, catalysts, and adsorbents for the removal of pollutants.

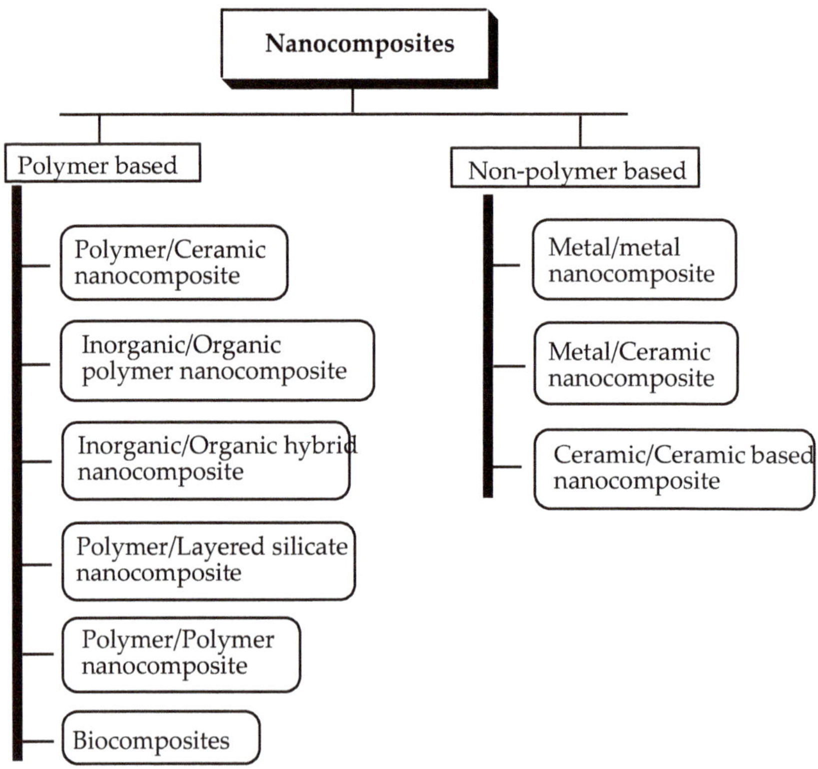

Scheme 4.1 Classification of nanocomposites based on matrices.

4.2.3.2 Carbon nanotube matrix–based nanocomposites

Carbon nanotubes (CNTs) can be classified as single-walled nanotubes (SWNTs) and multiwalled nanotubes (MWNTs). SWNTs consist of a cylindrical single sheet with a diameter between 1 and 3 nm and a length of several micrometers. They possess a cylindrical nanostructure formed by rolling up a single graphite sheet into a tube. MWNTs consist of a coaxial arrangement of concentric single nanotubes like rings of a tree trunk separated from one another by 0.34 nm. They usually have a diameter of about 2–20 nm. The production of SWNTs or MWNTs is highly dependent on the synthesis process and conditions.

Among all carbon products, carbon nanotubes have been of great interest and have found potential applications due to their mechanical and unique electronic properties. They also offer tremendous opportunities in the design of multifunctional materials systems and have been utilized in nanoelectronic devices, composites, chemical sensors, biosensors, environmental remediation, etc. (Scheme 4.2).

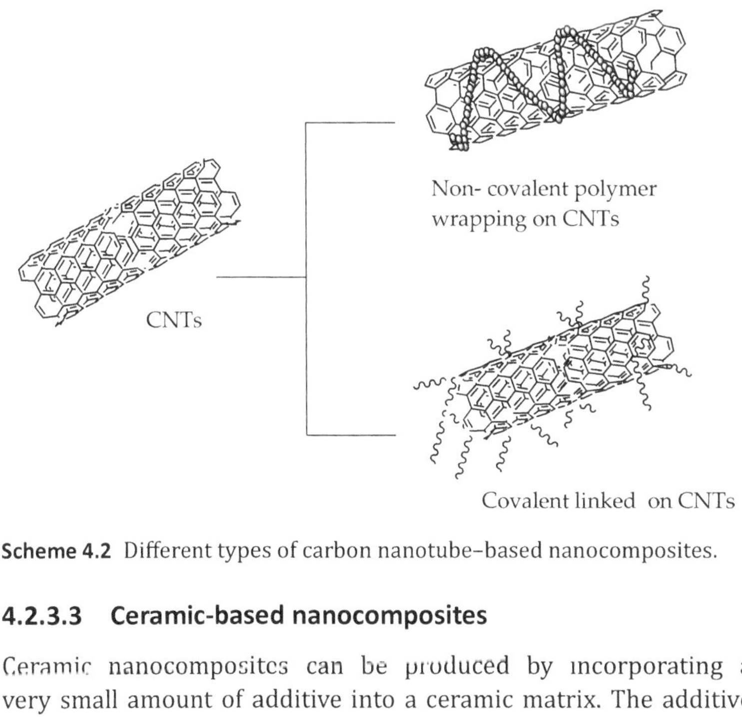

Scheme 4.2 Different types of carbon nanotube-based nanocomposites.

4.2.3.3 Ceramic-based nanocomposites

Ceramic nanocomposites can be produced by incorporating a very small amount of additive into a ceramic matrix. The additive segregates at the grain boundary with a gradient concentration or precipitates as molecular or cluster-sized particles within the grains or at the grain boundaries. In most nanocomposite materials, structural control at the molecular level could be obtained by optimized processing. They can be further classified as metal/ceramic, ceramic/ceramic and polymer/ceramic.

4.2.3.4 Polymeric nanocomposites

Polymer–nanoparticle composite (PCNs) materials have attracted the attention of a number of researchers owing to their synergistic and hybrid properties. These are the nanocomposites based on a polymer filler in any matrix. PCNs can be further classified as polymer-ceramic, polymer/layered silicate NCs, polymer/inorganic-inorganic NCs, polymer/polymer polymer/organic-organic NCs and bio-NCs. The fabrication of many devices has been carried out using different PNCs due to the ease of processability polymers combined with the better mechanical and optical properties of NPs. Dispersion of filler in the matrix is the most important step in fabrication of polymer-based nanocomposites. Various mechanochemical approaches, including sonication by ultrasound, can be used for this purpose. However, the reaggregation of the individual component limits the scope of this approach [13].

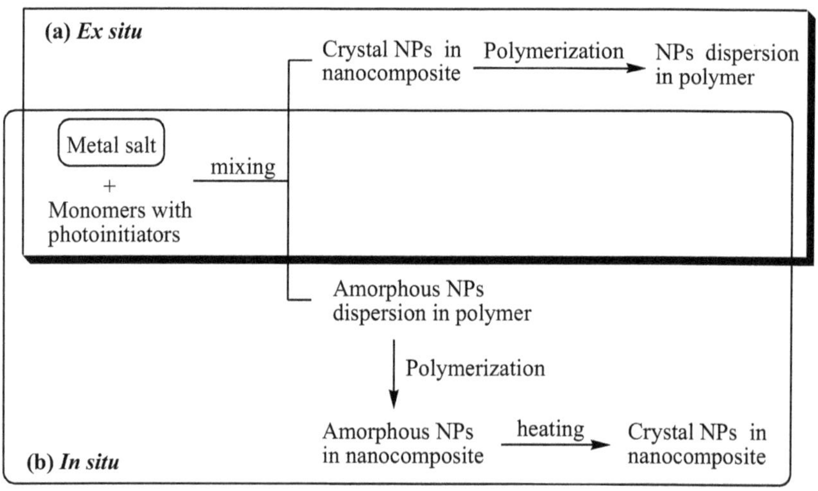

Scheme 4.3 Synthetic approaches for polymeric nanocomposites: (a) ex situ, (b) in situ.

PNCs can be synthesized using the ex situ and in situ approaches. In the ex situ method, NPs are dispersed directly into polymer to form composites. This method is more suitable for large-scale industrial applications than the in situ method (Scheme 4.3a). In the in situ methods, nanocomposites are generated inside a polymer matrix by precursors, which are transformed into the desired NPs by appropriate reactions (Scheme 4.3b). In situ approaches are currently getting much attention because of their obvious technological advantages over ex situ methods [11].

4.3 General Approaches to Nanocomposite Fabrication

Various techniques can be employed for the fabrication of NCs [14]. For example, NC systems such as ceramic supported metal NPs can be prepared by evaporating metal onto the chosen substrate or dispersal via solvent chemistry. However, novel processing techniques such as template synthesis, scanning probe electrochemical methods, and electrospinning are required for NCs having complex structures with coexisting ceramic and polymeric phases [15]. The progress of nanostructured materials requires a central structure called template within which a network is formed in such a way that removal of which creates a filled cavity with morphological and/or stereochemical features related to the template. Template synthesis requires the preparation of a variety of micro- and nanomaterials of desired morphology and therefore provides a route to enhancing nanostructure order. On the other hand, in case of nontemplate self-assembly, the larger structures are produced by the interaction of individual components without the assistance of external forces. However, for the synthesis of different nanostructures, a variety of approaches are available, but the methods capable of making pure, uniform, template-free nanostructures are required. Template-free methods of synthesizing nanostructures require only the mixing of components to achieve an ordered structure. They have several advantages, including simple synthesis, purification, and high efficiency. Also, uniform nanofibers are formed, which are easily scalable and reproducible and can be utilized as water-dispersible sensors, which facilitates eco-friendly processing and biological applications.

4.4 Applications of Nanocomposites in Pollution Control

Various nanocomposites can be used in environmental remediation by different ways such as sensing and detection, catalytic and redox degradation of contaminants and adsorption or removal of pollutants from the systems (Table 4.1).

Table 4.1 Different nanocomposites utilized for controlling pollution

Class of Nanocomposite	Nanocomposite	Pollutant
Metal oxides	α-Fe_2O_3/γ-C_3N_4 NCs	Ethanol gas
	CD-Fe_3O_4	β-Naphthol
	Ag-Fe_3O_4	4-mercaptobenzoic acid
	Zn_2SnO_4/C	Rhodamine B
	V_2O_5/TiO_2	Arsenic (III)
	Cu_xO/TiO_2	Volatile organic compounds
	TiO_2@WO_3/Au TiO_2-WO_3, P25, TiO_2@WO_3/Au	Aromatic pollutants
	M_xWO_3/ZnO (M = K, Rb, NH_4) NCs films and powders	Cr(VI)
		NO gas
	ZnO/γ-Fe_2O_3 NC	2,4-Dichlorophenoxyacetic acid
	$CoFe_2O_4$/CdS	Methylene blue, rhodamine B, and methyl orange
	BiOCl/HA	Oxytetracycline
	ZnO/SnO_2 NC	Methylene blue
	PbO_2-TiO_2	Methyl orange
	CNT/P-TiO_2	Methyl orange

Class of Nanocomposite	Nanocomposite	Pollutant
Carbon nanotubes	MWCNT/TiO_2-ceramic	*Escherichia coli* O157:H7
	GdO_2-MWCNT	Methylene blue
	MWCNT/TiO_2/SiO_2	Bisphenol A and carbamazepine
	Amine-functionalized magnetic CNTs	Cationic dyes
	Magnetic-modified corncobs	Congo red
	MWCNT/polyaniline	To remove methyl orange and Cr
	Magnetic/MWCNTs	Nitrofurans, such as furazolidone, nitrofurazone, nitrofurantoin and furaltadone
	Functionalized G/MNPs	Nitrite
Graphene Nanocomposites	BWO-RGO	Bisphenol A
	Co-rGO/ZnO	Benzyl alcohol
		Cr(VI)
		4-nitrophenol
		Methylene blue
	$Zn_{(1-x)}Mg_xO$ /rGO	Methylene blue
	TiO_2-GO	Butane
	G/TiO_2	Sodium pentachlorophenol and methylene blue
	G/Fe_3O_4	Methylene blue
	GO-COOH-CuS	Phenol and rhodamine B
	Ag-$CoFe_2O_4$-GO	*E. coli*, *B. subtilis* and *S. aureus*
	Fe_3O_4/GO	Congo red and methylene blue
	Fe_3O_4@nSiO_2@mSiO_2-TTRDNA	Hg^{2+}

(Continued)

Table 4.1 *(Continued)*

Class of Nanocomposite	Nanocomposite	Pollutant
Silica-based NCs	$Fe_3O_4@SiO_2$ NPs	Cu^{2+}
	$Fe_3O_4@SiO_2$-EDTA-Ni	Fluoride ion
	$Fe_3O_4@SiO_2$core ($Fe_3O_4@nSiO_2@mSiO_2$-Fe core–shell	DDT, DDD, DDE
	Fe_3O_4cluster@SiO_2@ TiO_2-N	Phenol
	$Fe_3O_4@SiO_2/TiO_2$	Methyl orange, methylene Blue, neutral red, bromocresol green and methyl red
	TiO_2/SiO_2/Ag triple	Ni, Pb and methylene blue
	$Fe_3O_4/mSiO_2$-C18	Methylene blue
	ZnO/MMT	Disperse red 54
Clay-based NCs	Hectorite–TiO_2	Formaldehyde
	Polycation–clay mineral	Trinitrophenol and trichlorophenol
	Linde type A zeolite (LTA)-goethite	Phosphate
	Fe_3O_4/bentonite	Methylene blue
	rectorite/chitosan	$CHCl_3$

4.4.1 Sensing and Detection of Pollutant

The α-Fe_2O_3/γ-C_3N_4 NCs were synthesized with a better gas-sensing performance, which possess excellent selectivity to ethanol gas, faster response and recovery time than the pure α-Fe_2O_3 and γ-C_3N_4. It could be attributed to the porous α-Fe_2O_3 nanotubes wrapped by lamellar γ-C_3N_4 nanostructures and the formation of heterojunction [16]. A series of new M_xWO_3/ZnO (M = K, Rb, NH_4) NCs films and powders was developed with good UV light absorption, visible light transmittance, and near infrared light shielding properties. Also, these films showed good toxic NO gas photo decomposition property, due to which they can be employed in energy saving and environmental cleanup [17]. NCs such as magnetic cyclodextrin NCs (CD-Fe_3O_4) could detect and remove β-naphthol from wastewater as a spectral probe [18]. A new magnetic adsorbent, 3-mercaptopropionic acid–coated 3-aminopropyl triethoxysilane–modified magnetic NPs (MNPs), is used for the extraction and preconcentration of arsenic ions in aqueous solutions with wide linear range and low detection limits are obtained [19]. Novel dipyridile-modified MNPs have been used for the extraction and determination of low levels of Pb^{2+} ions in various samples of rice, baking powder, wheat, and other foodstuffs [20]. Ag-Fe_3O_4 NCs are assembled into an orderly arrayed surface-enhanced Raman spectroscopy (SERS) substrate with an applied external magnetic field for detection of 4-mercaptobenzoic acid [21]. Fe_3O_4@Ag magnetic substrate with a portable SERS sensor has been utilized for the onsite monitoring and quantification of As speciation determination of As(III) and As(V) in complex media such as juice, wine, and soils [22].

Nickel oxide NP–decorated CNTs NCs (NiO·CNT NCs) were used as ultrasensitive and selective 4-aminophenol sensor that could be used for the major application of toxic agents in biological, green environmental, and health-care fields [23]. Multiwalled CNTs (MWCNTs) have been modified by MNPs with application for the preconcentration of Cd and Pb [24, 25]. MNPs have been used to construct a novel nitrite sensor by electropolymerization of alizarin red on the surface of glassy carbon electrode modified with Fe_3O_4-MWCNT composite nanofilm [26]. Functionalized graphene anchored with MNPs an electrochemical sensor loaded on a glassy carbon electrode (GCE) was developed for the detection of nitrite in tap, river, and rain water samples [27]. A regenerable and highly selective core–shell structured magnetic mesoporous silica NC with functionalization of thymine (T) and T-rich DNA (denoted as Fe_3O_4@nSiO_2@mSiO_2-TTRDNA NC) has been developed for simultaneous detection and removal of Hg^{2+} from wastewater. The Fe_3O_4@nSiO_2@mSiO_2-T-TRDNA NCs for Hg^{2+} could be easily regenerated using a simple acid treatment and resistance to nuclease digestion [28].

Polydiacetylene/silica NC was synthesized for use as a chemosensor. The polydiacetylene/silica NCs on irradiation with UV light acquire a blue color that could change to red on exposure of a variety of environmental perturbations, such as temperature, pH, and amphiphilic molecules [29]. Novel phosphatidylserine-functionalized Fe_3O_4@SiO_2 NPs and enzyme-encapsulated liposomes were used for the visual detection of as little as 0.1–0.5 M Cu^{2+} by employing phosphatidylserine and the enzymatic catalysis/oxidation of 3,3′,5,5′-tetramethyl benzidine sulfate (TMB) as a signal generator [30]. Fluorescent chemosensors containing MNPs have been developed for detecting anions such as ClO^- and SCN^- [31]. Highly fluorescent carbon quantum dot (CD) and magnetically separable nickel ethylenediaminetetraacetic acid (EDTA) complex bound-silica coated MNPs (Fe_3O_4@SiO_2-EDTA-Ni) are used as fluorophore and fluoride ion receptor. The fluoride sensors could be effectively used several times [32].

Aluminum-doped zinc oxide/polyaniline (Al/ZnO-PANI) NC was employed in the gas-sensing reaction. Al-doping effectively improves the gas-sensing properties of ZnO-PANI NC [33]. A three-dimensional nanocomposite was developed by the intercalation of a chitosan bilayer (a biopolymer) in Na^+-montmorillonite (MMT) through cationic exchange and H-bonding processes. The above reactions turn the resulting nanocomposite into an anionic exchanger, applied in the development of potentiometric sensors which show a remarkable selectivity toward monovalent anions due to the special arrangement of the biopolymer as a nanostructured bidimensional system [34, 35].

4.4.2 Catalytic and Redox Degradation of Contaminants

Novel Zn_2SnO_4/C NCs were prepared with high photocatalytic activity against rhodamine B (RhB) as a model organic pollutant. The above NC exhibits much better visible-light photocatalytic activity as compared to pure Zn_2SnO_4, NCs [36]. V_2O_5/TiO_2 NCs were employed as solar energy transducers to photo-oxidation of As (III) in aqueous solution. V_2O_5/TiO_2 with core–shell spheres morphology possesses excellent photocatalytic activity in the oxidation of arsenite onto the one with solid sphere nanostructure, because of its larger specific surface area, enhanced visible harvest ability, and improved charge separation efficiency [37]. Ag-$SrTiO_3$ NCs (Ag-STO) were synthesized with enhanced visible light-driven activity on NO removal without the generation of NO_2 in comparison with pristine $SrTiO_3$. The Ag-loading amount has a significant influence on light absorption properties of

Ag-STO, which further affects the photocatalytic efficiency [38]. The grafting of nanometer-sized Cu_xO clusters onto TiO_2 generates an excellent risk-reduction material in indoor environments. In the Cu_xO clusters, Cu(II) species enriches TiO_2 with efficient visible-light photo-oxidation of volatile organic compounds, whereas the Cu(I) species provides antimicrobial properties under dark conditions. By controlling the balance between Cu(I) and Cu(II) in Cu_xO, efficient decomposition and antipathogenic activity were achieved in the hybrid Cu_xO/TiO_2 NCs [39]. Double-shelled, positively and negatively charged, nanostructured hollow spheres with supported Au-NPs. TiO_2, WO_3, and Au NPs were coated successively onto the functionalized polystyrene (PS) template spheres. It is the rate of photodgradation activity for rhodamine B and trimesic acid (i.e., color and colorless aromatic pollutants) in decreasing order were $TiO_2@WO_3/Au > TiO_2\text{-}WO_3 >$ P25. $TiO_2@WO_3/Au$ was used for the photodegradation of aromatic pollutants and photoreduction of Cr(VI) [40]. New NC $MnO_2/Al_2O_3/Fe_2O_3$ photocatalyst was synthesized by the sol-gel method with high photoabsorption property in the visible light region [41]. The $ZnO/\gamma\text{-}Fe_2O_3$ catalyst was used for the photodegradation 2,4-dichlorophenoxyacetic acid (2,4-D) under UV irradiation. $ZnO/\gamma\text{-}Fe_2O_3$ NC exhibited enhanced photoactivity compared to pure ZnO and can be used as a photocatalyst in removing organic pollutants in wastewater [42]. A magnetic $CoFe_2O_4/CdS$ nanocomposite has been used for the sono/photocatalytic degradation of organic pollutant such as H_2O_2-assisted degradation of MB, rhodamine B, and methyl orange (MO) under ultrasonic irradiation. The NC displayed excellent sonocatalytic activity towards the degradation of all dyes examined and is a more efficient sonocatalyst than pure CdS [43]. The photocatalytic activity of hydroxyapatite supported BiOCl (BiOCl/HA) in the presence of H_2O_2 was reported using solar radiation during photocatalysis with good reusability. Both H_2O_2 and BiOCl/HA had a synergistic effect on oxytetracycline removal by simultaneous adsorption and photocatalysis [44]. ZnO/SnO_2 NC was synthesized using the co-precipitation method and used for the removal of methylene blue (MB) [45]. The electrocatalytic degradation of MO on $PbO_2\text{-}TiO_2$ nanocomposite electrodes has been reported. In comparison to PbO_2 electrodes, $PbO_2\text{-}TiO_2$ nanocomposite electrodes possess higher oxygen evolution over potential and higher COD removal efficiency and instantaneous current efficiency with MO degradation [46].

CNT/P-TiO_2 exhibits high photocatalytic efficiency for the degradation of MO dye both under UV irradiation and under visible light irradiation. The surface states of phosphorous, which generate the Ti-O-P linkage in the photocatalyst, and the presence of CNTs, promote the separation of photo-generated carriers [47]. MWCNT/TiO_2-ceramic under fluorescent light was employed for the removal of *Escherichia coli O*157:H7 during the filtration of water [48]. GdO_2-decorated MWCNT NCs were reported, with high photocatalytic degradation of MB than the MWCNT/TiO_2 NC and commercial titania and it could be reused for five times [49]. Another NC was also prepared using Gd, N, and S, tri-doped titania, decorated on oxidized MWCNTs for the degradation of naphthol blue black (NBB), in water under simulated solar light irradiation. The higher photocatalytic activity is attributed to the combined effect of improved visible light absorption and charge separation due to the synergistic effect of all the components of NCs. The prepared NC was fairly stable and could be reused for five times [50]. MWCNT/TiO_2/SiO_2 NC was used for photocatalytic treatment of pharmaceutical wastewater, in removal of bisphenol A and carbamazepine from water [51]. Novel strontium titanate nanocubes/reduced graphene oxide (STO/rGO) NCs were synthesized by the sol-precipitation process, which exhibits better adsorption and photocatalytic performance as compared to STO [52]. The uniform B_2WO_6-reduced graphene oxide (BWO-RGO) was reported for the photocatalytic degradation of pollutant bisphenol A, selective oxidation of benzyl alcohol, removal of Cr(VI), and selective reduction of 4-nitrophenol [53]. Cobalt activated ZnO NCs have been developed on reduced graphene oxide (rGO) nanosheets for the simultaneous enhancement of photocatalytic degradation of methylene blue (MB), and antibacterial activities. Compared with bare ZnO, the Co-rGO activated ZnO NCs exhibit significantly enhanced photocatalytic and antibacterial activities due to the synergetic effects of various mechanisms related to the Co and rGO incorporation [54]. Photocatalytic activity and photovoltaic and UV detector applications of $Zn_{(1-x)}Mg_xO$ (x = 0.02, 0.04, and 0.06) NPs and $Zn_{(1-x)}Mg_xO$/graphene NCs were reported. The $Zn_{(1-x)}Mg_xO$ /rGO NCs showed an excellent performance to remove methylene blue (MB) dye under natural sunlight illumination compared to the $Zn_{(1-x)}Mg_xO$ NPs [55]. A coupled $Au/Fe_3O_4/GO$ hybrid material improves the catalytic activity, stability, and separation capability of AuNPs and Hg^{2+} [56]. TiO_2-graphene oxide (TiO_2-GO) NC was prepared for the photocatalytic degradation of butane in the gas phase [57]. G/TiO_2 NC with high photocatalytic activity for degradation of sodium pentachlorophenol was synthesized, which showed higher photocatalytic activity efficiency than commercial $P_{25}TiO_2$. The improved photocatalytic activity may be attributed to the accelerated interfacial electron transfer process and the significantly prolonged lifetime of electron–hole pairs imparted by graphene sheets in the NCs [58]. TiO_2/graphene NC photocatalysts (GNP) was used for the photo-degradation of MB under visible light. GNP and graphene nanowire (GNW) have higher performance than their counterparts of pure TiO_2 NP and NW. The adsorptivity of GNW is significantly higher than GNP. As a result, the photocatalytic activity of GNW is much higher than GNP and pure NWs or NPs (Scheme 4.4) [59]. G/Fe_3O_4 NCs with different ratios were prepared by the solvothermal method for the removal of MB dye from aqueous solutions [60].

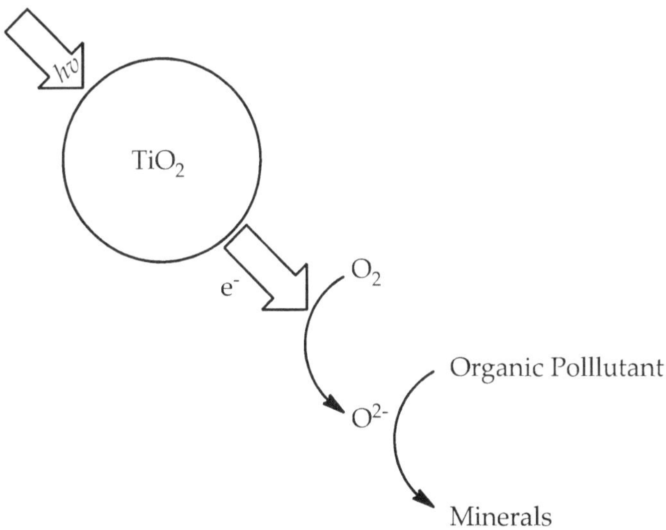

Scheme 4.4 Photocatalytic degradation of pollutants at TiO_2 surface.

Carboxylic acid-functionalized graphene oxide-copper(II) sulfide nanoparticle composite (GO-COOH-CuS) was reported with excellent photocatalytic degradation performance of phenol and rhodamine B, high antibacterial activity toward *E. coli* and *B. subtilis*, and good recovery and reusability [61].

A novel multifunctional microsphere with an iron oxide–improved mesoporous silica shell (Fe_3O_4@SiO_2core(Fe_3O_4@nSiO_2@mSiO_2-Fe core–shell) was utilized for the removal of 1,1,1-trichloro-2,2-bis(4-chlorophenyl)ethane (DDT) and its derivatives, i.e., 1,1-dichloro-2,2-bis(4-chlorophenyl)ethane (DDD) and 1,1-dichloro-2,2-bis(4-chlorophenyl)ethylene (DDE) [62]. Different nanocomposites such as pure anatase phase titania (TiO_2) NPs, nitrogen-doped titania (TiO_2-N), titania-coated magnetic silica (Fe_3O_4cluster@SiO_2@TiO_2 (FST)), and a novel magnetically recoverable TiO_2 nanocomposite photocatalyst containing nitrogen element (Fe_3O_4cluster@SiO_2@TiO_2-N (FST-N)) have been synthesized via a sol-gel process. Nitrogen-doped NCs such as TiO_2-N and FSTN NCs can be applied for the photocatalytic degradation of phenol under visible light irradiation, while undoped samples were almost inactive under same operating conditions [63]. TiO_2/SiO_2-based NC (Fe_3O_4@SiO_2/TiO_2) have been employed for removing dyes from the textile wastewater under UV irradiation. The NC could remove a large range of cationic, anionic and neutral dyes including MO, MB, neutral red (NR), bromocresol green (BG), and methyl red (MR). NR and BG have good results in reusing treatment. The NC could be separated and reused for three times [64]. NCs using organo-templated mesoporous silica and TiO_2 NPs with high molecular selective photocatalysis for the decomposition of organic molecules such as 4-nonylphenol (alkyl phenol) from water have been employed. The molecules possessing larger alkyl group undergo decomposition at a higher rate [65]. TiO_2/SiO_2/Ag triple NCs have been synthesized for the adsorption of Ni and Pb. The NC can also be employed for the efficient removal of MB with remarkable antibacterial activity from textile wastewater as compared to only TiO_2/SiO_2 [66]. A semi-crystalline NC cation exchanger has been synthesized by the sol-gel method at pH 1.0. The material was selective for Hg(II), Bi(III), Zr(IV), and Pb(II) ions. Besides the ion-exchanger, polyaniline/Ti(IV) as a composite material has been successfully applied for the photochemical degradation of industrial dye as well as a conducting material [67].

ZnO immobilized on MMT was used to synthesize ZnO/MMT NC for the photocatalytic ozonation of dispersed red 54 (DR54). The high decolorization efficiency could be attributed to the photocatalytic ozonation as compared to adsorption, single ozonation, catalytic ozonation, and photolysis [68]. The adsorption capacity and photocatalytic removal efficiency of formaldehyde using a hectorite–TiO_2 composite has been reported using different UV-sources. Hectorite–TiO_2 possesses 4.1 times higher formaldehyde uptake capacity than TiO_2 (Degussa P25). Removal efficiencies are proportional to the Ti content and the contact time but inversely proportional to relative humidity. Removal of formaldehyde also depends on the nature of UV source and complete elimination can be accomplished by using 254 + 185 nm UV source [69]. The photocatalytic oxidation of selected, recalcitrant contaminants was reported using Fe(VI)-TiO_2-UV system. The oxidation in the presence of ferrate increases significantly. The performance of the ferrate/photocatalysis system is strongly influenced by the reaction conditions, the pH, and dissolved oxygen concentration, arising from the complex nature of the interactions between the catalyst and the solution [70]. Hybrid photocatalysts consisting of titania (TiO_2) and an adsorbent such as mordenite have been employed for the efficient removal of toluene, an indoor air pollutant. The photocatalytic activities of hybrid photocatalysts in decomposing toluene are 1.33 times as high as pure P25 [71].

4.4.3 Adsorption of Pollutants

A Fe_3O_4/metal-organic frameworks (MOFs) composite adsorbs methylene blue (MB) and methyl violet from an aqueous solution and reused for several cycles [72]. Fe_3O_4@3-aminophenol-formaldehyde (Fe_3O_4@APF) core–shell resin polymer magnetic NCs are excellent adsorbent for the removal of MB [73]. Humic acid–coated MNPs remove MB from the aqueous solution [74]. The core–shell Ni@$BaTiO_3$ NCs were synthesized and possess high microwave absorption properties as compared to Ni NPs [75]. ZrO_2 NPs were immobilized in quaternary-aminated wheat straw to obtain an inexpensive, eco-friendly Ws-N-Zr NC. The biomass-based Ws-N-Zr exhibited higher preference toward phosphate than commercial anion exchanger IRA-900 when competing sulfate ions coexisted at relatively high levels and can be recycled without any significant loss of activity in contaminated waters [76]. Ni@$Mg(OH)_2$ NCs exhibit excellent heavy metal adsorption capacity and recyclable property. The first removal efficiency is almost 100% for target metals, and after five cycles, the adsorption capacity remains 95% [77]. Amine-functionalized magnetic CNTs effectively remove cationic dyes from binary systems [78]. Magnetic-modified corncobs show good adsorption performance for Congo red (CR) [79]. MWCNTs decorated with MNPs are modified by polyaniline to remove MO and CR [80]. Magnetic MWCNT NCs have been used for the removal of four nitrofurans, namely, furazolidone (FZD), nitrofurazone (NFZ), nitrofurantoin (NFT), and furaltadone (FTD) from aqueous solution [81]. Iron-nickel NPs deposited onto aluminum oxide (α-Al_2O_3) and carbon nanotubes (CNTs) to form NC materials Fe-Ni/Al_2O_3 and Fe-Ni/CNTs, respectively, were used as adsorbents [82].

Ag-$CoFe_2O_4$-GO NCs exhibit excellent antibacterial activity against *E. coli* and *S. aureus* compared with $CoFe_2O_4$, Ag-$CoFe_2O_4$, and $CoFe_2O_4$-GO composite, which can be attributed to the combination of GO nanosheets and AgNPs [83]. Fe_3O_4/GO NCs effectively remove CR as an anionic dye and MB as a cationic dye in aqueous solutions [84]. Hybrid materials consisting of polyacid brushes using poly(itaconic acid) (PIA) and poly(acrylic acid) (PAA) at the amine functional groups of chitosan and MNPs have been utilized for the reversible pH-responsive behavior and controlled adsorption/desorption of MB [85].

Novel mesoporous cetyl trimethyl ammonium bromide (CTAB)-functionalized magnetic microspheres with a core/shell structure (mesoporous Fe_3O_4@SiO_2@CTAB-SiO_2) effectively remove perfluorooctane sulfonates (PFOS) from water at acidic conditions (pH = 3) [86]. Mesoporous-silica-capped Fe_3O_4 mesoporous clusters exhibit excellent adsorption of MB from aqueous solution and could be regenerated easily by the photo-Fenton degradation reaction due to its high specific surface area, accessible pore channels, and good magnetic separation properties [87]. A C18-functionalized core–shell magnetic mesoporous silica composite

(Fe_3O_4/mSiO_2-C18) has excellent adsorption ability toward MB dye due to the large surface area and the abundant hydrophobic C18 groups [88]. Sunset yellow anionic dye is removed by amino propyltriethoxysilane-modified MNPs [89]. The NCs of electroactive polymers PANI or polypyrrole with ultrafine SiO_2 particles have been employed for metal uptake. The uptake of gold was comparatively faster and easier than palladium [90]. Organosilicate NC hexagonal mesostructure (NHMS) were synthesized from dodecylamine and tetraethyl orthosilicate (TEOS) to remove and retain chlorinated phenols (2,4-dichlorophenol) from aqueous solution [91]. Magnetic Fe_3O_4-chitosan NPs (m-Fe_3O_4-CNs) could adsorb bromothymol blue (BB) from aqueous solutions [92]. Fe_3O_4/chitosan/TiO_2 NCs show good adsorptive, photocatalytic, regenerated, and magnetic properties for the degradation of MB in wastewater [93]. By encapsulating zirconium phosphate (ZrP) NPs into three macroporous polystyrene resins with various surface groups, i.e., $-CH_2Cl$, $-SO_3-$, and $-CH_2N-(CH_3)_3$, three NC adsorbents (ZrP-Cl, ZrP-S, and ZrP-N) have been fabricated for lead removal from water. The presence of the charged functional groups ($-SO_3^-$ and $-CH_2N-(CH_3)_3$) is more favorable than the neutral $-CH_2Cl$ group to improve nano-ZrP dispersion (i.e., to achieve smaller ZrP NPs). ZrP-N and ZrP-S had higher capacity than ZrP-Cl for Pb removal. As compared to ZrP-N, ZrP-S exhibits higher preference toward lead ion at high calcium levels [94].

Polycation–clay mineral NCs were designed for the removal of pollutants such as trinitrophenol (PA) and trichlorophenol (TCP) from water. The adsorption kinetics of polydiallyl dimethylammonium chloride (PDADMAC) and poly-4-vinylpyridine-co-styrene (PVPcoS) on MMT was significantly faster than on sepiolite. Consequently, polycation–MMT composites were chosen to test pollutant adsorption. Both PA (anionic) and TCP (nonionic) showed higher affinity to the less charged polycation PVPcoS than to the highly charged polycation PDADMAC [95]. Linde type A zeolite (LTA)-goethite NC was synthesized for the adsorption of phosphate higher than 1.6 times the amount adsorbed on the mixture, indicating generation of synergistic effect in the LTA-goethite NC [96]. Fe_3O_4/bentonite NC has been utilized for the removal of MB from aqueous solutions. The adsorption capacity for the dye increases by increasing the contact time and initial pH of the solution. However, it decreased with increasing the adsorbent [97]. A novel NC bioadsorbent rectorite/chitosan was prepared by controlling different mass ratios of chitosan to rectorite using the water phase intercalation technique for adsorbent toward $CHCl_3$ from aqueous solution [98].

4.5 Future Prospects

Nanotechnology provides huge scope for the development of sustainable innovative materials for agriculture, water treatment, food production, processing, preservation and packaging applications. With the advancement of techniques, new nanocomposites have been developed using different matrix and fillers. This has opened a new area for research and development. However, the knowledge about the toxicity and biodegradability of NCs is very scarce. Nanotechnology could provide a platform for the development of various methodologies and techniques employed for solving the existing and pressing environmental issues such as detectors, sensors, pollution control, prevention and remediation. At the large scale, NCs have not yet been utilized for remediation projects and selectively used since economical and eco-friendly applications are not yet developed.

4.6 Conclusions

Nanocomposites are highly reactive because of their large surface area to volume ratio and presence of active sites. These properties of NCs can be applied for solving potential environmental issues such as air, water, and soil contamination and their remediation. NCs can be used for the removal of contaminants and deadly pollutants such as heavy metals, nitrates, oil spill, various gases, and toxins released by industries and combustion of fossil fuels from air, water, and soil. Metals NPs and metal oxides are used as fillers in composites, which imparts biocidal nature, and thus can be used in environmental remediation. NCs can also be used to prepare nano-sensors that can be used to detect heavy metal ions without any prior analysis, which can be handled easily. Nanobiosensors are used extensively for assessing pollution and water quality. Thus, nanocomposites provide opportunities and rewards creating fresh worldwide interest in these new materials.

References

1. Okpala, C. C. (2013). Nanocomposites-an overview, *Int. J. Eng. Res. Dev.*, **8**, pp. 17–23.
2. Okpala, C. C. (2014). The benefits and applications of nanocomposites, *Int. J. Adv. Eng. Tech.*, **1**, pp. 12–18.
3. Kaushik, A., Kumar, R., Arya, S. K., Nair, M., Malhotra, B. D., and Bhansali, S. (2015). Organic-inorganic hybrid nanocomposite-based gas sensors for environmental monitoring, *Chem. Rev.*, **115**, pp. 4571–4606.
4. Patil, M. R., and Shrivastava, V. S. (2015). Adsorption removal of carcinogenic acid violet 19 dye from aqueous solution by polyaniline-Fe_2O_3 magnetic nano-composite, *J. Mater. Environ. Sci.*, **6**, pp. 11–21.
5. Salavati-Niasari, M., Davar, F., and Fereshteh, Z. (2009). Synthesis and characterization of ZnO nanocrystals from thermolysis of new precursor, *Chem. Eng. J.*, **146**, pp. 498–502.
6. Salavati-Niasari, M., Bazarganipour, M., and Davar, F. (2011). Hydrothermal synthesis and characterization of bismuth selenide nanorods via a co-reduction route, *Inorg. Chim. Acta*, **364**, pp. 61–64.
7. Salavati-Niasari, M. (2009). Template synthesis and characterization of hexa aza macrocycles containing pyridine iron(II) complex nanoparticles dispersed within nanoreactors of zeolite-Y, *Inorg. Chem. Commun.*, **12**, pp. 359–363.
8. Cammarata, R. (2006). *Introduction to Nano Scale Science and Technology* Springer Publishers, USA.
9. Tan, O. K., Cao, W., Zhu, W., Chai, J. W., and Pan, J. S. (2003). Ethanol sensors based on nano-sized α-Fe_2O_3 with SnO_2, ZrO_2, TiO_2 solid solutions, *Sens. Actuators B*, **93**, pp. 396–401.
10. Guzman, M. G., Dille, J., Godet, S. (2009). Synthesis of silver nano-particles by chemical reduction method and their antibacterial activity, *Int. J. Chem. Bio. Eng.*, **2**, pp. 104–111.
11. Logvinenko, V., Polunina, O., Mikhailov, Y., Mikhailov, K., Bokhonov, B. (2007). Study of thermal decomposition of silver acetate, *J. Therm. Anal. Calorimetry*, **90**, pp. 813–816.

12. Fernando K. A. S., Smith, M. J., Harruff, B. A., Lewis W. K., Guliants, E. A., Bunker, C. E. (2009). Sonochemically assisted thermal decomposition of alane N, N′ dimethylethylamine with titanium(IV) isopropoxide in the presence of oleic acid to yield air stable and size selective aluminium core-shell nanoparticle, *J. Phys. Chem. C.*, **113**, pp. 500–503.

13. Guo, Q., Ghadiri, R., Weigel, T., Aumann, A., Gurevich, E. L., Esen, C., Medenbach, O., Cheng, W., Chichkov, B., and Ostendorf, A. (2014). Comparison of in situ and ex situ methods for synthesis of two-photon polymerization polymer nanocomposites, *Polymers*, **6**, pp. 2037–2050.

14. Tanahashi, M. (2010). Development of fabrication methods of filler/polymer nanocomposites: With focus on simple melt-compounding based approach without surface modification of nanofillers, *Materials*, **3**, pp. 1593–1619.

15. Kickelbick, G. (2007). *Hybrid Materials: Synthesis, Characterization, and Applications*: Wiley-VCH, USA.

16. Zhang, Y., Zhang, D., Guo, W., Chen, S. (2016). The α-Fe_2O_3/g-C_3N_4 heterostructural nanocomposites with enhanced ethanol gas sensing performance, *J. Alloys Compounds*, **685**, pp. 84–90.

17. Wua, X., Wanga, J., Zhanga, G., Katsumatab, K., Yanagisawac, K., Satod, T., Yind, S. (2017). Series of $MxWO_3$/ZnO (M = K, Rb, NH_4) nanocomposites: Combination of energy saving and environmental decontamination functions, *Appl. Catal. B: Environ.*, **201**, pp. 128–136.

18. Song, X. J., Qin, Z. Q., Wang, X. B., Yang, F., Fang, Q. L., She, C. G. (2016). β-Cyclodextrin modified with magnetic nanoparticles noncovalently for β-naphthol removal from wastewater, *Synth. React. Inorg. Met. Org. Nano. Met. Chem.*, **46**, pp. 143–146.

19. Pourghazi, K., Amoli-Diva, M., Beiraghi, A. (2015). Speciation of ultra-trace amounts of inorganic arsenic in water and rice samples by electrothermal atomic absorption spectrometry after solid-phase extraction with modified Fe_3O_4 nanoparticles, *Int. J. Environ. Anal. Chem.*, **95**, pp. 324–338.

20. Karimi, M., Aboufazeli, F., Lotfi Zadeh Zhad, H. R., Sadeghi, O., Najafi, E. (2014). Use of dipyridile amine functionalized magnetic nanoparticles for preconcentration and determination of lead ions in food samples, *J. AOAC Int.*, **97**, 1446–1451.

21. Zhu, S., Fan, C., Wang, J., He, J., Liang, E., Chao, M. (2015). Realization of high sensitive SERS substrates with one-pot fabrication of Ag-Fe_3O_4 nanocomposites, *J. Colloid Interface Sci.*, **438**, pp. 116–121.

22. Du, J., Cui, J., Jing, C. (2014). Rapid in situ identification of arsenic species using a portable Fe_3O_4@Ag SERS sensor, *Chem. Commun.*, **50**, pp. 347–349.

23. Hussain, M. M., Rahman, M. M., Asiri, A. M. (2017). Ultrasensitive and selective 4-aminophenol chemical sensor development based on nickel oxide nanoparticles decorated carbon nanotube nanocomposites for green environment, *J. Environ. Sci.*, **53**, pp. 27–38.

24. Ju, S., Yu, J., Ma, Y., Yang, Y., Liu, M. (2015). Rapid determination of cadmium and lead in maca (*Lepidium meyenii*) by magnetic solid-phase extraction and flame atomic absorption spectrometry, *Anal. Lett.*, **48**, pp. 2566–2580.

25. Wang, L., Hang, X., Chen, Y., Wang, Y., Feng, X. (2016). Determination of cadmium by magnetic multiwalled carbon nanotube flow injection preconcentration and graphite furnace atomic absorption spectrometry, *Anal. Lett.*, **49**, pp. 818–830.

26. Qu, J., Dong, Y., Wang, Y., Lou, T., Du, X., Qu, J. (2016). Novel nanofilm sensor based on poly-(alizarin red)/Fe_3O_4 magnetic nanoparticles-multiwalled carbon nanotubes composite material for determination of nitrite, *J. Nanosci. Nanotechnol.*, **16**, pp. 2731–2736.

27. Bharath, G., Madhu, R., Chen, S. M., Veeramani, V., Mangalaraj, D., and Ponpandian, N. (2015). Solvent-free mechanochemical synthesis of graphene oxide and Fe_3O_4-reduced graphene oxide nanocomposites for sensitive detection of nitrite, *J. Mater. Chem. A*, **3**, pp. 15529–15539.

28. He, D., He, X., Wang, K., Zhao, Y., and Zou, Z. (2013). Regenerable multifunctional mesoporous silica nanocomposites for simultaneous detection and removal of Mercury(II), *Langmuir*, **29**, pp. 5896–5904.

29. Su, Y. L. (2006). Preparation of diacetylene/silica nancomposite for use as a chemosensor, *React. Funct. Polym.*, **66**, pp. 967–973.

30. He, Y., Li, M., Jiang, W., Yang, W., Lin, L., Xu, L., and Fu, F. (2016). Phosphatidylserine-functionalized Fe_3O_4@SiO_2 nanoparticles combined with enzyme-encapsulated liposomes for the visual detection of Cu^{2+}, *J. Mater. Chem. B*, **4**, pp. 752–759.

31. Zhi, L., Wang, Z., Liu, J., Liu, W., Zhang, H., Chen, F., and Wang, B. (2015). White emission magnetic nanoparticles as chemosensors for sensitive colorimetric and ratiometric detection, and degradation of ClO^- and SCN^- in aqueous solutions based on a logic gate approach, *Nanoscale*, **7**, pp. 11712–11719.

32. Mohapatra, S., Sahu, S., Nayak, S., and Ghosh, S. K. (2015). Design of Fe_3O_4@SiO_2@carbon quantum dot based nanostructure for

fluorescence sensing, magnetic separation, and live cell imaging of fluoride ion, *Langmuir*, **31**, pp. 8111–8120.

33. Archana, R., Divya Priya, B., Monisha, M., Vinoth, T., Raji, P., and Sharma, R. K. (2015). Synthesis of Al/ZnO-PANI NCs for gas sensing application, *Karpagam J. Engineer. Res.* (KJER), II, *Special Issue on IEEE Sponsored International Conference on Intelligent Systems and Control* (ISCO'15) pp. 320–326.
34. Darder, M., Colilla, M., and Ruiz-Hitzky, E. (2003). Biopolymer-clay nanocomposites based on chitosan intercalated in montmorillonite, *Chem. Mater.*, **15**, pp. 3774–3780.
35. Colilla, M. D. M., and Ruiz-Hitzky, E. (2005). Chitosan-clay nanocomposites: Application as electrochemical sensors, *Appl. Clay Sci.*, **28**, pp. 199–208.
36. Jia, T., Fu, F., Long, F., Min, Z., Zhao, J., Chen, J., Li, J. (2016). Synthesis, characterization and enhanced visible-light photocatalytic activity of Zn_2SnO_4/C nanocomposites with truncated octahedron morphology, *Ceramics Int.*, **42**, pp. 13893–13899.
37. Xie, L., Liu, P., Zheng, Z., Weng, S., and Huang, J. (2016). Morphology engineering of V_2O_5/TiO_2 nanocomposites with enhanced visible light-driven photo functions for arsenic removal, *Appl. Cat. B: Environ.*, **184**, pp. 347–354.
38. Zhang, Q., Huang, Y., Xu, L., Cao, J., Ho, W., Lee, S. C. (2016). Visible-light-active plasmonic Ag-$SrTiO_3$ NCs for the degradation of NO in air with high selectivity, *ACS Appl. Mater. Interfaces*, **8**, pp. 4165–4174.
39. Qiu, X., Miyauchi, M., Sunada, K., Minoshima, M., Liu, M., Lu, Y., Li, D., Shimodaira, Y., Hosogi, Y., Kuroda, Y., Hashimoto, K. (2012). Hybrid Cu_xO/TiO_2 nanocomposites as risk-reduction materials in indoor environments, *ACS Nano*, **6**, pp. 1609–1618.
40. Cai, J., Wu, X., Li, S., Zheng, F. (2016). Synthesis of TiO_2@WO_3/Au nanocomposites hollow spheres with controllable size and high visible-light-driven photocatalytic activity, *ACS Sust. Chem. Eng.*, **4**, pp.1581–1590.
41. Logita, H. H., Tadesse, A., Kebede, T. (2015). Synthesis, characterization and photocatalytic activity of $MnO_2/Al_2O_3/Fe_2O_3$ nanocomposite for degradation of malachite green, *Afr. J. Pure Appl. Chem.*, **9**, pp. 211–222.
42. Mun, L. K., Abdullah, A. H., Hussein, M. Z., Zainal, Z. (2014). Synthesis and photocatalysis of ZnO/γ-Fe_2O_3 nanocomposite in degrading herbicide 2,4-dichlorophenoxyacetic acid, *Sains Malaysiana*, **43**, pp. 437–441.

43. Farhadi, S., and Siadatnasab, F. (2016). $CoFe_2O_4/CdS$ nanocomposite: Preparation, characterisation, and application in sonocatalytic degradation of organic dye pollutants, *Chinese J. Cat.*, **37**, 1487–1495.

44. Raizda, P., Gautam, S., Priya, B., Singh, P. (2016). Preparation and photocatalytic activity of hydroxyapatite supported BiOCl nanocomposite for oxytetracyline removal, *Adv. Mater. Lett.*, **7**, pp. 312–318.

45. Sabbaghi, S., Doraghi, F. (2016). Photo-catalytic degradation of methylene blue by ZnO/SnO_2 nanocomposite, *J. Water Environ. Nanotech.*, **1**, pp. 27–34.

46. Yao,Y. W., Cui, L. H., Li, Y., Yu, N. C., Dong, H. S., Chen, X., and Wei, F. (2015). Electrocatalytic degradation of methyl orange on PbO_2-TiO_2 nanocomposite electrodes, *Int. J. Environ. Res.*, **9**, pp. 1357–1364.

47. Wang, S., Zhou, S. (2011). Photodegradation of methyl orange by photocatalyst of CNTs/P-TiO_2 under UV and visible-light irradiation, *J. Hazard Mater.*, **185**, pp. 77–85.

48. Oza, G., Pandey, S., Gupta., A, Shinde, S., Mewada, A., Jagadale, P., Sharon, M. (2013). Photocatalysis-assisted water filtration: Using TiO_2-coated vertically aligned multi-walled carbon nanotube array for removal of Escherichia coli O157:H7, *Mater. Sci. Eng. C Mater.*, **3**, pp. 4392–4400.

49. Mamba, G., Mbianda. X. Y., and Mishra, A. K. (2014). Gadolinium nanoparticles decorated multiwalled carbon nanotube/titania NC for degradation of methylene blue in water under simulated solar light, *Environ. Sci. Poll. Res.*, **21**, pp. 5597–5609.

50. Mamba, G., Mbianda, X. Y., and Mishra, A. K. (2015). Photocatalytic degradation of diazo dye naphthol blue black in water using MWCNT/Gd, N, S-TiO_2 nanocomposites under simulated solar light, *J. Environ. Sci.*, **33**, pp. 219–228.

51. Czech, B., and Buda, W. (2015). Photocatalytic treatment of pharmaceutical waste water using new multiwall-carbon nanotubes/ TiO_2/SiO_2 nanocomposite, *Environ. Res.*, **137**, pp. 176–184.

52. Pan, X., Linn, S., Bi, K., Hao, Y., and Lei, M. (2016). A facile route to strontium titanate nanocubes/reduced graphene oxide nanocomposites and their enhanced adsorption and photocatalytic activity, *Mater. Lett.*, **185**, pp. 36–39.

53. Yang, J., Wang, X., Zhao, X., Dai, J., and Mo, S. (2015). Synthesis of uniform Bi_2WO_6-reduced graphene oxide NCs with significantly enhanced photocatalytic reduction activity, *J. Phys. Chem. C*, **119**, pp. 3068–3078.

54. Ravichandran, K., Nithiyadevi, K., Sakthivel, B., Arun, T., Sindhuja, E., and Muruganandam, G. (2016). Synthesis of ZnO: Co/rGO nanocomposites for enhanced photocatalytic and anti bacterial activities, *Ceram. Int.*, **142**, pp. 17539–17550.

55. Kharatzadeh, A., Jamali-Sheini, F., and Yousefi, R. (2016). Excellent photocatalytic performance of $Zn_{(1-x)}Mg_xO$/rGO nanocomposites under natural sunlight irradiation and their photovoltaic and UV detector applications, *Mater. Des.*, **107**, pp. 47–55.

56. Zhang, S., Li, H., Wang, Z., Liu, J., Zhang, H., Wang, B., and Yang, Z. (2015). A strongly coupled Au/Fe_3O_4/GO hybrid material with enhanced nanozyme activity for highly sensitive colorimetric detection, and rapid and efficient removal of Hg^{2+} in aqueous solutions, *Nanoscale*, **7**, pp. 8495–8502.

57. Stengl, V., Bakardjieva, S., Grygar, T. M., Bludska, J., and Kormunda, M. (2013). TiO_2-graphene oxide NC as advanced photocatalytic materials, *Chem. Cent. J.*, **7**, pp. 41–52.

58. Zhang, Y., Zhou, Z., Chen, T., Wang, H., Lu, W. (2014). Graphene TiO_2 nanocomposite with high photocatalytic activity for degradation of sodium pentachlorophenol, *J. Environ. Sci.*, **26**, pp. 2114–2122.

59. Farghali, M. A., Salah El-Din, T. A., Al-Enizi, A. M., and El Bahnasawy, R. M. (2015). Graphene/magnetite nanocomposite for potential environmental application, *Int. J. Electrochem. Sci.*, **10**, pp. 529–537.

60. Pan, X., Zhao, Y., Wang, S., and Fan, Z. (2013). TiO_2/graphene nanocomposite for photocatalytic application, *Mater. Processes Energy Commun. Curr. Res. Technol. Dev.* (A. Mendez-Vilas, ed.), pp. 913–920.

61. Yu, S., Liu, J., Zhu, W., Hu, Z. T., Lim, T. T., and Yan, X. (2015). Facile room-temperature synthesis of carboxylated graphene oxide copper sulfide nanocomposite with high photodegradation and disinfection activities under solar light irradiation, *Sci. Rep.*, **5**, pp. 1–12.

62. Tian, H., Chen, J., He, J., and Liu, F. (2015). Pd-loaded magnetic mesoporous nanocomposites: A magnetically recoverable catalyst with effective enrichment and high activity for DDT and DDE removal under mild conditions, *J. Colloid Interface Sci.*, **457**, 195–202.

63. Nakhjavani, S. H., Tavakoli, O., Akhlaghi, S. P., Salehi, Z., Esmailnejad-Ahranjani, P., and Arpanaei, A. (2015). Efficient photocatalytic degradation of organic pollutants by magnetically recoverable nitrogen-doped TiO_2 nanocomposite photocatalysts under visible light irradiation, *Environ. Sci. Pollut. Res.*, **22**, pp. 18859–18873.

64. Ahangar, L. E., Movassaghi, K., Emadi, M., and Yaghoobi, F. (2016). Photocatalytic application of TiO_2/SiO_2-based magnetic nanocomposite ($Fe_3O_4@SiO_2/TiO_2$) for reusing of textile waste water, *Nano. Chem. Res.*, **1**, pp. 33–39.

65. Inumaru, K., Kasahara, T., Yasui, M., et al. (2005). Direct nanocomposite of crystallite TiO_2 particles and mesoporous silica as a molecular selective and highly active photocatalyst, *Chem. Commun.*, **16**, pp. 2131–2133.

66. Mohseni, A., Malekina, L., Fazaeli, R., Ahmadi, E. (2013). Synthesis $TiO_2/SiO_2/Ag$ nanocomposite by sonochemical method and investigation of photo-catalyst effect in waste water treatment, *Nanocon,* **10**, pp.16–18.

67. Shahadat, M., Nabi, S. A., Bushra, R., Raeissi, A. S., Umara, K., and Ansari, M. O. (2012). Synthesis, characterization, photolytic degradation, electrical conductivity and applications of a nanocomposite adsorbent for the treatment of pollutants, *RSC Adv.*, **2**, pp. 7207–7220.

68. Khataee, A., Kiransan, M., Karaca, S., and Arefi-Oskoui, S. (2016). Preparation and characterization of ZnO/MMT nanocomposite for photocatalytic ozonation of a disperse dye, *Turk. J. Chem.*, **40**, pp. 546–564.

69. Kibanova, D., Sleiman, M., Cervini-Silva, J., and Destaillats, H. (2012). Adsorption and photocatalytic oxidation of formaldehyde on a clay-TiO_2 composite, *J. Hazard. Mater.*, **15**, pp. 211–212.

70. Sharma, V. K., Graham, N. J., Li, X. Z., and Yuan, B. L. (2010). Ferrate(VI) enhanced photocatalytic oxidation of pollutants in aqueous TiO_2 suspensions, *Environ. Sci. Pollut. Res. Int.*, **17**, pp. 453–461.

71. Mo, J., Zhang, Y., Xu, Q., and Yang, R. (2009). Effect of TiO_2/adsorbent hybrid photocatalysts for toluene decomposition in gas phase, *J. Hazard. Mater.*, **168**, pp. 276–281.

72. Li, L., Yuan, L. J., Hong, W., Fan, L., Mao, L. B., Liu, L. (2015). Hybrid Fe_3O_4/MOFs for the adsorption of methylene blue and methyl violet from aqueous solution, *Desalin. Water Treat.*, **55**, pp. 1973–1980.

73. Zhao, J., Luque, R., Qi, W., Lai, J., Gao, W., Gilani, M. R. H. S., Xu, G. (2015). Facile surfactant-free synthesis and characterization of Fe_3O_4@ 3-aminophenol-formaldehyde core-shell magnetic microspheres, *J. Mater. Chem. A*, **3**, pp. 519–524.

74. Chen, R. P., Zhang, Y. L., Wang, X. Y., Zhu, C. Y., Ma, A. J., Jiang, W. M. (2015). Removal of methylene blue from aqueous solution using humic-acid coated magnetic nanoparticles, *Desalin. Water Treat.*, **55**, pp. 539–548.

75. Shi, G., Li, Y., Ai, L., Shi, F. (2016). Two step synthesis and enhanced microwave absorption properties of polycrystalline $BaTiO_3$ coated Ni nanocomposites, *J. Alloys Compounds*, **680**, pp. 735–743.

76. Qiu, H., Liang, C., Zhang, X., Chen, M., Zhao, Y., Tao, T., Xu, H., Liu, G. (2015). Fabrication of a biomass-based hydrous zirconium oxide nanocomposites for preferable phosphate removal and recovery, *ACS Appl. Mater. Interfaces*, **7**, pp. 20835–20844.

77. Zhang, M., Song, W., Chen, Q., Miao, B., He, W. (2015). One-pot synthesis of magnetic $Ni@Mg(OH)_2$ core-shell nanocomposite as a recyclable removal agent for heavy metals, *ACS Appl. Mater. Interfaces*, **7**, pp. 1533–1540.

78. Mahmoodi, N. M., Bagherpour, F., and Nariyan, E. (2015). Amine functionalized magnetic carbon nanotube: Synthesis and binary system dye removal, *Desalin. Water Treat.*, **56**, pp. 107–120.

79. Dong, L., Zhipeng, Z., and Yigang, D. (2016). A Simple method to prepare magnetic modified corncobs and its application for Congo red adsorption, *J. Dispers. Sci. Technol.*, **37**, 73–79.

80. Zhao, Y., Chen, H., Li, J., and Chen, C. (2015). Hierarchical MWCNTs/Fe_3O_4/PANI magnetic composite as adsorbent for methyl orange removal, *J. Colloid Interface Sci.*, **450**, pp. 189–195.

81. Zhao, Z. Y., and Xiong, Z. H. (2016). Removal of four nitrofurans drugs from aqueous solution by magnetic multi-wall carbon nanotubes, *Fullerenes Nanotubes Carbon Nanostruct.*, **23**, pp. 640–648.

82. Wang, Y. S., Hsieh, S. H., Lee, C. H., and Horng, J. J. (2013). Adsorption of complex pollutants from aqueous solutions by nanocomposite, *Mate. Clean Soil Air Water*, **41**, 574–580.

83. Ma, S., Zhan, S., Jia, Y., and Zhou, Q. (2015). Highly efficient antibacterial and Pb(II) removal effects of Ag-$CoFe_2O_4$-GO nanocomposite, *ACS Appl. Mater. Interfaces*, **7**, pp. 10576–10586.

84. Kim, D. W., Bach, L. G., Hong, S. S., Park, C. L., and Kwon, T. A. (2015). Facile route towards the synthesis of Fe_3O_4/graphene oxide nanocomposites for environmental applications, *Mol. Cryst. Liquid Cryst.*, **599**, 43–50.

85. Dolatkhah, A., and Wilson, L. D. (2016). Magnetite/polymer brush nanocomposites with switchable uptake behavior toward methylene blue, *ACS Appl. Mater. Interfaces*, **8**, pp. 5595–5607.

86. Li, K., Zeng, Z., Xiong, J., Yan, L., Guo, H., Liu, S., Dai, Y., and Chen, T. (2015). Fabrication of mesoporous $Fe_3O_4@SiO_2@CTAB-SiO_2$

magnetic, microspheres with a core/shell structure and their efficient adsorption performance for the removal of trace PFOS from water, *Colloids Surf. A: Physicochem. Eng. Aspects*, **465**, pp. 113–123.

87. Tan, X., Lu, L., Wang, L., and Zhang, J. (2015). Facile synthesis of bimodal mesoporous Fe_3O_4@SiO_2 composite for efficient removal of methylene blue, *Eur. J. Inorg. Chem.*, **18**, pp. 2928–2933.

88. Zhang, X., Zeng, T., Wang, S., Niu, H., Wang, X., and Cai, Y. (2015). One-pot synthesis of C18-functionalized core-shell magnetic mesoporous silica composite as efficient sorbent for organic dye, *J. Colloid Interface Sci.*, **448**, pp. 189–196.

89. Rajabi, H. R., Arjmand, H., Hoseini, S. J., and Nasrabadi, H. (2015). Surface modified magnetic nanoparticles as efficient and green sorbents: Synthesis, characterization, and application for the removal of anionic dye, *J. Magn. Magn. Mater.*, **394**, pp. 7–13.

90. Neoh, K. G., Tan, K. K., Goh, P. L., Huang, S. W., Kang, E. T., and Tan, K. L. (1999). Electroactive polymer–SiO_2 nanocomposites for metal uptake, *Polymer*, **40**, pp. 887–893.

91. Abdel-Fattah, T. M., and Bishop, B. (2004). Organo-silicate nanocomposites for the removal of chlorinated phenols from aqueous media: Kinetics and environmental stability, *J. Environ. Sci. Health Part A Toxic/Hazard. Subs. Environ. Eng.*, **A39**, pp. 2855–2866.

92. Akýn, D., Yakar, A., and Gunduz, U. (2015). Synthesis of magnetic Fe_3O_4-chitosan nanoparticles by ionic gelation and their dye removal ability, *Water Environ. Res.*, **87**, pp. 425–436.

93. Xiang, Y., Wang, H., He, Y., and Song, G. (2015). Efficient degradation of methylene blue by magnetically separable Fe_3O_4/chitosan/TiO_2 nanocomposites, *Desalin. Water Treat.*, **55**, pp. 1018–1025.

94. Zhang, Q., Pan, B., Zhang, S., Wang, J., Zhang, W., and Lv, L. (2011). New insights into nanocomposite adsorbents for water treatment: A case study of polystyrene-supported zirconium phosphate nanoparticles for lead removal, *J. Nanopart. Res.*, **13**, pp. 5355–5364.

95. Ganigar, R., Rytwo, G., Gonen, Y., Radian, A., and Mishael, Y. G. (2010). Polymer-clay NCs for the removal of trichlorophenol and trinitrophenol from water, *Appl. Clay Sci.*, **49**, pp. 311–316.

96. Kugbe, J., Matsue, N., and Henmi, T. (2009). Synthesis of Linde type A zeolite-goethite nanocomposite as an adsorbent for cationic and anionic pollutants, *J. Hazard. Mater.*, **164**, pp. 929–935.

97. Hashem, F. S. (2013). Removal of methylene blue by magnetite covered bentonite nano-composite, *Eur. Chem. Bull.*, **2**, pp. 524–529.

98. Li, S., Zhou, P., and Ding, L. (2011). Adsorption application for removal of hazardous chloroform from aqueous solution by nanocomposites rectorite/chitosan adsorbent, *J. Water Resource Protect.*, **3**, pp. 448–455.

Chapter 5

Zinc Oxide-Based Nanocomposites for Photocatalytic Conversion of Organic Pollutants in Water

Sze-Mun Lam[a] and Jin-Chung Sin[b]

[a]*Department of Environmental Engineering, Faculty of Engineering and Green Technology, Universiti Tunku Abdul Rahman, Jalan Universiti, Bandar Barat, 31900 Kampar, Perak, Malaysia*
[b]*Department of Petrochemical Engineering, Faculty of Engineering and Green Technology, Universiti Tunku Abdul Rahman, Jalan Universiti, Bandar Barat, 31900 Kampar, Perak, Malaysia*

lamsm@utar.edu.my; sinjc@utar.edu.my

Environmental pollution caused by organic pollutants has become a major problem nowadays for the human society and seriously threatens the existence of terrestrial lives. The light-driven semiconductor-mediated photocatalysis, which can decompose organic pollutants into nontoxic molecules, is regarded as one of the most promising ways for the removal of polluted wastewater. Zinc oxide (ZnO), due to its exceptional properties, is widely used as significant photocatalytic material. With the UV light present in

Nanocomposites for Pollution Control
Edited by Chaudhery Mustansar Hussain and Ajay Kumar Mishra

ISBN 978-981-4774-45-1 (Hardcover), 978-1-315-14368-2 (eBook)
www.panstanford.com

natural sunlight, nanocrystalline ZnO photocatalytically degrades and eliminates various organic pollutants in water. The NiO/ZnO nanocomposites enhance the UV photocatalytic efficiency. The Ag_2O/ZnO and CuO/ZnO nanocomposites absorb visible light and photocatalytically degrade organic pollutants. WO_3/ZnO nanocomposites photocatalytically degrade the organic pollutants under sunlight effectively. ZnO-based nanocomposites reduce the photogenerated charge carriers leading to a large number of reactive oxygen species involved in the photocatalytic process. In this chapter, the mechanism of photocatalytic conversion of organic pollutants is discussed. Finally, the future prospects together with the challenges for ZnO-based nanocomposite photocatalysis on the organic pollutants conversion are summarized.

5.1 Introduction

5.1.1 What Is Zinc Oxide?

Zinc oxide (ZnO) is an oxidic compound naturally occurring as the rare mineral zincite, which crystallizes in three different structures: wurtzite, zincblende, and rocksalt. It usually appears as a white powder, nearly insoluble in water but soluble in acids and alkalis. ZnO is the most frequently used among the zinc compounds and commercially produced using the French process—the metallic zinc is vaporized in a large container by external heating. In an adjoining off-take pipe or combustion chamber, the vapor is burned off in the air to a fine ZnO powder [1] and American process—oxidized ores of roasted sulfide concentrates are mixed with anthracite coal (carbon additive) and smelted in a furnace. The coal together with the products of partial combustion mainly carbon monoxide reduced the ore to metallic zinc, which is released as vapor. The zinc vapor is then re-oxidized by lower temperature air and formed ZnO particulate. The purity of the ZnO produced by this process is normally rather inferior to that from the French process as it contained low levels of lead and sulfur [1, 2].

In general, ZnO crystallizes in a hexagonal wurtzite structure (Fig. 5.1a) with lattice parameters a = 3.25 Å and c = 5.20 Å at ambient pressure and temperature. The structure of ZnO can be simply described as a number of alternating planes composed of tetrahedrally coordinated O^{2-} and Zn^{2+} ions stacked alternately along the *c*-axis. The tetrahedral coordination in ZnO resulted in piezoelectric and pyroelectric due to the noncentral symmetric structure. Another important characteristic of ZnO is polar surfaces, which are the basal planes (0 0 0 1 and 0 0 0 1¯). One end of the basal polar plane (0 0 0 1) is terminated by Zn atoms and the other end (0 0 0 1¯) is terminated by oxygen atoms. The oppositely charged ions produced positively charged Zn-(0 0 0 1) and negatively charged O^- (0 0 0 1¯) surfaces, resulting in a normal dipole moment and spontaneous polarization along the *c*-axis as well as a divergence in surface energy. Thus, to maintain electrical neutrality, the polar surfaces typically have facet or exhibit massive surface reconstructions, but not in ZnO ± (0 0 0 1) surfaces that they are atomically flat, stable and with no reconstruction [3, 4]. Although the wurtzite structure of ZnO is the thermodynamically stable phase, two other structures of ZnO existed, which are known as zincblende (Fig. 5.1b) and rocksalt (Fig. 5.1c). These phases are metastable and only occurred under certain conditions such as through epitaxial growth of ZnO on a suitable cubic substrate to obtain the zincblende structure, while the rocksalt (NaCl-type) structure is observed when subjected to high pressures (~9 GPa at 300 K) [5]. For this reason, ZnO has a strong natural tendency to crystallize in the wurtzite structure. Depending on the optical absorption properties, the refractive index of ZnO (2.0) is smaller than that of TiO_2 (2.5–2.7), so ZnO hardly scatters light, thereby making it colorless and enhancing the transparency. In addition, ZnO has a considerably high thermal conductivity of 54 W/mK, which increases its appeal as a substrate for homoepitaxy or heteroepitaxy [6]. As an important semiconductor material, ZnO has been applied in many fields such as gas sensors, solar cells, nanolasers, light-emitting diodes (LEDs), electron field emitters, fertilizers, cosmetics, and photocatalysts for degradation and complete elimination of environmental pollutants [6–9]. Therefore, the development of ZnO with controllable nanoscale features in various applications has received enormous scientific interest recently.

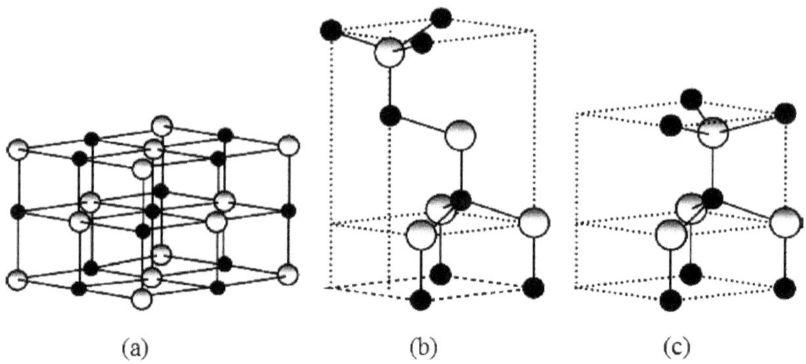

Figure 5.1 Stick-and-ball representation of ZnO crystal structures: (a) cubic rocksalt; (b) cubic zincblende and (c) hexagonal wurtzite. The shaded white and black spheres denote Zn and O atoms, respectively [5].

5.1.2 What Is Photocatalytic Conversion of Organic Pollutants?

Heterogeneous photocatalysis as an increase in the rate of a thermodynamically allowed ($\Delta G < 0$) reaction in the presence of catalysts with the increase originating from the creation of some new reaction pathways containing photogenerated species and a reduction of the activation energy. The catalysts in photocatalytic reactions are normally semiconductor materials, which can form electron (e^-) and hole (h^+) when exposed to light and the reactions are normally either oxidation or reduction.

The steps of a photocatalytic reaction involved the light absorption using a semiconductor material subsequent by e^- transferred from the valence band (VB) to the conduction band (CB) and creating h^+ in the VB. A band model is often used for schematic illustration of the electronic structures of semiconducting materials (Fig. 5.2). The photocatalytic reaction typically contains three main active species: h^+, hydroxyl (•OH) radical and superoxide anion (O_2•-) radical, where •OH radical is the principal oxidant in the photocatalytic conversion of the organic pollutant. The production of •OH radicals generally through two ways, (i) O_2 presented in the aqueous solution is reduced by photogenerated e^- to form O_2•- radicals, followed by reacting with H^+ (forming hydroperoxyl (HO_2•) radicals) and then further decomposition to produce •OH radicals and (ii) H_2O and

OH^- ion in the water environment are readily oxidized by photogenerated h^+ to form •OH radicals. It was also noted that the photogenerated e^- can easily recombine with h^+ after their generation in the absence of e^- or h^+ scavengers. In this regard, the presence of species scavengers is crucial for inhibiting the charge carrier recombination and for enhancing the efficiency of photocatalytic reaction. A series of chain oxidative and reductive reactions (Eqs. (5.1)–(5.11)) occurred at the photonic excitation of the catalyst is widely postulated as follows [10, 11]:

$$\text{Photoexcitation:} \quad SC + h\nu \rightarrow e^- + h^+ \tag{5.1}$$

$$\text{Oxygen ionosorption:} \quad O_2 + e^- \rightarrow O_2\bullet^- \tag{5.2}$$

$$\text{Ionization of water:} \quad H_2O \rightarrow OH^- + H^+ \tag{5.3}$$

$$\text{Oxidation of hydroxyl:} \quad OH^- + h^+ \rightarrow \bullet OH \tag{5.4}$$

$$\text{Protonation of superoxide:} \quad O_2\bullet^- + H^+ \rightarrow HO_2\bullet \tag{5.5}$$

$$\text{Co-scavenging of } e^-\text{:} \quad HO_2\bullet + e^- \rightarrow HO_2^- \tag{5.6}$$

$$\text{Formation of } H_2O_2\text{:} \quad HO_{2-} + H^+ \rightarrow H_2O_2 \tag{5.7}$$

$$\text{Formation of } \bullet OH\text{:} \quad H_2O_2 + e^- \rightarrow \bullet OH + OH^- \tag{5.8}$$

$$\text{Pollutant} + \bullet OH \rightarrow \text{degradation products} \tag{5.9}$$

$$\text{Pollutant} + h^+ \rightarrow \text{oxidation products} \tag{5.10}$$

$$\text{Pollutant} + e^- \rightarrow \text{reduction products} \tag{5.11}$$

where SC = semiconductor.

There has been a great deal of interest in ZnO for photocatalytic conversion of a wide variety of environmental contaminants lately, as seen from a surge of a relevant number of publications [7, 10, 12–19]. The interest in ZnO is fueled and fanned by its various exotic properties such as suitable band gap, high stability, high photosensitivity, high catalytic activity, non-toxic, low cost and environmental friendliness. According to the thermodynamic point of view, the pre-requisite for an efficient semiconductor catalyst is that the VB and CB of the catalyst should be positioned in such a way that the oxidation potential of the hydroxyl radical ($E^0(H_2O/\bullet OH)$ = +2.72 V vs. NHE) and the reduction potential of superoxide anion radical ($E^0(O_2/O_2\bullet^-)$ = −0.33 V vs. NHE) lay well within the band gap [7]. Table 5.1

illustrates the band positions of various semiconductors and their band gap energy.

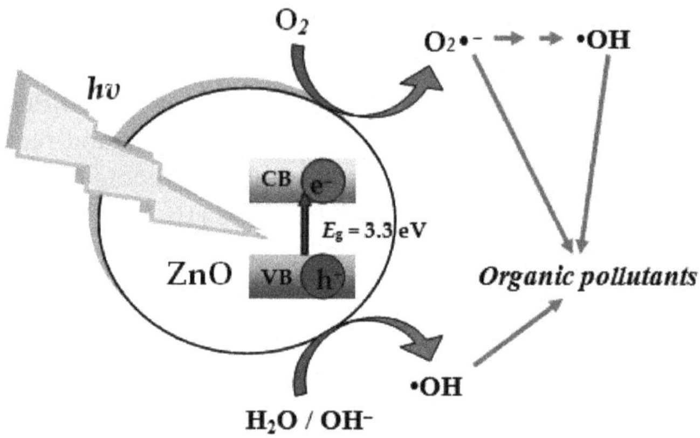

Figure 5.2 Schematic of the mechanism of photocatalytic conversion of organic pollutants.

Table 5.1 The CB and VB positions of some common semiconductor catalysts at pH 1 [20, 21]

Semiconductor	VB (eV vs. NHE ± 0.1 V)	CB (eV vs. NHE ± 0.1 V)	Band gap energy (eV)
ZnO	+2.8	−0.5	3.3
TiO_2	+2.8	−0.3	3.2
WO_3	+3.0	+0.2	2.8
ZrO_2	+4.0	−1.0	5.0
Fe_2O_3	+2.9	+0.6	2.3
SnO_2	+4.1	+0.3	3.8
ZnS	+1.4	-2.3	3.7
CdS	+2.1	-0.4	2.5
CdSe	+1.6	-0.1	1.7
GaAs	+1.0	-0.4	1.4

It is clear that TiO_2, ZnO, and ZrO_2 exhibited favorable band gap positions compared to other catalysts. TiO_2, which has a band gap of 3.2 eV, is the most preferred catalyst for the photocatalytic conversion of organic pollutants due to its photoactive, inert, and corrosion-resistant qualities. As an analogous to TiO_2, ZnO has a band gap energy of 3.3 eV, which is an alternative potential catalyst for the conversion of organic pollutants due to its high quantum efficiency; however, it photocorroded in acidic aqueous suspensions and suffered dissolution to form $Zn(OH)_2$ on the catalyst surface [7]. Some of the experimental results have shown that ZnO actually exhibited higher photocatalytic activities than TiO_2 and other semiconductor catalysts such as CdS, WO_3, Fe_2O_3, SnO_2, and ZrO_2 [15–18, 22, 23]. As the semiconductor for the reaction, ZnO is able to absorb over a larger portion of the solar spectrum compared to TiO_2. Solar energy is infinitely available, non-polluting, and appropriate for the photocatalytic conversion reaction. Thus, the utilization of solar energy for the photocatalytic conversion of organic pollutants could be considered as an ultimate solution to solve the crisis of energy shortage and environmental degradation.

5.2 Synthesis and Characterization of Zinc Oxide-Based Nanocomposite Photocatalysts

ZnO is a remarkable UV-activated photocatalyst for environmental pollution remediation. ZnO photocatalyst reviewed by our group and others exhibited excellent photoactivity for the conversion of organic dyes and phenolic compounds [10, 17, 24–26]. Despite these promissory findings, the application of an UV-activated photocatalyst is practically obstacle by a rapid e^-–h^+ recombination, which results in the competition with the separation of charge carriers and in turn reducing the supply of charge carriers to the photocatalyst surface. Moreover, its wide band gap structure also gives problem due to only absorption of UV component in the solar spectrum although it is paramount provision for environmental applications. To surmount the two shortfalls, two approaches according to the principle of photocatalysis have been proposed, including (i) encouraging the shuttle and separation of photogenerated e^-–h^+ pair and (ii) prolonging the wavelength of absorption to the visible light region. Till now, the fabrication of ZnO-based nanocomposite photocatalysts that improved the visible light photoactivity of ZnO has garnered great attraction as they can use a huge fraction of sunlight. A variety of semiconductor photocatalysts have been used for the synthesis of ZnO-based nanocomposites. They mainly include metal oxides (e.g., TiO_2 [27, 28], CuO [29, 30], Fe_2O_3 [31], NiO [32], WO_3 [24], CeO_2 [33], Ag_2O [34], SnO_2 [35, 36]), metal sulfides (e.g., ZnS [37], CdS [38]), metallates (e.g., $BiFeO_3$ [39], Bi_2WO_6 [40], $BiVO_4$ [41, 42]), and other nanomaterials (e.g., ZnSe [43], C_3N_4 [44]). The widely used synthetic strategies to prepare ZnO-based nanocomposite photocatalysts can be divided into four types: sol-gel, solution mixing, hydrothermal and/or solvothermal and microwave-assisted methods.

5.2.1 Solution Mixing

Solution mixing is the simplest method to prepare ZnO-based nanocomposite photocatalysts by mixing two or more metal precursors with desired concentrations in an aqueous solvent. During the blending, a hetero-interface between the semiconductors is formed and therefore the composites are formed when the aqueous solvent solution evaporated. For example, CeO_2/ZnO nanocomposite photocatalysts with different CeO_2 loadings were successfully prepared by vigorously mixing both precursors of zinc nitrate hexahydrate and cerium nitrate hexahydrate in ammonia solution at room temperature [33]. After that, H_2O_2 was added to the above mixture and the obtained products were centrifuged, washed with ethanol for several times, dried at 60°C for 12 h and finally calcined in air for 2 h at 500°C to form the composites. In another study, Dai et al. [29] reported the synthesis of CuO/ZnO nanocomposite photocatalysts embedded in mesoporous silica SBA-15 where the copper nitrate trihydrate and zinc nitrate hexahydrate were mixed in *n*-hexane solution containing SBA-15, subsequent by a calcination process in a muffle oven to obtain the interface interaction between the two nanoparticles supported on the SBA-15. Jeyasubramanian et al. [35] prepared the ZnO sol by hydrolysis of zinc acetate with NaOH and then the prepared tin (II) chloride solution was mixed with the sol to form the SnO_x ($x = 1, 2$)/ZnO nanocomposites after heat treatment under vacuum condition. This method was very appealing due to its simplicity. However, in most cases, the interaction between the semiconductors was weak since chemical bonding was not expected in this method. The intimate contact between the semiconductor photocatalysts was crucial for effective separation and transfer of e^-–h^+ pairs.

5.2.2 Sol-Gel Method

The sol-gel method is a wet-chemical technique widely used in the synthesis and fabrication of ZnO-based nanocomposite photocatalysts. It is chosen because of the high purity and homogeneity in the final products, low processing temperatures, low cost and easy synthetic process. The starting materials or precursors are metal alkoxides and metal salts which undergo hydrolysis and polycondensation reaction to form a colloid. The sol evolved with time towards the formation of an inorganic three-dimensional network forming another phase gel [45, 46].

The sol-gel synthesis was suitable for the preparation of ZnO-based nanocomposite photocatalysts due to the network of metal–OH which enabled to establish new bonds of oxygen bridges through hydrolysis–polycondensation reactions on the semiconductor surfaces. Accordingly, the resulting chemical bonding formed between the ZnO and the surface of coupled semiconductors. It was then followed by an annealing step to promote the crystallization of the nanocomposites. For example, Chen et al. [32] used sol-gel method to synthesize NiO/ZnO nanocomposite photocatalysts as follows: 3.6 g of zinc acetate was dissolved in 100 mL deionized water and stirred vigorously for 20 min, which was named solution A. Solution B was prepared by mixing the nickel nitrate hexahydrate solution with 200 mL ethanol and 5.4 g oxalic acid. Then, the solution A was added to solution B dropwise and the obtained colloidal solution was dried in an oven at 70°C for 24 h to form powder. It was finally annealed at 350–500°C for 1–4 h to obtain well-crystalline NiO/ZnO nanocomposite photocatalysts. The obtained NiO/ZnO nanocomposite photocatalysts had surface area of 17.1–35.3 m^2/g and the NiO particles closely dispersed on the surface of ZnO forming p-n junction photocatalysts. Kwiatkowski et al. [27] deposited TiO_2 layers via a sol-gel method onto the surface of ZnO nanorods by placing the ZnO samples in a vigorously stirred solution of titanium butoxide diluted in isopropanol. In the preparation process, TiO_2 was formed based on the hydrolytic reaction of titanium butoxide followed by a condensation process and finally annealed it to convert to crystalline anatase TiO_2 to yield TiO_2/ZnO nanocomposite photocatalysts. Typical SEM and TEM images in Fig. 5.3 clearly showed that the TiO_2 layer composed of nanoparticles with diameters in the range 3–8 nm were successfully dispersed on the ZnO nanorod surface.

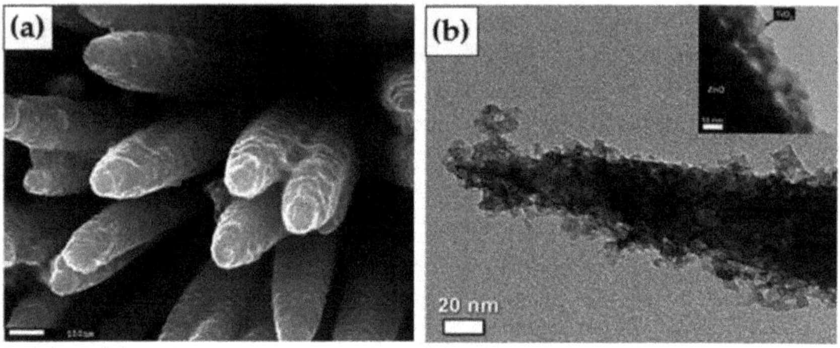

Figure 5.3 (a) SEM and (b) TEM images of TiO_2/ZnO nanocomposites [27].

5.2.3 Hydro(Solvo)Thermal Approach

The hydro(solvo)thermal approach has been widely applied in the preparation of ZnO-based nanocomposite photocatalysts due to its many advantages including relatively non-polluting set-up, high reactivity, low energy requirement and ease of controlling of the aqueous solution [46, 47]. Hydrothermal synthesis is defined as a reaction that is conducted within a closed system utilizing water as the reaction medium. A more general term "solvothermal" referred to a similar reaction in which a non-aqueous solvent (organic or inorganic) is utilized. These techniques are commonly carried out in an autoclave under controlled temperature and pressure. The reaction temperature is held above the water (or solvent) boiling point to self-produce saturated vapor pressure [48]. For example, Mansournia and Ghaderi [30] synthesized CuO/ZnO nanocomposite photocatalysts by a two-step hydrothermal treatment of hydrothermally synthesized ZnO nanoparticles and copper nitrate trihydrate in ammonia-water solvent. The mixture was sealed in a Teflon-lined autoclave and the temperature was kept at 70°C for 1 h. Finally the CuO/ZnO nanocomposite photocatalysts were collected by centrifugation, washing and drying at 75°C. Typical FESEM images of CuO/ZnO nanocomposite photocatalysts at different CuO loadings are shown in Fig. 5.4. The morphologies of the CuO/ZnO nanocomposite photocatalysts could be controlled by altering the loading of CuO. Cheng et al. [28] also used a two-step hydrothermal method to produce urchin-like TiO_2 nanospheres/ZnO nanospindles photocatalysts. The first step: three-dimensional urchin-like TiO_2 nanospheres were synthesized using 0.1 g Degussa P25 and 60 mL of 10 M NaOH solution in a Teflon-lined autoclave at 160°C for 2 h. The second step: one-dimensional ZnO nanospindles assembled on the surface of urchin-like TiO_2 nanospheres by adding 2.5 mg zinc nitrate hexahydrate and 5 mL of 0.05 M hexamethylene-tetramine solution to 30 mL deionized water. Then the homogenous solution was transferred into a stainless steel Teflon-lined autoclave and heated at 100°C for 3 h. The obtained nanocomposites were collected, washed with ethanol and deionized water for several times and finally dried in an oven at 60°C. Recently, ternary nanocomposites system was obtained by an one-pot hydrothermal process. Adhikari et al. [49] prepared Ag/ZnO/g-C_3N_4 from the aqueous solution of silver nitrate and zinc nitrate hexahydrate with g-C_3N_4 particles in an autoclave at 130°C for 3 h.

Rashad et al. [50] fabricated SnO_2/ZnO nanocomposites through a solvothermal method. In a typical experiment, a stoichiometric amount of tin chloride pentahydrate and zinc acetate was first dissolved in methanol. 5 M sodium hydroxide was then dissolved in methanol under stirring to obtain the pH 7 for complete precipitation of Zn and the mixture was transferred into a 100 mL Teflon-lined autoclave. The autoclave was heated at 150°C for 6 h to obtain the SnO_2/ZnO nanocomposites. In another study, $NiFe_2O_4$/ZnO composites were prepared when zinc nitrate tetrahydrate and ethanol were mixed with $NiFe_2O_4$ powder in an aqueous medium. It was then treated in an autoclave at 160–200°C [51]. The prepared $NiFe_2O_4$/ZnO composites extended absorptive capacity to the visible light region compared to pure ZnO. Using Planck's equation, the band gap energies of pure $NiFe_2O_4$, pure ZnO and $NiFe_2O_4$/ZnO (NNi1) were measured to be 1.64, 3.12 and 2.78 eV, respectively (as recorded in Fig. 5.5). The factors such as temperature, pH, treatment time, pressure and solvent type played crucial roles in this preparation method. Importantly, it was also possible to obtain a large quantity of nanocomposites at low cost in an environmental-friendly way.

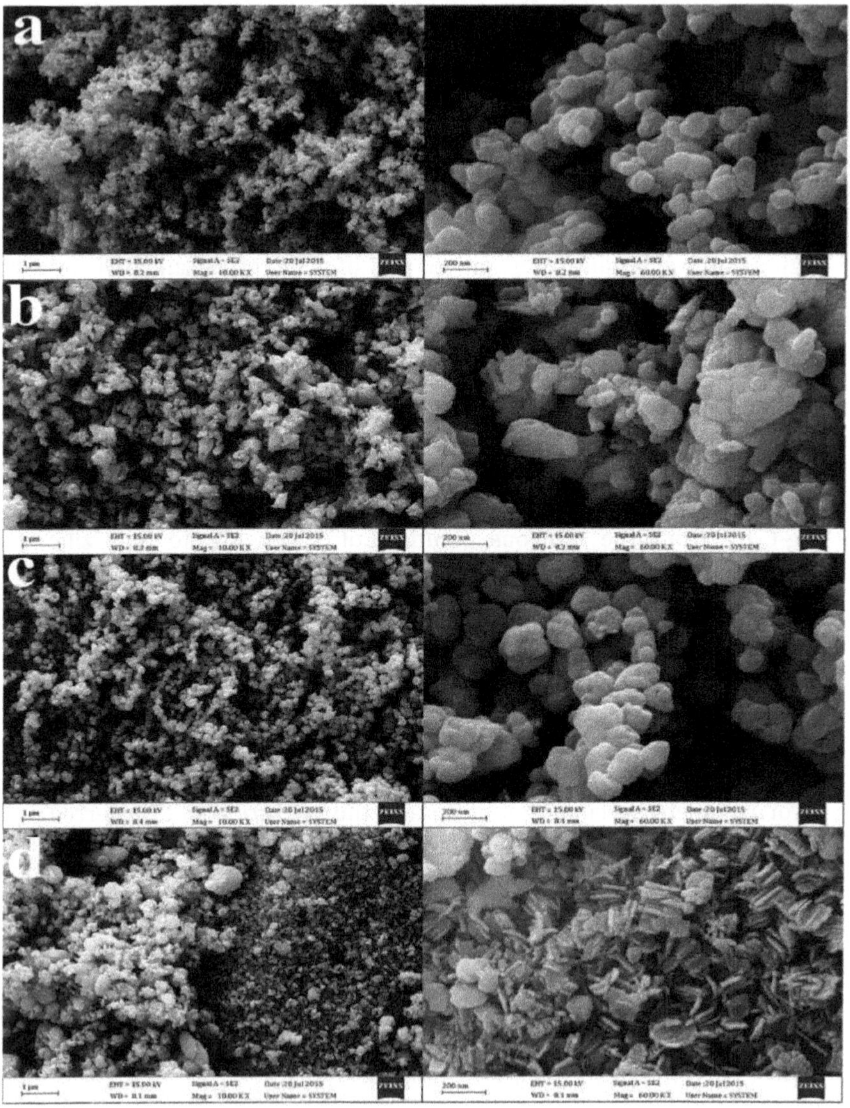

Figure 5.4 FESEM images with different magnifications of the CuO/ZnO nanocomposite photocatalysts: (a) 0.4% CuO/ZnO, (b) 2.0% CuO/ZnO, (c) 10% CuO/ZnO and (d) 50% CuO/ZnO [30].

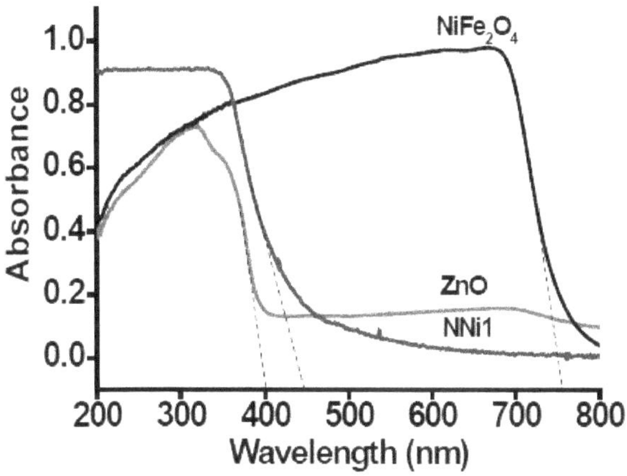

Figure 5.5 UV-vis diffuse reflectance spectra of pure $NiFe_2O_4$, pure ZnO and $NiFe_2O_4$/ZnO composites photocatalysts [51].

5.2.4 Microwave-Assisted Method

Microwave irradiation has recently received great attention as one of the important methods for the preparation of both organic and inorganic materials with controlled morphology and size. The microwave irradiation is principally based on microwave dielectric heating of the precursor solution which led to solvents and reagents volumetric heating. Due to the intense friction and the collision molecules produced by microwave irradiation, it can offer rapid and uniform heating properties to result in homogeneous nucleation and growth of ZnO-based nanocomposite photocatalysts [52, 53]. Sherly et al. [54] reported microwave-assisted synthesis of CuO/ZnO nanoparticles in aqueous medium. In that experiment, zinc nitrate and copper nitrate were dissolved in aqueous urea solution, and then treated in a microwave oven to obtain CuO/ZnO nanocomposite photocatalysts. The FESEM images in Fig. 5.6 show that the spherical-like CuO nanoparticles with sizes in the range of 15–20 nm were well dispersed on the surface of ZnO nanoparticles. Microwave-assisted methods have also been used to prepare ZnS/ZnO [37] and C/CdS/ZnO [38], TiO_2/ZnO [55] and ZnSe/ZnO [43] composites. Compared to the other synthetic methods, microwave-assisted synthesis offered selective activation of the target precursor to initiate nucleation and thus can lead to the generation of smaller and more uniform catalyst particles. In addition, the microwave-assisted method was able to be programmed and thus controlled the different catalyst synthesis steps.

Figure 5.6 FESEM images of (a, b) pure ZnO, (c, d) pure CuO and (e, f) CuO/ZnO nanocomposite photocatalysts [54].

5.3 Applications of Zinc Oxide-Based Nanocomposites for Photocatalytic Conversion of Organic Pollutants

Rapid industrial development has sparked a global crisis concerning the disposal of a large number of various toxic pollutants to the environment. Phenolic compounds, dyes and persistent organic pollutants (POPs) are the major organic contaminants in industrial wastewater and have received considerable attention worldwide due to their gravest threat to wildlife and human health [10, 24]. Some of these organic contaminants have been identified to exhibit endocrine-disrupting ability [56]. A developing research by the invention of various chemical, physical and biological treatment processes with varying levels of success has encouraged dramatic progress in the scientific society. Of major interest, ZnO photocatalytic technology with free radical reaction has currently become the most privileged and promising fundamental approach in environmental conservation. This technology is gradually being applied in various industries and even being commercialized in many countries [57, 58]. Significant decrease in the concentration of the contaminants has been observed when ZnO was used as a photocatalyst under irradiation of light [10, 19]. In this section, the modification techniques will be typically elucidated for encouraging the photoactivity according to the ZnO-based nanocomposites.

According to literature works [59–61], the NiO semiconductor has been garnered much attention because it possessed fascinating electrical, magnetic and optical properties for lithium ion batteries, electrochemical supercapacitors, magnetics, gas sensors and catalysis. NiO has been conventionally regarded as a poor candidate for photocatalytic application due to its wide band gap (~3.4 eV) and its instability to oxidize H_2O or OH^- groups. To address such a lapse, systematic investigation works have been conducted to apply NiO as a coupled semiconductor [62–64]. The NiO/ZnO system, where NiO served the role of photosensitizer for ZnO has garnered great attention. The CB of ZnO is more positive (–0.3 V) than that of NiO so that it could act as a pool for the photogenerated e^- [32]. By coupling NiO with ZnO, the photogenerated e^- can store on the ZnO while the photogenerated h^+ can store on the NiO due to the NiO/ZnO heterojunction formation. Based on Liu et al. [65] and Kanjwal et al. [60], NiO/ZnO system suppressed the charge carrier recombination rate, enabling better photocatalytic efficiency in organic pollutant conversion. For instance, NiO/ZnO nanofibers have also been studied as potential photocatalysts for the degradation of dairy effluent (DE) and methylene blue (MB) [60]. The NiO/ZnO nanocomposites were found to be more efficient than the pristine ZnO and NiO for photoconversion of the DE and dye under UV irradiation as illustrated in Fig. 5.7. Liu et al. [65] also fabricated similar system (NiO/ZnO) for photodecomposition of methyl orange (MO). Complete decomposition of the dye was obtained after 25 min of UV irradiation.

The revised sentence is "However, the development of this configuration for photocatalytic applications at a huge scale prevents its practicality owing to the wide band gap of both NiO and ZnO. As already mentioned, the improvement of photoactivity of ZnO under exposure of solar light irradiation recently moved to the coupling of narrow band gap oxide for ZnO. This coupled ZnO system can decrease the band gap, prolongs the absorbance range to visible light region and results in better separation of e^-–h^+ pair under exposure of irradiation.

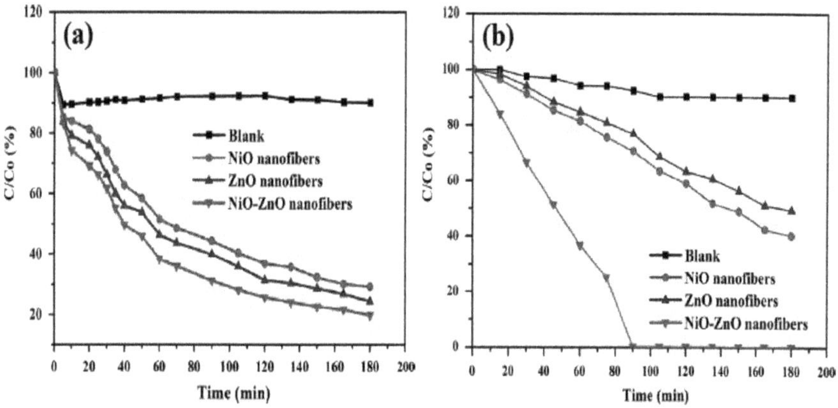

Figure 5.7 Photoconversion of (a) DE and (b) MB dye using blank, pristine ZnO, NiO nanofibers and NiO/ZnO nanofibers [60].

Currently, the small band gap oxides that have mostly been investigated are Ag_2O, CuO, Fe_2O_3, $BiFeO_3$, WO_3, Bi_2O_3 and MoO_3 [24, 30, 31, 34, 39, 66, 67]. Coupling of these semiconductors with ZnO has achieved outstanding photocatalytic activities compared to ZnO alone. Ag_2O with a small band gap of ~1.2 eV is one of the well-known semiconductors. Both the CB and VB of Ag_2O are higher than their neighbors in ZnO. The CB potentials of ZnO and Ag_2O are −0.5 and −1.3 eV, respectively [68–70]. Such proper alternation in the energy positions of the VB and CB bands permitted Ag_2O to play as a photosensitizer to absorb visible light irradiation and to form the e^- and h^+. Oppositely, the e^- is transported into the CB of the inactivated ZnO, leaving the h^+ behind in the VB of Ag_2O. The ZnO received e^- transported from the Ag_2O, which could effectively suppress the recombination of charge carrier in Ag_2O. Ag_2O/ZnO composite has been investigated by our group and other groups to be an efficient system with superior visible light photoactivity [34, 69]. In particular, the Ag_2O/ZnO composite developed by Umukoro et al. [69] has shown a great optical absorption of the composite from 370 to 800 nm (Fig. 5.8a) and owing to this absorption, a great enhancement of the visible light photoactivity was attained in the presence of visible light irradiation (λ > 420 nm), the decomposition of organic dye Acid Blue 74 utilizing the composite was about 2 times as high as that observed utilizing ZnO alone. The improvement in the photocatalytic ability was ascribed to two reasons: an enhancement in the photogenerated charge carrier separation efficiency as a consequence of the Ag_2O/ZnO heterojuncture and an increase in the formation of more reductive e^- through the charge carrier transfer process between the Ag_2O and ZnO (Fig. 5.8b).

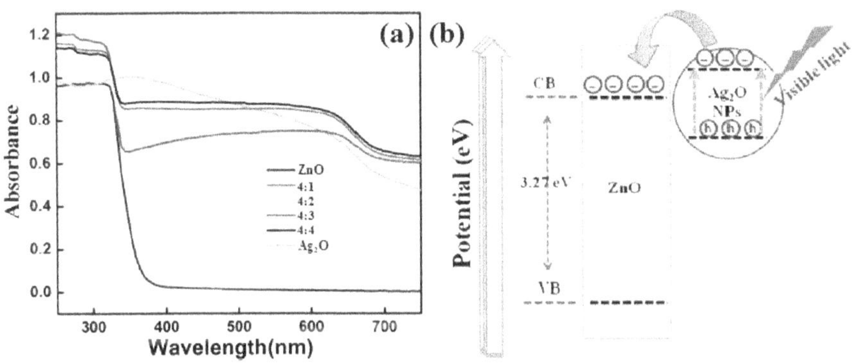

Figure 5.8 (a) UV-vis diffuse reflectance spectra of Ag_2O/ZnO at different Ag_2O loadings and (b) Schematic view for the charge carrier separation and energy band match of Ag_2O/ZnO under visible light irradiation [69].

CuO is another semiconductor that has also been extensively utilized as photosensitizer for ZnO photocatalysts. The main interest in utilizing this semiconductor is its photostability against corrosion and its small band gap (1.2–1.7 eV) that closely match the solar spectrum [71, 72]. The CB and VB potentials of ZnO are less negative and more positive, respectively, compared to the corresponding band of the CuO [73–75]. The CB position of CuO (−0.76 eV) is higher than the CB position of ZnO (−0.50 eV). The CuO-ZnO/pottery plate thin film heterojunction has been studied by Shavisi et al. [76] at different concentrations of CuO onto ZnO for photocatalytic degradation of synthetic wastewater-containing ammonia (Fig. 5.9). It has been reported in their work that the CuO absorbed a large portion of visible light and when the junction contained CuO, the absorbance commenced at

500 nm. By conducting a photocatalytic experiment, CuO (5 wt.%)/ZnO displayed superior ability to remove more than 77% of ammonia in 240 min solar irradiation time.

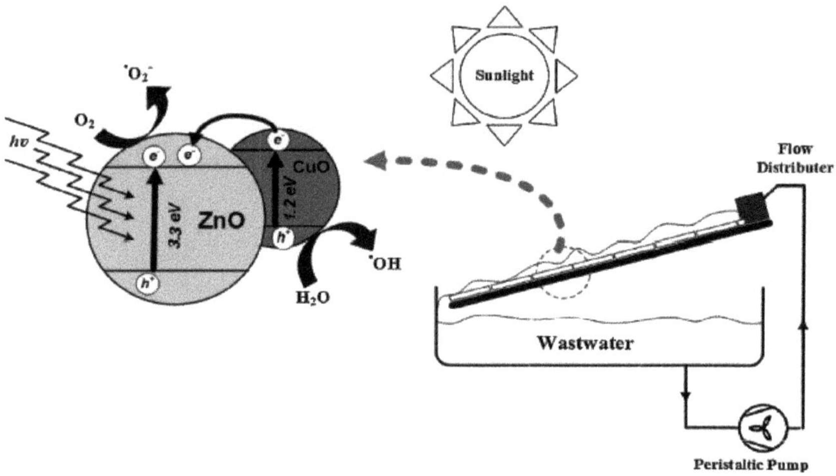

Figure 5.9 Schematic diagram of photoreactor system and photogenerated charge carrier in the CuO-ZnO system [76].

Nevertheless, success in the application of narrow band gap oxides as photocatalysts is very limited due to the fast charge carrier recombination. It has been reported by Vinu and Madras [77] that a good matching of the CB and VB of both metal oxides can ensure effective charge carriers transferred from one to another. If a suitable absorbate substrate is available to trap these e^- and h^+, recombination can be avoided and subsequent redox reactions can happen. For an ideal coupled photocatalyst, the redox potential of VB h^+ for both two metal oxides must be sufficiently positive to produce •OH radicals and that of the CB e^- must be sufficiently negative to produce O_2•- radicals. Hence, in developing the efficient coupled photocatalysts based on heterojunction structure, the levels of CB and VB of coupled photocatalysts permitting the occurrence of oxidative and reductive paths were regarded as the vital issue to enhance the photoactivity.

The WO_3/ZnO coupled system has garnered extensive attention in the area of photo-electrochemistry [78]. The band gap of WO_3 is ~2.8 eV and both the VB and CB levels of WO_3 are lower than those of ZnO. Under exposure of UV-vis light, WO_3 can be photoexcited and the photogenerated h^+ can be transferred from WO_3 to ZnO. The photogenerated e^- in the CB of ZnO could be easily received by WO_3 owing to the standard reduction potential of WO_3 was +0.5 V [79, 80]. The e^- at both sides of ZnO and WO_3 can then induce numerous reduction reactions to form the $O_2{\bullet}^-$, $HO_2{\bullet}$ and H_2O_2 as shown in Eqs. (5.12)–(5.14). On the other hand, considering that the CB position of WO_3 was lower than the standard redox potential $E^0(O_2/O_2{\bullet}^-)$(−0.33 V vs. NHE) and $E^0(O_2/HO_2{\bullet})$(−0.046 V vs. NHE), direct e^- shuttle to molecular O_2 to form the $O_2{\bullet}^-$ and $HO_2{\bullet}$ radicals will be difficult. Thus, the e^- in the CB of WO_3 will be transferred to O_2 species via the reactions stated in Eq. (5.14). The resulted H_2O_2 could then undergo several reactions to produce active •OH radicals to some extent [21, 81]. At the same time, the photogenerated h^+ in the WO_3 VB will be transferred to that of ZnO. The h^+ at the VBs of ZnO (2.80 V vs. NHE) and WO_3 (3.30 V vs. NHE) will further form more •OH radicals via Eqs. (5.15) and (5.16) that induced the degradation the organic molecules [82, 83].

$$O_2 + e^- \rightarrow O_2{\bullet}^-, \quad E^0 = -0.33\ \text{V} \tag{5.12}$$

$$O_2 + H^+ + e^- \rightarrow HO_2{\bullet}, \quad E^0 = -0.046\ \text{V} \tag{5.13}$$

$$O_2 + 2H^+ + 2e^- \rightarrow H_2O_2, \quad E^0 = +0.682\ \text{V} \tag{5.14}$$

$$OH^- + h^+ \rightarrow {\bullet}OH, \quad E^0 = +1.99\ \text{V} \tag{5.15}$$

$$H_2O + h^+ \rightarrow H^+ + {\bullet}OH, \quad E^0 = +2.72\ \text{V} \tag{5.16}$$

As we reported earlier [84], a WO_3/ZnO nanorod (NR) photocatalyst was developed via a simple hydrothermal-deposition route (Fig. 5.10). Construction of the photocatalyst structure was owing to the matching of the crystal planes as exhibited by the HRTEM image shown in Fig. 5.10b. The amount of WO_3 was adjusted by controlling the loading of WO_3 in the dispersion, which thereby affected the photoactivity performance. The optimum photoactivity of WO_3 (2 wt%)/ZnO NR in the presence of sunlight degradation of 2,4-dichlorophenoxyacetic acid (2,4-D) was 4.3 and 17.7 times higher than those of commercial TiO_2 and commercial WO_3 (Fig. 5.10c). Furthermore, a dramatic visible light photoactivity was also observed for the decomposition of organic pollutants (phenol, resorcinol and bisphenol A) from our group [24, 85]. This composite is one of the best visible light photocatalysts previously reported by our studies. The one-dimensional (1 D) catalyst structure together with the matched band level was regarded to be efficient for the photogenerated e^-–h^+ charge carrier separation in WO_3/ZnO NR and thus accountable for the improvement in the performance, which was evidenced by the prolonged lifetime of the photogenerated e^-–h^+ in the composites as analyzed through the photoluminescence (PL) spectra (Fig. 5.10d). The increase in the photoactivity of ZnO was also affirmed by Yu et al. [86] as a result of coupling with WO_3 for the decomposition of Acid Orange II. Their photocatalytic tests indicated that WO_3/ZnO photocatalysts displayed higher Acid Orange II decomposition than that of the similar system with pristine ZnO.

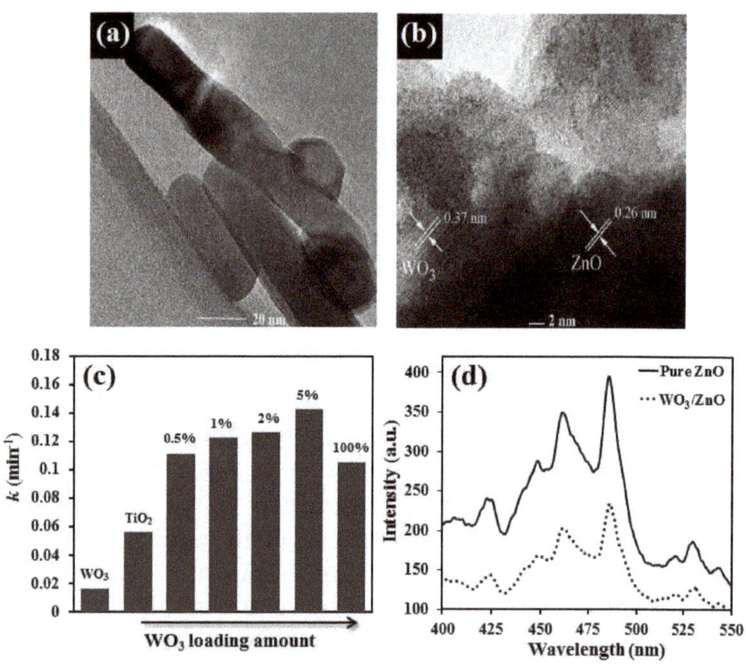

Figure 5.10 (a) TEM and (b) HRTEM images of WO_3 (2 at%)/ZnO nanorods composites, (c) photoconversion of 2,4-D under visible light using WO_3/ZnO samples at different WO_3 loadings, pure ZnO, pure WO_3 and commercial TiO_2 and (d) PL decay curves measured at λ_{325} nm for pure ZnO and WO_3 (2 at%)/ZnO nanocomposites [84].

The aforementioned findings suggested an intimate interaction between semiconductor photocatalysts and coupled metal oxides is beneficial for the visible light absorption and separation of charge carrier, resulting in the improved photocatalytic performance for organic pollutant conversion.

5.4 Degradation Mechanism of the Improved Photocatalytic Performance for Organic Pollutants Conversion

The mechanism of semiconductor-accommodated photocatalysis has been extensively described in Section 5.1.2. The photocatalytic process includes a series of photochemical reactions and following the first step of e^-–h^+ generation. The photogenerated VB h^+ can combine with adsorbed H_2O molecules or OH^- group to form •OH radical. In the presence of dissolved O_2, the CB e^- can be reacted with O_2 to form the $O_2\bullet^-$ and H_2O_2. These radicals have been deliberated to be the key active species accountable for the photocatalytic reaction of ZnO-based nanocomposites.

In order to identify the active species that partook in the degradation mechanism, several indirect detection techniques which were correlated to additional of a probe molecule in the medium were developed. The probe molecules chemically reacted with active species, thus leading to the production of a quantifiable signal over a longer time [85, 87]. These techniques contained electron paramagnetic resonance (EPR) [88], electron spin resonance (ESR) [89, 90] luminescence [91] and fluorescence [85, 87, 92]. Ye et al. [89] applied the ESR spin-trapping method to examine the active species formed in the solar light ZnO/CeO_2@halloysite nanotube composites. This technique used 5,5-dimethyl-1-pyrroline N-oxide (DMPO) as a spin trap that altered highly active species to numerous stable adducts, DMPO–•OH and DMPO–$O_2\bullet^-$. As shown in Fig. 5.11a, EPR emission spectrum of •OH radicals was intense than that of $O_2\bullet^-$ radicals. The result indicated that the •OH radicals and not $O_2\bullet^-$ radicals governed the photoactivity reaction. The production of •OH radical has also been studied over metal oxide nanoparticles (WO_3, CuO and Nb_2O_5)-coupled with ZnO nanorods in aerated aqueous solution using terephthalic acid (TA) as a fluorescence probe

[85, 93]. This method showed the TA conversion into a fluorescent product, 2-hydroxyterephthalic acid (HTA) via their reaction with •OH radicals that formed by the nanostructures as illustrated in Fig. 5.11b.

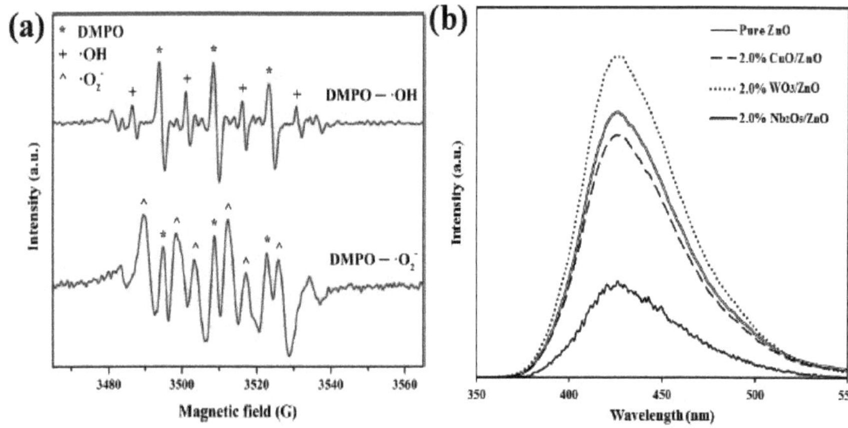

Figure 5.11 (a) DMPO spin-trapping ESR of the ZnO/CeO_2@halloysite nanotube under UV-vis light [89] and (b) PL spectra of the terephthalic acid solution with an excitation at 315 nm using different coupled photocatalysts [85, 93].

Radical scavengers have also been utilized to test the photocatalytic decomposition process of ZnO-based nanocomposites because they can provide information about the involvement of different active species, altering the kinetic profile of the reaction [82, 89, 94, 95]. Using Ag_3PO_4/ZnO composites as catalysts, the decomposition of RhB was significantly restrained when the addition of •OH radical scavenger, isopropanol (IPA), while $O_2\bullet^-$ radicals were not identified by trapping with benzoquinone (BQ) [82]. This suggested that the •OH radicals and not the $O_2\bullet^-$ radicals were the major active species responsible for visible light photocatalytic decomposition of MO dye. In another study, Wang et al. [96] found decreased tendency became more noticeable as increasing the *t*-butanol (*t*BuOH) concentration, which inferred that the •OH radicals were the predominant species in the visible light photodecomposition of RhB using Bi_2O_3/ZnO composites (Fig. 5.12). Moreover, they added that the decomposition was slightly retarded in the presence of h^+ scavenger (1 mM ethylenediaminetetraacetic acid, EDTA). This implied only minor extent of •OH radicals were originated from the h^+. This behavior can be explained by the band edge position plotted in Fig. 5.12d, the photogenerated e^- on the CB of Bi_2O_3 can interact with O_2, while h^+ on the VB of ZnO can react with H_2O. The photocatalytic decomposition of organic pollutants in the presence of ZnO-based nanocomposites was thereby relied on the positions of CB and VB of both metal oxides for the occurrence of redox reactions. Such hypothesis was also made by our group and some researchers for the photocatalytic conversion of other pollutants [85, 97, 98].

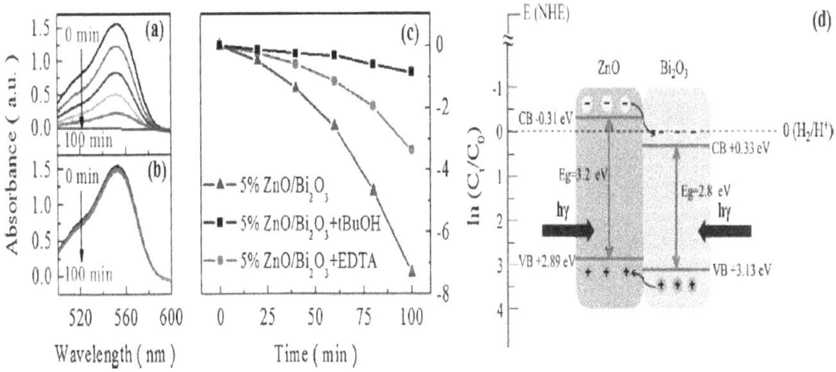

Figure 5.12 Absorbance changes of RhB solution at different irradiation times via (a) Bi_2O_3/ZnO + 1 mM EDTA and (b) Bi_2O_3/ZnO *t*BuOH. (c) Kinetic profiles of the decomposition of RhB solution. (d) Schematic diagram indicating the energy band structure and charge carrier separation in the Bi_2O_3/ZnO heterojunction [96].

Apart from the monitoring probing and radical trapping radicals during the photodecomposition process, identification of the reaction intermediates must be conducted to get a whole picture of the degradation mechanism. Gas chromatography/mass spectroscopy (GC/MS), high performance liquid chromatography (HPLC), high performance liquid chromatography/mass spectroscopy (HPLC/MS) and FTIR spectroscopy are the common characterization tools used to detect the reaction intermediates. In literatures, the photocatalytic experiments typically involved monitoring absorbance changes at a specific wavelength (λ_{max}), other complementary analyses must also be performed. These included the measurement of total organic carbon (TOC), chemical oxygen demand (COD), biological oxygen demand (BOD), CO_2 evolution and the release of inorganic ions. The temporal changes in these analyses will also reflect the involvement of active radicals during the course of photodecomposition process.

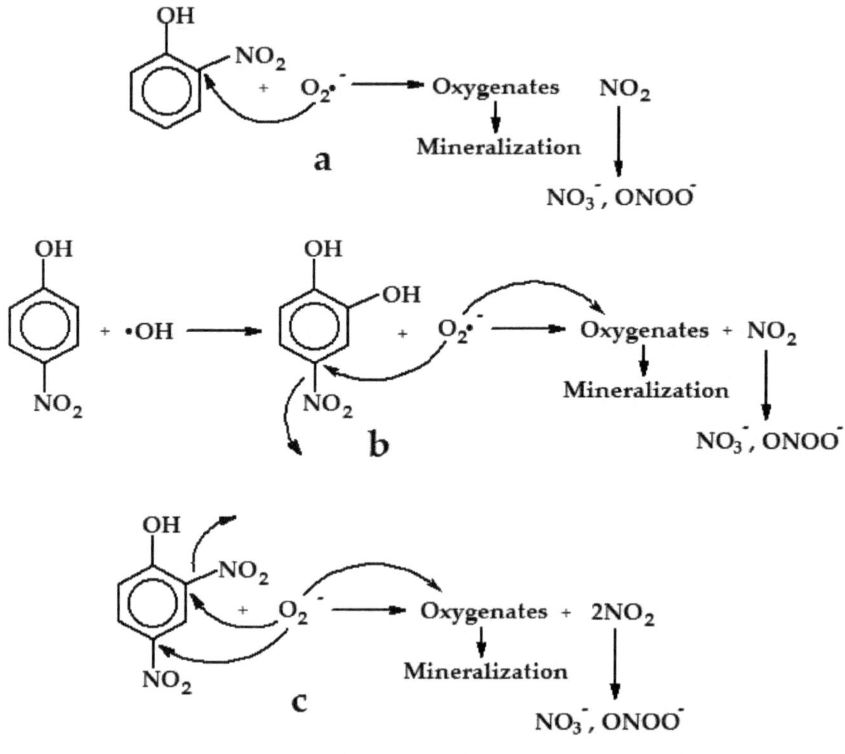

Figure 5.13 Degradation mechanisms of (a) 2-NP, (b) 4-NP and (c) 2,4-NP [99].

The photoactivity of the V_2O_5/ZnO composite was examined for the decomposition of three nitrophenol pollutants (2-nitrophenol (2-NP), 4-nitrophenol (4-NP) and 2, 4-dinitrophenol (2,4-NP)) [99]. The degradation mechanism of nitrophenols intermediates was investigated utilizing HPLC and GC/MS analyzers and illustrated in Fig. 5.13. The experimental results deduced that the coordination of nitro group in 2-NP and 4-NP considerably influenced the decomposition rate. The detection of OH^- group comprising intermediates in the decomposition of 4-NP verified the production and pivotal role of •OH radicals in the decomposition process. On the contrary, $O_2\bullet^-$ radicals were the main contributors in the photodecomposition of 2-NP and 2,4-DNP. They added that the release of N_2^-, N_3^- and N_4^+ ions upon the decomposition reaction aided in detecting the possible interaction sites. Using a ternary hybrid $TiO_2/ZnO/Bi_2O_3$ catalyst, Pei et al. [100] identified the degradation products of *o*-xylene through a GC/MS. They reported that the decomposition of *o*-xylene was initiated by the attack of •OH radicals on the methyl group, leading to generation of aromatic ring compounds such as $C_6H_4O_2$, C_7H_8 and C_6H_6. Subsequently, the aromatic ring rupture occurred to form carboxylic acids. The carboxylic acids can be directly converted to CO_2 and H_2O, which led to the certain mineralization degrees of *o*-xylene.

5.5 Summary and Perspectives

In summary, ZnO can be coupled with numerous semiconductors to form ZnO-based nanocomposite photocatalysts due to its exceptional chemical and physical properties, environmental stability and low cost. The exceptional properties of ZnO have opened up an alternative pathway to prepare high performance photocatalysts. In this chapter, we have highlighted the various synthetic routes such as solution-mixing, sol-gel, hydro(solvo) thermal and microwave-assisted strategy that have been applied for preparing the ZnO-based nanocomposites. These nanocomposites have demonstrated potential application in photocatalytic conversion of organic pollutants. In addition, the nanocomposites also demonstrated excellent photocatalytic conversion of organic pollutants over the currently available commercial semiconductor photocatalysts.

Despite the substantial progress that has been accomplished, there are some vital technical points that require to be further examined. The first challenge is the development of the heterojunction with proper CB and VB to encompass the oxidation and reduction potentials for the desirable catalytic reaction. ZnO is believed to be a better starting UV material to form nanocomposites with semiconductor photocatalysts. Therefore, the development of novel photocatalysts with high redox performance is necessary. The second one is the design and tailor-made of more efficient semiconductor photocatalysts. One of the virtues of ZnO incorporation into the nanocomposites is served to promote the separation of charge carriers and transport of photogenerated electrons and holes. The performance of photocatalysts is highly reliant on their structures of surface including the morphologies and surface states. Thus, precise controlled synthesis strategies have to be developed to prepare ZnO-based nanocomposites. Furthermore, coupling ZnO with other semiconductors designed to extend the absorption wavelength range to visible light region. The underlying mechanism of visible light-responsive ZnO-based nanocomposites still need more research works to investigate ranging from experimental using novel characterizations to computational techniques. A direct and systematic comparison of ZnO-based nanocomposites with other visible light catalyst materials is also indispensable to provide potential application for the next generation solar photocatalyst systems. Undeniably, there are still many challenges and prospects for ZnO-based nanocomposites and they are still anticipated to be developed as potential photocatalysts to circumvent various environmental and energy-related issues.

Acknowledgements

This work was supported by the Universiti Tunku Abdul Rahman (UTARRF/2015-C1/L02) and Ministry of Higher Education of Malaysia (FRGS/1/2015/TK02/UTAR/02/2 and FRGS/1/2016/TK02/UTAR/02/1).

References

1. Harper, C. A. (2001). *Guide to Ceramic Materials. Handbook of Ceramics, Glasses, and Diamonds* (McGraw-Hill, New York).
2. Porter, F. (1991). *Zinc Dust and Compounds. Zinc Handbook Properties, Processing, and Use in Deign* (Marcel Dekker, New York).
3. Wang, Z. L. (2004). Zinc oxide nanostructures: Growth, properties and applications, *J. Phys. Condens. Matter.*, **16**, pp. R829–R858.
4. Baruah, S., and Dutta, J. (2009). Hydrothermal growth of ZnO nanostructures, *Sci. Technol. Adv. Mater.*, **10**, pp. 1–18.

5. Özgür, Ü., Alivov, Y. I., Liu, C., Teke, A., Reshchikov, M. A., Doðan, S., Avrutin, V., Cho, S. J., and Morkoçd, H. (2005). A comprehensive review of ZnO materials and devices, *J. Appl. Phys.*, **98**, pp. 1–103.

6. Janotti, A., and Van de Walle, C. G. (2009). Fundamentals of zinc oxide as a semiconductor, *Rep. Prog. Phys.*, **72**, pp. 1–29.

7. Rehman, S., Ullah, R., Butt, A. M., and Gohar, N. D. (2009). Strategies of making TiO_2 and ZnO visible light active, *J. Hazard. Mater.*, **170**, pp. 560–569.

8. Lu, M. P., Song, J. H., Lu, M. Y., Chen, M. T., Gao, Y. F., Chen, L. J., and Wang, Z. L. (2009). Piezoelectric nanogenerator using p-type ZnO nanowire arrays, *Nano Lett.*, **9**, pp. 1223–1227.

9. Anta, J. A., Guillén, E., and Tena-Zaera, R. (2012). ZnO based dye-sensitized solar cells, *J. Phys. Chem. C*, **116**, pp. 11413–11425.

10. Lam, S. M., Sin, J. C., Abdullah, A. Z., and Mohamed, A. R. (2012). Degradation of wastewaters containing organic dyes photocatalysed by zinc oxide: A review, *Desalin. Water Treat.*, **41**, pp. 131–169.

11. Zhou, H. L., Qu, Y. Q., Zeid, T., and Duan, X. F. (2012). Towards highly efficient photocatalysts using semiconductor nanosrchitectures, *Energy Environ. Sci.*, **5**, pp. 6732–6743.

12. Sin, J. C., Lam, S. M., Satoshi, I., Lee, K. T., and Mohamed, A. R. (2014). Sunlight photocatalytic activity enhancement and mechanism of novel europium-doped ZnO hierarchical micro/nanospheres for degradation of phenol, *Appl. Catal. B: Environ.*, **148–149**, pp. 258–268.

13. Yuan, Y., Huang, G. F., Hu, W. Y., Xiong, D. N., and Huang, W. Q. (2016). Tunable synthesis of various ZnO architectural structures with enhanced photocatalytic activities, *Mater. Lett.*, **175**, pp. 68–71.

14. Hafez, H. S. (2012). High active ZnO rod-like nanomaterials: Synthesis, characterization and photocatalytic activity for dye removal, *Phys. E*, **44**, pp. 1522–1527.

15. Kansal, S. K., Ali, A. H., and Kapoor, S. (2010). Photocatalytic decolorization of biebrich scarlet dye in aqueous phase using different nanophotocatalysts, *Desalination*, **259**, pp. 147–155.

16. Gnanaprakasam, A., Sivakumar, V. M., Sivayogavalli, P. L., and Thirumarimurugan, M. (2015). Characterization of TiO_2 and ZnO nanoparticles and their applications in photocatalytic degradation of azo dyes, *Ecotoxicol. Environ. Saf.*, **121**, pp. 121–125.

17. Sin, J. C., Lam, S. M., Lee, K. T., and Mohamed, A. R. (2013). Self-assembly fabrication of ZnO hierarchical micro/nanospheres for enhanced

photocatalytic degradation of endocrine-disrupting chemicals, *Mater. Sci. Semicond. Process.*, **16**, pp. 1542–1550.

18. Pare, B., Jonnalagadda, S. B., Tomar, H., Singh, P., and Bhagwat, V. W. (2008). ZnO assisted photocatalytic degradation of acridine orange in aqueous solution using visible irradiation, *Desalination*, **232**, pp. 80–90.
19. Sin, J. C., Lam, S. M., Lee, K. T., and Mohamed, A. R. (2015). Surfactant-free solvothermal synthesis of ZnO nanorods for effective sunlight degradation of 2,4-dichlorophenol, *Mater. Lett.*, **140**, pp. 51–54.
20. Hernández-Alonso, M. D., Fresno, F., Suárez, S., and Coronado, J. M. (2009). Development of alternative photocatalysts to TiO_2: Challenges and opportunities, *Energy Environ. Sci.*, **2**, pp. 1231–1257.
21. Malato, S., Fernández-Ibánez, P., Maldonado, M. I., Blanco, J., and Gernjak, W. (2009). Decontamination and disinfection of water by solar photocatalysis: Recent overview and trends, *Catal. Today*, **147**, pp. 1–59.
22. Kansal, S. K., Singh, M., and Sud, D. (2007). Studies on photodegradation of two commercial dyes in aqueous phase using different photocatalysts, *J. Hazard. Mater.*, **141**, pp. 581–590.
23. Sakthivel, S., Neppolian, B., Shankar, M. V., Arabindoo, B., Palanichamy, M., and Murugesan, V. (2003). Solar photocatalytic degradation of azo dye: Comparison of photocatalytic efficiency of ZnO and TiO_2, *Solar Energy Mater. Solar Cells*, **77**, pp. 65–82.
24. Lam, S. M., Sin, J. C., Abdullah, A. Z., and Mohamed, A. R. (2012). ZnO nanorods surface-decorated by WO_3 nanoparticles for photocatalytic degradation of endocrine disruptors under a compact fluorescent lamp, *Ceram. Int.*, **39**, pp. 2343–2352.
25. Lam, S. M., Sin, J. C., and Mohamed, A. R. (2016). Fabrication of ZnO nanorods via a green hydrothermal method and their light driven catalytic activity towards the erasure of phenol compounds, *Mater. Lett.*, **167**, pp. 141–144.
26. Sin, J. C., Lam, S. M., Lee, K. T., and Mohamed, A. R. (2015). Preparation of cerium-doped ZnO hierarchical micro/nanospheres with enhanced photocatalytic performance for phenol degradation under visible light, *J. Mol. Catal. A: Chem.*, **409**, pp. 1–10.
27. Kwiatkowski, M., Bezverkhyy, I., and Skompska, M. (2015). ZnO nanorods covered with TiO_2 layer: Simple sol-gel preparation, and optical, photocatalytic and photoelectrochemical properties, *J. Mater. Chem. A*, **3**, pp. 12748–12760.

28. Cheng, P. F., Wang, Y. L., Xu, L. P., Sun, P., Su, Z. S., Jin, F. M., Liu, F. M., Sun, Y. F., and Lu, G. Y. (2016). High specific surface area urchin-like hierarchical ZnO-TiO_2 architectures: Hydrothermal synthesis and photocatalytic properties, *Mater. Letts.,* **175**, pp. 52–55.

29. Dai, P., Zhang, L. L., Zhang, G. T., Li, G., Sun, Z. Q., Liu, X. S., and Wu, M. Z. (2014). Characterization and photocatalytic activity of (ZnO-CuO)/SBA-15 nanocomposites synthesized by two-solvent method, *Mater. Res. Bull.,* **56**, pp. 119–124.

30. Mansournia, M., and Ghaderi, L. (2017). CuO@ZnO core shell nanocomposites: Novel hydrothermal synthesis and enhancement in photocatalytic property, *J. Alloy. Compd.,* **691**, pp. 171–177.

31. Liu, Y. J., Sun, L., Wu, J. G., Fang, T., Cai, R., and Wei, A. (2015). Preparation and photocatalytic activity of ZnO/Fe_2O_3 nanotube composites, *Mater. Sci. Eng. B,* **194**, pp. 9–13.

32. Chen, S. F., Zhao, W., Liu, W., and Zhang, S. J. (2009). Preparation, characterization and activity evaluation of p-n junction photocatalyst p-NiO/n-ZnO, *J. Sol-Gel Sci. Technol.,* **50**, pp. 387–396.

33. Lv, Z. J., Zhong, Q., and Ou, M. (2016). Utilizing peroxide as precursor for the synthesis of CeO_2/ZnO composite oxide with enhanced photocatalytic activity, *Appl. Surf. Sci.,* **376**, pp. 91–96.

34. Lam, S. M., Sin, J. C., Abdullah, A. Z., and Mohamed, A. R. (2013). Efficient photodegradation of resorcinol with Ag_2O/ZnO nanorods heterostructure under a compact fluorescent lamp irradiation, *Chem. Pap.,* **67**, pp. 1277– 1284.

35. Jeyasubramanian, K., Hikku, G. S., and Sivashakthi, M. (2016). Synthesis, characterization and photo-catalytic studies of mixed metal oxides of nano ZnO and SnO_x, *Mater. Sci. Semicond. Process.,* **51**, pp. 25–32.

36. Pascariu, P., Airinei, A., Olaru, N., Olaru, L., and Nica, V. (2016). Photocatalytic degradation of Rhodamine B dye using ZnO-SnO_2 electrospun ceramic nanofibers, *Ceram. Int.,* **42**, pp. 6775–6781.

37. Hu, Y., Qian, H. H., Liu, Y., Du, G. H., Zhang, F. M., Wang, L. B., and Hu, X. (2011). A microwave-assisted rapid route to synthesize ZnO/ZnS core-shell nanostructures via controllable surface sulfidation of ZnO nanorods, *CrystEngComm,* **13**, pp. 3438–3443.

38. Zhou, M. J., Hu, Y., Liu, Y., Yang, W. L., and Qian, H. S. (2012). Microwave-assisted route to fabricate coaxial ZnO/C/CdS nanocables with enhanced visible light-driven photocatalytic activity, *CrystEngComm,* **14**, pp. 7686–7693.

39. Loh, L., Briscoe, J., and Dunn, S. (2015). Chemical protection of ZnO nanorods at ultralow pH to form a hierarchical $BiFeO_3/ZnO$ core-shell structure, *ACS Appl. Mater. Interfaces,* **7**, pp. 152–157.
40. Min, Y. L., Zhang, K., Chen, Y. C., Zhang, Y. G., and Zhao, W. (2012). Synthesis of nanostructured ZnO/Bi_2WO_6 heterojuction for photocatalysis application, *Sep. Purif. Technol.,* **92**, pp. 115–120.
41. Balachandran, S., Prakash, N., Thirumalai, K., Muruganandham, M., Sillanpää, M., and Swaminathan, M. (2014). Facile construction of heterostructured $BiVO_4$–ZnO and its dual application of greater solar photocatalytic activity and self-cleaning property, *Ind. Eng. Chem. Res.,* **53**, pp. 8346–8356.
42. Fu, X. D., Xie, M. Z., Luan, P., and Jing, L. Q. (2014). Effective visible-excited charge separation in silicate-bridged $ZnO/BiVO_4$ nanocomposite and its contribution to enhanced photocatalytic activity, *ACS Appl. Mater. Interfaces,* **6**, pp. 18550–18557.
43. Cho, S. C., Jang, J. W., Lee, J. S., and Lee, K. H. (2012). Porous ZnO–ZnSe nanocomposites for visible light photocatalysis, *Nanoscale,* **4**, pp. 2066–2071.
44. Zhou, J. W., Zhang, M., and Zhu, Y. F. (2014). Preparation of visible light-drive g-C_3N_4@ZnO hybrid photocatalyst via mechanochemistry, *Phys. Chem. Chem. Phys.,* **16**, pp. 17627–17633.
45. Das, R. K., and Das, M. (2015). Catalytic activity of acid and base with different concentration on sol-gel kinetics of silica by ultrasonic method, *Ultrason. Sonochem.,* **26**, pp. 210–217.
46. Lam, S. M., Sin, J. C., Mohamed, A. R. (2016). A review on photocatalytic application of g-C_3N_4/semiconductor (CNS) nanocomposites towards the erasure of dyeing wastewater, *Mater. Sci. Semicond. Process.,* **47**, pp. 62–84.
47. Namratha, K., and Byrappa, K. (2013). Hydrothermal processing and in situ surface modification of metal oxide nanomaterials, *J. Supercrit Fluids,* **79**, pp. 251–260.
48. Byrappa, K., and Adschiri, T. (2007). Hydrothermal technology for nanotechnology, *Prog. Cryst. Growth Charact. Mater.,* **53**, pp. 117–166.
49. Adhikari, S. P., Pant, H. R., Kim, J. H., Kim, H. J., Park, C. H., and Kim, C. S. (2015). One pot synthesis and characterization of Ag-ZnO/g-C_3N_4 photocatalyst with improved photoactivity and antibacterial properties, *Colloids Surf. A: Physicochem. Eng. Aspects,* **482**, pp. 477–484.
50. Rashad, M. M., Ismail, A. A., Osama, I., Ibrahim, I. A., and Kandil, A. H. T. (2014). Photocatalytic decomposition of dyes using ZnO-

doped SnO_2 nanoparticles prepared by solvothermal method, *Arab. J. Chem.*, **7**, pp. 71–77.

51. Zulhadjri, R., Emriadi, N. J., and Arief, S. (2016). Synthesis of ZnO-$NiFe_2O_4$ magnetic nanocomposites by simple solvothermal method for photocatalytic dye degradation under solar light, *Orient. J. Chem.*, **32**, pp. 1411–1419.
52. Geng, J., and Song, G. H. (2013). One-pot fast synthesis of spherical ZnS/Au nanocomposites and their optical properties, *J. Mater. Sci.*, **48**, pp. 636–643.
53. Vijayalakshmi, K., and Sivaraj, D. (2015). Enhanced antibacterial activity of Cr doped ZnO nanorods synthesized using microwave processing, *RSC Adv.*, **5**, pp. 68461–68469.
54. Sherly, E. D., Vijaya, J. J., and Kennedy, L. J. (2015). Visible-light-induced photocatalytic performances of ZnO-CuO nanocomposites for degradation of 2,4-dichlorophenol, *Chinese J. Catal.*, **36**, pp. 1263–1272.
55. Arin, J., Thongtem, S., and Thongtem, T. (2013). Single-step synthesis of ZnO/TiO_2 nanocomposites by microwave radiation and their photocatalytic activities, *Mater. Lett.*, **96**, pp. 78–81.
56. Sin, J. C., Lam, S. M., Mohamed, A. R., and Lee, K. T. (2012). Degrading endocrine disrupting chemicals from wastewater by TiO_2 photocatalysis: A review, *Int. J. Photoenergy*, **2012**, p. 185159.
57. Tsuzuki, T. (2013). *Nanotechnology Commercialisation* (Pan Stanford Publishing, Boca Raton).
58. Suib, S. L. (2013). *New and Future Developments in Catalysis: Solar Photocatalysis* (Elsevier, Amsterdam).
59. Hameed, A., Montini, T., Gombac, V., and Fornasiero, P. (2009). Photocatalytic decolorization of dyes on NiO-ZnO nano-composites, *Photochem. Photobiol. Sci.*, **8**, pp. 677–682.
60. Kanjwal, M. A., Chronakis, I. S., and Barakat, N. A. M. (2015). Electronspun NiO, ZnO and composite NiO-ZnO nanofibers/photocatalytic degradation of dairy effluent, *Ceram. Int.*, **41**, pp. 12229–12236.
61. Belhadi, A., Boumaza, S., and Trari, M. (2011). Photoassisted hydrogen production under visible light over NiO/ZnO hetero-system, *Appl. Energy*, **88**, pp. 4490–4495.
62. Tian, F. S., and Liu, Y. L. (2013). Synthesis of p-type NiO/n-type ZnO heterostructure and its enhanced photocatalytic activity, *Scr. Mater.*, **69**, pp. 417–419.

63. Ku, Y., Lin, C. N., and Hou, W. M. (2011). Characterization of coupled NiO/TiO_2 photocatalyst for the photocatalytic reduction of Cr(VI) in aqueous solution, *J. Mol. Catal. A: Chem.*, **349**, pp. 20–27.

64. Song, L. M., and Zhang, S. J. (2010). A simple mechanical mixing method for preparation of visible-light-sensitive NiO-CaO composite photocatalyts with high photocatalytic activity, *J. Hazard. Mater.*, **174**, pp. 563–566.

65. Liu, Y. L., Li, G. Z., Mi, R. D., Deng, C. K., and Gao, P. Z. (2014). An environment-benign method for the synthesis of p-NiO/n-ZnO heterostructure with excellent performance for gas sensing and photocatalysis, *Sensor Actuat. B Chem.*, **191**, pp. 537–544.

66. Lam, S. M., Sin, J. C., Abdullah, A. Z., and Mohamed, A. R. (2013). Efficient photodegradation of endocrine-disrupting chemicals with Bi_2O_3-ZnO nanorods under a compact fluorescent lamp, *Water Air Soil Pollut.*, **224**, p. 1565.

67. Lam, S. M., Sin, J. C., Abdullah, A. Z., and Mohamed, A. R. (2013). Investigation on visible-light photocatalytic degradation of 2,4-dichlorophenoxyacetic acid in the presence of MoO_3/ZnO nanorod composites, *J. Mol. Catal. A: Chem.*, **370**, pp. 123–131.

68. Xu, L. L., Wei, B., Liu, W. L., Zhang, H. L., Su, C. Y., and Che, J. X. (2013). Flower-like $ZnO–Ag_2O$ composites: Precipitation synthesis and photocatalytic activity, *Nanoscale Res. Lett.*, **8**, p. 536.

69. Umukoro, E. H., Peleyeju, M. G., Ngila, J. C., and Arotiba, O. A. (2016). Photocatalytic degradation of acid blue 74 in water using $Ag–Ag_2O–ZnO$ nanostructures anchored on graphene oxide, *Solid State Sci.*, **51**, pp. 66–73.

70. Kadam, A., Dhabbe, R., Gophane, A., Sathe, T., and Garadkar, K. (2016). Template free synthesis of ZnO/Ag_2O nanocomposites as a highly efficient visible active photocatalyst for detoxification of methyl orange, *J. Photochem. Photobiol. B: Biol.*, **154**, pp. 24–33.

71. Li, B. X., and Wang, Y. F. (2010). Facile synthesis and photocatalytic activity of ZnO–CuO nanocomposite, *Superlattices Microstruct.*, **47**, pp. 615–623.

72. Meshram, S. P., Adhyapak, P. V., Mulik, U. P., and Amalnerkar, D. P. (2012). Facile synthesis of CuO nanomorphs and their morphology dependent sunlight driven photocatalytic properties, *Chem. Eng. J.*, **204–206**, pp. 158–168.

73. Wei, S. Q., Chen, Y. Y., Ma, Y. Y., and Shao, Z. C. (2010). Fabrication of CuO/ZnO composite films with cathodic co-electrodeposition and

their photocatalytic performance, *J. Mol. Catal. A: Chem.*, **331**, pp. 112–116.

74. Ghosh, A., and Mondal, A. (2016). Fabrication of stable, efficient and recyclable p-CuO/n-ZnO thin film heterojunction for visible light driven photocatalytic degradation of organic dyes, *Mater. Lett.*, **164**, pp. 221–224.

75. Samad, A., Furukawa, M., Katsumata, H., Suzuki, T., and Kaneco, S. (2016). Photocatalytic oxidation and simultaneous removal of arsenite with CuO/ZnO photocatalytic, *J. Photochem. Photobiol. A: Chem.*, **325**, pp. 97–103.

76. Shavisi, Y., Sharifinia, S., and Mohamadi, Z. (2016). Solar-light-harvesting degradation of aqueous ammonia by CuO/ZnO immobilized on pottery plate: Linear kinetic modeling for adsorption and photocatalysis process, *J. Environ. Chem. Eng.*, **4**, pp. 2736–2744.

77. Vinu, R., and Madras, G. (2010). Environmental remediation by photocatalysis, *J. Indian Inst. Sci.*, **90**, pp. 189–231.

78. Liu, Y., He, H. Z., Li, J., Li, W. Z., Yang, Y. H., Li, Y. M., and Chen, Q. Y. (2015). ZnO nanoparticle-functionalized WO_3 plate with enhanced photoelectrochemical properties, *RSC Adv.*, **5**, pp. 46928–46934.

79. Li, D., and Haneda, H. (2004). Enhancement of photocatalytic activity of sprayed nitrogen-containing ZnO powders by coupling with metal oxides during the acetaldehyde decomposition, *Chemosphere*, **54**, pp. 1099–1100.

80. Gupta, S. M., and Tripathi, M. (2012). An overview of commonly used semiconductor nanoparticles in photocatalysis, *High Energy Chem.*, **46**, pp. 1–9.

81. Gaya, U. I., and Abdullah, A. H. (2008). Heterogeneous photocatalytic degradation of organic contaminants over titanium dioxide: A review of fundamentals, progress and problems, *J. Photochem. Photobio. C: Photochem. Rev.*, **9**, pp. 1–12.

82. Liu, W., Wang, M. L., Xu, C. X., Chen, S. F., and Fu, X. L. (2013). Ag_3PO_4/ZnO: An efficient visible light-sensitized composite with its application in photocatalytic degradation of Rhodamine B, *Mater. Res. Bull.*, **48**, pp. 106–113.

83. Liu, W., Wang, M. L., Xu, C. X., Chen, S. F., and Fu, X. L. (2013). Significantly enhanced visible-light photocatalytic activity of g-C_3N_4 via ZnO modification and the mechanism study, *J. Mol. Catal. A: Chem.*, **368–369**, pp. 9–15.

84. Lam, S. M., Sin, J. C., Abdullah, A. Z., and Mohamed, A. R. (2015). Sunlight responsive WO_3/ZnO nanorods for photocatalytic degradation and

mineralization of chlorinated phenoxyacetic acid herbicides in water, *J. Colloid Interface Sci.*, **450**, pp. 34–44.

85. Lam, S. M., Sin, J. C., Abdullah, A. Z., and Mohamed, A. R. (2014). Transition metal oxide loaded ZnO nanorods: Preparation, characterization and their UV-vis photocatalytic activities, *Sep. Purif. Technol.*, **132**, pp. 378–387.

86. Yu, C. L., Yang, K., Shu, Q., and Li, X. (2011). Preparation of WO_3/ZnO composite photocatalyst and its photocatalytic performance, *Chinese J. Catal.*, **32**, pp. 555–565.

87. Zheng, L. R., Chen, C. Q., Zheng, Y. H., Zhan, Y. Y., Cao, Y. N., Lin, X. Y., Zheng, Q., Wei, K., and Zhu, J. F. (2014). Photocatalytic activity of $ZnO/Sn_{1-x}Zn_xO_{2-x}$ nanocrystals: A synergistic effect of doping and heterojunction, *Appl. Catal. B: Environ.*, **148–149**, pp. 44–50.

88. Dodd, A., McKinley, A., Takuya, T., and Saunders, M. (2009). Mechanochemical synthesis of nanoparticles ZnO-$ZnWO_4$ powders and their photocatalytic activity, *J. Eur. Ceram. Soc.*, **29**, pp. 139–144.

89. Ye, Z. F., Li, J. Z., Zhou, M. J., Wang, H. Q., Ma, Y., Huo, P. W., Yu, L. B., and Yan, Y. S. (2016). Well-dispersed nebula-like ZnO/CeO_2@HNTs heterostructure for efficient photocatalytic degradation of tetracycline, *Chem. Eng. J.*, **304**, pp. 917–933.

90. Yang, Y. W., Que, W. X., Zhang, X. Y., Yin, X. T., Xing, Y. L., Que, M. D., Zhao, H. Y., and Du, Y. P. (2017). High-quality Cu_2ZnSnS_4 and $Cu_2ZnSnSe_4$ nanocrystals hybrid with ZnO and $NaYF_4$: Yb, Tm as efficient photocatalytic sensitizers, *Appl. Catal. B: Environ.*, **200**, pp. 402–411.

91. Chauhan, I., Nigam, S., Sudarsan, V., and Vatsa, R. K. (2016). Hetero-junction assisted improved luminescence of Eu^{3+} doped ZnO-SnO_2 nanocomposite, *J. Lumin.*, **176**, pp. 124–129.

92. Qamar, M. T., Aslam, M., Rehan, Z. A., Soomro, M. T., Basahi, J. M., Ismail, I. M. I., Almeelbi, T., and Hameed, A. (2017). The influence of p-type Mn_3O_4 nanostructures on the photocatalytic activity of ZnO for the removal of bromo and chlorophenol in natural sunlight exposure, *Appl. Catal. B: Environ.*, **201**, pp. 105–118.

93. Lam, S. M., Sin, J. C., Satoshi, I., Abdullah, A. Z., and Mohamed, A. R. (2014). Enhanced sunlight photocatalytic performance over Nb_2O_5/ZnO nanorod composites and the mechanism study, *Appl. Catal. A: Gen.*, **471**, pp. 126–135.

94. Wu, D., An, T. C., Li, G. Y., Wang, W., Cai, Y. C., and Yip, H. Y. (2015). Mechanistic study of the visible-light-driven photocatalytic

inactivation of bacteria by grapheme oxide-zinc oxide composite, *Appl. Surf. Sci.*, **358**, pp. 137–145.

95. Gohari, M. S., and Yangjeh, A. H. (2016). Photosentization of Fe_2O_3/ZnO by AgBr and Ag_3PO_4 to fabricate novel magnetically recoverable nanocomposites with significantly enhanced photo-catalytic activity under visible-light irradiation, *Ceram. Int.*, **42**, pp. 15224–15234.

96. Wang, X., Ren, P. R., and Fan, H. Q. (2015). Room-temeparture solid state synthesis of ZnO/Bi_2O_3 heterojunction and their solar light photocatalytic performance, *Mater. Res. Bull.*, **64**, pp. 82–87.

97. Lamba, R., Umar, A., Mehta, S. K., Anderson, W. A., and Kansal, S. K. (2015). Visible-light-driven photocatalytic properties of self assembled cauliflower-like AgCl/ZnO hierarchical nanostructures, *J. Mol. Catal. A: Chem.*, **408**, pp. 189–201.

98. Xiang, Z., Zhong, J. B., Huang, S. T., Li, J. Z., Chen, J. F., Wang, T., Li, M. J., and Wang, P. (2016). Efficient charge separation of Ag_2CO_3/ZnO composities prepared by a facile precipitation approach and its dependence on loading content of Ag_2CO_3, *Mater. Sci. Semicond. Process.*, **52**, pp. 62–67.

99. Aslam, M., Ismail, M. I., Almeelbi, T., Salah, N., Chandrasekaran, S., and Hameed, A. (2014). Enhanced photocatalytic activity of V_2O_5–ZnO composites for the mineralization of nitrophenols, *Chemosphere*, **117**, pp. 115–123.

100. Pei, C. C., and Leung, W. W. F. (2015). Photocatalytic oxidation of nitrogen monoxide and o-xylene by TiO_2/ZnO/Bi_2O_3 nanofibers: Optimization, kinetic modeling and mechanisms, *Appl. Catal. B: Environ.*, **174–175**, pp. 515–525.

Chapter 6

Polymer Nanocomposites as Nanoadsorbents for Environment Remediation

Priyanka Ghanghas,[a] Kavita Poonia,[a] and Dinesh Kumar[b]

[a]*Department of Chemistry, Banasthali University, Rajasthan 304022, India*
[b]*School of Chemical Sciences, Central University of Gujarat, Gandhinagar, Gujarat 382030, India*

dsbchoudhary2002@gmail.com

Recently, researchers have been highly concerned about the extinction of various water resources. Among various treatments, advanced processes in nanomaterial sciences have drawn scientists' attention. Nanomaterials are playing a key role in environmental remediation. They have many shapes, such as nanoparticles, fibers, wires, and tubes. They may also be capable of working as catalysts and adsorbents and their composites with polymers are used for detecting and clearing away various gases such as SO_2, NO_x, and CO, contaminated chemicals which are toxic arsenic, manganese, nitrate, iron, heavy metals, as well as organic pollutants (both aliphatic and aromatic hydrocarbons). They show excellent physical and chemical properties which are

Nanocomposites for Pollution Control
Edited by Chaudhery Mustansar Hussain and Ajay Kumar Mishra

ISBN 978-981-4774-45-1 (Hardcover), 978-1-315-14368-2 (eBook)
www.panstanford.com

unique in their function caused by the unusually large surface-area-to-volume ratios, and high interfacial reactivity of the nanofillers. This chapter provides an overview of polymer nanocomposites (PNCs) for environment application, PNC fabrication methods, and strategies for the implication of PNC materials for controlling contaminants and pollutant sensing and identification.

6.1 Introduction

Nanoparticles (NPs) with a size of approximately 1–100 nm have a significant impact in many scientific fields, including chemistry, electronics, medicine, biology, and material sciences. The physical and chemical properties of NPs are directly related to their intrinsic compositions, apparent sizes, and extrinsic surface structures. Thus, the design, synthesis, characterization, and applications of nanostructures are critical aspects of the emerging field of nanomaterials.

Nanocomposites are a mixture of two or more nanoparticle components in which one component is present in a nanoscale range which ranges from 1 to 100 nm. The main motive of producing nanocomposites is to add nanomaterial in the form, single entity with other components to produce a synergistic effect. The most suitable example of targeting those composites in which each component structure and their properties combine distinctly to give hybrid component which show multifunctions in both structures as well as properties. Based on previous researchers' examination and interest, this chapter focuses on nanoadsorbents of polymer-based multifunctional nanocomposites as well as magnetic nanocomposites.

Polymer nanocomposites (PNCs) are generally made up of a polymer or copolymer having nanoparticles or nanofillers scattered in the polymer matrix. They play a vital role in removing toxicity and heavy metals remediation from water and wastewater.

Nanofibres, hollow nanofibres, core–shell nanofibres, nanorods and nanotubes produced have an excellent potential for a wide range of applications mainly homogeneous and heterogeneous catalysis, sensors, filter applications, and optoelectronics.

The magnetic nanocomposite materials are responsible for the discovery of the outstanding new process, having potential applications in the field of science and technology. Magnetic nanoparticles (MNPs) have drawn attention due to their internal magnetic properties, which make them an excellent source of drug delivery, useful in magnetic resonance imaging devices, anticancer agent adsorbents and magnetically recoverable catalysts. The major class of nanomaterials includes metallic and bimetallic nanoparticles, metal oxides, ferrite and superparamagnetic iron oxide nanoparticles (SPIONs) (McCarthy et al., 2008; Maleki et al., 2012). The magnetic nanocomposites are mainly made up of magnetic nanoparticles which are present inside either nonmagnetic or magnetic matrix (Gupta et al., 2001; Huang et al., 2001). Likewise, magnetic nanoparticles scattered in composites basically have a strong tendency to give rise agglomerates for the elimination of energy present with high surface area-to-volume ratio of the nanosized particles. To avoid collection of magnetic nanoparticles, protective ways have been made to chemically stabilise the visible magnetic nanoparticles by grafting of or coating with organic species. This chapter focuses on various properties, synthesis, and applications of magnetic nanoparticles and their polymer nanocomposites that contain surfactants or polymers or are covered with an inorganic layer, such as carbon and silica. Conductive polymer nanostructures or metal oxide are prepared from pseudoactive materials in a carbon matrix which has high conductivity and a large surface area (Hughes et al., 2002; Zhu et al., 2012); thus, it can potentially increase the capacitance of the resulting composite electrodes. Nanocomposites have high biocompatibility, light weight, anticorrosion, relatively easy fabrication and processing, and low cost. Many properties can be introduced into the composite entity to suit many steady biological functions.

Environmental remediation refers to reducing radiation exposure, for example, from contaminated soil, groundwater or surface water. The purpose is more than just eliminating radiation sources; it is about protecting people and the environment from the potentially harmful effects of exposure to ionising radiation. This process is well regulated and evaluated for human health and environmental risks.

The targeting remediation is mainly focused on nanocomposite solid wastes, wastewater, heavy metals, arsenic, and organic dyes. Finally, this chapter concludes that the multifunctional nanocomposites are important for the sustainability of environmental remediation. Although this chapter does not comprehensively cover every aspect of multifunctional nanocomposites, it provides selective reports what we have achieved till date.

Polymer-based nanoadsorbents: Polymer-based adsorbents are hybrid materials that achieve nanoscale particle immobilisation by the nanopore template inside the host matrix and improved hydraulic conductivity due to large particle properties, as well as the affinities of inorganic nanoparticles towards heavy metals. Therefore, it is feasible to use the inorganic adsorbent for the increased capture of target pollutants. Trace heavy metal purification, including Pb(II), Cd(II), As(V), etc., and the utilisation of polymer-based nanocomposites have also been explained in this chapter.

Polymer-based nanoadsorbents have received immense attention in sample pre-treatment procedures. Inorganic polymers, either as a system in which nanosized inorganic materials can be inserted for improving their various properties like chemical, mechanical, thermal, and sorption or as a bed or as a template to synthesise and grow nanoparticles, are being considered as the adsorbents of interest. Nanopolymer spheres also work as desirable adsorbents for separation purposes. The most important and beneficial fact of acquiring the nanoscale inorganic particles inside polymeric beads is that these small particles cannot incorporate easily in fixed-bed columns or any plug-flow-type system only because of extravagant pressure drops and bad durability. Inorganic nanoscale hybrid adsorbents are more amenable than the single one for the regeneration and reuse of the polymer. A few important advantages of nanoscale polymer–inorganic hybrids are excellent adsorption capacity and best chemical stability in a wide range pH due to ease of functionalization between different polymeric units such as epoxide and epoxy. Nevertheless, the specific resistance of the polymeric groups and their linkage to acid and base hydrolysis are an additional benefit of using them as ligand-carrying polymers (Kaya et al., 2011. Hu et al., 2006) stated that a type of magnetite (γ-Fe_2O_3) used for the eliminating of Cr(VI) ions from wastewater. Although this adsorbent could be distinguished promptly from the post-treatment wastewater system, its maximum adsorption capacity was only about 17.0 mg g^{-1}. In order to reach the highest adsorption capacity, common adsorbents have been upgraded by embedding them into a polymeric shell with many functional groups, which mainly includes phosphate amide, carboxylate, hydroxyl, sulphate, and amino groups. Magnetic nanoparticles have been tailored by chemical alteration of polymeric shells; by selective adsorption, various metal ions were attained (Zhao et al., 2010). Nanoporous polymers can be categorised into three main categories: (1) Cross-linked polymers prepared by the process of polymerization in the presence of the porogen, whose surface area is up to 1000 m^2g^{-1} and they are extensively used as the separation media; (2) hyper-cross-linked polymers, which has a surface area of ≥ 2000 m^2g^{-1}; and (3) the polymers that have basic microporosity with a surface area up to 1000 m^2g^{-1} (Germain et al., 2008). The sorption of the toxic metals on polymer-based nanoadsorbents is based on the chemistry of the adsorbents surface. The analysis of pH point of zero charges (PZC) for glycidyl methacrylate-based polymer (GMD) and ferric oxide of nanosized loaded glycidyl methacrylate-based polymer (GMDFe) was carried out (Kaya et al., 2011). GMD adsorbent showed PZC at pH 6.3, which was shifted to pH 6.86 for GMDFe adsorbent. The surface of adsorbents was positively charged at GMDFe and GMD and these were eliminated anions from the solution under these conditions. At the time of loading of iron nanoparticles onto the surface, there was an increase in the positive potential of the particle surface.

6.2 Preparation of Polymer-Based Nanoadsorbents

Various methods have been used for nanopolymer sphere fabrication, for instance, precipitation, spraying, phase separation, and emulsion. An emulsification solvent diffusion (ESD) method in nanosized particles of PLGA or PLA (polylactide) was achieved by using a mixture of acetone and water The particles are prepared by the emulsification process in the water phase and consequent solvent-evaporation with balanced mechanical stirring.

The formation of polymer nanoadsorbent is described in the following steps:

(1) The polymeric solution was added to the aqueous phase to obtain emulsion droplet.
(2) Acetone was quickly diffused out from each emulsion droplet, thereby decreasing its size to nano-order.
(3) Consequently, by the process of "solvent-evaporation," the remaining solvent was eliminated from the system by solidifying the droplets to obtain final polymeric nanoparticles.

This method has been used by Murakami et al. (1999), and sometimes it results in serious aggregation in particle formation because of the huge quantity of residual dichloromethane as solvent. They recommended an improved method in which a mixture of two water-miscible as well as water-immiscible organic solvents with low hydrolysation and polymerization category of polyvinyl alcohols (PVAs) was used as a quasi-emulsifier. The combination of binary organic solvents like acetonitrile and alcohol results in the formation of nanoparticles without aggregation. The suspension polymerization technique widely has been used for preparing polymeric templates, which was applied in polymer-based nano-adsorbents. Ethylene glycol methacrylate (EGDMA) copolymer beads were prepared by the co-precipitation method, which was used to form nanosized ferric oxide in synthesising polymer-based hybrid nanoadsorbent (Kaya et al., 2011). EGDMA copolymer beads were further improved by using diethylamine and then by cross-linking amine-containing beads fraternised with chloroacetone by drenching in the solution of chloroacetone/dimethylformamide. The yield of quaternization was followed by the analysis of chlorine ions in the product, which was done by mercuric thiocyanate method. In a hypothermal condition, the co-precipitation of ferric to ferrous ions in sodium hydroxide solution is done for the preparation of a magnetic hybrid sorbent. The chemical equation can be written as follows:

$$Fe^{2+} + 2Fe^{3+} + 8OH^{-} \rightarrow Fe_3O_4 + 4H_2O$$

According to above equation, the initial molar ratio of ferric to ferrous ions is (2:1), which is required for the formation of Fe_3O_4. As a result, nanosized ferric-oxide-loaded glycidyl-methacrylate-based polymeric (GMDFe) was attained by adding glycidyl methacrylate-based polymer GMD beads into an aqueous solution having $FeCl_3$ and $FeCl_2$ under effective stirring.

6.3 Applications of Polymer-Based Nanoadsorbents

6.3.1 Removal of Organic Compound

Polymer-based clay nanocomposites for the elimination of trichlorophenol and trinitrophenol from water have been used by Ganigar et al. (2010). Two nanosized clay minerals, MMT and sepiolite (SEP), were loaded on two polycations, which are poly diallyl dimethyl ammonium chloride (PDADMAC) and poly-4-vinylpyridine-co-styrene (PVPcoS). Polycations are formed from clay mineral nanocomposites, which prefer two phenolic pollutants, trichlorophenol (TCP) and trinitrophenol, in lakes, river and water reservoirs. The concentrations of PA and TCP were analysed by UV–vis spectroscopy. The adsorption kinetics of PDADMAC and PVPcoS on MMT were significantly faster (2–4 h) than on sepiolite (3–4 days), which was elucidated by the latter's porous structure. Consequently, PDADMAC–MMT and PVPcoS–MMT were chosen as optimal adsorbents for the elimination of phenol derivatives from the water. Both PA (anionic) and TCP (nonionic) showed higher affinity to the less charged polycation PVPcoS (40% of PA and the monomers charged) than to the highly charged polycation PDADMAC. However, PA elimination by the PVPcoS–MMT composite was nearly complete while TCP removal reached 40–60% of the added amounts. The adsorption isotherms of the pollutants suggested that the binding to PVPcoS–MMT was directed not only mainly by hydrophobic interactions but also by electrostatic interactions in the case of PA. Differences were also seen in the binding kinetics PA and TCP to dry and wet composites. The hydration property of PA enhanced its binding to wet composites, whereas the property of hydrophobicity of TCP enhanced its binding to dehydrated composites. The results of this study emphasise the importance of a better understanding of pollutant–adsorbent interactions to enable more efficient tailoring of polymer–clay mineral composites for water treatment.

6.3.2 Removal of Heavy Metals

Zirconium phosphate (ZrP) is a highly active and most efficient ion-exchanger. It helps in trapping heavy metals, including Pb^{2+}, Zn^{2+}, and Cd^{2+} from solution (Zhang et al., 2010; Pan et al., 2007). ZrP NPs were scattered onto a porous polymeric cations exchanger D-001 and attained a new hybrid nanomaterials ZrP-001. The lead broke through suddenly onto D-001 due to its nonspecific selectivity toward the lead. On the contrary, breakthrough results were observed for ZrP-001 under otherwise identical conditions and ZrP-001 retention resulted in a dramatic lead decrease from 40 to less than 0.1 mg/L, such satisfactory performance was ascribed to the "Donnan membrane effects," i.e. the negatively charged polymeric matrix D-001 would highly increase lead permeation and preconcentration from solution to the active sorption sites, and lead concentrations around ZrP particles were much higher than in solution. Similar results were also obtained by Sarkar et al. (2005) and De Marco et al. (2003) and co-workers for hydrated ferric oxides impregnated onto an anion exchanger for arsenic removal.

Besides, the thiol derivative (ZrPS) of ZrP has also shown preferable sorption towards heavy metals, and the column adsorption for trapping trace heavy metals from drinking water has also shown that ZrPS-001 can effectively decrease three toxic heavy metals to below their respective drinking water standards (Zhang et al., 2008).

Hydrated manganese oxides (Tripathy et al., 2005; Ren et al., 2011; Stroes-Gascoyne et al., 1987) can selectively bind heavy metals and anionic ligands (e.g. arsenate, phosphate) through inner–sphere complex formation. However, like ZrP, they cannot be directly used in columns or any other flow-through systems due to the fine or ultrafine particles. Polymer-supported hydrated manganese oxides were then developed to overcome the bottleneck (Su et al., 2010; Wan et al., 2010).

A complete effluent history of a separate fixed-bed column packed with either HMO-001 or D-001 for a feeding solution containing lead ion and competing cations like Na^+, Ca^{2+} and Mg^{2+}. Lead broke through quickly on D-001 due to its poor selectivity towards the lead. On the contrary, satisfactory breakthrough results were observed for HMO-001 even when the competing cations were about 200 times more than the lead ion in mass concentration. The lead concentration in the effluent reduces dramatically from 1.0 to less than 0.01 mg/L, which is the allowance of drinking water suggested by WHO. Moreover, the exhausted ZrP/ZrPS-001 beads have efficiently been regenerated by 2M HCl solution at 303 K for repeated use without any significant capacity loss.

Similarly, other inorganic nanoparticles exhibiting specific affinity toward heavy metal ions, namely, hydrated zirconium oxides (HFOs) $Ti(HPO_4)_2$, hydrated ferric oxide (HFO), can also be anchored onto porous cation exchanger to obtain hybrid nanomaterials (Jia et al., 2009; Pan et al., 2009). Biopolymers as abundant, biodegradable and renewable materials have high capacity to adsorb extensive variety of molecules and ions by chemical and physical interactions. Chitosan is known as a hydrophilic, non-toxic, biocompatible, and biodegradable polysaccharide with antibacterial activity which exhibits high adsorption capacity for the removal of various heavy metal ions from wastewater (Lakouraj et al., 2014).

6.4 Magnetic Polymer-Based Nanocomposites

Magnetic polymer nanocomposites are of particular interest because of the combination of excellent magnetic properties, stability, and good biocompatibility. Organic–inorganic magnetic nanocomposites can be prepared by different processes in situ and ex situ by co-precipitation, melt blending, microwave reflux and ceramic–glass processing and plasma polymerization techniques. These nanocomposites have been exploited for in vivo imaging, as superparamagnetic or negative contrast agents, drug carriers, heavy metal adsorption, and magnetically recoverable photocatalyst for the degradation of organic pollutants.

A polymer and the plastic magnet is a nonmetallic magnet which is made from an organic polymer. This is a new class of magnetic materials which has gained significant interest of researchers. Poly(1,3,5-triaminobenzene) was synthesised when oxidised with iodine by Torrance et al. (1998) to show a ferromagnetic phase up to 400°C. Rajca et al. (2001) synthesised organic pi-conjugated polymer with a very large magnetic moment and magnetic order even at low temperatures below 10 K. Zaidi et al. (2009) synthesized a novel magnetic polymer PANICNQ produced from polyaniline (PANI) and tetracyanoquinodimethane (TCNQ) which was an acceptor molecule, the first magnetic polymer to function at room temperature. PANICNQ combines a fully conjugated nitrogen-carrying backbone with molecular charge transfer side groups, and this combination produces a stable polymer with a high density of localised spins which are expected to produce coupling. Crayston et al. (2000) synthesised organic magnets and other observed developments in the field of organic magnets.

6.5 Magnetic Nanocomposites

There are mainly three types of magnetic nanocomposites: (i) core–shell inorganic nanocomposites, (ii) self-assembled nanocomposites, and (iii) organic–inorganic nanocomposites.

6.5.1 Core–Shell Inorganic Nanocomposites

The combination of two nanoscale entities into a single hybrid particle has attracted much attention due to the numerous prospective of applications (Wang et al., 2009). Silica NPs have been employed as a very common and promising material for coating and encapsulating magnetic NPs. There are many reasons for choosing SiO_2 as coating or matrix or shell material in magnetic composite materials. First of all, it reveals silanol surface groups which may easily be derivatized with a variety of functional groups, providing a convenient platform for further functionalization. SiO_2-based magnetic composites have been studied in view of various biomedical and environmental applications due to their biocompatibility, stability against degradation and hydrophilic character. Magnetic SiO_2 composite materials have been prepared by several procedures, including aerosol routes (e.g. aerosol pyrolysis, spray-drying), microemulsion polymerization, and sol–gel processes (Behrens et al., 2011).

6.5.2 Self-Assembled Colloidal Nanocomposites

The self-assembly of small building blocks (e.g. atoms, molecules, and nanoparticles) into well-arranged macroscopic superstructures have been an important concern in various fields of chemistry, biology and materials science. 2-D and 3-D self-assembly of NPs and their superlattices which have a high degree of translational order have drawn considerable attention (Talapin et al., 2010). These super crystals showed not only a translational order but also an orientational order with a crystallographic arrangement. The assembly of NPs of different materials into defined colloidal crystals or quasicrystals provides a general path to a large variety of composite materials with new collective properties which are arising from the interaction of the different nanocrystals in the assembly.

6.5.3 Organic–Inorganic Nanocomposites

This is the classical type of a nanocomposite, where the isolated NPs are finely scattered in an organic polymer. Originally, agglomerated NPs are scattered in a polymer matrix. Functional organic-inorganic nanocomposites with improved physical properties allow various applications (e.g. in biomedical, micro-optics, electronics, energy conversion, or storage). The resulting composite flow behaviour limits huge solid loadings and therefore property adjustment due to restrictions in shaping. Trim rate and temperature-dependent as well as oscillatory theological investigations are therefore necessary for a detailed description of the composite flow properties early to shape forming (Mezger et al., 2006).

6.6 Synthesis of Magnetic Nanoparticles

The synthesis of MNPs of different compositions and phases, which mainly includes iron oxides, like Fe_3O_4 and α-Fe_2O_3 (Neveu et al., 2002; Kumar et al., 2009; Kumar et al., 2013), pure metals such as cobalt and iron (Park et al., 2000; Puntes et al., 2001), spinel ferrites such as $MnFe_2O_4$, $MgFe_2O_4$, $CoFe_2O_4$, $NiFe_2O_4$, and $ZnFe_2O_4$ (Chen et al., 1999; Park et al., 2004), alloys such as $FePt_3$ and $CoPt_3$, dilute magnetic semiconductors, and polymer magnets (Zaidi et al., 2009; Sun et al., 2000; Shevchenko et al., 2002; Sharma et al., 2010) have been described in the literature. During the past few years, researchers have developed efficient synthetic ways for the synthesis of shape-controlled, highly stable, and monodisperse magnetic nanoparticles. A variety of methods that have been used for the formation of high-quality magnetic nanoparticles mainly include hydrothermal techniques: sol–gel processing, surfactant-assisted synthesis, co-precipitation, microemulsion techniques, and solution combustion.

6.6.1 Synthesis of Magnetic Polymer Nanocomposites

There are many methods available for preparing magnetic polymer nanocomposites, such as plasma polymerization, layer-by-layer method, and seed precipitation. The plasma polymerization technique is a room-temperature and environmentally benign process. The main principle of the plasma polymerization technique is that the ionised and excited molecules and radicals created by the electrical field bombardment react with the surface of the substrate. A versatile approach for the gas-phase synthesis of magnetic polymer nanocomposites is the application of microwaves for plasma generation. This approach was developed by Vollath et al. (1999). The inorganic cores are formed by homogeneous nucleation, and the organic shell of hybrid NPs condenses by the process of heterogeneous nucleation and polymerises outside of the plasma zone on the cores synthesised in the plasma. A similar approach was used by others (Lamparth et al., 2002; Srikanth et al., 2002) to synthesise magnetic core–shell nanocomposites. Layer-by-layer accumulations of NPs and polymers were carried out for the formation of highly homogeneous thin composite films. The controlled synthesis of novel magnetic core–shell nanocomposites that consist of alternative layers of magnetic nanoparticles and polyelectrolytes as a shell and polystyrene (PS) as a centre was developed by Caruso et al. (1999). The core–shell polymer template was first made by the accumulation of three layers of poly (allylamine hydrochloride) (PAH) onto PS particle surface. Magnetic NPs (10–15 nm) were then adsorbed onto the polymer template through electrostatic complexation. Alternating deposition of PAH and magnetic NPs emerged in the formation of multilayer shells on the template surface. Nevertheless, this method is time consuming because of sequential polyelectrolyte and magnetic NP depositions and purification cycles. Additionally, the leaching problem of magnetic NPs from the inner shell layer and the stability of the shell layers to pH and electrolyte changes may still be of concern. Seed precipitation polymerization has been developed for the preparation of core–shell composite microparticles. The hydrophilic temperature-sensitive magnetic latexes prepared by Sauzedde et al. (1999). Seed magnetic NPs were prepared by adsorption of negatively charged Fe_3O_4 NPs onto positively charged colloidal template [PS/poly(isopropyl acrylamide)]. The encapsulation process executes via copolymerization of N-isopropyl acrylamide and N,N-methylene bisacrylamide, the itaconic acid in the presence of the seed particles. The resulting particles had a diameter of about 500–800 nm but with narrow size distribution. Composite particles via seed precipitation polymerization of methacrylic acid and hydroxyethyl methacrylate in the presence of tris(hydroxyl methyl) aminomethane hydroxide-coated magnetic NPs were prepared (Zaitsev et al., 1999). The resulting particles had a diameter of about 150 nm but with broad size distribution. Vinyl terminated magnetic NP-encapsulated PS composite prepared (Gu et al., 2009). Nevertheless, in the seed precipitation polymerization, the encapsulation process is quite difficult to control (Deng et al., 2003).

6.6.2 Application of Magnetic Polymer-Based Nanocomposites

6.6.2.1 Removal of heavy metals

Industrial wastes from various industries, such as paints and pigments, mining and extraction, glass production, plating, and battery manufacturing plants introduce various metals into water bodies. These toxicants are not biodegradable, and their presence in water causes bioaccumulation in living organisms, resulting in diseases such as cancer, kidney failure, metabolic acidosis, oral ulcer, and renal failure. Many of them are not toxic, but their presence in non-permissible doses can cause health problems. Heavy metal ions as cadmium, chromium, copper, lead, zinc, arsenic, and mercury, generated from natural geological or industrial sources, cause hazardous water pollution (Dean et al., 1972). Fan et al. (2016) reported magnetic removal of non-magnetic heavy metal substances from the soil. An effective elimination of undesirable metals from water systems is very important but still a challenging task for environmental researchers. A number of traditional as well as modern technologies—such as chemical precipitation (Wang et al., 2005), ion exchange, membrane filtration (Walsh et al., 1994), reverse osmosis (Bodalo-Santoyo et al., 2003), electrochemical treatment (Ersahin et al., 2012), coagulation (Zhang et al., 2003), extraction (Rykowska et al., 2008), irradiation (Batley et al., 1978), biological methods and adsorption (Srivastava et al., 2011)—have been employed for the removal of heavy metals from wastewater. Moreover, adsorption (Zamboulish et al., 2011) is regarded as the most promising technique because of its high efficiency, simplicity, and cost-effectiveness. There are many types of materials, such as activated carbons, clay minerals, biopolymers, chelating materials, and zeolites, which have been exploited to adsorb metal ions from aqueous solutions. If the centre of adsorbent is magnetic, then separation is done. The removal of metal ions using magnetic adsorbents has recently grabbed attention for wastewater treatment. The main adsorption criterion of magnetite polyacrylamide amino-amidoxime magnetite nanocomposites as adsorbents for methylene blue, Co^{2+} and Ni^{2+} cations from water. MB contains highly electro-negative heteroatoms such as sulphur and nitrogen that can form hydrogen bonds with water to promote the diffusion of MB into adsorbents. The electrostatic interaction mechanism is favoured as the explanation of the removal of MB using magnetite polyionic composites from water. Magnetite nanocomposites can be used as an adsorbent to remove toxic materials (Atta et al., 2016).

Cr(VI) is a toxic hexavalent ion produced from a broad range of anthropogenic sources which include pigment, chrome plating, wood preservation, stainless steel, textile dying, and anti-corrosion coatings (Zhu et al., 2012; Bhaumik et al., 2011). Frequently, it is found in groundwater with a broad range of pH at industrial sites. Upon excessive exposure, it leads to cancers and skin lesions (Wise et al., 2009). The maximum limit of 0.1 mg/L for total chromium in drinking water suggested by USEPA (Zhang et al., 2010). Cr(VI) must be removed from the polluted water to meet above level from different water sources.

Various methods have been developed for the environmental cleanup and removal of chromium from polluted water, including adsorption (Amuda et al., 2007), electrochemical precipitation (Kongsricharern et al., 1995), ion exchange, reduction, membrane filtration, reverse osmosis and biomaterials (Park et al., 2008). Reduction of Cr(VI) to Cr(III) is an active means of removing the Cr(VI) from the environment (Cui et al., 2010), is the best method among all above-mentioned methods, and can be achieved from the oxidation of several reductants, including zero-valent iron (Lai et al., 2008), organic compounds (Wittbrodt et al., 1995), biomaterials (Park et al., 2004), and hydrogen sulphide (Kim et al., 2001). After reduction, the Cr(III) can be removed from the solution via precipitation by adjusting the solution's pH (Gu et al., 2013). Currently, the conductive polymer polyaniline (PANI) has been described and used for Cr(VI) removal (Kumar et al., 2007; Samamni et al., 2010). PANI has three distinct structures: leucoemeraldine (LEB), emeraldine (EB), and pernigraniline (PB) (Ogoshi et al., 2011). These PANI powders have rough surfaces and look like good candidates for Cr(VI) removal due to their large specific surface area, bulk density production and low cost (Guo et al., 2011). Moreover, recycling the PANI powders after Cr(VI) treatment remains a challenge and the reaction mechanisms between PANI and Cr(VI) need be understood. PANI powder nanocomposite can be used to reduce and remove very toxic Cr(VI) in aqueous solutions, to less toxic Cr(III). The reduced Cr(III) entered the inside of the nanocomposites alternatively being adsorbed on the surface. Simultaneously, the coated PANI was moderately converted into its oxidised state after treatment with Cr(VI). Cr(VI) concentration has been appropriately controlled to within EPA needs, and the synthesised nanocomposites were recycled easily by simply using a permanent magnet and then acid to regenerate. It has been actively reused for further Cr(VI) removal.

6.6.2.2 Removal of oil

Wastewater contains oils in the form of fats, lubricants, petroleum products, and cutting oils. This is one of the significant environmental concerns at present. Oil spills in water bodies pose threat to aquatic life (Wang et al., 2010). Jiang et al. reported facile removal of oils from water surfaces through highly hydrophobic and magnetic polymer nanocomposites. (Jiang et al., 2014). Oil spills in the water bodies pose a threat to aquatic life. Though oil spills are normal accidents, they are on rise in world as per various reports from international agencies. The oils present in the emulsion form are more problematic to tackle because of high stability in the aqueous medium (Annunciado et al., 2005). The floating barriers are the most commonly used technique to control the spread of oil, whereas sorbents are utilised to eliminate final traces of oil or in areas that cannot be achieved by skimmers. Techniques such as adsorption (Ahmad et al., 2005; Husseien et al., 2008), coagulation (Yang, 2007), coalescence (Ji et al., 2009) and membrane filtration (Ahmad et al., 2003) have been utilised by researchers worldwide to separate oils from water. Natural organic adsorbents such as fibres, peat moss and cellulose have also been utilised for the adsorption of oil. In recent times, advanced functional nanomaterials comprising of organic and inorganic counterparts have gained importance (Zhu et al., 2010; Zhang et al., 2011). The common techniques are expensive and separation is not quite easy. Magnetic materials having an affinity for oil are an excellent alternative. However, magnetic materials such as Fe_3O_4 and ferrites have no affinity for oils. They are modified by species which can make the adsorption of oil possible. The Fe_3O_4 nanoparticles coated with polystyrene (Fe_3O_4@PS) were synthesised (Chen et al., 2013) as a material with a higher affinity for oil and showed water repellent action. Due to a hydrophobic organic counterpart in organic–inorganic magnetic nanocomposites, oil gets easily absorbed and magnetically removed or recovered due to magnetic core inside the main member of magnetic nanoparticle. Polymers show various properties due to their large surface areas, high chemical stability, hydrophobicity, and high functionality. Therefore, composites based on Fe_3O_4 and polymers have been most exploited for oil removal applications. In a similar way, magnetic polystyrene nanocomposites (Tempest et al., 2014) and Fe_3O_4 in alkyl resin (de-Sauza et al., 2010) were used for the recovery of oil from oil-contaminated water. Magnetic Fe_3O_4 nanoparticles (MNPs) were coated with a polystyrene layer successfully to form water-repellent and oil-absorbing surfaces, which could float on water and selectively absorb lubricating oil up to 3 times the particles' weight while completely repelling water. Oils could be removed from the surfaces of nanocomposites by a simple treatment and the nanocomposites still kept highly hydrophobic and superoleophilic characteristics, so the nanocomposites have an excellent recyclability in the oil-absorbent capacity (Chen et al., 2013).

6.6.3 Multifunctional Nanocomposites

Multifunctional magnetic nanocomposites have found a variety of applications in electronics, magnetic sensing and environmental remediation. Compared to their metallic counterpart, polymer-based multifunctional nanocomposites have the advantages of light weight, flexibility, processability, and anticorrosion. The high specific area of these structured materials provides increased adsorption capacity, while their magnetic property makes possible the recycling of the new family of adsorbents. This unique magnetic property enables easy recycling of magnetic nanocomposites adsorbents for environmental remediation, leaving no secondary pollution, and simultaneously improving the economic value of magnetic nanocomposites.

Magnetoresistance (MR) is a process in which the resistance of a material changes when an external magnetic field is applied. Giant magnetoresistance (GMR) is the large resistance change that occurs when the relative orientation of the magnetic domains in adjacent layers is adjusted from anti-parallel to parallel under an applied magnetic field. It is defined as the ratio $(R\text{AP}-R\text{P})/R\text{P}$, where RP and RAP are the resistances of materials for parallel and antiparallel alignments, respectively. The GMR aspect was first discovered in multilayered structural materials, in which ferromagnetic Fe metal layers were separated by a nonmagnetic Cr layer in 1988. This is intriguing, in particular, for conducting polymer-based multifunctional nanocomposites, as the conducting polymer is a conjugated structure with electrons being exchanged and transported along the backbone in the case of oxidative remediation of heavy metal species.

6.6.4 Applications of Multifunctional Nanocomposites

6.6.4.1 Removal of heavy metals

Rapid industrialisation has led to an increased discharge of wastes into the environment, posing potential harm to society and human health—heavy metal species being found in underground and surface water as well as solid plastic wastes accumulating in landfills and other locations. The main threats are associated with the consumption of elements such as arsenic, lead, chromium, mercury, antimony, cadmium and nickel, some of which appear in the form of soluble oxy-ions in natural water. Among these heavy metal species, Cr(VI) is a commonly identified contaminant because of its high toxicity and mobility (Hsu et al., 2010; Xu et al., 2007). Traditional removal methods for heavy metals are classified as relatively selective (chemical coagulation; filtration, adsorption) and non-selective (nanofiltration and reverse osmosis). Among these methods, adsorption is considered as one of the most promising methods because metal-loaded adsorbents are more compact and generally form stronger bonds. For this reason, the use of consumable adsorbents is nowadays the dominant trend, since it is the simplest removal method. The qualification of the proper adsorbent for an individual heavy metal is based on a number of conditions defined by the uptake mechanism of its species. High chemical affinity, stabilisation of positive or negative surface charge and incorporation of ion or electron exchange potential are described as possible directions of optimisation. A large variety of nanostructured materials, usually in the form of inorganic engineered nanoparticles, have been studied as adsorbents for the removal of heavy metals. Depending on the established risk of each heavy metal and its concentration, iron-based nanoparticles are the most widely applied systems for the uptake of heavy metals in water. The combination of properties like chemical affinity to targeted oxy-ions, surface charge and redox potential together with their stability and low-cost enable their use for various cases. Arsenic is known for its toxicity and carcinogenicity to human beings (Kitchen et al., 2009; Chandra et al., 2010; Nata et al., 2011). Drinking water contaminated with arsenic has become a serious concern throughout the world. Long-term exposure to arsenic can cause cancers of the bladder, lungs, skin, kidney, liver, and prostate (Navas-Acien et al., 2005). Many countries have set strict regulations to limit arsenic level in drinking water, which adds to the technical challenges as well as places a financial burden on water facilities to reach the 10 ppb requirements set by the USEPA. Leila et al. (2015) synthesised new multifunctional nanocomposite polymer nanocomposites for cadmium removal from industrial wastewaters. Thus, there is an urgent need for efficient and economic methods to remove heavy metal species and arsenic.

6.6.5 Solid Waste Treatment

With the industrial advancement, solid waste treatment has become a major problem with the increased use of plastics. There are mainly three ways to deal with solid wastes: incineration, landfilling, and recycling. Incineration may cause secondary pollution by generating toxic gases, and land filling is also not practical considering the limited land resources and possible leaching of hazardous substances from landfill sites. Hence, inclusive recycling has become the best method for the treatment of solid wastes. An example is PS/Ni@NiO PNC recycling (Zhu et al., 2011), where magnetic core@shell structured Ni@PNCs, as well as liquid fuel products, were generated. The former is an excellent adsorbent that can be used for environmental remediation due to the high specific surface area of the carbon shell and the high magnetisation of the Ni core. Liquid fuels produced from the PNCs contain more saturated components than those from PS, which is attributed to catalytic hydrogenation by the Ni nanofiller during the pyrolysis process.

A large number of technologies for wastewater treatment and remediation, such as cyanide treatment (Monser et al., 2002), electrochemical precipitation (Kongsricharoem et al., 2004), reverse osmosis (Hafez, 2004; Modrzejewska et al., 1999), ion exchange (Rengaraj et al., 2001, 2003) and adsorption (Zhang et al., 2010; Mohan et al., 2005; Sharma et al., 2008; Hu et al., 2005; Aydln et al., 2009; Zhao et al., 2011; Sawalha et al., 2005; Parsons et al., 2002), have been developed. Among these technologies, chlorination of cyanides can result in highly toxic intermediates and other toxic organochlorines. These compounds together with residual chlorine create additional environmental problems.

Precipitation is considered to be the most applicable and economical way, but this technique may produce a large amount of precipitate sludge that needs additional processes for further treatment. Although reverse osmosis can finally decrease metal ions, its applications are limited due to high costs and low pH value (Modrzejewska et al., 1999). There are very limited studies on the removal of Cr(III) by ion exchange (Pansini et al., 1991), and its operation cost is also higher compared to other techniques. In addition, Cr(VI) reduction by bulk zero-valence Fe (Melitas et al., 2001; Wei et al., 2011), Fe(II) (Deng et al., 1996; Jardine et al., 1999), hydrogen (Powell et al., 1995; Loyaux-Lawniczak et al., 1999), dissolved organic compounds (Reardon, 1995) and sulphur compounds (Gangoly et al., 1975) was also reported to have reached satisfactory removal capacity over an extended treatment time. Currently, nanostructured iron and iron oxides have been proved highly useful for heavy metal removal through either reduction or adsorption (Zhang, 2003; Pondar et al., 2000; Hu et al., 2008). Despite this, using these nanomaterials still has two extreme challenges: One is easy oxidation/dissolution of pure Fe NPs particularly in acidic solutions and the other is the recovery of minute NPs, mainly in a continuous flow system. To prevent oxidation/dissolution of NPs, a chemically inert shell structure is usually introduced around the NP core. The shell material may include silica (Lu et al., 2002; Wang et al., 1999; Zhu et al., 2011), polymer (Zhu et al., 2012; Pellegrino et al., 2004), carbon (Lin et al., 2008) and noble metals (Cho et al., 2005; Lu et al., 2005; Guo et al., 2005). From these substitutes, carbon shell has proved to be feasible due to its low cost and high specific surface area (Oliveira et al., 2002; Vaughan et al., 2005).

In general, inorganic arsenic is more toxic than the organic form, and arsenic trivalent is more mobile and toxic than arsenate pentavalent (Schnoor, 1996). There have been various efforts for the remediation of arsenic from the disposal scene, including ion exchange (Kim et al., 2004), reverse osmosis (Kang et al., 2000), and adsorption. Compared with others methods, adsorption has the advantages of low cost and the low technology need for operation and maintenance. Magnetic composites with core–shell Fe-Fe_2O_3 have been used for arsenic removal from the disposal scene, along with ion exchange and reverse osmosis (Zhu et al.,

2012). NPs also showed great potential in the remediation of arsenic species. The removal rate of arsenic reached 7 mg/g after a 2 h interaction at room temperature when the starting concentration of As(III) was 4 ppm and the pH 7. Tan et al. reported selective removal of Pb(II) and Cr(VI) on multifunctional nanocomposite Fe_3O_4@SiO_2–mPD/SP (Tan et al., 2014). Yu et al. (2016) synthesised Fe_3O_4@SiO_2-EDTA nanocomposites for the removal of Pb(II) and Cu(II) from aqueous solution.

6.6.6 Removal of Dye

Dyes are mainly applicable in many industries such as paper, textile, pulp, pharmaceutical and bleaching. Residual dyes discharged into natural water resources consist of serious organic pollutants, posing health and environmental concerns. The main source of dye pollutants is textile industries due to which it produced in significant amounts. Degradation products of these dyes such as benzidine, naphthalene and other aromatic compounds are cogent to cause carcinogenic or mutagenic effects in various life forms. Therefore, treatment of dye-polluted water is very important for environmental remediation. Because of the excellent performance of multifunctional nanocomposites in the remediation of heavy metal species and arsenic, they could be a potent alternative for organic dye removal from wastewater, as the dyes are mainly organic molecules, which are anionic, nonionic, or cationic. Thereby adjusting the pH, porosity, and combined functionalities of multifunctional nanocomposite adsorbents, organic dyes can be eliminated, and their often reuse is expected in the near future. Tian et al. (2016) prepared multifunctional palygorskite/carbon/Ag nanocomposite for the removal of dye. Magnetic polydopamine (PDA)–LDH (MPL) multifunctional material, which was fabricated by an easy and green approach for the removal of toxic metals and anionic dyes. PDA and LDHs were expected to capture toxic metals and anionic dyes pollutants. In a mono-component system, the removal behaviours showed heterogeneous removal capacities (Li et al., 2016). A stable magnetic nanocomposite of collagen and superparamagnetic iron oxide nanoparticles (SPIONs) was prepared by a simple process utilizing protein wastes from leather industry. Molecular interaction between helical collagen fibres and spherical SPIONs was proved through calorimetric, microscopic and spectroscopic techniques. This nanocomposite exhibited selective oil absorption (Thanikaivelan et al., 2012).

6.7 Summary

This chapter highlights nanocomposites such as polymer-based nanocomposites, magnetic-based nanocomposites and multifunctional nanocomposites, their applications in environment remediation, and the excellent performance of these synergistic materials in the remediation of waste species such as for chromium and arsenic elimination. Specifically, compared to the widely accepted activated carbon, their great potential is shown for application in other types of pollutants, which include organic dyes that are discharged into the environment.

Acknowledgment

We gratefully acknowledge support from the Ministry of Human Resource Development Department of Higher Education, Government of India, under the scheme of Establishment of Centre of Excellence for Training and Research in Frontier Areas of Science and Technology (FAST), for providing the necessary financial support to carry out this study vide letter No, F. No. 5-5/201 4-TS. Vll.

References

Ahmad AL, Ismail S, Bhatia S (2003). Water recycling from palmoil mill effluent (PzME) using membrane technology, *Desalination*; **157**: 87–95.

Ahmad AL, Sumathi S, Hameed BH (2005). Adsorption of residue oil from palm oil mill effluent using powder and flake chitosan: equilibrium and kinetic studies, *Water Res*; **39**: 2483–2494.

Amuda OS, Giwa AA, Bello IA (2007). Removal of heavy metal from industrial wastewater using modified activated coconut shell carbon, *Biochem Eng J*; **36**: 174–181.

Annunciado R, Sydenstricker THD, Amico SC (2005). Experimental investigation of various vegetable fibres as sorbent materials for oil spills, *Mar Pollut Bull*; **50**: 1340–1346.

Atta AM, Al-Lohedan HA, Ezzat AO, Issa ZA, Oumi AB (2016). Synthesis and application of magnetite polyacrylamide amino-amidoxime nano-composites as adsorbents for water pollutants, *J Polym Res*; **23**: 69.

Aydln YA, Aksoy ND (2009). Adsorption of chromium on Chitosan: Optimization, kinetics and thermodynamics, *Chem Eng J*, **151**: 188–194.

Batley GE, Farrar YJ (1978). Irradiation techniques for the release of bound heavy metals in natural waters and blood, *Anal Chim Acta*; **99**: 283–292.

Behrens S (2011). Preparation of functional magnetic nanocomposites and hybrid materials: recent progress and future directions, *Nanoscale*; **3**: 877–892.

Bhaumik MA, Maity VV, Srinivasu, Onyango MS (2011). Enhanced removal of Cr(VI) from aqueous solution using polypyrrole/Fe_3O_4 magnetic nanocomposites, *J Hazard Mater*; **190**: 381–390.

Bódalo-Santoyo A, Gómez-Carrasco JL, Gómez-Gómez E, Máximo-Martín F, Hidalgo-Montesinos AM (2003). Application of reverse osmosis to reduce pollutants present in industrial wastewater, *Desalination*; **155**: 101–108.

Caruso F, Susha AS, Giersig M, Möhwald H. (1999). Magnetic core–shell particles: Preparation of magnetite multilayers on polymer latex microspheres, *Adv Mater*; **11**: 950–953.

Chandra V, Park J, Chun Y, Lee JW, Hwang IC, Kim KS (2010). Water-dispersible magnetite-reduced graphene oxide composites for arsenic removal, *ACS Nano*; **4**: 3979–3986.

Chen M, Jiang W, Wang F, Shen P, Ma P, Gu J, Mao J, Li F. (2013). Synthesis of highly hydrophobic floating magnetic polymer nanocomposites for removal of oils from water surface, *Appl Surf Sci*; **286**: 249–256.

Chen Q, Rondinone AJ, Chakoumakos BC, Zhang ZJ (1999). Synthesis of superparamagnetic $MgFe_2O_4$ nanoparticles by coprecipitation, *J Magn Magn Mater*; **194**: 1–7.

Cho SJ, Idrobo JC, Olamit J, Liu K, Browning ND, Kauzlarich SM (2005). Growth mechanisms and oxidation-resistance of Au-coated Fe nanoparticles, *Chem Mater*; **17**: 3181–3186.

Crayston JA, Devine JN, Walton JC (2000). Conceptual and synthetic strategies for the preparation of organic magnets, *Tetrahedron*; **56**: 7829–7857.

Cui H, Fu M, Yu S, and Wang MK (2011). Reduction and removal of Cr(VI) from aqueous solutions using modified byproducts of beer production, *J Hazard Mater*; **186**: 1625–1631.

De Souza Jr FG, Jessica AM, Cezar HMR, Jose CP (2010). A magnetic composite for cleaning of oil spills on water, *Macromol Mater Eng*; **295**: 942–948.

Dean JG, Bosqui FL, Lanouette KH (1972). Removing heavy metals from waste water, *Environ Sci Technol*; **6**: 518–522.

DeMarco MJ, SenGupta AK, Greenleaf JE (2003). Arsenic removal using a polymeric/inorganic hybrid sorbent, *Water Res*; **37**: 164–176.

Deng B, Stone AT (1996). Surface-catalyzed chromium (VI) reduction: Reactivity comparisons of different organic reductants and different oxide surfaces, *Environ Sci Technol*; **30**: 2484–2494.

Deng YH, Yang WL, Wang CC, Fu SK (2003). A novel approach for preparation of thermo responsive polymer magnetic microspheres with core–shell structure, *Adv Mater*; **15**: 1729–1732.

Ersahin ME, Ozgun H, Dereli RK, Ozturk I, Roest K (2012). A review on dynamic membrane filtration: Materials, applications and future perspectives, *Bioresour Technol*; **122**: 196–206.

Fan L, Song J, Bai W, Wang S, Zeng M, Li X, Zhou Y, Li H, Lu H (2016). Chelating capture and magnetic removal of non-magnetic heavy metal substances from soil, *Sci Rep*; **6**: 21027.

Gangoli N, Markey DC, Thodos G (1975). Removal of heavy metal ions from aqueous solutions with fly ash. 2nd National Conference on Complete Water Reuse, Chicago, USA, Proceedings, pp. 270–275.

Ganigar R, Rytwo G, Gonen Y, Radian A, Mishael YG (2010). Polymer–clay nanocomposites for the removal of trichlorophenol and trinitrophenol from water, *Appl Clay Sci*; **49**: 311–316.

Germain J, Rolandi M, Backer SA, Fréchet JM (2008). Sulfur as a novel nanopatterning material: An ultrathin resist and a chemically addressable template for nanocrystal self-assembly, *Adv Mater*; **20**: 4526–4529.

Gu J, Jiang W, Wang F, Chen M, Mao J, Xie T (2014). Facile removal of oils from water surfaces through highly hydrophobic and magnetic polymer nanocomposites, *Appl Sur Sci*; **301**: 492–499.

Gu H, Rapole SB, Huang Y, Cao D, Luo Z, Wei S, Guo Z (2013). Synergistic interactions between multi-walled carbon nanotubes and toxic hexavalent chromium, *J Mater Chem A*; **1**: 2011–2021.

Gu S, Shiratori T, Konno M (2003). Synthesis of monodisperse, magnetic latex particles with polystyrene core, *Colloid Polym Sci*; **281**: 1076–1081.

Guo X, Fei GT, Su H, De Zhang L (2011). High-performance and reproducible polyaniline nanowire/tubes for removal of Cr(VI) in aqueous solution, *J Phys Chem C*; **115**: 1608–1613.

Guo Z, Kumar CS, Henry LL, Doomes EE, Hormes J, Podlaha EJ (2005). Displacement synthesis of Cu shells surrounding Co nanoparticles, *J Electrochem Soc*; **152**: D1–D5.

Gupta S, Ranjit R, Mitra C, Raychaudhuri P, Pinto R (2001). Enhanced room-temperature magnetoresistance in $La_{0.7}$ $Sr_{0.3}$ MnO_3-glass composites, *Appl Phys Lett*; **78**: 362–364.

Hafez A, El-Mariharawy S (2004). Design and performance of the two-stage/two-pass RO membrane system for chromium removal from tannery wastewater, *Desalination*; **165**: 141–151.

Hsu LC, Wang SL, Lin YC, Wang MK, Chiang PN, Liu JC, Kuan WH, Chen CC, Tzou YM (2010). Cr(VI) removal on fungal biomass of Neurospora crassa—the importance of dissolved organic carbons derived from the biomass to Cr(VI) reduction, *Environ Sci Technol*; **44**: 6202–6208.

Hu J, Chen G, Lo IMC (2005). Removal and recovery of Cr(VI) from wastewater by maghemite nanoparticles, *Water Res*; **39**: 4528–4536.

Hu J, Chen GH, Lo IMC (2006). Selective removal of heavy metals from industrial wastewater using maghemite nanoparticle: Performance and mechanisms, *J Environ Eng*; **132**: 709.

Hu JS, Zhong LS, Song WG, Wan LJ (2008). Synthesis of hierarchically structured metal oxides and their application in heavy metal ion removal, *Adv Mater*; **20**: 2977–2982.

Huang YH, Chen X, Wang ZM, Liao CS, Yan CH, Zhao HW, Shen BG (2002). Enhanced magnetoresistance in granular $La_{2/3}Ca_{1/3}MnO_3$/polymer composites, *J Appl Phys*; **91**: 7733–7735.

Hughes M, Chen GZ, Shaffer MSP, Fray DJ, Windle AH (2002). Electrochemical capacitance of a nanoporous composite of carbon nanotoubes and polypyrrole, *Chem Mater*; **14**: 1610–1613.

Husseien M, Amer AA, El-maghraby A (2008). Experimental investigation of thermal modification influence on sorption qualities of barley straw, *J Appl Sci Res*; **4**: 652–657.

Jardine PM, Fendorf SE, Mayes MA, Larsen IL, Brooks SC, Bailey WB (1999). Fate and transport of hexavalent chromium in undisturbed soil, *Environ Sci Technol*; **33**: 2939–2944.

Ji F, Li C, Dong X, Li Y, Wang D (2009). Separation of oil from oily wastewater by sorption and coalescence technique using ethanol grafted polyacrylonitrile, *J Hazard Mater*; **164**: 1346–1351.

Jia K, Pan B, Lv L (2009). Impregnating titanium phosphate nanoparticles onto a porous cation exchanger for enhanced lead removal from waters, *J Colloid Interface Sci*; **331**: 453–457.

Kang M, Kawasaki M, Tamada S, Kamei T, Magara Y (2000). Effect of pH on the removal of arsenic and antimony using reverse osmosis membranes, *Desalination*; **131**: 293.

Kaya IGB, Duranoglu D, Beker U, Senkal BF (2011). Noval calyx[4] arene network resin for Cr(IV) ions remediation: A response surface approach, *Clean-Soil Air Water*; **39**: 980.

Kim J, Benjamin MM (2004). Modeling a novel ion exchange process for arsenic and nitrate removal, *Water Res*; **38**: 2053.

Kim CQ, Zhou B, Deng EC, Thornton, Xu H (2001). Chromium(VI) reduction by hydrogen sulfide in aqueous media: Stoichiometry and kinetics, *Environ Sci Technol*; **35**: 2219–2225.

Kitchin KT, Conolly R (2009). Arsenic-induced carcinogenesis-oxidative stress as a possible mode of action and future research needs for more biologically based risk assessment, *Chem Res Toxicol*; **23**: 327–335.

Kongsricharoern N, Polprasert C (1995). Electrochemical precipitation of chromium from an electroplating wastewater, *Water Sci Technol*; **31**: 109–117.

Kongsricharoern N, Polprasert C (1996). Chromium removal by a bipolar electro-chemical precipitation process, *Water Sci Technol*; **34**: 109–116.

Kumar PAS, Chakraborty, Ray M (2008). Removal and recovery of chromium from wastewater using short chain polyaniline synthesized on jute fiber, *Chem Eng J*; **141**: 130–140.

Kumar R, Gautam S, Hwang IC, Lee JR, Chae KH, Thakur N (2009). Preparation and characterization of α-Fe_2O_3 polyhedral nano-crystals via annealing technique, *Mater Lett*; **63**: 1047–1050.

Kumar P, Singh RK, Rawat N, Barman PR, Katyal SC, Jang H, Lee HL, Kumar R (2013). A novel method for controlled synthesis of nanosize hematite (α-Fe_2O_3) thin film on liquid-vapor interface, *J Nanopart Res*; **15**: 1532(1)–1532(13).

Lai KCK, Lo IMC (2008). Removal of chromium (VI) by acid-washed zero-valent iron under various groundwater geochemistry conditions, *Environ Sci Technol*; **42**: 1238–1244.

Lakouraj MM, Hasanzadeh F, Zare EN (2014). Nanogel and superparamagnetic nanocomposite of thiacalix[4]arene functionalized chitosan: Synthesis, characterization and heavy metal sorption, *Iran Polym J*; **23**: 933–945.

Lamparth I, Szabó DV, Vollath D (2002). Ceramic nanoparticles coated with oligomers based on acrylic derivatives, *Macromol Symp*; **181**: 107–112.

Li J, Fan Q, Wu Y, Wang X, Chen C, Tang Z, Wanga X (2016). Magnetic polydopamine decorated with Mg-Al LDH nanoflakes as a novel bio-based adsorbent for simultaneous removal of potentially toxic metals and anionic dyes, *J Mater Chem A*; **4**: 1737–1746.

Lin CAJ, Sperling RA, Li JK, Yang TY, Li PY, Zanella M, Chang WH, Parak WJ (2008). Design of amphiphilic polymer for nanoparticle coating and functionalization, *Small*; **4**: 334–341.

Loyaux-Lawniczak S, Refait P, Ehrhardt JJ, Lecomte P, Génin JMR (1999). Trapping of Cr by formation of ferrihydrite during the reduction of chromate ions by Fe(II)–Fe(III) hydroxysalt green rusts, *Environ Sci Technol*; **34**: 438–443.

Lu Z, Prouty MD, Guo Z, Golub VO, Kumar CSSR, Lvov YM (2005). Magnetic switch of permeability for polyelectrolyte microcapsules embedded with Co@Au nanoparticles, *Langmuir*; **21**: 2042–2050.

Lu Y, Yin Y, Mayers BT, Xia Y (2002). Modifying the surface properties of superparamagnetic iron oxide nanoparticles through a sol-gel approach, *Nano Lett*; **2**: 183–186.

Maleki H, Simchi A, Imami M, Costa BFO (2012). Size-controlled synthesis of superpara magnetic iron oxide nanoparticles and their surface coating by gold for biomedical applications, *J Magn Magn Mater*; **324**: 3997–4005.

McCarthy JR, Weissleder R (2008). Multifunctional magnetic nanoparticles for targeted imaging and therapy, *Adv Drug Deliv Rev*; **60**: 1241–1251.

Melitas N, Chuffe-Moscoso O, Farrell J (2001). Kinetics of soluble chromium removal from contaminated water by zerovalent iron media: corrosion inhibition and passive oxide effects, *Environ Sci Technol*; **35**: 3948–3953.

Mezger TR (2006). *The Rheology Handbook: For Users of Rotational and Oscillatory Rheometers*, Vincentz Network, Hannove; **155**: 1397.

Modrzejewsk Z, Kaminski W (1999). Separation of Cr(VI) on chitosan membranes, *Ind Eng Chem Res*; **38**: 4946–4950.

Mohan D, Singh KP, Singh VK (2005). Removal of hexavalent chromium from aqueous solution using low-cost activated carbons derived from agricultural waste materials and activated carbon fabric cloth, *Ind Eng Chem Res*; **44**: 1027–1042.

Monser L, Adhoum N (2002). Modified activated carbon for the removal of copper, zinc, chromium and cyanide from wastewater, *Sep Purif Technol*; **26**: 137–146.

Murakami H, Kobayashi M, Takeuchi H, Kawashima Y (1999). Preparation of poly(DL-lactide-co-glycolide) nanoparticles by modified spontaneous emulsification solvent diffusion method, *Int J Pharm*; **187**: 143.

Nata IF, Sureshkumar M, Lee CK (2011). One-pot preparation of amine-rich magnetite/bacterial cellulose nanocomposite and its application for arsenate removal, *RSC Adv*; **1**: 625–631.

Navas-Acien A, Sharrett AR, Silbergeld EK, Schwartz BS, Nachman KE, Burke TA, Guallar E (2005). Arsenic exposure and cardiovascular disease: a systematic review of the epidemiologic evidence, *Am J Epidemiol*; **162**: 1037–1049.

Neveu S, Bee A, Robineau M, Talbot D (2002). Size-selective chemical synthesis of tartrate stabilized cobalt ferrite ionic magnetic fluid, *J Colloid Interface Sci*; **255**: 293–298.

Ogoshi T, Hasegawa Y, Aoki T, Ishimori Y, Inagi S, Yamagishi T (2011). Reduction of emeraldine base form of polyaniline by pillar[5] arene based on formation of poly (pseudorotaxane) structure, *Macromolecules*; **44**: 7639–7644.

Oliveira LCA, Rios RVRA, Fabris J, Garg V, Sapag K, Lago RM (2002). Activated carbon/iron oxide magnetic composites for the adsorption of contaminants in water, *Carbon*; **40**: 2177–2183.

Pan B, Wu J, Pan B (2009). Development of polymer-based nanosized hydrated ferric oxides (HFOs) for enhanced phosphate removal from waste effluents, *Water Res*; **43**: 4421–4429.

Pan B, Zhang QR, Zhang WM (2007). Highly effective removal of heavy metals by polymer-based zirconium phosphate: A case study of lead ion, *J Colloid Int Sci*; **310**: 99–105.

Pansini M, Colella C, Gennaro M (1991). Chromium removal from water by ion exchange using zeolite, *Desalination*; **83**: 145–157.

Park J, An K, Hwang Y, Park JG, Noh HJ, Kim JY, Park JH, Hwang NM, Hyeon T (2004). Ultra-large-scale syntheses of monodisperse nanocrystals, *Nat Mater*; **3**: 891–895.

Park SJ, Kim S, Lee S, Khim, Char K, Hyeon T (2000). Synthesis and magnetic studies of uniform iron nanorods and nanospheres, *J Am Chem Soc*; **122**: 8581–8582.

Park D, Yun YS, Lee HW, Park JM (2008). Advanced kinetic model of the Cr(VI) removal by biomaterials at various pHs and temperatures, *Bioresour Technol*; **99**: 1141–1147.

Park D, Yun YS, Park JM (2004). Reduction of hexavalent chromium with the brown seaweed ecklonia biomass, *Environ Sci Technol*; **38**: 4860–4864.

Parsons JG, Hejazi M, Tiemann KJ, Henning J, Gardea-Torresdey JL (2002). An XAS study of the binding of copper(II), zinc(II), chromium(III) and chromium(VI) to hopbiomass, *Microchem J*; **71**: 211–219.

Pellegrino T, Manna L, Kudera S, Liedl T, Koktysh D, Rogach AL, Keller S, Rädler J, Natile G, Parak WJ (2004). Hydrophobic nanocrystals coated with an amphiphilic polymer shell: a general route to water soluble nanocrystals, *Nano Lett*; **4**: 703–707.

Ponder SM, Darab JG, Mallouk TE (2000). Remediation of Cr (VI) and Pb(II) aqueous solutions using supported, nanoscale zero-valent iron, *Environ Sci Technol*; **34**: 2564–2569.

Powell RM, Puls RW, Hightower SK, Sabatini DA (1995). Coupled iron corrosion and chromate reduction: Mechanisms for subsurface remediation, *Environ Sci Technol*; **29**: 1913–1922.

Puntes VF, Krishan KM, Alivisatos AP (2001). Colloidal nanocrystals shape and size control: The case of cobalt, *Science*; **291**: 2115–2117.

Rajca A, Wongsriratanakul J, Rajca S (2001). Magnetic ordering in an organic polymer, *Science*; **294**: 1503–1505.

Reardon EJ (1995). Anaerobic corrosion of granular iron measurement and interpretation of hydrogen evolution rates, *Environ Sci Technol*; **29**: 2936–2945.

Ren Y, Yan N, Wen Q (2011). Graphene/δ-MnO_2 composite as adsorbent for the removal of nickel ions from wastewater, *Chem Eng J*; **175**: 1–7.

Rengaraj S, Joo CK, Kim Y, Yi J (2003). Kinetics of removal of chromium from water and electronic process wastewater by ion exchange resins 1200H, 1500H and IRN97H, *J Hazard Mater*; **102**: 257–275.

Rengaraj S, Yeon KH, Moon SH (2001). Removal of chromium from water and wastewaters by ion exchange resins, *J Hazard Mater*; **87**: 273–287.

Rykowska I, Wasiak W, Byra J (2008). Extraction of copper ions using silica gel with chemically modified surface, *Chem Pap*; **62**: 255–259.

Samani MR, Borghei SM, Olad A, haichi MJ (2010). Removal of chromium from aqueous solution using polyaniline-Poly ethylene glycol composite, *J Hazard Mater*; **184**: 248–254.

Sarkar S, Gupta A, Biswas RK (2005). Well-head arsenic removal units in remote villages of Indian subcontinent: Field results and performance evaluation, *Water Res*; **39**: 2196–2206.

Sauzedde F, Elaissari A, Pichot C (1999). Hydrophilic magnetic polymer latexes. 2. Encapsulation of adsorbed iron oxide nanoparticles, *Colloid Polym Sci*; **277**: 1041–1050.

Sawalha MF, Gardea-Torresdey JL, Parsons JG, Saupe G, Peralta-Videa JR (2005). Determination of adsorption and speciation of chromium species by saltbush (atriplex canescens) biomass using a combination of XAS and ECP-OES, *Microchem J*; **81**: 122–132.

Schnoor JL (1996). *Environmental Modeling: Fate and Transport of Pollutants in Water, Air, and Soil.* New York: John Wiley & Sons.

Sharma PK, Dutta RK, Pandey AC (2010). Alteration of magnetic and optical properties of ultrafine dilute magnetic semiconductor ZnO: Co^{2+} nanoparticles, *J Colloid Interface Sci*; **345**: 149–153.

Sharma YC, Singh B. Agrawal A, Weng CH (2008). Removal of chromium by riverbed sand from water and wastewater effect of important parameters, *J Hazard Mater*; **151**: 789–793.

Shevchenko EV, Talapin DV, Rogach AL, Kornowski A, Haase M, Weller H (2002). Colloidal synthesis and self-assembly of $CoPt_3$ nanocrystals, *J Am Chem Soc*; **124**: 11480–11485.

Srikanth H, Hajndl R, Chirinos C, Sanders J, Sampath A, Sudarshan TS (2001). Magnetic studies of polymer-coated Fe nanoparticles synthesized by microwave plasma polymerization, *Appl Phys Lett*; **79**: 3503–3505.

Srivastava V, Weng CH, Singh VK, Sharma YC (2011). Adsorption of nickel ions from aqueous solutions by nano alumina: Kinetic, mass transfer, and equilibrium studies, *J Chem Eng Data*; **56**: 1414–1422.

Stroes-Gascoyne S, Kramer JR, Snodgrass WJ (1987). Preparation, characterization and aging of δ-MnO_2, for use in trace metal speciation studies, *Appl Geochem*; **2**: 217–226.

Su Q, Pan B, Pan B (2009). Fabrication of polymer-supported nanosized hydrous manganese dioxide (HMO) for enhanced lead removal from waters, *Sci Total Environ*; **407**: 5471–5477.

Su Q, Pan B, Wan S (2010). Use of hydrous manganese dioxide as a potential sorbent for selective removal of lead, cadmium, and zinc ions from water, *J Colloid Interface Sci*; **349**: 607–612.

Sun S, Murray CB, Weller D, Folks L, Moser A (2000). Monodisperse Fe Pt nanoparticles and ferromagnetic Fe Pt nanocrystal superlattices, *Science*; **287**: 1989–1992.

Talapin DV, Lee JS, Kovalenko MV, Shevchenko EV (2010). Prospects of colloidal nanocrystals for electronic and optoelectronic applications, *Chem Rev*; **110**: 389–458.

Tan L, Xu J, Xue X, Lou Z, Zhu J, Baig SA, Xu X (2014). Multifunctional nanocomposite Fe_3O_4@SiO_2–mPD/SP for selective removal of Pb(II) and Cr(VI) from aqueous solutions, *RSC Adv*; **4**: 45920–45929.

Tempest P, Bonini M, Ridi F, Bagiloni P (2014). Magnetic polystyrene nanocomposite for separation of oil and water, *J Mater Chem A*; **2**: 15268–15272.

Thanikaivelan P, Narayanan NT, Pradhan BK, Ajayan PM (2012). Collagen based magnetic nanocomposites for oil removal applications, *Sci Rep*; **2**: 230.

Tian G, Wang W, Zong L, Kang Y, Wang A (2016). From spent dye-loaded palygorskite to a multifunctional palygorskite/carbon/Ag nanocomposite, *RSC Adv*; **6**: 41696–41706.

Torrance JB, Bagus PS, Johhannsen I, Nazzal AI, Parkin SSP, Batail P (1998). Ferromagnetic interactions in organic-solids—an overview of theory and experiment, *J Appl Phys*; **63**: 2962–2965.

Tripathy SS, Kanungo SB (2005). Adsorption of Co^{2+}, Ni^{2+}, Cu^{2+} and Zn^{2+} from 0.5 M NaCl and major ion sea water on a mixture of δ-MnO_2 and amorphous FeOOH, *J Colloid Interface Sci*; **284**: 30–38.

Vaughan RL, Reed BE (2005). Modeling As(V) removal by a iron oxide impregnated activated carbon using the surface complexation approach, *Water Res*; **39**: 1005–1014.

Vollath D, Szabó DV (1999). Coated nanoparticles: A new way to improved nanocomposites, *J Nanopart Res*; **1**: 235–242.

Vollath D, Szabó DV, Fuchs J (1999). Synthesis and properties of ceramic-polymer composites, *Nanostruct Mater*; **12**: 433–438.

Walsh FC, Reade GW (1994). Electrochemical techniques for the treatment of dilute metal-ion solutions, *Stud Environ Sci*; **59**: 3–44.

Wan S, Zhao X, Lv L (2010). Selective adsorption of Cd(II) and Zn(II) ions by nano-hydrous manganese dioxide (HMO) encapsulated cation exchanger, *Ind Eng Chem Res*; **49**: 7574–7579.

Wang G, Harrison A (1999). Preparation of iron particles coated with silica, *J Colloid Interface Sci*; **217**: 203–207.

Wang G, Sun Q, Zhang Y, Fan J, Ma L (2010). Sorption and regeneration of magnetic exfoliated graphite as a new sorbent for oil pollution, *Desalination*; **263**: 183–188.

Wang C, Xu C, Zeng H, Sun S (2009). Recent progress in syntheses and applications of dumbbell-like nanoparticles, *Adv Mater*; **21**: 3045–3052.

Wang LK, Vaccari DA, Li Y, Shammas NK (2005). Chemical precipitation physicochemical treatment processes, *Handbook Environ Eng*; **6**: 3141–3197.

Wei S, Wang Q, Zhu J, Sun L, Lin H, Guo Z (2011). Multifunctional composite core-shell nanoparticles, *Nanoscale*; **3**: 4474–4502.

Wise SS, Shaffiey F, LaCerte C, Goertz CEC, Dunn JL, Gulland FMD, Aboueissa AM, Zheng T, Wise JP (2009). Particulate and soluble hexavalent chromium are cytotoxic and genotoxic to Steller sea lion lung cells, *Aquat Toxicol*; **91**: 329–335.

Wittbrodt PR, CD Palmer (1995). Reduction of Cr(VI) in the presence of excess soil fulvic acid, *Environ Sci Technol*; **29**: 255–263.

Xing Y, Chen X, Wang D (2007). Electrically regenerated ion exchange for removal and recovery of Cr (VI) from wastewater, *Environ Sci Technol*; **41**: 1439–1443.

Xu Y, Zhao D (2007). Reductive immobilization of chromate in water and soil using stabilized iron nanoparticles, *Water Res*; **4**: 2101–2108.

Yang C (2007). Electrochemical coagulation for oily water demulsification, *Sep Purif Technol*; **54**: 388–395.

Yu L, Ruiqi F, Yue S, Xiaoxin Z, Baig SA, Xinhua X (2016). Multifunctional nanocomposites Fe_3O_4@SiO_2-EDTA for Pb(II) and Cu(II) removal from aqueous solutions, *Appl Sur Sci*; **369**: 267–276.

Zaidi NA, Giblin SR, Terry I, Monkman AP (2004). Room temperature magnetic order in an organic magnet derived from polyaniline, *Polymer*; **45**: 5683–5689.

Zaitsev VS, Filimonov DS, Presnyakov IA, Gambino RJ, Chu B (1999). Physical and chemical properties of magnetite and magnetite-polymer nanoparticles and their colloidal dispersions, *J Colloid Interface Sci*; **212**: 49–57.

Zamboulis D, Peleka EN, Lazaridis NK, Matis KA (2011). Metal ion separation and recovery from environmental sources using various flotation and sorption techniques, *J Chem Technol Biotechnol*; **86**: 335–344.

Zhang W (2003). Nanoscale iron particles for environmental remediation: an overview, *J Nanopart Res*; **5**: 323–332.

Zhang QR, Du W, Pan BC (2008). A comparative study on Pb^{2+}, Zn^{2+} and Cd^{2+} sorption onto zirconium phosphate supported by a cation exchanger, *J Hazard Mater*; **152**: 469–475.

Zhang P, Hahn HH, Hoffmann E (2003). Different behavior of iron (III) and aluminum (III) salts to coagulate silica particle suspension, *Acta Hydrochim Hydrobiol*; **31**: 145–151.

Zhang Q, Pan B, Pan B (2008). Selective sorption of lead, cadmium and zinc ions by a polymeric cation exchanger containing nano-$Zr(HPO_3S)_2$, *Environ Sci Technol*; **42**: 4140–4145.

Zhang J, Seeger S (2011). Polyester materials with super wetting silicone nano filaments for oil/water separation and selective oil absorption, *Adv Funct Mater*; **21**: 4699–4704.

Zhang D, Wei S, Kaila C, Su X, Wu J, Karki AB, Young DP, Guo Z (2010). Carbon-stabilized iron nanoparticles for environmental remediation, *Nanoscale*; **2**: 917–919.

Zhang D, Wei S, Kaila C, Su X, Wu J, Karki A, Young DP, Guo Z (2013). Carbon-stabilized iron nanoparticles for environmental remediation, *Nanoscale*; **2**: 917–919.

Zhao Y, Peralta-Videa JR, Lopez-Moreno ML, Ren M, Saupe G, Gardea-Torresdey JL (2011). Kinetin increases chromium absorption, modulates its distribution, and changes the activity of catalase and ascorbate peroxidase in Mexican Palo Verde, *Environ Sci Technol*; **45**: 1082–1087.

Zhao YG, Shen HU, Pan SD, Hu MQ Xia QH (2010). Template-induced covalent assembly of hybrid particles for the facile fabrication of magnetic Fe_3O_4-polymer hybrid hollow microspheres, *J Mater Sci*; **45**: 5291.

Zhu J, Chen M, Qu H, Zhang X, Wei H, Luo Z, Colorado HA, Wei S, Guo Z (2012). Electrochromic polyaniline/graphite oxide nanocomposites with endured electrochemical energy storage, *Polymer*; **53**: 5953–5964.

Zhu J, Sadu R, Wei S, Chen D, Haldolaarachchige N, Luo Z, Gomes J, Young DP, Guo Z (2012). Magnetic graphene nanoplatelet composites toward arsenic removal, *ECS J Solid State Sci Technol*; **1**: M1–M5.

Zhu Q, Tao F, Pan Q (2010). Fast and selective removal of oils from water surface via highly hydrophobic core-shell Fe_2O_3@C nanoparticles under magnetic field, *ACS Appl Mater Interfaces*; **2**: 3141–3146.

Zhu J, Wei S, Gu H, Rapole SB, Wang Q, Luo Z, Haldolaarachchige N, Young DP, Guo Z (2012). One-pot synthesis of magnetic graphene nanocomposites decorated with core@double-shell nanoparticles for fast chromium removal, *Environ Sci Technol*; **46**: 977–985.

Zhu J, Wei S, Haldolaarachchige N, Young DP, Guo Z (2011). Electromagnetic field shielding polyurethane nanocomposites reinforced with core-shell Fe silica nanoparticles, *J Phys Chem C*; **115**: 15304–15310.

Zhu J, Wei S, Lee IY, Park S, Willis J, Haldolaarachchige N, Young DP, Luo Z, Guo Z (2012). Silica stabilized iron particles toward anti-corrosion magnetic polyurethane nanocomposites, *RSC Adv*; **2**: 1136–1143.

Zhu J, Wei S, Li Y, Pallavkar S, Lin H, Haldolaarachchige N, Luo Z, Young DP, Guo Z (2011). Comprehensive and sustainable recycling of polymer nanocomposites, *J Mater Chem*; **21**: 16239–16246.

Zoleikani L, Issazadeh H, ZareNezhad B (2015). Preparation of new conductive polymer nanocomposites for cadmium removal from industrial wastewaters, *J Chem Technol Met*; **50**: 71–80.

Chapter 7

Role of New-Generation Technology in Remediating Environmental Pollution

Sapna[a] **and Dinesh Kumar**[b]

[a]*Department of Chemistry, Banasthali University, Rajasthan 304022, India*
[b]*School of Chemical Sciences, Central University of Gujarat, Gandhinagar, Gujarat 382030, India*

dsbchoudhary2002@gmail.com

Heavy metal pollution in the environment is a major problem. Therefore, contamination with these heavy metal ions has increased public concerns because of their toxicities in relatively low concentration. Hence, it is vital to develop nanomaterials for the treatment of these contaminants in the environment. In this chapter, the use of carbon nanotubes (CNTs), photocatalyst, TiO_2 nanoparticles, fullerene, dendrimers, zeolite, iron nanoparticles is considered with special focus on the removal of heavy metals as well as air pollutants.

Nanocomposites for Pollution Control
Edited by Chaudhery Mustansar Hussain and Ajay Kumar Mishra

ISBN 978-981-4774-45-1 (Hardcover), 978-1-315-14368-2 (eBook)
www.panstanford.com

7.1 Introduction

In the past few decades, environmental pollution has emerged as a major concern not only in developing countries but also in developed countries. This occurs as air, water, soil, noise, and light. The term "pollution" has several definitions, one being "the presence of a substance in the environment whose chemical composition or quantity adversely affects the natural processes as well as human health." The technological advancement has also given rise to new pollutants which imbalance the whole biodiversity. Hence the emergency to develop new technologies that would reduce the pollution to risk-free levels quickly and in an easy manner. Although, many technologies have been developed, nanomaterial-based techniques have been explored to provide a new solution to remediate the environmental pollution and improve the performance of conventional technologies. Nowadays, nanomaterials have drawn much attention because of their characteristic property as its large surface-to-volume ratio. Various kinds of popular nanomaterials are currently used for controlling the environmental pollution, such as carbon nanotubes (CNTs), iron nanoparticles (Fe NPs), dendrimers, TiO_2, fullerenes, nanoscale zeolite, and metal oxides.

7.2 CNTs for Remediation

CNTs were first discovered by Iijima et al. (1991). These are synthesised by various methods, such as arc discharge, laser ablation, and chemical vapour deposition (Bethune et al., 1993). CNTs fundamentally include both single-walled carbon nanotubes and multi-walled carbon nanotubes (Merkoci et al., 2006). They are considered an allotrope of carbon after graphite, fullerene and diamond. CNTs have attracted immense attention due to its distinctive properties in the field of remediation of environmental pollution owing to their unique mechanical and physical, electrochemical properties (Avouris et al., 2002; Nuzzo et al., 2006). They offer superior environmental technologies as sensors and sorbent materials to sense and treat existing contaminants such as gas pollutants, harmful microbes, waterborne pathogens, organic dyes, toxic heavy metals, and pesticides (Goering et al.,

2008; Hyung et al., 2008; Li et al., 2002, 2003). These are used to prevent new pollution caused by many other anthropogenic activities (Chen et al., 2006). CNT-based sensors have been used to detect various environmental contaminants not only in the soil but also in water and air because of their high adsorption efficiency and better regeneration potential. CNTs have been utilized as a strengthening material in green nanocomposites. Besides their remediation property, they find wide application in electrochemical devices of energy green nanocomposites. CNTs show excellent electronic, conductive, and mechanical properties and have been utilized in the manufacture of biosensors and hydrogen storage systems (Eswaramoorthy et al., 1999). Because of their amazing sorption of organic and inorganic contaminations, the present research is now gaining more attention to incorporate CNTs into several devices so that they can work as adsorbents and recognition matrices. Furthermore, these nanomaterials' surface functionalization with negatively charged functional groups can appreciably improve the sorption capacity. Further, CNTs have likewise found a few different applications, including environmental remediation. They are among strong sorbents reported by many researchers, including Fagan et al. (2004), who reported the use of CNTs for 1,2-dichlorobenzene adsorption. Lu et al. (2005) demonstrated potential removal of trihalomethanes from water by CNTs. Long et al. (2001) studied the sorption of highly toxic dioxin and found that CNTs have more sorption capacity than activated carbon, confirmed by the temperature-programmed desorption technique (TPD). Yang et al. (2006) observed the adsorption of toxic polycyclic aromatic compounds such as pyrene, phenanthrene, and naphthalene onto CNTs. Wang et al. (2008) found capability of adsorbing dissolved organic matter such as peptone and α-phenylalanine. Zhou et al. (2006) investigated the potential application to remove dichlorodiphenyltrichloroethane (DDT) from aqueous. Shi et al. (2010) showed the effect of humic acid and surfactant on the surface of CNTs for the adsorption of atrazine. Hilding et al. (2001) established a good adsorption capacity of butane onto the surface of carbon nanotubes, and other polar and non-polar organic chemicals. Yu et al. (2012) oxidized MWCNTs by different concentrations of NaOCl and found that 3% NaOCl-functionalized MWCNTs had maximum adsorption capacity for toluene, ethylbenzene and xylene isomers in an aqueous solution. Li et al. (2003) reported that MWCNTs absorb toxic heavy metal ions, such as, Pb(II), Cu(II) and Cd(II) and the affinity order was Pb(II) > Cu(II) > Cd(II). Kandah et al. (2007) synthesized MWCNTs by chemical vapour decomposition and further oxidized with concentrated nitric acid. They concluded that oxidized CNT possessed higher adsorption capacity 49.261 mg/g for nickel ion. Peng et al. (2005) synthesized novel adsorbent by grafting CeO_2 onto the surface of the CNTs. Adsorbent regenerated by using NaOH solution, and the efficiency was found over 94%. Additionally, CNTs have been used for the sorption of copper(II), lead(II), trace gold(III), mercury(II), nickel(II) (Bystrzejewski et al., 2011; Tian et al., 2010; Xu et al., 2008; Afzali et al., 2010; Luo et al., 2010; Lu et al., 2006). CNTs also have a huge potential to remove biothreat agents, which makes them very useful in tackling environment pollution. CNTs showed high bacterial adsorption capacity (Tan et al., 2008; Brady et al., 2008; Srivatsava et al., 2004; Deng, 2008). Nellore et al. (2015) synthesized the PGL antimicrobial peptide and glutathione conjugated CNTs bridged three-dimensional porous graphene oxide membrane, in which the highly porous membrane nature of CNTs help in the capturing the *E. coli* O157:H7 bacteria from the wastewater. Upadhyayula et al. (2008) observed that the hydrophobic nature of SWCNTs show efficient adsorption capacity against the waterborne pathogens such as *S. aureus* SH 1000 and *E. coli* pKV-11. Olivi et al. 2013 diagnosed the antimicrobial characteristic of CNTs against many pathogenic microorganisms which are responsible for serious infections in the human body such as *Staphylococci*, *Streptococci*, *Salmonella*, and *Candida*. Recently, nanomaterials have found prominence in research to develop more efficient nanostructured reactive membranes and filter for wastewater treatment and desalination. CNT-based filters have been synthesized consisting of hollow cylinders united with radially aligned CNT walls. These filters have the ability to remove microbes at nanometer scale size such as Poliovirus (25 nm) and bacteria (*E. coli* and *S. aureus*) from contaminated water. Among commercially used membranes like ceramic- and polymer-based separation membranes, carbon nanotube-fabricated membranes have outstanding performance due to their thermal and mechanical stability, their high surface area, and ease of and cost-effective fabrication of nanotubes membranes. These filters can be cleaned through autoclaving or ultrasonication and reused for further application (Kang et al., 2009; Srivastava et al., 2004).

7.3 Role in Controlling Air Pollution

Due to increasing industrialization, various toxic gases spread in the surrounding areas, which makes the environment polluted and adversely affects the life on earth as well as human health. Therefore, researchers have to make an effort to solve this problem. Bai et al. (2007) developed gas sensors using conducting polymer to control air pollution, which are highly sensitive, easy to fabricate and work at room temperature, but they suffer from low selectivity and long-term stability problems. Penza et al. (2009) reported a highly selective method for sensing NO_2 gas by decorating CNTs with different metals such as Pt, Ru, and Ag. Further, Hoa et al. (2009) functionalized CNTs with metal oxides, such as SnO_2, WO_3, and TiO_2. Among these, SnO_2, the multi-walled hybrid material, showed higher selectivity towards higher sensitivity than a pure SnO_2 thin film gas sensor to ultra-low concentrations of NO_2 at ppb level.

7.3.1 Photocatalysis

Photocatalysis is a highly developed oxidation process that is active in the area of detoxification of drinking water and decomposes the sewage as well as industrial effluent and is particularly used for the oxidative elimination of micropollutants and microbial pathogens (Friedmann et al., 2010). As reported, mostly organic pollutants have been destroyed by heterogeneous photocatalysts. Because of its high accessibility, less harmful nature, cost efficiency, and well-known material properties, TiO_2 is extensively used as a photocatalyst (Chong et al., 2010; Gaya et al., 2008; Qu et al., 2013). When TiO_2 is exposed to an ultraviolet light with a suitable wavelength in the range of 200–400 nm, electrons get photoexcited and are transferred into the conduction band. Therefore, at the time of photoexcitation, electron–hole pairs are formed, leading to a complex chain of oxidative-reductive reactions. Titanium dioxide is a semiconductor, known as titania or TiO_2, which shows excellent catalytic properties. Basically, it is found in nature in three phases: anatase (tetragonal), rutile (tetragonal), and brookite (orthorhombic). Among these phases, the anatase phase is the most photoreactive (Srinivasan et al., 2003). Because of its smaller particle sizes below 50 nm, high concentrations of surface hydroxyl (OH) groups, and a high surface area, it shows high photocatalytic activity. However, the band gap of anatase TiO_2 is 3.2 eV; so it is capable of absorbing light only below 400 nm and electron–hole pairs generated on the photocatalyst surface (Hoffmann et al., 1995; Al et al., 2001). Dyes, impurities and dopants can sensitize TiO_2 to react to a much larger visible light region and only be activated under UV light irradiation with a wavelength of 387 nm. This high-energy band gap rules out the use of solar energy as the photoactivity source. The UV source requires large quantities of electric energy, which would result in high costs in practical applications. Anatase titanium dioxide has been proved to be the most effective and suitable photocatalyst in the process of degradation of toxic environment pollutants. The photocatalytic activity of TiO_2 depends on its particle size, lattice impurities, crystallinity, crystal phase, density of surface hydroxyl groups, and surface area. It is widely used in environmental applications as a photocatalyst for oxidation reactions of an organic compound as well as in the fields of photovoltaics, photoelectrochromic, ceramics, and sensors (Kitano et al., 2007; Klosek et al., 2001; Guo et al., 2011, 2012 2013; Bae et al., 2007; Aran et al., 2008). Since the discovery of the phenomenon of the photocatalytic splitting of water on a TiO_2 electrode under UV light, huge efforts devoted to titania research have led to the most promising semiconductor photocatalysts, and TiO_2-based materials are therefore expected to play the main role in tackling the serious issue of environmental pollution.

7.4 Methods for the Synthesis of TiO_2 Nanoparticles

Synthesis of titania nanoparticles has been done by various methods, such as sol-gel, hydrothermal, solvothermal, and chemical vapour deposition (Chen et al., 2007; Kiri et al., 2010; Zaki et al.,

2010). These techniques decide titania's shape, size, and many other photochemical properties. TiO_2 has gained much attention for use as an environmental remediation catalyst to complete the disposal of organic and inorganic pollutants because of its excellent characteristics. It possesses high thermal stability, high photocatalytic activity, high resistance to chemicals and photocorrosion, non-toxicity, and dielectric properties as well as is inexpensive (Ahn et al., 2007; Neppolian et al., 2005; Liu et al., 2003; Bessekhouad et al., 2003; Tian et al., 2009). TiO_2 nanoparticles are among the emerging and most promising photocatalysts for water purification (Adesina et al., 2004; Li et al., 2008). These photocatalysts involve the formation of highly reactive oxidants, such as OH radicals, for the elimination of microorganisms, bacteria, fungi, algae, viruses, etc. (Einaga et al., 1999; Fujishima et al., 2000; Shephard et al., 2002; Ibanez et al., 2003; Li et al., 2003; Cho et al., 2004, 2005; Hajkova et al., 2007; Zan et al., 2007). TiO_2, after 8 h of simulated solar exposure, reduced the feasibility of numerous waterborne pathogens, such as, fungi, *E. coli*, protozoa, and *P. aeruginosa* (Lonnen et al., 2005). The complete inactivation of fecal coliform done by TiO_2 photocatalyst and no further growth of bacteria were observed in the presence of UV light; it shows less photocatalytic activity and adsorption capacity compared with the visible-light region (Gelover et al., 2006; Ni et al., 2007). So to extend its photocatalytic activity and adsorption capacity, doping has been done by several methods, such as self-doping, non-metal doping, transitional metal doping, and rare-earth metal doping (Liu, et al., 2014, 2015; Zu et al., 2012; Xing et al., 2014; Marschall et al., 2014; Song et al., 2013; Yu et al., 2012; Angelis et al., 2007; Borlaf et al., 2014). Moreover, anionic non-metal doping such as nitrogen doped with TiO_2 a via hydrothermal method to enhance the absorption range of from the UV region to visible light carbon, sulphur or fluorine into TiO_2 (Asahi, et al., 2001; Burda et al., 2003; Irie et al., 2003; Diwald et al., 2004; Yang et al., 2004; Khan et al., 2002; Sakthivel et al., 2003 Noworyta et al., 2004; Umebayashi et al., 2002; Ohno et al., 2003; Yamamoto et al., 2004; Yu et al., 2002). Recently, Ag doping of TiO_2 resulted in enhanced bacterial inactivation by complete removal of *E. coli* or reduced the inactivation time thereby enhancing disinfection under UV wavelengths and solar radiations (Sokmen et al., 2001; Sung-Suh et al., 2004; Vamathevan et al., 2004; Zhang et al., 2005). As highlighted, the synthesis of visible-light-activated TiO_2 nano-particles has attracted considerable interest, and TiO_2 nanoparticles and nanocrystalline irradiated with UV-Vis light exhibited strong bactericidal activity against *E. coli* (Page et al., 2007; Bae et al., 2003; Ireland et al., 1993; Wei et al., 1994; Bekbolet et al., 1996; Maness et al., 1999). Metal-doped TiO_2 nanoparticles, sulphur, and iron have shown strong antibacterial effects against *E. coli* (Yu et al., 2005; Egerton et al., 2006).

7.3.2 Fullerenes

Fullerene, an allotrope of carbon, was discovered in 1985 by contribution of many scientists—Richard Smalley, Robert Curl, James Heath, Sean O'Brien, and Harold Kroto at Rice University. They were awarded the Nobel prize in 1996 for this novel discovery (Kroto et al., 1985; Anton et al., 1996). Fullerene finds application in several technological fields due to its chemical, optical, and structural properties. The structure of fullerene is composed of carbon molecule, and it is found in various shapes like sphere, ellipsoid, and tube. Fullerenes have cage-like appearance and a highly symmetrical structure and are composed of 12 pentagons and 20 hexagons with sp^3, and sp^2 bond character, which is responsible for the angle strain. The isomers with the adjoining pentagons also show lower stability and relative abundance than the isomers with inaccessible five-member ring in which π bonds are delocalized due to resonance over the fullerene structure (Kroto et al., 1987; Campbell et al., 1996). The degree of delocalization of charge is still an unresolved question for scientists; the chemical behaviour of fullerene (C-60) is between an aromatic molecule and a straight-chained alkene (Taylor et al., 1993). Such type of balance between stability and reactivity classifies Buckminsterfullerene, C-60, from the degenerate C-20 fullerenes and inert, planar graphite fullerene consists of an icosahedral symmetry, which is stabilized by its resonating structures, electronic state, and bonding geometry for each carbon atom (Johnson et al., 1990, 1992). Fullerene transforms into various structures such as covalent, supramolecular, and endohedral forms and permits molecular manipulation and formation of polymeric material for specific environmental applications. However, C-60 does not undergo oxidation easily due to its lowest unoccupied molecular orbital (LUMO), which has the tendency to accommodate only up to six electrons. This leads to its possible structural scaffolding for reactive adducts. Bonifazi et al. (2007) provided an excellent review of the extensive supramolecular chemistry of fullerene. The functionalization of fullerene has been done by supramolecular techniques, including molecular self-assembly, immobilisation of fullerene derivatives via non-covalent interactions van der Waals, electrostatic, and hydrophobic interactions intrinsic to fullerenes or their complementary reactants. These supramolecular techniques enhance the biomedical and environmental applications, increase the solubility of an organic molecule, and reduce the aggregation of hydrophobic fullerene molecules by surface modification, and conformation (Diederich et al., 1996; Lehn et al., 1996; Stone et al., 1986). A wonderful application of fullerene works as a nanoadsorbent due to their large surface-to-volume ratio and its manipulated surface chemistry helps in the removal of toxic metals and organic molecules from water and wastewater (Falcao et al., 2007; Cheng et al., 2004; Bagrovskaya et al., 2012; Yang et al., 2006). Many studies have demonstrated the sorption of metal ions using C-60. Gallego et al. (1994) investigated the analytical potential of C-60 for the pre-concentration of metal traces by using neutral chelating ligand such as ammonium pyrrolidine dithiocarbamate to check the lead traces in water where C-60 fullerene was demonstrated to perform better in terms of both accuracy and selectivity. Bagrovskaya et al. (2012) synthesized the composite fullerene, modifying it with polystyrene film. They were found that the sorption capacity for Cu(II) was four times higher than the unmodified composite compared with Ni(II), Zn(II), and Cd(II). The selectivity of sorption on the composite for these metals is Cu(II) > Cd(II) > Ni(II) > Zn(II). Samonin et al. (2008) reported metal cation sorption capacity by the modification of fullerene with activated carbon. Cheng et al. reported the elimination of organic hydrophobic contaminant such as naphthalene and 1,2-dichlorobenzeneby nano-fullerene. Serrano et al. (2006) tested C-60 fullerene's sorptive capacity for benzene, toluene, ethylbenzene, and xylene isomers compared with other sorbents such as Tenax TA, and RP-C18. C-60 fullerenes have been the best choice in terms of selectivity, sensitivity, precision, and reusability. Polycyclic hydrocarbons (PAHs) are very toxic; such organic toxicants have been discharged into the soil due to anthropogenic activities. They possess an aromatic structure, low polarity, and elevated hydrophobicity. PAHs tend to adhere to soil particulates on their surface when they persist and accumulate in the soil for a long time, which causes serious environmental concern and risk to human and ecological system due to their acute toxic nature. Many researchers have found that carbonaceous nanomaterials, including C-60 fullerenes, may efficiently sequester PAHs and reduce their bio-accessibility and extractability (Towell et al., 2011). To remove toxic organic PAHs and heavy metal ions, fullerenes have been used as an efficient adsorbent.

7.3.3 Dendrimers

Dendrimers consist of repetitive tree-like branches, nano-sized, and radially symmetric molecules with well-defined, homogeneous, and monodisperse macromolecular structure (Srinivasa et al., 2007). These are an artificially manufactured and have been synthesized by using various approaches (Buhleier et al., 1978; Tomalia et al., 1984). These are basically a polymer made up of many repeating units of monomers. Dendrimers have a wide range of applications in environment pollution remediation due to their macromolecular chemistry. Dendritic polymers are given opportunities to develop effective ultra-filtration processes for detoxification of toxic metal ions, contaminated water by radionuclides, organic-inorganic solutes, and various waterborne pathogens such bacteria and viruses (Chechik et al., 2000). The wonderful nature of such filtration membranes is that after filtration, the loaded ions can be recovered, and the dendritic polymer can be recycled. Additionally, dendrimer-based membranes filter requires minimum operating pressure and lower energy consumption. Sun et al. (2006) prepared dendrimers by anchoring low-generation diethanolamine into cross-linked polystyrene-supported to show good adsorption capacities for Ag(I), Cu(II) and Hg(II) ions, and the adsorption mechanism was characterised by XPS. Shahbazi et al. (2011) utilised SBA-15 mesoporous silica material to functionalize amine ($-NH_2$) and melamine-based dendrimer amines (MDA) dendrimers for the removal of metal ions for from an aqueous system. Anbia et al. synthesized mesoporous carbon nitride functionalized with melamine-based dendrimer amine for the removal of Pb(II) and Cu(II). They achieved good adsorption efficiency for Pb(II) and Cu(II) at 196.34 and 199.75 mg/g, respectively (Anbia et al., 2015). Dendritic polymers made out of arbitrary hyper-expanded polymers, dendrigraft polymers, dendrons and dendrimers are generally monodisperse and exceedingly stretched macromolecules with the controlled piece and engineering comprising of three parts: a centre, inside branch cells and terminal branch cell.

Diallo et al. (2004, 2005) found that poly(amidoamine) (PAMAM) with an ethylene diamine (EDA) acts as the core dendrimer for the elimination of copper toxic metal ion from the aqueous system. Arkas et al. (2006) developed the organic-inorganic hybrid filters by using functionalized poly (propylene imine) dendrimer, poly (ethylene imine) hyper branched polymer, and beta-cyclodextrin derivatives. The filters appeared to be more efficient for the removal of PAHs above 95% than to other pollutants such as monoaromatic hydrocarbons as benzene, toluene, xylene (BTX), and pesticides (simazine) removed less than 80%. Afterwards, filters were tested for the purification of water by continuous filtration experiments, employing a variety of water pollutants.

7.3.4 Zeolite

Zeolites are naturally occurring microporous materials whose pore size is less than 2 nm. They belong to the mineral class tectosilicates. Zeolites mainly contain aluminosilicate minerals with a different composition. They show a wide range of applications in environment remediation due to their unique physicochemical properties such as high chemical and mechanical resistance and high sorption capacity, which provide the basis for catalysis, ion exchange, separation, non-toxic nature, and low cost. Various zeolites have been widely applied in the elimination of heavy metals, such as Cr(III), Ni(II), Zn(II), Cu(II), Fe(II), Pb(II) and Cd(II) from industrial effluent water as well as the domestic wastewater (Allabashi et al., 2007). Various methods have been used for eliminating dissolved toxic heavy metals in drinking water, such as sorption, precipitation, ultrafiltration, reverse osmosis, and electrodialysis (Ahluwalia et al., 2007; Babel et al., 2003). With all these methods recently, zeolites have extensively been used in the removal of heavy metal ions. The structures of zeolites contain three-dimensional frameworks of SiO_4 and AlO_4 tetrahedral and isomorphic substitution of Si(IV) by Al(III) making its lattice completely negative charged, which is balanced by exchangeable cations (sodium, potassium, or calcium) (Khachatryan et al., 2014). These cations are further transferrable with heavy metal ions, which are found in wastewater as well in industrial effluents, such as lead, cadmium, zinc, and manganese (Barer et al., 1987). In fact, the ion exchange property of zeolites makes them particularly suitable for removing undesirable heavy metal ions from industrial effluent water. The first zeolite mineral, stilbite, was discovered in 1756 by Swedish chemist and mineralogist Axel Fredicka Cronstedt. There are various natural zeolites found in nature: clinoptilolite mordenite, phillipsite, chabazite, stilbite, analcime and laumontite, offretite, paulingite, barrerite, and mazzite. Among all the zeolites, clinoptilolite is the most abundant natural zeolite which is widely used in the remediation, and its chemical formula is $Na_{0.1}K_{8.57}Ba_{0.04}(Al_{9.31}Si_{26.83}O_{72})\cdot 19.56\ H_2O$ (Brunner et al., 1993). The sorption capacity of natural zeolite (clinoptilolite) for inorganic cations has been reported by many authors. Zamzow et al. reported selective order of inorganic metal ions by the clinoptilolite in the sodium form as Pb(II) > Cd(II) > Cs(I) > Cu(II) > Co(II) > Cr(III) > Zn(II) > Ni(II) > Hg(II) (Zamzow et al., 1990). Mier et al. (2001) Studied the sorption of Cu(II), Pb(II), Fe(III) and Cr(III) on natural clinoptilolite and disclosed that equilibrium is favourable for Pb, unfavourable for Cu, and of sigmoid shape for Cr(III) and Fe(III). Other than clinoptilolite, many other modified forms of zeolites have been studied in the removal of heavy metals. Sorption capacity by the zeolite depends on the various parameters like (S/L), a value which indicates the ratio of zeolite and wastewater, contact time, temperature, the initial concentration of metal ions, and solution pH (Ibrahimi et al., 2015). The sorption capacity of modified zeolite was impregnated by NaY and NaP, observed higher 83.2 mg/g than natural zeolite mordenite was 3.5 mg/g (Covarrubias et al., 2006).

Further, another natural Brazilian zeolite showed adsorption capacity of 3–14.5 mg/g and Greek adsorbents showed 4.1 mg/g for the removal of the Cr(III) ion (Bosco et al., 2005; *Alvarez-*Ayuso et al., 2003). The efficiency improved by a higher extent by using modified zeolite 4A and NaP1 from coal fly ash was 56.4 mg/g (Hui et al., 2005). Das et al. (2013) reported the removal of toxic and non-biodegradable cadmium heavy metal from wastewater by zeolite synthesized from fly ash at pH 6, 40°C and at an adsorbent dosage of 2 g/100 mL. Martýnez et al. (2004) reported the removal of cadmium toxic metal from aqueous systems by the modification of zeolite with surfactant hexadecyltrimethylammonium (HDTMA) bromide with the sorption efficiency 15.86 mg/g at pH 7. A naturally modified Jordanian zeolite tuff and NJ zeolite have been used to eliminate cadmium and their efficiency was 25.9 for Cd(II) (Almjadleh et al., 2014; Taamneh et al., 2016). The removal efficiency of natural zeolite for the metal Pb(II) was 78.7 mg/g at 70°C, and pH 4.5 (Curkovi et al., 1997; Bekta et al., 2004). Further, the adsorption efficiency was increased up to 91.2 mg/g of Croatia natural zeolite by base treatment with a 2 M NaCl solution at 70°C for the one-day time. Jha et al. (2008) reported that Pb(II) uptake efficiency was 228 mg/g by varying Pb(II) stock solution range from 100 to 2,000 mg/L at 25°C for 24 h using composites of activated carbon/zeolite prepared from coal fly ash as an adsorbent. Ismael et al. (2012) studied the removal of lead from wastewater and found that natural zeolite has less tendency toward removal than the modified and treated form of zeolite with acid and alkali solutions.

7.5 Iron Nanoparticles

Numerous kinds of nanomaterials have been developed in active research. In between all iron oxide, nanoparticles have also remarkably attracted researchers' attention in controlling pollution (Pragnesh et al., 2014). Due to their unique properties such as small size, high surface area, and magnetic property, they are widely used in various fields (Oh et al., 2011; Laurent et al., 2008). These nanoparticles have been synthesized mainly by three methods: chemical method, physical method, and biological method. The chemical method is the most acceptable method for the synthesis of iron oxide nanoparticles. It is further divided into various types, such as chemical co-precipitation, thermal decomposition, sol-gel, and electrochemical methods. The small size and high surface area of iron oxide nanoparticles make them an ideal adsorbent. Heavy metals in the solution easily disperse on the active surface of Fe_3O_4 nanoparticles. The surface area of magnetic nanoparticles plays a vital role in the adsorption process. To increase the adsorption efficiency and avoid the aggregation of particles, these have been modified with various suitable functional groups such as carboxylic acids, phosphoric acid, silanol, thio, and amine as well as small organic molecules and biomolecules.

Water pollution by toxic arsenic metal ion causes serious environmental pollution. It causes severe diseases, such as cancer, neurological disorder, nausea, hyperkeratosis, and muscular weakness (Mandal et al., 2002). Arsenic contamination in drinking water has been discovered locally which can be related to various kinds of sickness. Around 45 to 57 million people in Bangladesh and 13 million people in the United States have been affected by unsafe arsenic contaminated water reported by WHO (2006) (An et al., 2011). According to USEPA guidelines, the amount of arsenic in the drinking water should be 0.010 ppm or lesser (Yang et al., 2010). Hence, it is essential to have a potent technique to drive out arsenic from the consumption water and wastewater. Iron oxide nanoparticles have shown outstanding performance for the decontamination of arsenic from the water (Yavuz et al., 2010; Jeong et al., 2007; Lin et al., 2012). Various forms of iron such as magnetite, hematite, and maghemite nanoparticles have been developed by using diverse chemical methods to remove As(III) and As(V) from wastewater (Ma et al., 2013; Cui et al., 2006).

The main characteristic properties of magnetic nanoparticles are small size and high surface area attributed to the suitable and eco-friendly adsorbent. Due to their high surface area, heavy metals in the solution easily disperse on the dynamic surface of Fe_3O_4 nanoparticles. The surface area of magnetic nanoparticles plays a major role in the adsorption process. Due to large surface of nanoparticles, chances of aggregation increase in the aqueous system, which could decrease their effectiveness; so it is necessary to modify the surface of iron nanoparticles to enhance their efficacy (Mohammad-Beigi et al., 2011; Kango et al., 2013; Feng et al., 2012). Chalasani et al. (2012) synthesized β-cyclodextrin-functionalized magnetic iron oxide nanoparticles (CMβCD) with average nanoparticles size 10 nm by thermal decomposition and post grafting methodology. These CMCD Fe_3O_4 have been used for the removal of As(III), As(V), 2-naphthol, and naphthalene. Mou et al. (2011) reported the adsorption capacity of As(V) 143.12 m^2/g by Fe_2O_3 chestnut-like amorphous-core/γ-phase shell hierarchical nanostructure by one-step template-free method. There are numerous applications of magnetic nanomaterials to remove arsenic salts from water.

Copper has various applications in various industries. It is utilized as a part of electroplating, paint, electrical, and manure industries. Because of the extensive variety of utilization of copper, it can be collected in the earth because it directly disposes of in water streams without any treatment, which makes water more polluted and the aquatic as well as terrestrial life is affected. Hao et al. (2010) carried out 98% removal of copper from a polluted river, tap water using surface modified magnetite nanoparticles. The maximum adsorption capacity was 25.77 mg/g at pH 6 and 298 K. Li et al. (2013) reported the maximum adsorption capacity of 22.4 mg/g by amine-functionalized magnetic nanoparticles prepared with 3-aminopropyltriethoxysilan by using the sol-gel method. The adsorption capacity of these composites increased with increasing the pH due to the presence of amine group on its surface and strong complexation took place between Cu(II) and the free amine group. Further adsorption capacity decreased at lower pH due to the protonation of amine. Banerjee and Chen reported fast adsorption rate of Fe_3O_4-gum-arabic nanocomposite for the elimination of Cu(II). The coupling between a carboxylic acid group of gum Arabic and hydroxyl group of nanoparticles was responsible for the attachment of gum Arabic on the surface of Fe_3O_4. The size of the particles obtained was in the range 13–67 nm and 5.1 wt% of Arabic gum was immobilized. Strong complexation between Cu(II) and the amine group of Arabic gum confirmed fast adsorption rate and short equilibrium was attained within 2 min. The effect of pH on adsorption was also examined at pH < 2; no adsorption took place. The adsorption increased when the pH was increased from 2 to 6. The adsorption isotherm fitted well to the Langmuir equation with maximum adsorption capacity of 27 mg/g. GA-MNP, too, has good reusability and regeneration capability. The acid solution used for regeneration and the maximum adsorption capacity of GA-MNP was 28.12, 27.64, and 27.18 mg/g. Subsequently Banerjee et al. (2007 and Zhou et al. (2009) demonstrated the removal of Cu(II) from the aqueous system. Chitosan-coated maghemite nanoparticle was modified with a biodegradable and green reagent α-ketoglutaric acid. Techniques such as XRD, FT-TR, TEM, VSM, and EDS were used to characterize the magnetic-based nanoadsorbent. The TEM showed the particles size of about 30 nm and confirms that no aggregation occurred in the synthesized nanoadsorbent chitosan-coated magnetic nanoparticles (CCMNPs) bearing α-ketoglutaric acid. VSM and EDS analysis also confirmed the superparamagnetic nature of iron nanoparticle and confirms the adsorption of Cu(II) on the surface of iron nanoparticles. The characteristic peak around 589 cm^{-1} and 1644 cm^{-1} in the FT-IR spectra indicates successful coating of γ-Fe_2O_3 and confirms the vibration of the primary amine, which vanishes after the immobilization of α-ketoglutaric acid, and the new peak around 1618 cm^{-1} is the vibration peak of secondary amine that also confirmed the coating of α-ketoglutaric acid to magnetic CCMNPs. The other bands were 1402 and 1718 cm^{-1} of C–H stretching vibration of CH_2 from α-ketoglutaric acid and amide functional groups, respectively. The crystalline structure of γ-Fe_2O_3 was confirmed by the XRD pattern. The XRD peaks became broader and low intensity, which indicated that KA uniformly modifies CCMNPs. The saturation magnetization of KA-CCMNPs was 33.5 emu/g and showed that they are supermagnetic in nature and highly susceptible to the external magnetic field. The adsorption efficiency was 96.15 mg/g, which followed the Langmuir isotherm. Similarly, chitosan supported on the surface of magnetite nanoparticles to enhance the removal efficiency. Chang et al. (2005) synthesized novel monodisperse chitosan-bound Fe_3O_4 magnetic nanoadsorbent, employed for the removal of Cu(II). Chitosan covalently attached to the Fe_3O_4 nanoparticles surface through carbodiimide activation. TEM showed the diameter of nanoadsorbent around 13.5 nm. Nanoadsorbent showed better efficacy above pH 2. The pH and temperature studies demonstrated that the adsorption capacity increases with pH within the range 2–5. The maximum adsorption capacity was 21.5 mg/g, which followed the Langmuir equation. Yan et al. (2012) successfully synthesized magnetic composite microspheres bearing Fe_3O_4 nanoparticles and poly(acrylic) acid-chitosan by using the chemical co-precipitation method for the removal of Cu(II). Subsequently the nanoadsorbent could be further used for adsorption due to their reusability and regeneration tendency. Tseng et al. (2009) found that CS/PPA-Fe_3O_4 microspheres exhibit more adsorption capacity than CS-Fe_3O_4 microspheres for the copper ions in an aqueous system. The Fe_3O_4-polyvinyl acetate-iminodiacetic acid contains EDTA, which was also a useful adsorbent for the removal of Cu(II). Badruddoza et al. (2011) modified iron oxide nanoparticles with carboxmethyl-β-cyclodextrin. The modified magnetite nanoparticles (CMβCD-Fe_3O_4) showed maximum adsorption capacity of 47.2 mg/g at 25°C for the removal of Cu(II) ions from an aqueous system. CMβCD grafted on the surface of magnetic nanoparticles was confirmed by FTIR, TEM, TGA and XPS analyses. The maximum adsorption efficiency of this magnetic nanoadsorbent was due to the presence of various hydroxy and carboxyl groups. Zhu et al. (2012) synthesized an efficient magnetic-graphene nanocomposite by one-pot thermal decomposition method for the removal of Cr(VI) from wastewater. The morphology of magnetic NPs was confirmed by TEM images. Various effects were studied to check their removal efficiency for Cr(VI), such as the contact time, adsorbent dosage, and pH. The requirement of less contact time and low pH to remove Cr(VI) made them a promising adsorbent. Yuan et al. (2009) synthesized montmorillonite clay-supported Fe_3O_4 nanoparticles by co-precipitation and hydrosol methods, which showed better adsorption capacity for the removal of Cr(VI) than Fe_3O_4. The high efficiency was due to the porous nature of clay, which facilitates better dispersion of magnetite nanoparticles inside and less aggregation. The average size of nanoadsorbent without and with support was around 20 and 15 nm, respectively. Tellez et al. (2011) developed the nanoadsorbent by using a low-cost biomaterial, orange peel. The magnetic nanoparticles were supported on the orange peel pith by using the redox precipitation method. Tubular shaped nanoparticles with size around 20 to 80 nm and octahedral crystals around 20 to 40 nm were observed. A strong adsorption capacity was demonstrated compared with unmodified magnetic nanoparticles for the removal of Cr(VI). Wang et al. (2012) synthesized magnetite-polypyrrole composite microspheres using Fe_3O_4 microspheres as a chemical template under sonication. Fe_3O_4/PPy, which showed higher adsorption capacity of around 209.2 mg/g for Cr(VI). Fe_3O_4 nanoadsorbents have been demonstrated good performance for the removal of Pb(II) ions from wastewater using a batch adsorption technique. Various influencing factors such as the effect of temperature, pH, and coexisting ions which affect the rate of adsorption were sudied in the adsorption experiments. Further adsorption of Pb(II) reached equilibrium in a short duration of around half an hour. The adsorption rate of lead on magnetic nanoparticle increased with raising the temperature that indicated endothermic adsorption and there was no influence of the co-existing cation on adsorption. The thermodynamics studies of Pb(II) adsorption onto the magnetic nanoadsorbent revealed that the adsorption was spontaneous, endothermic, and physical in nature. The maximum adsorption capacity of Pb(II) is 36 mg/g as confirmed by the Langmuir and Freundlich adsorption isotherms (Nassar et al., 2010). Cheng et al. (2012) synthesized the (γ-Fe_2O_3) nanoparticles with a diameter of 60 nm by a simple co-precipitation method for the selective decontamination of toxic heavy metals from electroplating wastewater. The developed nanoadsorbent was characterized by XRD, EDX, SEM, and EDX. The adsorption of Pb(II) attained equilibrium within 15 min. The adsorption equilibrium data fitted well with the Langmuir model. The rate of adsorption of Pb(II) increased with pH. Xu et al. (2012) synthesized magnetic nanoparticles and calcium alginate beads impregnated with *Phanerochaete chrysosporium* for the removal of toxic Pb(II) from aqueous solution. Adsorption-desorption studies were carried out by using HCl solution up to five times and the percentage removal of Pb(II) about 90% was maintained after performing the fifth cycle. Madrakian et al. (2013) synthesized an efficient adsorbent by anchored blue-19 dye on to surface of magnetic nanoparticles for the removal of Pb(II) from water samples. Wang et al. (2012) used the hydrothermal method for the preparation of efficient water-soluble magnetite nanoparticles. The modified adsorbent exhibited more solubility and higher stability and showed a high degree of magnetization, which allowed magnetic separation from wastewater. It showed high affinity for Pb(II) and Cr(VI) than water-insoluble magnetite nanoparticles. These water-soluble magnetic nanoparticles can remove 90% of Pb(II) in 2 min. Cao et al. (2012) synthesized α-Fe_2O_3 nanostructures with flower-like morphology through the template-free microwave-assisted solvothermal method. The maximum adsorption capacity for As(V) and Cr(V) was 51 and 30 mg/g, respectively. The high adsorption efficiency and inexpensive nature make them promising adsorbent for removal of As(V) and Cr(V) from wastewater. Further, Wei et al. (2013) synthesized α-Fe_2O_3 nanoparticles with different hollow nest like morphology via the template-free solvent thermal method and the glycerol-mediated microwave-assisted method. The prepared adsorbent was characterized by XRD, FESEM, and TEM. It has been used for the removal of As(V) and Cr(V) as well as organic pollutants from the water. The adsorption capacity of adsorbent for As(V) and Cr(V) was 75.3 and 58.5 mg/g, respectively. Roy et al. (2012) developed maghemite nanotubes by microwave irradiation method. The nanoadsorbent has the capability to detoxify more than two toxic heavy metal ions such as Cu(II), Zn(II), and Pb(II). The BET surface area and magnetic saturation obtained were 321.638 m^2/g and 68.7 emu/g, respectively. The maximum adsorption capacities of tubular maghemite adsorbents found with respect to various metal ions such as Cu(II), Pb(II), and Zn(II) were 111.11, 71.42, and 84.95 mg/g, respectively. The adsorption equilibrium of heavy metal ions followed Langmuir isotherm model. The kinetic data of adsorption followed by a pseudo–second–order equation thereby indicating their chemical adsorption. Zhang et al. (2012) coated thiol functional group on the surface of CNTs/Fe_3O_4 with the help of a binder, which is 3-mercaptopropyltriethoxysilane. They found the BET surface area of functionalized superparamagnetic carbon nanoadsorbent to be 97.16 m^2/g. The maximum adsorption capacity for the elimination of Pb(II) and Hg(II) was 65.40 and 65.52 mg/g, respectively, well fitted with the Langmuir isotherm. Inductively coupled plasma optical emission spectroscopy (ICP-OES) technique was used to check the effect of contact time, adsorbent dosage, and initial metal concentration during the adsorption experiment. The synthesized nanocomposite was highly pH dependent and optimal pH value 6.5 was reported. Parham et al. (2012) synthesized an efficient adsorbent by functionalizing the magnetic nanoparticle with 2-mercaptobenzthiazole. It works as an efficient adsorbent compared with unmodified magnetic nanoparticles for the removal of toxic metals such as Hg(II). Unmodified nanoparticles were able to remove 43.47% of 50 mg/mL of Hg(II) from polluted water, while modification with MBT enhanced the removal capacity up to 98.6% in the same concentration within 4 min. The change in the pH range and NaCl concentration did not significantly affect the adsorption efficiency of the synthesized adsorbent for the toxic mercury ion. Xin et al. (2012) synthesized amine-functionalized mesoporous magnetite (AF-Fe_3O_4) nanoparticles by using a cost-effective and eco-friendly method. AF-Fe_3O_4 was prepared by the hydrothermal method in which iron salt $FeCl_3 \cdot 6H_2O$ was added to ethylene glycol followed by sodium acetate and ethane diamine and sealed in a Teflon-lined stainless-steel autoclave and heated at 200°C for 8 h. The mesoporous morphology was confirmed by TEM pictures. The pore volume and the BET surface were 0.1833 cm^3/g and 25.94 m^2/g, respectively. The X-ray diffraction pattern proved that it is a magnetite phase structure. The zeta-potential value decreased with increase of pH. The adsorption rate of AF-Fe_3O_4 followed the pseudo second-order kinetic model. The equilibrium was achieved within 120 min at pH 7. According to the Langmuir isotherm, the maximum adsorption capacity for Pb(II), Cd(II), and Cu(II) was 369, 446.4, and 523 mg/g, respectively. Thermodynamic studies revealed that the AF-Fe_3O_4 adsorbent was endothermic and spontaneous. Ge et al. (2012) developed efficient functionalized magnetic nanoparticles by the copolymer of acrylic acid and crotonic acid, which was further modified with 3-aminopropyltroethoxysilane for the decontamination of heavy metals, such as Cu(II), Pb(II), Zn(II), and Cd(II) from aqueous solutions. The maximum adsorption efficiency observed for these metals was 126.9, 166.1, 43.4, and 29.6 mg/g, respectively, which was confirmed by the Langmuir isotherm. The effects of pH, function of time, reusability, and absorption-desorption studies were observed to check the adsorption capacity. The maximum adsorption was found at pH 5.5. Liu et al. (2008) synthesized magnetite nanoparticles coated with low-cost and eco-friendly humic acid via the co-precipitation method. The saturation magnetization of 79.6 emu/g of Fe_3O_4-HA enabled easy separation from the water at low magnetic field within a few minutes. The equilibrium was reached in 15 min. The maximum adsorption capacity obtained was 46.3 to 97.7 mg/g. The magnetic nanoparticles were stable in tap as well as in natural water and showed minimum leach-out with the acid-base treatment of 0.1M HCl and 2M NaOH solutions. The adsorbent could remove more than 99% of Hg(II) and Pb(II) and 95% of Cu(II) and Cd(II) in natural and tap water. Singh et al. (2011) reported a novel adsorbent by modifying Fe_3O_4 with succinic acid, ethylene diamine, and 2,3-dimercaptosuccinic, which have the capability to remove heavy metal ions as well as pathogens. The adsorbent possessed a high degree of magnetization at room temperature. The acid- and thiol-functionalized Fe_3O_4 was spherical in shape with average sizes of 10 nm and 6 nm, as confirmed by the TEM analysis, while amine-functionalized nanoparticles showed distinct morphology with a mesoporous nature and with diameter of 40 nm. Due to the magnetization property, the adsorbent can be easily separated from the aqueous solution. These developed adsorbents demonstrated good capacity for the removal of heavy metals, such as Cr(III), Ni(II), Co(II), Cu(II), Cd(II), Pb(II), and As(III). The adsorption capacity increased with increase in all three different types of adsorbents and due to strong complexation that took place between the surface of adsorbents and metals ions and weak electrostatic interaction with surface. The adsorption capacities of all three adsorbents increased with increased dosage of adsorbents because the active site was available more in number on the surface. Among the three adsorbents, thiol-MNP was more efficient due to the presence of the thiol group, which was considered as soft Lewis base, which has more tendency to interact with the soft Lewis acid such as heavy metal ions. The adsorbent could be regenerated after the desorption experiment was performed using 10 mL 0.1M HCl solution. The maximum adsorption efficiency observed above 85% for the removal of heavy metals from the aqueous solution. Badruddoza et al. (2013) synthesized magnetic nanoadsorbents by grafting carboxymethyl-β-cyclodextrin polymer on to the surface of Fe_3O_4. The maximum adsorption was shown for selective heavy metals, such as Pb(II), Cd(II), and Ni(II). The adsorption capacity of Pb(II),

Cd(II), and Ni(II) was 64.5, 27.7, and 13.2 mg/g, respectively. It was found that adsorption was pH- and temperature dependent. Modified nanoadsorbents showed maximum adsorption capacity due to the interaction between carboxyl and hydroxyl group of used polymers with the metal ion. Gong et al. (2012) synthesised magnetic nanoparticles by coating of biodegradable shellac layer on to the surface of Fe_3O_4. TEM analysis illustrated the core–shell structure of the modified (SCMN) Fe_3O_4, considered as core and shellac resin as shell with the diameter of 20 and 15 nm. FTIR results showed the interaction between the carboxyl group of resin and cadmium ions. The maximum adsorption capacity for the removal of Cd(II) was 18.80 mg/g. Yantasee et al. (2007) synthesized superparamagnetic iron oxide nanoparticles whose surface was modified with dimercaptosuccinic acid (DMSA). They are demonstrated for the removal of toxic metals, such as, Hg(II), Ag(I), Pb(II), Cd(II), and Tl(I). The ligand DMSA effectively bound to the metal ions. DMSA-Fe_3O_4 exhibited a capacity of 227 mg/g of mercury, which was 30-fold higher than conventional resin-based sorbents (GT-73). The chemical affinity, capacity, kinetics, and stability of these adsorbents were compared with GT-73, activated carbon, and non-porous silica in ground water, river water, seawater, human blood and plasma. MNP-DMSA was efficiently removed 99 wt% of 1 mg/L Pb in 1 min. However, other adsorbents such as GT-73 and chelex-100 took 120 and 10 min to remove 96% of Pb. Tan et al. developed an efficient nanoadsorbent for the removal of Pb(II) by the immobilization of 1,6-hexanediamine to the surface of Fe_3O_4. Synthesized nanoadsorbent had the tendency to remove up to 98% of Pb(II) from the industrial water. The maximum adsorption capacity of Pb(II) was 40.10 mg/g by MNPs-NH_2, as determined by the Langmuir adsorption isotherm model (Tan et al., 2012). Shen et al. (2009) prepared Fe_3O_4 magnetic nanoparticles with an average size of 8, 12 and 35 nm via three diverse methods: co-precipitation, surface decoration process, and the polyol process. These three were the important adsorbents for the removal of heavy metals, such as, Ni(II), Cd(II), Cr(II), and Cu(II) from wastewater. The maximum adsorption efficiency to remove heavy metals was observed 5.15, 7.45, and 35.46 mg/g, respectively. Zhang et al. (2011) used Fe_3O_4-NH_2/PEI-EDTA with an average size of 60 nm for the removal of heavy metal ions, such as,

Cu(II), Cd(II), and Pb(II). It was feasible at 5.5 pH. The adsorbent show about 98% of Pb(II) from the aqueous solutions. The higher efficiency and reusability make it a promising adsorbent. Kim et al. (2013) synthesized hierarchical MnO_2-coated with magnetic nanocomposite (Fe_3O_4/MnO_2) with flower-like morphology via hydrothermal process. They used it to remove heavy metal ions from the water. The higher adsorption capacity was the result of high surface area and high magnetic saturation. The removal efficiency of Fe_3O_4/MnO_2 was compared with unmodified Fe_3O_4 nanoparticles for heavy metals such Cd(II), Cu(II), Pb(II), and Zn(II). The maximum adsorption capacity of Cd(II) obtained was 53.2 mg/g, as determined by the Langmuir equilibrium sorption model. Various solution conditions did not affect the adsorption capacity of the modified adsorbent. They retained adsorption capacity over 80% countered in natural waters. High magnetization in the presence of external magnetic field makes this adsorbent reusable. Singh et al. (2013) synthesized ZnO magnetic semiconductor composite by fabrication of Fe_3O_4 on to its surface. They observed about 100% removal of heavy metals. The high surface area and highly porous nature make it more efficient. It also acts as a photocatalyst which degrades organic dye in presence of UV light and captures pathogens rapidly. Ma et al. (2013) developed $MnFe_2O_4$@Mn-Co oxide-based magnetic adsorbents which had the maximum adsorption capacity of 481.2, 386.2, and 345.5 mg/g, respectively, to eliminate Pb(II), Cu(II), and Cd(II) and also had good regeneration and reusability capabilities. Huang et al. (2009) prepared polyacrylic acid covalently coated on the surface of Fe_3O_4, which was further functionalized with amine group using diethylenetriamine. It was used an adsorbent to remove the Cu(II) and Cr(VI). The maximum adsorption ability demonstrated 12.43 and 11.24 mg/g, respectively, well fitted with Langmuir isotherms. Mahdavian et al. (2010) prepared 10 to 23 nm magnetic nanoparticles via co-precipitation followed by the modification of Fe_3O_4 with polyacrylic acid with the help of binder aminopropyltriethoxy silane. The prepared nanoadsorbent was used for the removal of heavy metals such as cadmium, lead, nickel, and copper. Karami et al. (2013) synthesized magnetite nanorods through pulsed current electrochemical method with average diameter 60 nm prepared. The developed nanorods were then utilized for the separation of heavy metals such as Pb(II), Zn(II), Cu(II), Pb(II), Ni(II), and Cd(II). The Langmuir maximum adsorptions of these metals were 127, 107.27, 79.10, 112.86, 95.42, and 88.38 mg/g, respectively.

Acknowledgement

We gratefully acknowledge support from the Ministry of Science and Technology and Department of Science and Technology, Government of India, under the scheme of Establishment of Women Technology Park, for providing the necessary financial support to carry out this study vide letter No, F. No SEED/WTP/063/2014.

References

Adesina AA (2004). Industrial exploitation of photocatalysis: Progress, perspectives and prospects. *Catal Surv Asia*, **4**: 265–273.

Afzali D, Ghaseminezhad S, Taher MA (2010). Separation and preconcentration of trace amounts of gold(III) ions using modified multiwalled carbon nanotube sorbent prior to flame atomic absorption spectrometry determination. *J AOAC Int*, **93**: 1287–1292.

Ahluwalia SS, Goyal D (2007). Microbial and plant derived biomass for removal of heavy metals from wastewater. *Bioresour Technol*, **98**: 2243–2257.

Ahn WY, Sheeley SA, Rajh T, Cropek DM (2007). Photocatalytic reduction of 4-nitrophenol with arginine-modified titanium dioxide nanoparticles. *Appl Catal B*, **74**: 103–110.

Al QS, Salman SR (2001). Photocatalytic degradation of methyl orange as a model compound. *J Photochem Photobiol A Chem*, **148**: 161–168.

Allabashi R, Arkas M, Hormann G, Tsiourvas D (2007). Removal of some organic pollutants in water employing ceramic membranes impregnated with cross-linked silylated dendritic and cyclodextrin polymers. *Water Res*, **41**(2): 476–486.

Almjadleh M, Alasheh S, Raheb I (2014). Use of natural and modified jordanian zeolitic tuff for removal of cadmium(II) from aqueous solutions. *Jordan J Civ Eng*, **8**: 1–8.

Alvarez-Ayuso E, Garcia-Sanchez A, Querol X (2003). Purification of metal electroplating waste waters using zeolites. *Water Res*, **37**: 4855–4862.

An B, Liang Q, Zhao D (2011). Removal of arsenic(V) from spent ion exchange brine using a new class of starch-bridged magnetite nanoparticles. *Water Res*, **45**: 1961–1972.

Anbia M, Haqshenas M (2015). Adsorption studies of Pb(II) and Cu(II) ions on mesoporous carbon nitride functionalized with melamine-based dendrimer amine. *Int J Environ Sci Technol*, **12**: 2649–2664.

Angelis F, Selloni A, Gratzel M, Nazeeruddin MK (2007). Influence of the sensitizer adsorption mode on the open-circuit potential of dye-sensitized solar cells. *Nano Lett*, **7**: 3189–3195.

Anton WJ, Stephen RW, David IS (1996). *Biological Applications of Fullerenes*. New York City, NY: Elsevier Science.

Aran Ja, Alonso AP, Rodriguez JMD, Melian JAH, Diaz OG, Pena JP (2008). Comparative study of MTBE photocatalytic degradation with TiO_2 and Cu-TiO_2. *Appl Catal B*, **78**: 355–363.

Arkas M, Allabashi R, Tsiourvas D, Mattausch EM, Perfler R (2006). Organic/inorganic hybrid filters based on dendritic and cyclodextrin "nanosponges" for the removal of organic pollutants from water. *Environ Sci Technol*, **40**: 2771–2777.

Asahi R, Morikawa T, Ohwaki, Aoki K, Taga Y (2001). Visible-light photocatalysis in nitrogen-doped titanium oxides. *Science*, **293**: 269–271.

Avouris P (2002). Molecular electronics with carbon nanotubes. *Acc Chem Res*, **35**: 1026–1034.

Babel S, Kurniawan TA (2003). Low-cost adsorbents for heavy metals uptake from contaminated water: A review. *J Hazard Mater*, **97**: 219–243.

Badruddoza AZM, Tay ASH, Tan PY, Hidajat K, Uddin MS (2011). Carboxymethyl-β-cyclodextrin conjugated magnetic nanoparticles as nano-adsorbents for removal of copper ions: Synthesis and adsorption studies. *J Hazard Mater*, **185**: 1177–1186.

Badruddoza AZM, Shawon ZB, Tay WJ, Hidajat K, Uddin MS (2013). Fe_3O_4/cyclodextrin polymer nanocomposites for selective heavy metals from industrial wastewater. *Carbohydr Polym*, **91**: 322–332.

Bae, SW, Borse PH, Hong SJ, Jang JSL, Jeong JS, Hong ED, Yoon TE, Jin JH (2007). Photophysical properties of nanosized metal-doped TiO_2 photocatalyst working under visible light. *J Korean Phys Soc*, **51**: 22–26.

Bae E, Choi W (2003). Highly enhanced photoreductive degradation of perchlorinated compounds on dye-sensitized metal/TiO_2 under visible light. *Environ Sci Technol*, **37**: 147–152.

Bagrovskaya NA, Alekseeva OV (2012). Sorption properties of fullerene-modified polystyrene. *Protect Metals Phys Chem Surf*, **48**: 217–220.

Bai H, Shi G (2007). Gas sensors based on conducting polymers. *Sensors*, **7**: 267–307.

Banerjee SS, Chen DH (2007). Fast removal of copper ions by gum arabic modified magnetic nano-adsorbent. *J Hazard Mater*, **147**: 792–799.

Barer RM (1987). *Zeolites and Clay Minerals as Sorbent and Molecular Sieves.* Academic Press, New York.

Bekbolet M, Araz CV (1996). Inactivation of *Escherichia coli* by photo-catalytic oxidation. *Chemosphere*, **32**: 959–965.

Bekta N, Kara S (2004). Removal of lead from aqueous solutions by natural clinoptilolite: Equilibrium and kinetic studies. *Sep Purif Technol*, **39**: 189–200.

Bethune DS, Kiang CH, Vries MSd, Gorman G, Savoy R, Vazquez J, Beyers R (1993). Cobalt-catalyzed growth of carbon nanotubes with single-atomic-layer walls. *Nature*, **363**: 605–607.

Bessekhouad Y, Robert D, Weber JV (2003). Synthesis of photocatalytic TiO_2 nanoparticles: Optimization of the preparation conditions. *J Photochem Photobiol A Chem*, **157**: 47–53.

Bonifazi D, Enger O, Diederich F (2007). Supramolecular [60] fullerene chemistry on surfaces. *Chem Soc Rev*, **36**(2): 390–414.

Borlaf M, Colomer MT, Andres Ade, Cabello F, Serna R, Moreno R (2014). TiO_2/Eu^{3+} thin films with high photoluminescence emission prepared by electrophoretic deposition from nanoparticulate sols. *Eur J Inorg Chem*, **2014**: 25152–5159.

Bosco SMD, Jimenez RS, Carvalh WA (2005). Removal of toxic metals from wastewater by Brazilian natural scolecite. *J Colloid Interf Sci*, **281**: 424–431.

Brady EAS, Kang S, Elimelech M (2008). A single-walled-carbon-nano-tube filter for removal of viral and bacterial pathogens. *Small*, **4**: 481–484.

Brunner GO. (1993). Quantitative zeolite topology can help to recognize erroneous structures and to plan syntheses. *Zeolites*, **13**: 88–91.

Buhleier E, Wehner W, Vogtle F (1978). Cascade- and "nonskidchain-like" syntheses of molecular cavity topologies. *Synthesis(Stuttg)*, **2**: 155–158.

Burda C, Lou Y, Chen X, Samia ACS, Stout J, Gole JL (2003). Enhanced nitrogen doping in TiO_2 nanoparticles. *Nano Lett*, **3**: 1049–1051.

Bystrzejewski M, Pyrzynska (2011). Kinetics of copper ions sorption onto activated carbon, carbon nanotubes and carbon-encapsulated magnetic nanoparticles. *Colloids Surf A*, **377**: 402–408.

Campbell EEB, Fowler PW, Mitchell D, Zerbetto F (1996). Increasing cost of pentagon adjacency for larger fullerenes. *Chem Phys Lett*, **250**: 544–548.

Cao CY, Qu J, Yan WS, Zhu JF, Wu ZY, Song WG (2012). Low-cost synthesis of flowerlike-Fe_2O_3 nanostructures for heavymetal ion removal: adsorption property and mechanism. *Langmuir*, **28**: 4573–4579.

Chalasani R, Vasudevan S (2012). Cyclodextrin functionalized magnetic iron oxide nanocrystal: A host-carrier for magnetic separation of non-polar molecules and arsenic from aqueous media. *J Mater Chem*, **22**: 14925–14931.

Chang YC, Chen DH (2005). Preparation and adsorption properties of monodisperse chitosan-bound Fe_3O_4 magnetic nanoparticles for removal of Cu(II) ions. *J Colloid Interface Sci*, **283**: 446–451.

Chechik V, Crooks RM (2000). Dendrimer-Encapsulated Pd Nanoparticles as Fluorous Phase-Soluble Catalysts. *J Am Chem Soc*, **122**: 1243–1244.

Chen X, Mao SS (2007). Titanium dioxide nanomaterials: Synthesis, properties, modifications and applications. *Chem Rev*, **107**: 2891–2959.

Chen CL, Wang XK (2006). Adsorption of Ni(II) from aqueous solution using oxidized multiwall carbon nanotubes. *Ind Eng Chem Res*, **45**: 9144–9149.

Cheng XK, Kan AT, Tomson MB (2004). Naphthalene adsorption and desorption from aqueous C-60 fullerene. *J Chem Eng*, **49**: 675–683.

Cheng Z, Tan ALK, Tao Y, Shan D, Ting KE, Yin XJ (2012). Synthesis and characterization of iron oxide nanoparticles and applications in the removal of heavy metals from industrial wastewater. *Int J Photoenergy*, **2012**: 1–5.

Cho M, Chung H, Choi W, Yoon J (2004). Linear correlation between inactivation of *E. coli* and OH radical concentration in TiO_2 photocatalytic disinfection. *Water Res*, **38**: 1069–1077.

Cho M, Chung H, Choi W, Yoon J (2005). Different inactivation behaviors of MS-2 phage and *Escherichia coli* in TiO_2 photocatalytic disinfection. *Appl Environ Microbiol*, **71**: 270–275.

Chong MN, Jin B, Chow CW, Saint C (2010). Recent developments in photo catalytic water treatment technology: A review. *Water Res*, **44**: 2997–3027.

Cortes-Martýnez R, Martýnez-Miranda V, Solache-Rýos M, Garcýa-Sosa I (2004). Evaluation of natural and surfactant modified zeolites in the removal of cadmium from aqueous solutions. *Separ Sci Tech*, **39**: 2711–2730.

Covarrubias C, Garcýa R, Arriagada R, Yanez J, Garland MT (2006). Cr(III) exchange on zeolites obtained from kaolin and natural mordenite. *Micropor Mesopor Mater*, **88**: 220–231.

Cui S, Shen X, Lin B (2006). Surface organic modification of Fe_3O_4 nanoparticles by silane-coupling agents. *Rare Metals*, **25**: 426–430.

Curkovi L, Cerjan-stefanovic S, Filipan T (1997). Metal ion exchange by natural and modified zeolites. *Water Res*, **31**: 1379–1382.

Das G, Pradhan NC, Madhu GM, Preetham HS (2013). Removal of cadmium from aqueous streams by zeolite synthesized from fly ash. *J Mater Environ Sci*, **4**(3): 410–419.

Deng S, Upadhyayula VKK, Smith GB, Mitchell MC (2008). Adsorption equilibrium and kinetics of single-walled carbon nanotubes. *IEEE Sens J*, **8**: 954–962.

Diallo MS, Christie S, Swaminathan P, Balogh L, Shi X, Um W, Papelis C, Goddard WA, Johnson JH (2004). Dendritic chelating agents 1. Cu(II) binding to ethylene diamine core poly(amidoamine) dendrimers in aqueous solutions. *Langmuir*, **20**: 2640–2651.

Diallo MS, Christie S, Swaminathan P, Johnson JH, Goddard WA (2005). Dendrimer-enhanced ultrafiltration. Recovery of Cu(II) from aqueous solutions using Gx–NH_2 PAMAM dendrimers with ethylene diamine core. *Environ Sci Technol*, **39**: 1366–1377.

Diederich F, Thilgen C (1996). Covalent fullerene chemistry. *Science*, **271**: 317–323.

Diwald O, Thompson TL, Zubkov T, Goralski EG, Walck SD, Yates JT, Jr (2004). Photochemical activity of nitrogen-doped rutile TiO_2 (110) in visible light. *J Phys Chem B*, **108**: 6004–6008.

Egerton TA, Kosa SAM, Christensen PA (2006). Photoelectrocatalytic disinfection of *E. coli* suspensions by iron doped TiO_2. *Phys Chem Chem Phys*, **8**: 398–406.

Einaga H, Futamura S, Ibusuki T (1999). Photocatalytic decomposition of benzene over TiO_2 in a humidified air stream. *Phys Chem Chem Phys*, **20**: 4903–4908.

Eswaramoorthy M, Sen R, Rao CNR (1999). A study of micropores in single-walled carbon nanotubes by the adsorption of gases and vapors, *Chem Phys Lett*, **304**: 207–210.

Fagan SB, Souza Filho A, Lima J, Filho JM, Ferreira OP, Mazali IO, Alves OL, Dresselhaus MS (2004). 1,2-dichlorobenzene interacting with carbon nanotubes. *Nano Lett*, **4**: 1285–1288.

Falcao EHL, Wudl F (2007). Carbon allotropes beyond graphite and diamond. *J Chem Technol Biotechnol*, **82**: 524–531.

Feng L, Cao M, Ma X, Zhu Y, Hu C (2012). Superparamagnetic high-surface-area Fe_3O_4 nanoparticles as adsorbents for arsenic removal. *J Hazard Mater*, **217–218**: 439–446.

Friedmann D, Mendive C, Bahnemann D (2010). TiO_2 for water treatment: Parameters affecting the kinetics and mechanisms of photocatalysis. *Appl Catal B Environ*, **99**: 398–406.

Fujishima A, Rao TN, Tryk DA (2000). Titanium dioxide photocatalysis. *J Photochem Photobiol C Photochem Rev*, **1**: 1–21.

Goering J, Kadossov E, Burghaus U (2008). Adsorption kinetics of alcohols on singlewall carbon nanotubes: An ultrahigh vacuum surface chemistry study. *J Phys Chem C*, **112**: 10114–10124.

Gallego M, Pena YPd, Valcarcel M (1994). Fullerenes as sorbent materials for metal preconcentration. *Anal Chem*, **66**: 4074–4078.

Gaya UI, Abdullah AH (2008). Heterogeneous photocatalytic degradation of organic contaminants over titanium dioxide: A review of fundamentals, process and problems. *J Photochem Photobiol A Chem*, **9**: 1–12.

Ge F, Li MM, Ye H, Zhao BX (2012). Effective removal of heavy metal ions Cd^{2+}, Zn^{2+}, Pb^{2+}, Cu^{2+} from aqueous solution by polymer-modified magnetic nanoparticles. *J Hazard Mater*, **211–212**: 366–372.

Gelover S, Gomez LA, Reyes K, MT Leal (2006). A practical demonstration of water disinfection using TiO_2 films and sunlight. *Water Res*, **40**: 3274–3280.

Gong J, Chen L, Zeng G, Long F, Deng J, Niu Q, He X (2012). Shellac-coated iron oxide nanoparticles for removal of cadmium(II) ions from aqueous solution. *J Environ Sci*, **24**: 1165–1173.

Guo CF, Cao S, Zhang J. Tang H, Guo S, Tian Y, Liu Q (2011). Topotactic transformations of superstructures: From thin films to two-dimensional networks to nested two-dimensional networks. *J Am Chem Soc*, **133**: 8211–8215.

Guo CF, Zhang J, Tian Y, Liu Q (2012). A general strategy to superstructured networks and nested self-similar networks of bismuth compounds. *ACS Nano*, **6**: 8746–8752.

Guo CF, Zhang J, Wang M, Tian Y, Liu Q (2013). A Strategy to prepare wafer scale bismuth compound superstructures. *Small*, **9**: 2394–2398.

Hajkova P, Spatenka P, Horsky J, Horska I, Kolouch A (2007). Photocatalytic effect of TiO_2 films on viruses and bacteria. *Plasma Process. Polym*, **43**: 97–401.

Hao YM, Man C, Hu ZB (2010). Effective removal of Cu(II) ions from aqueous solution by amino-functionalized magnetic nanoparticles. *J Hazard Mater*, **184**: 392–399.

Hilding J, Grulke EA, Sinnott SB, Qian D, Andrews R, Jagtoyen M (2001). Sorption of butane on carbon multiwall nanotubes at room temperature. *Langmuir*, **17**: 7540–7544.

Hoa ND, Quy NV, Kim D (2009). Nanowire structured SnOx-SWNT composites: High-performance sensor for NOx detection. *Sens Actuators B Chem*, **142**: 253–259.

Hoffmann MR, Martin ST, Choi W, Bahnemannt DW (1995). Environmental applications of semiconductor photocatalysis. *Chem Rev*, **95**: 69–96.

Huang SH, Chen DH (2009). Rapid removal of heavy metal cations and anions from aqueous solutions by an amino functionalized magnetic nano-adsorbent. *J Hazard Mater*, **163**: 74–179.

Hui, KS, Chao CY, Kot SC (2005). Removal of mixed heavy metal ions in wastewater by zeolite 4A and residual products from recycled coal fly ash. *J Hazard Mater*, **127**: 89–189.

Hyung H, Kim JH (2008). Natural organic matter (NOM) adsorption to multi-walled carbon nanotubes: Effect of NOM characteristics and water quality parameters. *Environ Sci Technol*, **42**: 4416–4421.

Ibanez JA, Litter MI, Pizarro RA (2003). Photocatalytic bactericidal effect of TiO_2 on Enterobacter cloacae. Comparative study with other Gram (-) bacteria. *J Photochem Photobiol A Chem*, **157**: 81–85.

Ibrahimi MM, Sayyadi AS (2015). Application of natural and modified zeolites in removing heavy metal Cations from aqueous media: an overview of including parameters affecting the process. *Int J Geol Agri Environ Sci*, **3**: 1–7.

Iijima S (1991). Helical microtubules of graphitic carbon. *Nature*, **354**: 56–58.

Ireland JC, Klostermann P, Rice EW, Clark RM (1993). Inactivation of *Escherichia coli* by titanium dioxide photocatalytic oxidation. *Appl Environ Microbiol*, **59**: 1668–1670.

Irie H, Watanabe Y, Hashimoto K (2003). Nitrogen-concentration dependence on photocatalytic activity of $TiO_{2-x}N_x$ powders. *J Phys Chem B*, **107**: 5483–5486.

Ismael IS, Melegy A, Kratochvil T (2012). Lead removal from aqueous solution by natural and pretreated zeolites. *Geotech Geol Eng*, **30**: 253–262.

Jeong Y, Fan M, Singh S, Chuang CL, Saha B, Leeuwen JHV (2007). Evaluation of iron oxide and aluminium oxide as potential arsenic(V) adsorbents. *Chem Eng Process*, **46**: 1030–1039.

Jha, VK, Matsuda M, Miyake M (2008). Sorption properties of the activated carbon-zeolite composite prepared from coal fly ash for Ni^{2+}, Cu^{2+}, Cd^{2+} and Pb^{2+}. *J Hazard Mater*, **160**: 148–153.

Johnson R, Bethune D, Yannoni C (1992). Fullerene structure and dynamics—A magnetic-resonance potpourri. *Acc Chem Res*, **25**: 169–175.

Johnson, RD, Meijer G, Bethune DS (1990). C-60 has icosahedral symmetry. *J Am Chem Soc*, **112**: 8983–8984.

Kandah MI, Meunier JL (2007). Removal of nickel ions from water by multi-walled carbon nanotubes. *J Hazard Mater*, **146**: 283–288.

Kang S, Mauter MS, Elimelech M (2009). Microbial cytotoxicity of carbon-based nanomaterials: Implications for river water and wastewater effluent. *Environ Sci Technol*, **43**: 2648–2653.

Kango S, Kalia S, Celli A, Njuguna J, Habibi Y, Kumar R (2013). Surface modification of inorganic nanoparticles for the development of organic-inorganic nanocomposites. *Prog Polym Sci*, **38**: 1232–1261.

Karami H (2013). Heavy metal removal from water by magnetite nanorods. *Chem Eng J*, **219**: 209–216.

Khachatryan SV (2014). Heavy metal adsorption by armenian natural zeolite from natural aqueous solutions. *Chem Biol*, **2**: 31–35.

Khan SUM, Al-Shahry M, Ingler JW (2002). Efficient photochemical water splitting by a chemically modified n-TiO_2. *Science*, **297**: 2243–2245.

Kim EJ, Lee CS, Chang YY, Chang YS (2013). Hierarchically structured manganese oxide-coated magnetic nanocomposites aqueous systems. *ACS Appl Mater Interface*, **19**: 9628–9634.

Kiri P, Hyett G, Binions R (2010). Solid state thermochromic materials. *Adv Mater Lett*, **1**: 86–105.

Kitano M, Matsuoka M, Ueshima M, Anpo M (2007). Recent developments in titanium oxide-based photocatalysts. *Appl Catal A*, **325**: 1–14.

Klosek S, Raftery D (2001). Visible light driven V-doped TiO_2 photocatalyst and its photooxidation of ethanol. *J Phys Chem B*, **105**: 2815–2819.

Kroto HW (1987). The stability of the fullerenes. C-24, C-28, C-32, C-36, C-50, C-60 and C-70. *Nature*, **329**: 529–531.

Kroto HW, Heath JR, Obrien SC, Curl RF, Smalley RE (1985). C-60—Buckminster fullerene. *Nature*, **318**: 162–163.

Laurent S, Forge D, Port M, Roch A, Robic C, Elst LV, Robert N (2008). Magnetic iron oxide nanoparticles: Synthesis, stabilization, vectorization, physicochemical characterizations and biological applications. *Chem Rev*, **108**: 2064–2110.

Lehn JM, Atwood JL, Davies JED, Nicol MD, Vogtle, F (1996). *Comprehensive Supramolecular Chemistry*; Pergamon: New York.

Li YH, Ding J, Luan ZK, Di ZC, Zhu YF, Xu CL, Wu DH, Wei BQ (2003). Competitive adsorption of Pb^{2+}, Cu^{2+} and Cd^{2+} ions from aqueous solutions by multiwalled carbon nanotubes. *Carbon*, **41**: 2787–2792.

Li Q, Mahendra S, Lyon DY, Brunet L, Liga MV, Li D (2008). Antimicrobial nanomaterials for water disinfection and microbial control: Potential applications and implications. *Water Res*, **42**: 4591–4602.

Li YH, Wang SG, Wei JQ, Zhang XF, Xu CL, Luan ZK, Wu DH, Wei BQ (2002). Lead adsorption on carbon nanotubes. *Chem Phys Lett*, **357**: 263–266.

Li H, Xiao DL, He H, Lin R, Zuo PL (2013). Adsorption behavior and adsorption mechanism of Cu (II) ions on aminofunctionalized magnetic nanoparticles. *Trans Nonferrous Met Soc China*, **23**: 2657–2665.

Li M, Zhang S, Peng Y, Lv L, Pan B (2015). Enhanced visible light responsive photocatalytic activity of TiO_2-based nanocrystallites: Impact of doping sequence. *RSC Adv*, **5**: 7363–7369.

Lin S, Lu D, Lu Z (2012). Removal of arsenic contaminants with magnetic γ-Fe_2O_3 nanoparticles. *Chem Eng*, **211–212**: 46–52.

Liu L, Chen XB (2014). Titanium dioxide nanomaterials: Self-structural modifications. *Chem Rev*, **114**: 9890–9918.

Liu SX, Liu JL, Li XS, Zhu X, Zhu AM (2015). Gliding arc plasma synthesis of visible light active C-doped titania photocatalysts. *Plasma Process Polym*, **12**: 422–430.

Liu HL, Yang TCK (2003). Photocatalytic inactivation of *Escherichia coli* and Lactobacillus helveticus by ZnO and TiO_2 activated with ultraviolet light. *Process Biochem*, **39**: 475–481.

Liu H, Yang W, Ma Y, Cao Y, Yao J, Zhang J, Hu T. (2003). Synthesis and characterization of titania prepared by using a photoassisted sol-gel method, *Langmuir*, **19**: 3001–3005.

Liu JF, Zhao ZS. Jiang GB (2008). Coating Fe_3O_4 magnetic nanoparticles with humic acid for high efficient removal of heavy metals in water. *Environ Sci Technol*, **42**: 6949–6954.

Long RQ, Yang RT (2001). Carbon nanotubes as a superior sorbent for dioxin removal. *J Am Chem Soc*, **123**: 2058–2059.

Lonnen J, Kilvington S, Kehoe SC, Al-Touati F, Mc Guigan KG (2005). Solar and photocatalytic disinfection of protozoan, fungal and bacterial microbes in drinking water. *Water Res*, **39**: 877–883.

Lu C, Chung YL, Chang KF (2005). Adsorption of trihalomethanes from water with carbon nanotubes. *Water Res*, **39**: 1183–1189.

Lu C, Liu C (2006). Removal of nickel (II) from aqueous solution by carbon nanotubes. *J Chem Technol Biotechnol*, **81**: 1932–1940.

Luo G, Yao H, Xu M, Cui X, Chen W, Gupta R, Xu Z (2010). Carbon nanotube-silver composite for mercury capture and analysis. *Energy Fuels*, **24**: 419–426.

Ma Z, Zhao D, Chang Y, Xing S, Wu Y, Gao Y (2013). Synthesis of $MnFe_2O_4$@Mn-Co oxide core-shell nanoparticles and their excellent performance for heavy metal removal. *Dalton Trans*, **42**: 14261–14267.

Ma J., Zhu Z, Chen B, Yang M, Zhou H, Li C, Yu F, Chen J (2013). One-pot, large-scale synthesis of magnetic activated carbon nanotubes and their applications for arsenic removal. *J Mater Chem A*, **1**: 4662–4666.

Madrakian T, Afkhami A, Ahmadi M (2013). Simple in situ functionalizing magnetite nanoparticles by reactive blue-19 and their application to the effective removal of Pb^{2+} ions from waters samples. *Chemosphere*, **90**: 542–547.

Mahdavian AR, Mirrahimi MAS (2010). Efficient separation of heavy metal cations by anchoring polyacrylic acid on superparamagnetic magnetite nanoparticles through surface modification. *Chem Eng J*, **159**: 264–271.

Marschall R, Wang L (2014). Non-metal doping of transition metal oxides for visible-light photocatalysis. *Catal Today*, **225**: 111–135.

Mandal BK, Suzuki KT (2002). Arsenic round the world: A review. *Talanta*, **58**: 201–235.

Maness PC, Smolinski S, Blake DM, Huang Z, Wolfrum EJ, Jacoby WA (1999). Bactericidal activity of photocatalytic TiO_2 reaction: Toward an understanding of its killing mechanism. *Appl Environ Microbiol*, **65**: 4094–4098.

Merkoci A (2006). Carbon nanotubes in analytical sciences. *Microchem Acta*, **152**: 157–174.

Mier MV, Callejas RL, Gehr R, Cisneros BEJ, Alvarez PJJ. (2001). Heavy metal removal with mexican clinoptilolite: Multicomponent ionic exchange. *Water Res*, **35**: 373–378.

Mohammad-Beigi H, Yaghmaei S, Roostaazad R, Bardania H, Arpanaei A (2011). Effect of pH, citrate treatment and silanecoupling agent concentration on the magnetic, structural and surface properties of functionalized silica-coated iron oxide nanocomposite particles. *Phys E*, **44**: 618–627.

Mou F, Guan J, Xiao Z, Sun Z, Shi W, Fan AX (2011). Solvent-mediated synthesis of a magnetic Fe_2O_3 chestnutlike amorphous-core/γ-phase-shell hierarchical nanostructures with strong As(V) removal capability. *J Mater Chem*, **21**: 5414–5421.

Nassar NN (2010). Rapid removal and recovery of Pb(II) from wastewater by magnetic nanoadsorbents. *J Hazard Mater*, **184**: 538–546.

Nellore BPV, Kanchanapally RF, Sinha SS, Pramanik A, Hamme AT, Arslan Z, Sardar D, Ray PC (2015). Bio-Conjugated CNT-Bridged 3D porous graphene oxide membrane for highly efficient disinfection of pathogenic bacteria and removal of toxic metals from water. *ACS Appl Mater Interfaces*, **7**: 19210–19218.

Neppolian B, Yamashita H, Okada Y, Nishijima H, Anpo M (2005). Preparation of unique TiO_2 nanoparticle photocatalysts by a multi-gelation method for control of the physicochemical parameters and reactivity. *Catal Lett*, **105**: 111–117.

Ni M, Leung MKH, Leung DYC, Sumathy K (2007). A review and recent developments in photocatalytic water-splitting using TiO_2 for hydrogen production. *Renew Sustainable Energy Rev*, **11**: 401–425.

Noworyta K, Augustynski J (2004). Spectral photoresponses of carbon-doped TiO_2 film electrodes. *Electrochem Solid-State Lett*, **7**: 31–33.

Nuzzo JB (2006). The biological threat to U.S. water supplies: Towards national water security policy. *Biosecur Bioterror*, **4**: 147–159.

Oh JK, Park JM (2011). Iron oxide-based superparamagnetic polymeric nanomaterials: Design, preparation, and biomedical application. *Prog Polym Sci*, **36**: 168–189.

Ohno T, Mitsui T, Matsumura M (2003). Photocatalytic activity of S-doped TiO_2 photocatalyst under visible light. *Chem Lett*, **32**: 364–365.

Olivi M, Zanni E, Bellis GD, Talora C, Sarto MS, Palleschi C, Flahaut E, Monthioux M, Rapino S, Uccelletti D, Fiorito S (2013). Inhibition of microbial growth by carbon nanotubes networks. *Nanoscale*, **5**: 9023–9029.

Page K, Palgrave RG, Parkin IP, Wilson M, Savin SLP, Chadwick AV (2007). Titania silver-titania composite films on glass—potent antimicrobial coatings. *J Mater Chem*, **17**: 95–104.

Parham H, Zargar B, Shiralipour R (2012). Fast and efficient removal of mercury from water samples using magnetic iron oxide nanoparticles modified with 2-mercaptobenzothiazole. *J Hazard Mater*, **205–206**: 94–100.

Peng X, Luan Z, Ding J, Di Z, Li Y, Tian B (2005). Ceria nanoparticles supported on carbon nanotubes for the removal of arsenate from the water. *Mater Lett*, **59**: 399–403.

Penza M, Rossi R, Alvisi M, Signore MA, Cassano G, Dimaio D, Pentassuglia R, Piscopiello E, Serra E, Falconieri M (2009). Characterization of metal-modified and vertically-aligned carbon nanotube films for functionally enhanced gas sensor applications. *Thin Solid Films*, **517**: 6211–6216.

Pragnesh ND, Lakhan VC (2014). Application of iron oxide nanomaterials for the removal of heavy metals. *J Nanotechnol*, **2014**: 1–14.

Qu X, Alvarez PJ, Li Q (2013). Applications of nanotechnology in water and wastewater treatment. *Water Res*, **47**: 3931–3946.

Roy A, Bhattachary J (2012). Removal of Cu(II), Zn(II) and Pb(II) from water using microwave-assisted synthesized maghemite nanotubes. *Chem Eng J*, **211–212**: 493–500.

Sakthivel S, Kisch H (2003). Daylight photocatalysis by carbon-modified titanium dioxide. *Angew Chem Int Ed Engl*, **42**: 4908–4911.

Samonin VV, Nikonova VY, Podvyaznikov ML (2008). Sorption properties of fullerene-modified activated carbon with respect to metal ions. *Protect Metals*, **44**: 190–192.

Serrano A, Gallego M (2006). Fullerenes as sorbent materials for benzene, toluene, ethylbenzene, and xylene isomers preconcentration. *J Sep Sci*, **29**: 33–40.

Shahbazi A, Younesi H, Badiei A (2011). Functionalized SBA-15 mesoporous silica by melamine-based dendrimer amines for adsorptive characteristics of Pb(II), Cu(II) and Cd(II) heavy metal ions in batch and fixed bed column. *Chem Eng J*, **168**(2): 505–518.

Shen YF, Tang J, Nie HZ, Wang YD, Ren Y, Zuo L (2009). Preparation and application of magnetic Fe_3O_4 nanoparticles for wastewater purification. *Sep Purif Technol*, **68**: 312–319.

Shephard GS, Stockenstrom S, Villiers Dd, Engelbrecht WJ, Wessels GFS (2002). Degradation of microcystin toxins in a falling film photocatalytic reactor with immobilized titanium dioxide catalyst. *Water Res*, **36**: 140–146.

Shi B, Zhuang X, Yan X, Lu J, Tang H (2010). Adsorption of atrazine by natural organic matter and surfactant dispersed carbon nanotubes. *J Environ Sci (China)*, **22**: 1195–1202.

Singh S, Barick KC, Bahadur D (2011). Surface engineered magnetic nanoparticles for removal of toxic metal ions and bacterial pathogens. *J Hazard Mater*, **192**: 1539–1547.

Singh S, Barick KC, Bahadur D (2013). Fe_3O_4 embedded ZnO nanocomposites for the removal of toxic metal ions organic dyes and bacterial pathogens. *J Mater Chem A*, **10**: 3325–3333.

Sokmen, Candan M, Sumer Z (2001). Disinfection of *E. coli* by the Ag-TiO_2/UV system: lipid peroxidation. *J Photochem Photobiol A Chem*, **143**: 241–244.

Song K, Han X, Shao G (2013). Electronic properties of rutile TiO_2 doped with 4*d* transition metals: First-principles study. *J Alloy Compd*, **551**: 118–124.

Srinivasan C, Somasundaram N (2003). Bactericidal and detoxification effects of irradiated semiconductor catalyst TiO_2. *Curr Sci*, **85**: 1431–1436.

Srinivasa GS, Yarema KJ (2007). *Nanotechnologies for the Life Sciences: Dendrimers in Cancer Treatment and Diagnosis*, 7. New York, Wiley.

Srivatsava A, Srivatsava ON, Talapatra S, Vajtai R, Ajayan PM (2004). Carbon nanotube filters. *Lett Nat*, **3**: 610–614.

Stone AJ, Wales DJ (1986), Theoreticalstudies of icosahedral C-60 and some related species. *Chem Phys Lett*, **128**: 501–503.

Sun C, Qu R, Ji C, Wang C, Sun Y, Yue Z, Cheng G (2006). Preparation and adsorption properties of crosslinked polystyrene-supported low generation diethanolamine-typed dendrimer for metal ions. *Talanta*, **70**(1): 14–19.

Sung-Suh HM, Choi JR, Hah HJ, Koo SM, Bae YC (2004). Comparison of Ag deposition effects on the photocatalytic activity of nanoparticulate TiO_2 under visible and UV light irradiation. *J Photochem Photobiol A Chem*, **16**: 337–344.

Taamneh Y, Sharadqah S (2016). The removal of heavy metals from aqueous solution using natural Jordanian zeolite. *Appl Water Sci*, 1–8.

Tan Y, Chen M, Hao Y (2012). High efficient removal of Pb(II) by amino-functionalized Fe_3O_4 magnetic nanoparticles. *Chem Eng J*, **191**: 104–111.

Tan X, Fang M, Chen C, Yu S, Wang X (2008). Counterion effects of nickel and sodium dodecylbenzene sulfonate adsorption to multiwalled carbon nanotubes in aqueous solution. *Carbon*, **46**: 1741–1750.

Taylor R, Walton DRM (1993). The chemistry of fullerenes. *Nature*, **363**: 685–693.

Tellez GL, Dýaz CEB, Hernandez PB, Morales GR, Bilyeu B (2011). Removal of hexavalent chromium in aquatic solutions by iron nanoparticles embedded in orange peel pith. *Chem Eng J*, **173**: 480–485.

Tian G, Fu H, Jing L, Tian C (2009). Synthesis and photocatalytic activity of stable nanocrystalline TiO_2 with high crystallinity and large surface area. *J Hazard Mater*, **161**: 1122–1130.

Tian X, Zhou S, Zhang Z, He X, Yu M, Lin D (2010). Metal impurities dominate the sorption of a commercially available carbon nanotube for Pb(II) from the water. *Environ Sci Technol*, **44**: 8144–8149.

Tomalia DA, Baker H, Dewald J, Hall M, Kallos G, Martin S, Roeck J, Ryder J, Smith P (1984). A new class of polymers: starburst-dendritic macromolecules. *Polym J*, **17**: 117–132.

Towell MG, Browne LA, Paton GI, Semple KT (2011). The impact of carbon nanomaterials on the behaviour of C-14-phenanthrene and C-14-benzo-[a] pyrene in the soil. *Environ Pollut*, **159**: 706–715.

Tseng JY, Chang CY, Chang CF, Chiang PC (2009). Kinetics and equilibrium of desorption removal of copper from magnetic polymer adsorbent. *J Hazard Mater*, **171**: 370–377.

Umebayashi T, Yamaki T, Itoh H, Asai K (2002). Band gap narrowing of titanium dioxide by sulfur doping. *Appl Phys Lett*, **81**: 454–456.

Upadhyayula VKK, Deng S, Mitchell MC, Smith GB, Nair VS, Ghoshroy S (2008). Adsorption kinetics of Escherichia coli and Staphylococcus aureus on single-walled carbon nanotube aggregates. *Wat Sci Technol*, **58**: 179–84.

Vamathevan V, Amal R, Beydoun D, Low G, McEvoy S (2004). Silver metallisation of titania particles: Effects on photoactivity for the oxidation of organics. *Chem Eng J*, **98**: 127–139.

Wang LJ, Jiang Q, Zhao L (2012). Water-soluble Fe_3O_4 nanoparticles with a high solubility for removal of heavymetal ions from wastewater. *Dalton Trans*, **41**: 4544–4551.

Wang X, Lu J, Xing B (2008). Sorption of organic contaminants by carbonnanotubes: Influence of adsorbed organic matter. *Environ Sci Technol*, **12**: 3207–3212.

Wang Y, Zou B, Gao T, Wu X, Lou S, Zhou S (2012). Synthesis of orange-like Fe_3O_4/PPy composite microspheres and their excellent Cr(VI) ion removal properties. *J Mater Chem*, **22**: 9034–9040.

Wei C, Lin WY, Zalnal Z, Williams NE, Zhu K, Kruzic AP, Smith RL, Rajeshwar K (1994). Bactericidal activity of TiO_2 photocatalyst in aqueous media: Toward a solar-assisted water disinfection system. *Environ Sci Technol*, **28**: 934–938.

Wei L, Shizhen H, Wenzhe C (2010). An MWCNT-doped SnO_2 thin film NO_2 gas sensor by RF reactive magnetron sputtering. *J Semicond*, **31**: 1–7.

Wei Z, Xing R, Zhang X, Liu S, Yu H, Li P (2013). Facile template-free fabrication of hollow nestlike-Fe_2O_3 nanostructures for water treatment. *ACS Appl Mater Interface*, **5**: 598–604.

Xin X, Wei Q, Yang J, Yan L, Feng R, Chen G, Du B, Li H (2012). Highly efficient removal of heavy metal ions by amine-functionalized mesoporous Fe_3O_4 nanoparticles. *Chem Eng J*, **184**: 132–140.

Xing M, Li X, Zhang J (2014). Synergistic effect on the visible light activity of Ti^{3+} doped TiO_2 nanorods/boron doped graphene composite. *Sci Rep*, **4**: 5493–5499.

Xu, Y, Hou T, Sun X, Li Y (2012). The influence of defects on Mo-doped TiO_2 by first-principles studies. *Chem Phys Chem*, **13**: 1514–1521.

Xu D, Tan X, Chen C, Wang X (2008). Removal of Pb(II) from aqueous solution by oxidized multiwalled carbon nanotubes. *J Hazard Mater*, **154**: 407–416.

Xu P, Zeng GM, Huang DL, LC, Zhao MH, Wei Z, Li NJ, Huang C, Xie GX (2012). Adsorption of Pb(II) by iron oxide nanoparticles immobilizedphanerochaetechrysosporium: Equilibrium, kinetic, thermodynamic and mechanisms analysis. *Chem Eng J*, **203**: 423–431.

Yamamoto T, Yamashita F, Tanaka I, Matsubara E, Muramatsu A (2004). Electronic states of sulfur doped TiO_2 by first principles calculations. *Mater Trans*, **45**: 1987–1990.

Yang S, Gao L (2004). New method to prepare nitrogen-doped titanium dioxide and its photocatalytic activities irradiated by visible light. *J Am Ceram Soc*, **87**: 1803–1805.

Yan H, Yang L, Yang Z, Yang H, Li A, Cheng R (2012). Preparation of chitosan/poly (acrylic acid) magnetic composite microspheres and applications in the removal of copper (II) ions from aqueous solutions. *J Hazard Mater*, **229–230**: 371–380.

Yang W, Kan AT, Chen W, Tomson MB (2010). PH dependent effect of zinc on arsenic adsorption to magnetite nanoparticles. *Water Res*, **44**: 5693–5701.

Yang K, Zhu LZ, Xing BS (2006). Adsorption of polycyclic aromatic hydrocarbons by carbon nanomaterials. *Environ Sci Technol*, **40**: 1855–1861.

Yantasee W, Warner CL, Sangvanich T, Addleman RS, Carter TG, Wiacek RJ, Fryxell GE, Timchalk C, Warner MG (2007). Removal of heavy metals from aqueous systems with thiol functionalized superparamagnetic nanoparticles. *Environ Sci Technol*, **41**: 5114–5119.

Yavuz CT, Mayo JT, Suchecki C, Wang J, Ellsworth AZ, Dcouto H, Quevedo E, Prakash A, Gonzalez L, Nguyen C, Kelty C, Colvin VL (2010). Pollution magnet: Nano-magnetite for arsenic removal from drinking water. *Environ Geochem Health*, **32**: 327–334.

Yu JC, Ho W, Yu J, Yip H, Po KW, Zhao J (2005). Efficient visible-light-induced photocatalytic disinfection on sulfur-doped nanocrystalline titania. *Environ Sci, Technol*, **39**: 1175–1179.

Yu F, Ma J, Wu Y (2012). Adsorption of toluene, ethylbenzene and xylene isomers on multi-walled carbon nanotubes oxidized by different concentration of NaOCl. *Front Environ Eng Sci*, **6**: 320–329.

Yu X, Tong S, Ge M, Zuo J, Cao C, Song W (2013). One-step, synthesis of magnetic composites of cellulose@iron oxide nanoparticles for arsenic removal. *J Mater Chem A*, **1**: 959–965.

Yu JC, Yu J, Ho W, Jiang Z, Zhang L (2002). Effects of F-doping on the photocatalytic activity and microstructures of nanocrystalline TiO_2 powders. *Chem Mater*, **14**: 3808–3816.

Yuan P, Fan M, Yang D, He H, Liu D, Yuan A, Zhu J, Chen T (2009). Montmorillonite-supported magnetite nanoparticles for the removal of hexavalent chromium Cr(VI) from aqueous solutions. *J Hazard Mater*, **166**: 821–829.

Zaki MI, Mekhemer GAH, Fouad NE, Jagadale TC, Ogale SB (2010). Surface texture and specific adsorption sites of sol-gelsynthesized anatase TiO_2 nanoparticles. *Mater Res Bull*, **45**: 1470–1475.

Zamzow MJ, Eichbaum BR, Sandgren KR, Shanks DE (1990). Removal of heavy metals and other cations from wastewater using zeolites. *Sep Sci Technol*, **25**: 1555–1569.

Zan L, Fa W, Peng T, Gong ZK (2007). Photocatalysis effect of nanometer TiO_2 and TiO_2-coated ceramic plate on Hepatitis B virus. *J Photochem Photobiol B Biol*, **86**: 165–169.

Zhang C, Sui J, Li J, Tang Y, Cai W (2012). Efficient removal of heavy metal ions by thiol-functionalized super magnetic carbon nanotubes. *Chem Eng J*, **210**: 45–52.

Zhang X, Zhou M, Lei L (2005). Preparation of an Ag-TiO_2 photocatalyst coated on activated carbon by MOCVD. *Mater Chem Phys*, **91**: 73–79.

Zhang F, Zhu Z, Dong Z, Cui Z, Wang H, Hu W, Zhao P, Wang P, Wei S, Li R, Ma J (2011). Magnetically recoverable facile nanomaterials: Synthesis, characterization and application in remediation of heavy metals. *Microchem J*, **98**: 328–333.

Zhou YT, Nie HL, Branford WC, He ZY, Zhu LM (2009). Removal of Cu^{2+} from aqueous solution by chitosan-coated magnetic nanoparticles modified with α-ketoglutaric acid. *J Colloid Interface Sci*, **330**: 29–37.

Zhou Q, Xiao J, Wang W (2006). Using multi-walled carbon nanotubes as solid phase extraction adsorbents to determine dichlorodiphenyltrichloroethane and its metabolites at trace level in water samples by high-performance liquid chromatography with UV detection. *J Chromatogr A*, **1125**: 152–158.

Zhu J, Wei S, Gu H, Rapole SB, Wang Q, Luo Z, Haldolaarachchige N, Young DP, Guo Z (2012). One-pot synthesis of magnetic graphene nanocomposites decorated with core@double-shell nanoparticles for fast chromium removal. *Environ Sci Technol*, **46**: 977–985.

Zu F, Bozhilov K, Dillon RJ, Wang L, Smith P, Zhao X, Bardeen C, Feng PY (2012). Active facets on titanium (III)-doped TiO_2: An effective strategy to improve the visible light photocatalytic activity. *Angew Chem Int Ed*, **124**: 6327–6330.

Part 2

Carbon Nanomaterial-Based Nanocomposites for Pollution Control

Chapter 8

Carbon Nanostructures: Applications and Perspectives for a Green Future

Ankita Dhillon, Ritu Painuli, and Dinesh Kumar

Department of Chemistry, Banasthali University, Rajasthan 304022, India

dsbchoudhary2002@gmail.com

8.1 Introduction

Nanomaterials are actively considered as contaminants of emerging concern by environmental researchers. Studying the negative environmental consequences of engineered nanomaterials highlights the necessity of designing benign nanomaterials. In this regard, the excellent chemical, electrical and physical properties of carbon nanomaterials prompt advanced answers to sustained ecological disputes.

The utilisation of nanotechnology in biomedical applications has paved a path for equivalent advancements in both environmental science and engineering. Advancement in carbon-based nanomaterials has revolutionised the nanotechnology area. Carbon-based nanomaterials and their functional derivatives have been continuously utilised in the drugs delivery via

Nanocomposites for Pollution Control
Edited by Chaudhery Mustansar Hussain and Ajay Kumar Mishra

ISBN 978-981-4774-45-1 (Hardcover), 978 1 315 14308-2 (eBook)
www.panstanford.com

dense tissues (Martin et al., 2003), in targeting cancerous cells (Kam et al., 2005), and functionalized derivatives as artificial transmembrane pores (Lopez et al., 2004). The environmental applications of carbon-based nanomaterials have been outlined in the present chapter. First, a brief discussion is done on carbon-based nanomaterials and their analogous properties. After that the application of carbon-based nanomaterials in wastewater treatment, air pollution monitoring, biotechnologies, renewable energy technologies, and green nanocomposites are discussed. Then, the potential hazards of these nanomaterials for large-scale application are discussed. The distinctive properties of carbon-based nanomaterials when these are utilised within a sustainable frame have facilitated their excellent environmental applications with limited allegations.

8.2 Carbon-Based Nanomaterials

Carbon-based nanomaterials, comprising fullerenes and single- and multi-walled nanotubes of carbon nanoparticles, are presently the smart nanomaterials from the application point of view. These nanomaterials present excellent chemical, physical, and electronic properties compared with their bulk forms (Tenne, 2006).

The unique hybridization characteristics of carbon and the ability of structural alterations in them during synthesis conditions have resulted in high-degree tailored manipulation not so far achieved by inorganic nanocomposites. The present chapter critically reviews various carbon-based nanomaterials for their environmental application.

All the chemical, physical, mechanical, and electronic properties of carbon-based nanomaterials are strictly related to the structural conformation of carbon and, so, its hybridization state (Ajayan, 1999). The ground-state orbital configuration of carbon is $1s^2$, $2s^2$, $2p^2$. In the excited state, the low-energy orbital electron gets promoted to the higher-energy p orbital as a consequence of low energy gap between the 2s and 2p electron shells. These promotions can result in various types of hybridization of carbon such as sp, sp^2, sp^3 depending on bonding relationships of carbon with the neighbouring atoms.

The higher energy excitation energy is compensated by the energy gained from covalent bonding with neighbouring atoms (Hu et al., 2007).

These variable hybridization states explain the presence of a wide variety of organic compounds and the significant variations among bulk configurations of carbon. Carbon assumes the sp^3 configuration of diamond at elevated temperatures or pressures, which are the thermodynamically favourable configurations at such high temperature. However, carbon attains the planar sp^2 conformation at lower temperature conditions and results in the formation of monolayer sheets having three sigma bonds and a single π bond. Some weak more interplane forces involve van der Waals forces and the interaction between parallel sheets overlapping orbitals. These weak forces are disturbed by mild shear forces, physical separation, and chemical modification and therefore resulted in the slipping of graphite's planes.

Graphite sheets' three-dimensional structures are thermodynamically stable at the nanoscale level. The curvature formed results in the generation of strain energy which is balanced by the lowering in a number of unfavourable dangling bonds (Falcao et al., 2007). The resulting structures are fullerenes and nanotubes, which contain a lot of graphite's features, but also demonstrate different and tunable characteristics as a result of quantum effects at the nano level, the presence of bonds having enhanced sp^3 character, quantum restriction in one or more dimensions (Dresselhaus et al., 2001), and closed topology. Therefore, carbon-based nanomaterials demonstrate the same bonding configurations as bulk carbon configurations; however, their properties and morphologies are governed by resonance structures stability as compared to the bulk averages of their crystalline forms.

Fullerenes. The unique chemical, optical, and structural properties of fullerenes are determined by their electronic structure. The perfectly structured fullerenes have cage-like structures which are made up of 12 five-member rings and an indefinite number of six-member rings. The structure having fewer six-member rings have higher sp^3 bonding character, greater strain energies, and more reactivity of carbon sites. The isomers having adjoining five-member rings exhibit lesser stability and lower relative abundance than the isomers having discrete five-member rings where as a result of resonance, π bonds delocalization occurs over the fullerene structure (Kroto, 1987; Campbell et al., 1996). Even though the degree of charge delocalization is still questionable, the chemical behaviour of fullerenes holds between a straight chain alkene and an aromatic molecule (Taylor et al., 1993).

Such type of balance between stability and reactivity discriminates Buckminster fullerene with that of degenerate C-20 fullerenes and inert, planar graphite. Fullerene contains icosahedral (Ih) symmetry and resonance stabilised structures that result in equivalent electronic states and bonding geometries around each carbon atom (Johnson et al., 1990, 1992). Transformations in covalent, supramolecular states allow manipulation at the molecular level and formation of polymeric material for various environmental applications. Although fullerene does not undergo oxidation easily, it can accommodate up to six electrons in its lowest unoccupied molecular orbital (LUMO). Such kind of behaviour leads to covalent fullerene chemistry and fullerenes utilisation as structural scaffolds for reactive adducts.

Utilisation of supramolecular pathways for the structural modifications in fullerenes has led to advancement in their significant properties appropriate for environmental applications. The increase in solubility and reduced aggregation of hydrophobic fullerene molecules can be achieved using supramolecular techniques and such kind of behaviour is of particular significance in the field of biomedical and environmental applications (Lehn et al., 1996; Bonifazi et al., 2007). Therefore, combined utilisation of covalent and supramolecular techniques has led to the development of a range of tailored structures (Diederich et al., 1999).

In addition to above-discussed conformations, carbon-based nanomaterial research has also described some additional fullerene conformations comprising large spherical fullerenes (Shinohara et al., 1991; Wang et al., 2001) and concentric fullerenes named carbon onions (Ugarte, 1992). Such kind of fullerene conformations has excellent reactivity which is due to the symmetry and bonding conformations. On the theoretical basis, large fullerenes and carbon onions adopt a faceted structure whereas, direct imaging, shows that these structures preserve the spherical geometry. This type of ambiguity suggests the presence of Stone-Wales defects in the standard bonding structure (Bates et al., 1998). These are the kind of defects that describe the four six-member rings rearrangement into a 5-7-7-5 conformation (Stone et al., 1986). The resultant conformation has decreased resonance and higher strain energy that enhanced the probability of nucleophilic attack, presenting one possible description for higher reactivity.

Fullerenes have the ability to form stable crystalline nanoparticles of diameter ~25–500 nm in various types of solutions, including water at appropriate pH and electrolyte concentrations (Fortner et al., 2005; Chen et al., 2006; Scrivens et al., 1994; Andrievsky et al., 1995). Aggregation of fullerenes occurs without humic acid and obeys DLVO theory of colloidal stability (Chen et al., 2006) and the average aggregate diameter has been found to a function of pH, ionic strength, mixing behaviour, initial fullerene concentration, and contact time (Duncan et al., 2008; Brant et al., 2005, 2006). On the other hand, in the presence of humic acid, macromolecules get adsorb to the fullerene nanoparticles that lead to steric stabilisation and consequently reduced aggregation rates (Chen et al., 2007).

Nanotubes. Carbon nanotubes (Iijima, 1991) are the one-dimensional analogues of 0-dimensional fullerene molecules. A nanotube is graphene sheet of micrometre size that is turned into a cylinder having a nanosize diameter and capped with a spherical fullerene. Such a nanoscale diameter tube having a micro- to centimetre length leads to the development of CNT structures of extraordinary aspect ratio (Xia et al., 2003; Hu et al., 1999, Dresselhaus et al., 2001).

There are two important aspects by which nanotubes differ from graphene planes. First is the 1-D tubule's geometry (Dresselhaus et al., 2001) and second is the development of structural indeterminacy upon graphene's 2-D hexagonal carbon conformation confinement to one dimension only. In spite of bulk graphite that exhibit semiconductor behaviour, isolated single-walled nanotubes having small diameter can show metallic, semimetallic, or semiconducting behaviour depending on the hexagonal lattice's orientation (Saito et al., 1992).

Double-walled and multiwalled nanotubes (MWNTs) are the 1-D analogues of carbon onions. These nanotubes possess many similar properties of the SWNTs because of weak coupling across the 0.34 nm interlayer distance (Bandow et al., 2001) of MWNTs. However, they do not show the metallic properties of SWNTs, but exhibit semiconductor properties of bulk graphite.

Strong forces of attraction between nanotubes make difficult CNT purification and manipulation. Such kind of forces explains the presence of tightly bundled CNTs conformation (Thess et al., 1996) and negligible dispersion in both polar and non-polar solvents. As a result, another process of physical dispersion like sonication is needed for the unbundling of nanotube sheaths. Other techniques like supramolecular techniques are can be efficiently utilised for the dispersion of CNT bundles, but, they have limited applicability in systems having pristine nanotubes like properties. Although great advancement has been achieved in functionalization and supramolecular modifications (Sun et al., 2002; Holzinger et al., 2001; Andreas, 2002; Georgalikas et al., 2002; Banerjee et al., 2005), the range and management of molecular modification is quite difficult in the case of nanotubes. Further lowering in reactivity of the carbons occurs due to reduced strain energy of the sidewall hexagons and therefore necessitates stronger reagents for modification.

Commercially existing single walled nanotubes and multiwalled nanotubes are normally associated with defects, contaminants, and other physical defects. Although, numerous purification methods have been developed (Chiang et al., 2001; Shelimov et al., 1998; Shi et al., 1999), but differentiation between samples is a matter of future research for the implications and large scale CNTs applications (Plata et al., 2008).

8.3 Unique Properties of Carbon Nanomaterials

Carbon-based nanomaterials have the combined properties of sp^2 hybridised carbon bonds and the extraordinary nanoscale properties of physics and chemistry. Single nanotubes offer good electrical conductivity and bulk nanomaterials offer good adsorptive capacity. Therefore, both single nanotubes and bulk molecules present excellent advancement regarding environmental applications. Surface area, size, and shape; molecular relationships and sorption properties; and optical, electronic, and thermal properties are some of the unique properties of carbonaceous nanomaterials that are usually referred to in environmental applications.

Size and Surface Area: The structure and molecular conformations in the fullerene cage of a material can be controlled using molecular manipulations. Even though existing fabrication practices for carbon-based nanomaterials are not uniform and precise, correlations among growth conditions and properties of developed product tell the production of tuned nanomaterials. A precise combination of fabrication conditions like temperature, pressure, catalyst, and electron field, can results in pure nanomaterial structure (Dresselhaus et al., 2001; Bandow et al., 1998; Cassell et al., 1999; Jost et al., 2004). Such as, the diameter of SWNT is a function of synthesis procedure, nanotubes between 0.7 and 1 nm can be developed using HiPCO synthesis procedure. On the other hand, laser ablation and other graphitic-based procedures results in the development of somewhat larger nanotubes (1–2 nm).

The properties and applications of fullerene and nanotubes structures are strongly correlated to their diameter. In the case of fullerenes, the number of carbons in the spherical molecule can be referred to as the diameter. The small diameters of SWNT lead to elevated strain energies, σ and π bonds and mixing, and rehybridization of electron orbital (Popo, 2004).

As a result of smaller diameter, nanotubes have novel application in separation, and size exclusion procedures comprising nanowire synthesis (Han et al., 1997) and membrane filtration (Holt et al., 2006). However, due to smaller diameters, CNT have limited applications due to complicated purification methods for removing bulk nanotube's growth catalysts (Rinzler et al., 1998).

The presence of high surface area to volume ratio in the case of nanomaterials separates them from their bulk materials across the spectrum of carbonaceous nanomaterials. The transfer of bulk materials into nanoscale leads to the increase in higher percentages of surface atoms. Such surface atoms present good adsorptive properties and broad applications (Lin et al., 2006; Krishna et al., 2005; Kongkanand et al., 2006; Anson et al., 2006).

Molecular Interactions and Sorption Properties. Carbon-based nanomaterials properties are usually studies using conventional physical-chemical models and theories comprising electrostatics (Keblinski et al., 2002), adsorption (Furmaniak et al., 2006), hydrophobicity (Walther et al., 2004), and Hansen solubility parameters (Ham et al., 2005). The interpretation of physical-chemical processes that are difficult to get using experimental techniques has been provided by molecular modelling (Fernandez, 2003).

The classic Lennard–Jones (LJ) continuum model is used to describe the potential energies of interaction between carbon nanomaterials (Girifalco et al., 2000; Girifalco, 1991). The rate of aggregation is reduced by functionalization of nanomaterials via covalent or supramolecular techniques and hydrophilicity in the otherwise hydrophobic nanoparticles is achieved by introducing polar functional groups (Taylor et al., 1993; Diederich et al., 1999; Chen et al., 2007; Terashima et al., 2007).

The adsorption studies of carbon nanomaterials show fast equilibrium rates, good adsorption capacity, lower pH dependency, negligible hysteresis (Béguin et al., 2006), and good correlation with conventional Langmuir, BET, or Freundlich isotherms (Hilding et al., 2001; Chen et al., 2007). However, these studies are complicated as a result of distinctive adsorption characteristics in micropores. More specifically, the high dispersion energy and the overlapping force fields by neighbouring carbon walls enhance sorbent-sorbates and sorbent-sorbent interactions, thus resulting in condensation inside the nanotube (Hilding et al., 2001). Such type of nanotube's filling may be a reason why various adsorption models illustrate the adsorption phenomenon on the surface area of a CNT (Maddox et al., 1995; Inoue et al., 1998; Li et al., 2004). Adsorption capacity has been found to be an important criterion in various environmental applications such as remediation of contaminant and hydrogen storage.

Electronic, Optical, and Thermal Properties. The unique conductive, optical, and thermal properties of fullerenes and nanotubes are due to their bonding configurations that present quite promising applications in the field of the electronic sector (Avouris, 2002). Additionally, they have broad applications in environmental sensing devices and competent power production in modern solar cells (Kamat et al., 2004). Lastly, photo-oxidation of persistent organics using fullerenes is also demonstrated as an ecological remediation technique (Bonchio et al., 2004; Zhu et al., 2007; Narita et al., 2007; Colvin, 2007). The notable optical activity of fullerenes in both the UV and visible ranges are due to strong electron transitions from bonding to antibonding orbitals and from HOMO and LUMO bands in the visible range, respectively. The strong absorption of the fullerene molecule in UV region also brings about photoexcitation from the ground level state to a singlet excited state that emits that has the ability to generate different reactive oxygen species (ROS), comprising singlet oxygen molecules, hydroxide radicals, and superoxide radical anions in common environmental aqueous media (Lee et al., 2007). SWNTs demonstrate excellent electronic properties including good current conductivity, reduced ionisation potential, and high field emission properties (Saito et al., 2000). Most of such properties are as a result of restricted electron flow in 1-D nanotubes. These properties of nanotubes are strongly related to the chirality, diameter, length, and the number of concentric tubules (Hamada et al., 1992). However, the heterogeneities in MWNTs produce imperfection in scattering and diffusive electron movement (Ebbesen et al., 1996; Tans et al., 1997; Langer et al., 1996), that result in their limited applications in environmental and customer products.

The electronic arrangement of the outermost tubules determines the conduction behaviour in MWNTs which is quite similar to the electronic behaviour of graphite.

Metallic SWNTs, on the other hand act as quantum wires, having electron arrangement along the radial route thus, quantizing the conduction bands into distinct energy states (Tans et al., 1997; Bockrath et al., 1997). Resonant tunnelling of electrons occurs in the nanotube through such distinct energy states that are then further delocalized over nanotubes length. Such kind of spatial distribution of charge improves the conductivity and current capacity, and also lessens the defects in the nanotube sidewalls. Quantum wires conductivity is quite sensitive to adsorption by nearby liquid or gaseous molecules, and, such kind of behaviour has found numerous applications as environmental sensors (Lim et al., 2006; Pejcic et al., 2007; Sinha et al., 2006; Wang, 2005). SWNTs have very low ionisation potential as compared to many common field emitters presently employed in the electronics sector. Additional lowering of ionisation potential is seen in the presence of some adsorbates, like water (Maiti et al., 2001). The nanotubes with reduced ionisation potential and higher field emission efficiency show efficient applications in green designing of future generation devices (Berber et al., 2000; Kim et al., 2001; Cahill et al., 2003). Further advancement in nanotube's joining can result in the development of long fibres having excellent thermal conductivity.

8.4 Carbon Nanomaterials as Sorbents

Sorption of ecological contaminants to sorbents, for example, NOM, clay and actuated carbon represent a noteworthy drop in common and engineered environmental systems. In order to remove the organic and inorganic contaminants, traditional drinking water treatment depends on physicochemical sorption processes. (Pignatello et al., 1996; Allen-King et al., 2002; Bailey et al., 1999; Luthy et al., 1997; Weber et al., 1991; Schwarzenbach et al., 1981). The various factors which limit the sorption capacity of conventional carbonaceous sorbents are activation energy of sorptive bonds density of surface active sites, the slow kinetics, and the mass transport rate to the sorbent surface. An extensive dimension of conventional sorbents also bounds their transportation via low porosity environments and obscure efforts in subsurface remediation. Owing to their elevated surface area to volume ratio, controlled pore size distribution, and manipulatable surface chemistry, carbonaceous nanosorbent, beat a significant number of these inherent limitations. By utilising carbon-based nanomaterials sorption studies demonstrates rapid equilibrium rates, effectiveness over a wide pH range, high adsorption capacity, consistency with BET, Langmuir, and Freundlich isotherms (Lu et al., 2005; Yang et al., 2006, 2006; Cheng et al., 2004, 2005; Cai et al., 2003).

π-π electron–donor-acceptor (EDA) interactions with aromatic sorbates (Long et al., 2001; Gotovac et al., 2006), reduced heterogeneity of adsorption energies (Papirer et al., 1999), and the absence of pore diffusion as an intermediate mechanism in adsorption (Peng et al., 2003) are the various attributes accredited to the elevated equilibrium rates of carbonaceous nanosorbents over activated carbon. The work of Yang et al. (2006) strengthened this conclusion by comparing an assortment of carbonaceous nanosorbents including C-60, nC-60 nanoparticles, single-walled carbon tubes, and varying dimensions of multi-walled carbon tubes. The virtual nonappearance of hysteresis between adsorption and desorption isotherms for liquids and gases under atmospheric pressure is the additional advantage linked with carbonaceous nanosorbents (Yang et al., 2007; Neimark et al., 2003; Punyamurtula et al., 2007). In water and wastewater treatment, the traditional applications of activated carbon comprise of declining in natural contaminants, residual taste, or odour. The cost and possible lethality of carbonaceous nanomaterials have prevented the broad research in extensive utilisation for water treatment. Till date, the majority of the research on the ecological applications of nanosorbents has been focused for the removal of particular hazardous contaminants for, e.g. trihalomethanes (Lu et al., 2005), polycyclic aromatic hydrocarbons (Yang et al., 2006), and naphthalene (Cheng et al., 2004). For targeting specific micropollutants, removing low concentration contaminants (Yan et al., 2006), or improving subsurface mobility, functionalized nanosorbents may provide an optimised approach. For instance, on comparing with activated carbon, hydrophilic OH and COOH groups functionalized CNTs possessed higher sorption of low molecular weight and polar compounds (Lu et al., 2005).

In contrast to the relatively nonspecific, hydrophobic sorption mechanisms describing organic sorption, inorganic sorption to carbonaceous nanostructures is characterised by specific complexation reactions. Rather than total surface area, surface functional group density becomes the primary determinant of inorganic sorption capacity. Competing for complexation reactions or metal speciation also limits the sorption capacity which is sensitive to changes in pH (Li et al., 2002; Wang et al., 2005; Li et al., 2003). Lastly, sorption of arsenic (an oxyanion metal contaminant) changes by the concentration of divalent cations in solution. This because of the cationic bridging of the metal to deprotonated oxygen groups on the surface of sorbent via a ternary surface complex reaction (Peng et al., 2005). Carbon nanomaterials have additionally been utilised as a high surface area scaffold for oxides or macromolecules with inherent sorbent capacity. For the removal of contaminants, the current instances of nanomaterials as scaffolding agents involve functionalization of

CNT by CeO_2 nanoparticles for AsO_4^{3-} and Cr^{4+} removal (Peng et al., 2005; Di et al., 2006), amorphous alumina for fluoride adsorption (Li et al., 2001) etc. The commercial ecological applications of nanomaterials sorptive capacity are not confined to the removal of ordinary contaminants.

8.5 Carbonaceous Nanomaterials as Antimicrobial Agents

The distinctive features and nanoscale dimensions of nanotubes and fullerenes have increased anxiety amid environmental scientists and toxicologists (Colvin, 2003; Nel et al., 2006; Lam et al., 2006). Toxicity may rely upon structural and physiochemical characteristics, for example, surface chemistry length, functional group density, residual catalyst contamination, and diameter (Jia et al., 2005; Karakoti et al., 2006; Worle-Knirsch et al., 2006; Sayes et al., 2006; Lyon et al., 2006; Tang et al., 2007; Kang et al., 2007, 2008). Various researchers are excited to use these observed antimicrobial properties in ecology and public health applications (Kang et al., 2007, 2008; Badireddy et al., 2007).

For the control of pathogens, the most potent method is deliberate perturbation of cell membranes. Novel antimicrobial surface coatings that utilise the intrinsic vulnerability of bacteria for carbon tubes may suggest solutions to various challenging problems of bacterial colonisation, biofilm development in medical implant devices, drinking water systems, and other submerged surfaces (Tong et al., 2007).

Various research groups are exploring the applications of antimicrobial and antiviral nanoparticles for treatment of water and distribution systems. Carbon tubes have additionally been projected as scaffolding agents for antimicrobial Ag nanoparticles (Morones et al., 2005) or semiconducting photocatalysts such as TiO_2 (Krishna et al., 2005).

If the obstacles to immobilisation and partition of the nanomaterials are surmounted (Wiesner, 2006), the deactivation of the pathogen by fullerol and nC-60 (Lyon et al., 2008) may also be relevant to water and wastewater treatment. Fullerols antiviral activity is supposed to act via the production of superoxide in the existence of both ultraviolet and an electron-donating molecule or singlet oxygen in the existence of ultraviolet light (Badireddy et al., 2007). Stable nC-60 suspensions possess potent antibacterial activity for diverse range bacteria over various environmental conditions (Lyon et al., 2008).

8.6 Carbon Tubes in Air Pollution

The remarkable electrochemical electrical, and optical properties of carbon tubes stimulated the curiosity of researchers to discover the possible applications of carbon tubes as the sensing elements to detect and monitor the concentration of lethal gases released in the environment (Wei et al., 2006; Van et al., 2008; Francia et al., 2009; Bondavalli et al., 2009; Lu et al., 2009; Penza et al., 2009, 2009; Zhang et al., 2009). The metallic or semi-conductivity of CNTs is notably affected by their 1D cylindrical structure. Gas sensors which are based on CNT present various preferences over traditional metal oxide semiconductor gas sensors (Endo et al., 2008). Carbon tube-based gas sensors have been utilised in numerous researchers for the detection of nitrogen oxide (Ueda et al., 2008, 2008), nitrogen dioxide (Kong et al., 2000; Cantalini et al., 2003; Valentini et al., 2004; Cho et al., 2006; Moon et al., 2008), ammonia (Nguyen et al., 2006; Quang et al., 2006; Nguyen et al., 2007), and sulphur dioxide (Suehiro et al., 2005) at room temperature. In comparison to gas sensors, the results demonstrated high and prompt response but the tedious recovery represents a challenge. Several strategies have been proposed, for the improvement of gas desorption from the sensor. Efforts have additionally been made to enhance the affinity and sensitivity of CNT-based gas sensors via polymer functionalization (Lu et al., 2009). The integration of CNTs in conducting polymers, for instance, polypyrrole and polyaniline results in an increment in sensitivity of the sensor for selected gases or vapours. Qi et al. (2003) investigated a polymer coating with polyethyleneimine (PEI) CNT-based gas sensor showed a high affinity for NO_2 detection without any ammonia interference. This is because of the reduced binding affinity and sticking coefficient of ammonia on the electron-rich carbon tubes.

In contrast to the PEI-coated sensor, carbon tube-based gas sensor coated with Nafion permits superior selectivity for the sensing of NH_3. On the basis of carbon tubes/poly(methyl methacrylate) composite chemo resistor, Abraham et al. (2004) fabricated a compact wireless gas sensor. Despite the fact that carbon tube-based gas sensors display good competition for conventional metal oxide sensors, continuous developments have been done to enhance the traditional gas sensor. To overcome the limitation of high working temperature, an integration of carbon tubes in metal oxide sensors has been done. To detect pollutant gases, for instance, NH_3 and NO_2 at room temperature a remarkable improvement in sensitivity and response time of carbon tubes/ metal oxide sensors were observed. The enhanced performance was credited to the change in conductance upon adsorption of the gas provided by carbon tubes and the efficient accessing of nanopassages by the gas.

8.7 Carbon Tubes in Biotechnology

There has been hasty development in biotechnology over the years in which living beings are utilised to make products or processes for particular uses. This is because of the escalating interest for innovative, environmentally friendly technologies. The emergence of biotechnology provides a possibility for the participation of CNTs, especially in biofuel cells. Biofuel cells are the fuel cells that depend on biocatalytic activity to produce electric power. In general, they are categorising as microbial fuel cells (MFCs) or enzymatic biofuel cells (EFCs). MFCs utilise the catabolic activities of microorganisms to produce electric energy. They have been regarded as the upcoming alternative in waste water treatment (Logan et al., 2006; Watanabe, 2008). However, because of low execution and shortage of technical experience, the technology is not feasible practically.

A lot of efforts have been centred on utilising and reforming carbon tubes as electrodes to intensify power production in MFC owing to their enhanced conductivity and huge surface area (Morozan et al., 2007; Qiao et al., 2007; Sharma et al., 2008; Tsai et al., 2009). As compared with a non-CNT coated electrode, the presence of CNTs achieved an upgraded power density to 250%. The optimisation of the power density was further acquired by employing the biocompatibility of carbon tubes with microorganisms in MFCs. Although the cytotoxicity behaviour of CNTs diminished after their modification and functionalization (Sayes et al., 2006). Morozan et al. (2007) examined the biocompatibility of *Staphylococcus aureus* with CNTs. It was demonstrated that there was a quick growth of a microorganism in CNT modified cell culture media. If applied to MFC anodic design, this leads to the potential of dropping the loss of power generation. EFCs utilise enzyme or protein catalysis for the conversion of chemical energy into electrical energy. Till now, owing to the poor enzyme stability, short lifetime, and low power density, the EFCs application as a power source for communication devices, low power sensors and medical implants has been impeded (Kim et al., 2006; Minteer et al., 2007). CNTs have also demonstrated great performance in EFCs by providing a sturdy platform for enzyme immobilisation (Fischback et al., 2006; Asuri et al., 2007; Li et al., 2008; Zhao et al., 2009). Moreover, the stability of enzyme could increase by nanoscale environment provided by CNTs. Enzymes that adhere covalently onto the carbon tubes were found to have a high degree of stability (Asuri et al., 2007; Govardhan, 1999; Sheldon, 2007). The curvature of carbon tubes was enhanced as the distance among enzyme molecules, thereby decreasing detrimental interaction among the enzymes. This leads to increase in enzyme stability (Asuri et al., 2006). The stability of the enzyme in EFCs is imperative as it gives a splendid operational stability, which is expected to make ready for expanding the power density and delaying the lifetime of EFCs.

8.8 Carbon Tubes in Energy Conversion

Solar energy has an incredible potential to satisfy the necessity of reliable clean energy sources (Wijewardane, 2009). On the basis of thin films processing new generation solar cells involve dye-sensitized solar cells, and polymer organic solar cells etc. Due to their lessened production expense as well as huge energy alteration performance, they are reliable photovoltaic devices. Broad research has been done to acquire enhanced photoconversion efficacy of solar cells. This can be done by integrating CNTs into the device, for example, nanoarchitecture on the working electrode surface (Sawatsuk et al., 2009; Imahori et al., 2009), carbon tubes composite as counter electrode etc. (Mei et al., 2009; Ramasamy et al., 2008). The incorporation of carbon tubes into the solar cells electrode had demonstrated an expanded photoconversion proficiency, which was ascribed to the powerful task of carbon tubes to encourage the transfer of charge over the photoactive layer, to lessen the speed of charge recombination and, in this manner, to enhance the photoconversion efficacy of solar cells. Li ions batteries are turning into the primary power source, because of their enhanced energy density, high operating voltage etc (Yang et al., 2009). In lithium ion batteries the preparation of TiO_2/multiwalled carbon tubes composites as electrode materials has been described since SnO_2 displayed an enhanced capacity (Du et al., 2010). The enhanced performances of SnO_2/MWCNT composites were accredited to carbon tubes as the intercalation host for SnO_2 nanoparticles that lessened the absolute volume change and prohibited the agglomeration of SnO_2 nanoparticles. Recently, carbon tubes have been investigated as electrode materials for hybrid supercapacitors. Electrochemical conducting polymers as well as transition metal oxides, for instance, MnO_2 and ZnO_2 are deposited onto Carbon-tubes to enhance the total capacitance via faradaic pseudocapacitance effects. Graphene nanosheet/carbon nanotube/polyaniline (GNS/CNT/PANI) composite possesses increased specific capacitance and brilliant cycle stability. With carbon tubes as the conductive filler in the electrode, higher energy capacity and cycle efficacy of electrochemical supercapacitors can be attained.

8.9 Carbon Tubes in Environmental Monitoring and Wastewater Treatment

Air pollutants have been the main reason causative to the depletion in the quality of air that results in an undesirable impact on the ecology and living wellbeing. Traditional solid-state gas sensors are extremely sensitive and selective, but they undergo from the ill effects of superior working temperatures and an unending stability issues (Bai et al., 2007; Barsan et al., 2007). CNTs have been revealed as capable gas-sensing nanomaterials (Wei et al., 2010). Pristine CNTs based on gas sensors are investigated to demonstrate excellent responses to ammonia, sulphur dioxide and nitrogen oxide (Goldoni et al., 2003, 2004). MWCNT-doped Tin oxide, thin film nitrogen dioxide gas sensor, displayed an enhanced sensitivity to extra low concentrations of NO_2 in the parts per billion range in comparison to the pure tin oxide thin film gas sensor (Wei et al., 2010). To serve as sensing elements carbon tubes functionalization by metal oxides had displayed enhanced sensitivity and selectivity for various air pollutants. Pathogens are generally uncoloured, insipid and odourless, that results in the difficulty to remove pathogens (Nuzzo, 2006). Carbon-tubes have shown promising applications for removal and detection of pathogens with its enhanced bacterial adsorption capacity. Single-walled Carbon tubes had enhanced affinity for adsorption in both *E. coli* and *S. aureus* bacterial cells (Upadhyayula et al., 2008, 2008). Owing to the unique fibrous mesoporous structure of SWCNT, these bacteria demonstrate a high adsorption and selectivity. In addition, a lot of researchers are concentrating on the functionalization of carbon tubes to develop a biosensor to detect pathogens. As carbon tubes exhibit antimicrobial properties, the biosensor will have the tendency to concentrate and detect pathogens. Dyes are colour organic compounds, which have been utilised in numerous industries. The presence of dyes in water bodies is objectionable as some of which are hazardous to public health and ecology (Crini, 2006). For the removal of methylene blue, CNTs have shown brilliant potential. Additionally, the functionalization of CNTs with biomaterials has demonstrated a huge affinity for dyes. For the removal of Congo red, the impregnation of carbon tubes with chitosan hydrogel beads has manifested to be an efficient bio sorbent material (Chatterjee et al., 2010). Heavy metals are the main lethal pollutants and are deleterious to human health and eco-friendly environment (Rao et al., 2007). The adsorption efficacy of CNTs for divalent heavy metal ions was significantly improved after the carbon tubes oxidation with acid solutions. The adsorption mechanism was accredited to the chemical relations among metal ions and surface functional groups of carbon tubes. Thus, the adsorption capacity of CNTs was mainly influenced by the surface total acidity of CNTs.

8.10 Pollution Prevention via Molecular Manipulation

This chapter has secured ecological applications seeing that nanomaterials are straightforwardly connected towards difficulties confronted by natural researchers, architects, and policy creators. Those who embrace the ideas of life-cycle evaluation, in any case, perceive that the utilisation stage constitutes just a small amount of an innovation's advancement from configuration through transfer. Unusual uses of carbon nanomaterials are ready to make critical commitments in these alternate periods of innovative advancement. Carbon nanomaterials empower a suite of composite materials with enhanced execution attributes. At the point when utilizations of a material are joined into product design, the potential exists for items with a more drawn out life span and more noteworthy material productivity. Supplanting current structural materials with high-quality nanotube composites, for instance, could diminish the volume of concrete (Li et al., 2007) or the mass of high-quality steels (Ajayan et al., 2001). Minimization of material streams crosswise over life-cycle stages will be a noteworthy commitment of nanotechnology to the field of green configuration. The commitments that nanomaterials make towards reduced material stream must not be "fixed" by asset serious or earth hazardous procedures for combining CNTs. Greener synthesis pathways, solvent substitution in refinement methods, and bottom-up manufacturing plans are being investigated to lessen the ecological weight of the nanomaterials themselves (McKenzie et al., 2004; Eckelman et al., 2008). Styrene synthesis utilising carbon tubes as a substitution for conventional iron-oxide catalyst changed over the response from endothermic to exothermic, diminished the temperature of reaction and enhanced the selectivity of the reaction. This procedure yields net energy funds of almost 50% and diminishes substantial metal discharges by 75% (Steinfeldt et al., 2007). Green nanotechnology ought to likewise consolidate end-of-life considerations in item outline, assembling, and applications improvement. Nanomaterial sorbents, for the case, may find another use in basic support or street construction. Finally, green item outline must consider disposal and regulation methodologies to minimise the dispersal of nanomaterials in the earth. This is a region ready for multidisciplinary joint effort between item engineers, environmental researchers, and toxicologists. In any case, hindrances exist to the acknowledgement of pollution prevention practices in nano-based item development, implementation, and life-cycle plan. Challenges arise when scaling up advancements from seat to modern level processes. The excellence, uniformity, and valuing of carbon nanomaterials are still in flux. Existing waste management infrastructure frameworks are not intended to accommodate new nano-based items. Researchers about on the fate, transport, and danger of carbon tubes and fullerenes report conflicting results. Also, open view of danger is highly unstable for new advancements (Fischhoff et al., 2002). Efforts to win open trust through responsible risk communication outlets are one important part of a comprehensive strategy to strengthen nanotechnology's status as a conduit for contamination avoidance and green outline.

8.11 Carbon Nanotubes in Green Nanocomposites Design

The world's economic growth is proportional to the waste generation. Wastes, especially synthetic polymer waste, have been playing a foremost role in environmental pollution. Thus, in order to overcome this complexity, a waste management concept has been suggested by the European Community. That is based on two strategies: increase the reuse and reprocess of waste which focuses on life-cycle assessment (LCA) and avoiding waste by improving product design (Baillie, 2004). For solving waste disposal issues, the future perspective point of view leads a challenge to produce green nanocomposites by utilising biodegradable polymers (Mojumdar et al., 2005; Wang et al., 2005). Biodegradable polymers because of their degradability can be utilised as a substitute to supplant non-renewable petroleum-based polymers. The low heat distortion temperatures and poorer mechanical properties of the biodegradable polymers restrict their use in broad applications. The various factors which determine the quality of the biodegradable polymer/CNT nanocomposites are the alignment of CNT, the CNT-biodegradable polymer adhesion, and the dispersion of CNT in the biodegradable polymer matrix (Grossiord et al., 2005; Ray et al., 2006; Vaudreuil et al., 2007). Following to the incorporation with CNTs there was an improvement by about 10°C in the thermal stability of poly(butylenes succinate) (PBSU) (Song et al., 2009).

Owing to the degradability of the biodegradable polymer, the additional advantage of the green nanocomposites is their capacity to reuse the incorporated CNTs. Biodegradable polymer can be degraded by means of enzymatic degradation or microbial degradation under particular conditions of pH and temperature. The reprocesses of CNTs could decrease the waste disposal and also decrease commercial for material processing.

8.12 Significant Applications of Carbon Nanotubes

To be applied in various environmental fields CNTs have been hailed as the potential materials. The foremost affair that limits the application of CNTs on a higher scale is the cost effective synthesis. Through the decomposition of hydrocarbons via catalytic chemical vapour deposition method, CNTs with a fabrication rate of 595 kg/h were reported (Agboola et al., 2007). Zhang et al. (2007) demonstrated that high purity CNTs can be effectively produced at the higher scale at a low price by utilising ceramic sphere as substrate and the liquefied petroleum gas as the carbon source material.

The disposed-of CNTs may enter into the living body through inhalation or via the food chain and leads to harmful effects on human health and the environment. The harmful effects arise as a function of their physical dimension (Magrez et al., 2006; Wick et al., 2007), the presence of impurities and chemical treatment (Dumortier et al., 2006). From the in vivo studies of CNTs, it was observed that CNTs could persuade lung cellular proliferation (Muller et al., 2005), pulmonary inflammation (Lam et al., 2004) and halted the heart muscle growth in rats. The lethality of agglomerated CNTs is more than of well dispersed CNTs (Wick et al., 2007). Thus, owing to the enhanced dispersion functionalized CNTs have a less toxic effect on mammalian cell viability (Dumortier et al., 2006). Thus far, still, insufficient learning regarding effect of CNTs on the human health and environmental. Thus, the excellent approach to avoid the direct contact of human beings and environment to CNTs is source reduction (Reijnders, 2006).

8.13 Conclusions

The proactive to retroactive uses of CNTs in the ecological frame have been investigated. Inferable from their splendid electrical, mechanical and physical properties, CNTs assume a chief part in air contamination risk and wastewater treatment. In waste water treatment, CNTs present as antimicrobial specialists, having nanofilter sorbents to pull out natural and inorganic pollutants, as well as lethal microbes. In confirming of air contamination, upgrading of CNT-based gas sensors delivers about enhanced affectability with brief sensor reaction for poison gases.

Moreover, CNTs are likewise illustrated as one of the decisive difficulties in generating "green" vitality, which constitutes clean combustion. The making of electricity by "green" energy innovations based on biomass catalytic activity, renewable sources such as photovoltaic gadget and hydrogen fuel cell inhibits the release of noxious gases to the atmosphere and blocks the overture for fossil fuel.

Be that as it may, these modernisations are still not yet been financially declining to earth since they are still in progressing stage and additional time is recommended to carry out specialised chrysalis. Beside from this, the preferable efficacy thickness given by Carbon tube-based supercapacitors has been seen as an optional way to substitute customary batteries. In green materials, the involvement of carbon tubes in green nanocomposite design embraces the LCA concept, which encourages reduction, reuse and recycle-capability of raw materials. For substantial scale applications, continuous creation of carbon tubes from low-cost sources has stepped to conquering the problem of the high cost of the union. Considering the welfare exhortation because of prevailing use of CNTs, an ideal approach to perpetuate a strategic distance from human and natural exposition to CNTs source decrease.

Acknowledgement

We gratefully acknowledge support from the Ministry of Human Resource Development Department of Higher Education, Government of India, under the scheme of the Establishment of Centre of Excellence for Training and Research in Frontier Areas of Science and Technology (FAST), for providing the necessary financial support to carry out this study vide letter No, F. No. 5-5/201 4-TS.Vll.

References

Abraham JK, Philip B, Witchurch A, Varadan VK, Reddy CA (2004). Compact wireless gas sensor using a carbon nanotube/PMMA thin film chemiresistor, *Smart Mater Struct*; **13**: 1045–1049.

Agboola AE, Pike RW, Hertwig TA, Lou HH (2007). Conceptual design of carbon nanotube processes, *Clean Technol Environ*; **9**: 289–311.

Ajayan PM (1999). Nanotubes from carbon, *Chem Rev*; **99**: 787–1800.

Ajayan PM (1999). Applications of carbon nanotubes, *Carbon Nanotubes*; **80**: 391–425.

Allen-King RM, Grathwohl P, Ball WP (2002). New modelling paradigms for the sorption of hydrophobic organic chemicals to heterogeneous carbonaceous matter in soils, sediments, and rocks, *Adv Water Resour*; **25**: 985–1016.

Andreas H (2002). Functionalization of single-walled carbon nanotube's, *Angew Chem Int Ed*; **41**: 1853–1859.

Andrievsky GV, Kosevich MV, Vovk OM, Shelkovsky VS, Vashchenko LA (1995). On the production of an aqueous colloidal solution of fullerenes, *J Chem Soc Chem Commun*; **12**: 1281–1282.

Anson A, Lafuente E, Urriolabeitia E, Navarro R, Benito AM, Maser WK, Martinez MT (2006). Hydrogen capacity of palladium-loaded carbon materials, *J Phys Chem B*; **110**: 6643–6648.

Asuri P, Bale SS, Pangule RC, Shah DA, Kane RS, Dordick JS (2007). Structure, function, and stability of enzymes covalently attached to single-walled carbon nanotube's, *Langmuir*; **23**: 12318–12321.

Asuri P, Karajanagi SS, Yang H, Yim TJ, Kane RS, Dordick JS (2006). Increasing protein stability through control of the Nanoscale environment, *Langmuir*; **22**: 5833–5836.

Avouris P (2002). Carbon nanotube electronics, *Chem Phys*; **281**: 429–445.

Badireddy AR, Hotze EM, Chellam S, Alvarez P, Wiesner MR (2007). Inactivation of bacteriophages via photosensitization of fullerol nanoparticles, *Environ Sci Technol*; **41**: 6627–6632.

Bai H, Shi G (2007). Gas sensors based on conducting polymers, *Sensors*; **7**: 267–307.

Bailey SE, Olin TJ, Bricka RM, Adrian DD (1999). A review of potentially low-cost sorbents for heavy metals, *Water Res*; **33**: 2469–2479.

Baillie C (2004). *Green Composites: Polymer Composites and the Environment.* CRC Press, New York.

Bandow S, Asaka S, Saito Y, Rao A, Grigorian L, Richter E, Eklund P (1998). Effect of the growth temperature on the diameter distribution and chirality of single-wall carbon nanotube's, *Phys Rev Lett*; **80**: 3779–3782.

Bandow S, Takizawa M, Hirahara K, Yudasaka M, Iijima S (2001). Raman scattering study of double-wall carbon nanotube's derived from the chains of fullerenes in single-wall carbon nanotubes, *Chem Phys Lett*; **337**: 48–54.

Banerjee S, Hemraj-Benny T, Wong SS (2005). Covalent surface chemistry of single-walled carbon nanotube's, *Adv Mater*; **17**: 17–29.

Barsan N, Koziej D, Weimar U (2007). Metal oxide-based gas sensor research: How to? *Sens. Actuators B Chem.*; **121**: 18–35.

Bates KR, Scuseria GE (1998). Why are bucky onions round, *Theor. Chem. Acc.*; **99**: 29–33.

Béguin F, Flahaut E, Linares-Solano A, Pinson J (2006). Surface properties, porosity, chemical and electrochemical application, in *Understanding Carbon Nanotubes*, Springer; 495–549.

Berber S, Kwon YK, Tomanek D (2000). Unusually high thermal conductivity of carbon nanotube's, *Phys. Rev. Lett.*; **84**: 4613–4616.

Bockrath M, Cobden DH, McEuen PL, Chopra NG, Zettl A, Thess A, Smalley RE (1997). single-electron transport in ropes of carbon nanotube's, *Science*; **275**: 1922–1925.

Bonchio M, Carraro M, Scorrano G, Bagno A (2004). Photooxidation water by new hybrid molecular photocatalysts integrating an organic sensitizer with a polyoxometalate core, *Adv. Synth. Catal.*; **346**: 648–654.

Bondavalli P, Legagneux P, Pribat D (2009). Carbon nanotubes based transistors as gas sensors: State of the art and critical review, *Sens Actuator B Chem*; **140**: 304–318.

Bonifazi D, Enger O, Diederich F (2007). Supramolecular [60] fullerene chemistry on surfaces, *Chem. Soc. Rev.*; **36**: 390–414.

Brant JA, Labille J, Bottero JY, Wiesner MR (2006). Characterizing the impact of preparation method on fullerene cluster structure and chemistry, *Langmuir*; **22**: 3878–3885.

Brant J, Lecoanet H, Hotze M, Wiesner M (2005). Comparison of electrokinetic properties of colloidal fullerenes (n-C60) formed using two procedures, *Environ Sci Technol*; **39**: 6343–6351.

Cahill DG, Ford WK, Goodson KE, Mahan GD, Majumdar A, Maris HJ, Merlin R, Phillpot SR (2003). Nanoscale thermal transport, *J Appl Phys*; **93**: 793–818.

Cai YQ, Jiang GB, Liu JF, Zhou QX (2003). Multiwalled carbon nanotubes as a solid-phase extraction adsorbent for the determination of bisphenol a, 4-n-nonylphenol, and 4-tertoctylphenol, *Anal Chem*; **75**: 2517–2521.

Campbell EEB, Fowler PW, Mitchell D, Zerbetto F (1996). Increasing the cost of pentagon adjacency for larger fullerenes, *Chem Phys Lett*; **250**: 544–548.

Cantalini C, Valentini L, Lozzi L, Armentano I, Kenny JM, Santucci S (2003). NO_2 gas sensitivity of carbon nanotubes obtained by plasma enhanced chemical vapour deposition, *Sens Actuator B Chem*; **93**: 333–337.

Cassell AM, Raymakers JA, Kong J, Dai HJ (1999). Large scale CVD synthesis of single-walled carbon nanotube's, *J Phys Chem B*; **103**: 6484–6492.

Chatterjee S, Lee MW, Woo SH (2010). Adsorption of Congo red by chitosan hydrogel beads impregnated with carbon nanotubes. *Bioresour Technol*; **101**: 1800–1806.

Chen W, Duan L, Zhu DQ (2007). Adsorption of polar and nonpolar organic chemicals to carbon nanotube's, *Environ Sci Technol*; **41**: 8295–8300.

Chen KL, Elimelech M (2006). Aggregation and deposition kinetics of fullerene (C60) nanoparticles, *Langmuir*; **22**: 10994–11001.

Chen KL, Elimelech M (2007). Influence of humic acid on the aggregation kinetics of fullerene (C60) nanoparticles in monovalent and divalent electrolyte solutions, *J Colloid Interface Sci*; **309**: 126–134.

Cheng XK, Kan AT, Tomson MB (2004). Naphthalene adsorption and desorption from aqueous C-60 fullerene, *J Chem Eng Data*; **49**: 675–683.

Cheng XK, Kan AT, Tomson MB (2005). Uptake and sequestration of naphthalene and 1,2-dichlorobenzene by C-60, *J Nanopart Res*; **7**: 555–567.

Chiang IW, Brinson BE, Smalley RE, Margrave JL, Hauge RH (2001). Purification and characterization of single-wall carbon nanotube's, *J Phys Chem B*; **105**: 1157–1161.

Cho WS, Moon SI, Paek KK, Lee YH, Park JH, Ju BK (2006). Patterned multiwall carbon nanotube films as materials of NO_2 gas sensors. *Sens Actuator B—Chem*; **119**: 180–185.

Colvin VL (2003). The potential environmental impact of engineered nanomaterials, *Nat Biotechnol*; **21**: 1166–1170.

Colvin VL (2007). Clean water from small materials, in *Nanotechnology in the Environment*, MRS: Boston, MA.

Crini G (2006). Non-conventional low-cost adsorbents for dye removal: A review. *Bioresour Technol*; **97**: 1061–1085.

Di ZC, Ding J, Peng XJ, Li YH, Luan ZK, Liang J (2006). Chromium adsorption by aligned carbon nanotube's supported ceria nanoparticles, *Chemosphere*; **62**: 861–865.

Diederich F, Gomez-Lopez M (1999). Supramolecular fullerene chemistry, *Chem Soc Rev*; **28**: 263–277.

Dresselhaus, MS, ed (2001). *Carbon Nanotubes: Synthesis, Structure, Properties, and Applications*, Springer: Berlin, NY.

Dresselhaus M, Endo M (2001). Relation of carbon nanotubes to other carbon materials, *Top Appl Phys*; **80**: 11–28.

Du G, Zhong C, Zhang P, Guo Z, Chen Z, Liu H (2010). Tin dioxide/ carbon nanotube composites with high uniform SnO_2 loading as anode materials for lithium ion batteries, *Electrochim Acta*; **55**: 2582–2586.

Dumortier H, Lacotte S, Pastorin G, Marega R, Wu W, Bonifazi D, Briand JP, Prato M, Muller S, Bianco A (2006). Functionalized carbon nanotubes are non-cytotoxic and preserve the functionality of primary immune cells. *Nano Lett*; **6**: 1522–1528.

Duncan LK, Jinschek JR, Vikesland PJ (2008). C-60 colloid formation in aqueous systems: Effects of preparation method on size, structure, and surface charge, *Environ Sci Technol*; **42**: 173–178.

Ebbesen TW, Lezec HJ, Hiura H, Bennett JW, Ghaemi HF, Thio T (1996). The electrical conductivity of individual carbon nanotubes, *Nature*; **382**: 54–56.

Eckelman M, Zimmerman J, Anastas P (2008). Toward green nano: Atom economy and *E*-factor analysis of several nanomaterials syntheses, *J Ind Ecol*; **12**: 316–328.

Endo M, Strano M, Ajayan P (2008). Potential applications of carbon nanotubes, *Carbon Nanotubes*; 13–62.

Falcao EHL, Wudl F (2007). Carbon allotropes: Beyond graphite and diamond, *J Chem Technol Biotechnol*; **82**: 524–531.

Fernandez A (2003). What caliber pore is like a pipe? Nanotubes as modulators of ionic gradients, *J Chem Phys*; **119**: 5315–5319.

Fischhoff B, Bostrom A, Quadrel M (2002). In *Oxford Textbook of Public Health, 4th ed*; Detels R, McEwen J, Reaglehole R, Tanaka H, eds; Oxford University Press: Oxford, UK.

Fischback MB, Jong KY, Zhao X, Wang P, Hyun GP, Ho NC, Kim J, Ha S (2006). Miniature biofuel cells with improved stability under continuous operation, *Electroanalysis*; **18**: 2016–2022.

Fortner JD, Lyon DY, Sayes CM, Boyd AM, Falkner JC, Hotze EM, Alemany LB, Tao YJ, Guo W, Ausman KD, Colvin VL, Hughes JB (2005). C-60 in water: Nanocrystal formation and microbial response, *Environ Sci Technol*; **39**: 4307–4316.

Francia GD, Alfano B, La Ferrara V (2009). Conductometric gas nanosensors, *J Sens*; **2009**: 1–18.

Furmaniak S, Terzyk AP, Gauden PA, Rychlicki G (2006). Simple models of adsorption in nanotube's, *J Colloid Interface Sci*; **295**: 310–317.

Girifalco LA (1991). Interaction potential for C-60 molecules, *J Phys Chem*; **95**: 5370–5371.

Girifalco LA, Hodak M, Lee RS (2000). Carbon nanotubes, buckyballs, ropes, and a universal graphitic potential, *Phys Rev B*; **62**: 13104–13110.

Georgakilas V, Kordatos K, Prato M, Guldi DM, Holzinger M, Hirsch A (2002). Organic functionalization of carbon nanotube's, *J Am Chem Soc*; **124**: 760–761.

Goldoni A, Larciprete R, Petaccia L, Lizzit S (2003). Single-wall carbon nanotube interaction with gases: Sample contaminants and environmental monitoring, *J Am Chem Soc*; **125**: 11329–11333.

Goldoni A, Petaccia L, Gregoratti L, Kaulich B, Barinov A, Lizzit S, Laurita A, Sangaletti L, Larciprete R (2004). Spectroscopic characterization of contaminants and interaction with gases in single-walled carbon nanotubes, *Carbon*; **42**: 2099–2112.

Gotovac S, Hattori Y, Noguchi D, Miyamoto J, Kanamaru, M, Utsumi S, Kanoh H, Kaneko K (2006). Phenanthrene adsorption from solution on single wall carbon nanotube's, *J Phys Chem B*; **110**: 16219–16224.

Govardhan CP (1999). Crosslinking of enzymes for improved stability and performance, *Curr Opin Biotechnol*; **10**: 331–335.

Grossiord N, Loos J, Koning CE (2005). Strategies for dispersing carbon nanotubes in highly viscous polymers, *J Mater Chem*; **15**: 2349–2352.

Ham HT, Choi YS, Chung IJ (2005). An explanation of dispersion states of single-walled carbon nanotubes in solvents and aqueous surfactant solutions using solubility parameters, *J Colloid Interface Sci*; **286**: 216–223.

Hamada N, Sawada S-I, Oshiyama A (1992). New one-dimensional conductors: Graphitic microtubules, *Phys Rev Lett*; **68**: 1579–1581.

Han W, Fan S, Li Q, Hu Y (1997). Synthesis of gallium nitride nanorods through a carbon nanotube-confined reaction, *Science*; **277**: 1287–1289.

Hilding J, Grulke EA, Sinnott SB, Qian D, Andrews R, Jagtoyen M (2001). Sorption of butane on carbon multiwall nanotubes at room temperature, *Langmuir*; **17**: 7540–7544.

Holt J, Park H, Wang Y, Stadermann M, Artyukhin A, Grigoropoulos C, Noy A, Bakajin O (2006). Fast mass transport through sub-2-nanometer carbon nanotube's, *Science*; **312**: 1034–1037.

Holzinger M, Vostrowsky O, Hirsch A, Hennrich F, Kappes M, Weiss R, Jellen F (2001). Sidewall functionalization of carbon nanotubes, *Angew Chem Int Ed*; **40**: 4002–4005.

Hu JT, Odom TW, Lieber CM (1999). Chemistry and physics in one dimension: Synthesis and properties of nanowires and nanotubes, *Acc Chem Res*; **32**: 435–445.

Hu Y, Shenderova O, Brenner D (2007). Carbon nanostructures: Morphologies and properties, *J Comput Theor Nanosci*; **4**: 199–221.

Imahori H, Umeyama T (2009). Donor-acceptor nonarchitecture on semiconducting electrodes for solar energy conversion, *J Phys Chem C*; **113**: 9029–9039.

Inoue S, Ichikuni N, Suzuki T, Uematsu T, Kaneko K (1998). Capillary condensation of N-2 on multiwall carbon nanotube's, *J Phys Chem B*; **102**: 4689–4692.

Iijima S (1991). Helical microtubules of graphitic carbon, *Nature*; **354**: 56–58.

Jia G, Wang HF, Yan L, Wang X, Pei RJ, Yan T Zhao YL, Guo XB (2005). Cytotoxicity of carbon nanomaterials: Single wall nanotube, multi-wall nanotube, and fullerene, *Environ Sci Technol*; **39**: 1378–1383.

Johnson R, Bethune D, Yannoni C (1992). Fullerene structure and dynamics: A magnetic-resonance potpourri, *Acc Chem Res*; **25**: 169–175.

Johnson RD, Meijer G, Bethune DS (1990). C-60 has icosahedral symmetry, *J Am Chem Soc*; **112**: 8983–8984.

Jost O, Gorbunov A, Liu XJ, Pompe W, Fink J (2004). Single-walled carbon nanotube diameter, *J Nanosci Nanotechnol*; **4**: 433–440.

Kam NWS, O'Connell M, Wisdom JA, Dai HJ (2005). Carbon nanotubes as multifunctional biological transporters and near-infrared agents for selective cancer cell destruction, *Proc Nat Acad Sci U S A*; **102**: 11600–11605.

Kamat PV, Haria M, Hotchandani S (2004). C60 Cluster as an electron shuttle in a Ru(II)-polypyridyl sensitizer-based photo chemical solar cell, *J Phys Chem B*; **108**: 5166–5170.

Kang S, Herzberg M, Rodrigues DF, Elimelech M (2008). Antibacterial effects of carbon nanotubes: Size does matter! *Langmuir*; **24**: 6409–6413.

Kang S, Pinault M, Pfefferle LD, Elimelech M (2007). Single-walled carbon nanotube's exhibit strong antimicrobial activity, *Langmuir*; **23**: 8670–8673.

Karakoti AS, Hench LL, Seal S (2006). The potential toxicity of nanomaterials: The role of surfaces, *JOM*; **58**: 77–82.

Keblinski P, Nayak SK, Zapol P, Ajayan PM (2002). Charge distribution and stability of charged carbon nanotubes, *Phys Rev Lett*; **89**: 255503.

Kim J, Jia H, Wang P (2006). Challenges in biocatalysis for enzyme-based biofuel cells, *Biotechnol Adv*; **24**: 296–308.

Kim P, Shi L, Majumdar A, McEuen PL (2001). Thermal transport measurements of individual multiwalled nanotubes, *Phys Rev Lett*; **8721**: 215502.

Kong J, Franklin NR, Zhou C, Chapline MG, Peng S, Cho K, Dai H (2000). Nanotube molecular wires as chemical sensors, *Science*; **287**: 622–625.

Kongkanand A, Kuwabata S, Girish KG, Kamat P (2006). Single-wall carbon nanotubes supported platinum nanoparticles with improved electrocatalytic activity for oxygen reduction reaction, *Langmuir*; **22**: 2392–2396.

Krishna V, Pumprueg S, Lee SH, Zhao J, Sigmund W, Koopman B, Moudgil BM (2005). Photocatalytic disinfection with titanium dioxide coated multi-wall carbon nanotubes, *Process Safety Environ Prot*; **83**: 393–397.

Kroto HW (1987). The stability of the fullerenes C-24, C-28, C-32, C-36, C-50, C-60 and C-70, *Nature*; **329**: 529–531.

Lam CW, James JT, McCluskey R, Hunter RL (2004). Pulmonary toxicity of single-wall carbon nanotubes in mice 7 and 90 days after intractracheal instillation, *Toxicol Sci*; **77**: 126–134.

Lam CW, James JT, McCluskey R, Arepalli S, Hunter RL (2006). A review of carbon nanotube toxicity and assessment of potential occupational and environmental health risks, *Crit Rev Toxicol*; **36**: 189–217.

Langer L, Bayot V, Grivei E, Issi JP, Heremans JP, Olk CH, Stockman L, Van Haesendonck C, Bruynseraede Y (1996). Quantum transport in a multiwalled carbon nanotube, *Phys Rev Lett*; **76**: 479–482.

Lee J, Fortner JD, Hughes JB, Kim JH (2007). Photochemical production of reactive oxygen species by C-60 in the aqueous phase during UV irradiation, *Environ Sci Technol*; **41**: 2529–2535.

Lehn JM, Atwood JL, Davies JED, MacNicol DD, Vögtle F (1996). *Comprehensive Supramolecular Chemistry*; Pergamon: New York.

Li YH, Ding J, Luan ZK, Di ZC, Zhu YF, Xu CL, Wu DH, Wei BQ (2003). Competitive adsorption of Pb^{2+}, Cu^{2+} and Cd^{2+} ions from aqueous solutions by multiwalled carbon nanotube's, *Carbon*; **41**: 2787–2792.

Li Z, Pan Z, Dai S (2004). Nitrogen adsorption characterization of aligned multiwalled carbon nanotubes and their acid modification, *J Colloid Interface Sci*; **277**: 35–42.

Li YH, Wang S, Wei J, Zhang X, Xu C, Luan Z, Wu D, Wei B (2002). Lead adsorption on carbon nanotube's, *Chem Phys Lett*; **357**: 263–266.

Li GY, Wang PM, Zhao X (2007). Pressure-sensitive properties and microstructure of carbon nanotube reinforced cement composites. *Cem Concr Compos*; **29**: 377–382.

Li X, Zhou H, Yu P, Su L, Ohsaka T, Mao L (2008). A Miniature glucose/O_2 biofuel cell with single-walled carbon nanotubes-modified carbon fiber microelectrodes as the substrate, *Electrochem Commun*; **10**: 851–854.

Lim TC, Ramakrishna S (2006). A conceptual review of nanosensors, *Z Naturforsch*; **61**: 402–412.

Lin YH, Cui XL, Bontha J (2006). Electrically controlled anion exchange based on polypyrrole and carbon nanotubes nanocomposites for perchlorate removal, *Environ Sci Technol*; **40**: 4004–4009.

Logan BE, Hamelers B, Rozendal R, Schröder U, Keller J, Freguia S, Aelterman P, Verstraete W, Rabaey K (2006). Microbial fuel cells: Methodology and technology, *Environ Sci Technol*; **40**: 5181–5192.

Long RQ, Yang RT (2001). Carbon nanotube's as a superior sorbent for dioxin removal, *J Am Chem Soc*; **123**: 2058–2059.

Lopez CF, Nielsen SO, Moore PB, Klein ML (2004). Understanding nature's design for a nanosyringe, *Proc Nat Acad Sci U S A*; **101**: 4431–4434.

Lu CS, Chung YL, Chang KF (2005). Adsorption of trihalomethanes from water with carbon nanotube's, *Water Res*; **39**: 1183–1189.

Lu J, Kumar B, Castro M, Feller JF (2009). Vapour sensing with conductive polymer nanocomposites (CPC): Polycarbonate-carbon nanotubes transducers with hierarchical structure processed by spray layer by layer, *Sens Actuator B Chem*; **140**: 451–460.

Luthy RG, Aiken GR, Brusseau ML, Cunningham SD, Gschwend PM, Pignatello JJ, Reinhard M, Traina SJ, Weber WJ, Westall JC (1997). Sequestration of hydrophobic organic contaminants by geosorbents, *Environ Sci Technol*; **31**: 3341–3347.

Lyon DY, Adams LK, Falkner JC, Alvarez PJJ (2006). Antibacterial activity of fullerene water suspensions: Effects of preparation method and particle size, *Environ Sci Technol*; **40**: 4360–4366.

Lyon DY, Brown DA, Alvarez PJJ (2008). Implications and potential applications of bactericidal fullerene water suspensions: Effect of nC60 concentration, exposure conditions and shelf life, *Water Sci Technol*, **57**: 1533–1538.

Maddox MW, Gubbins KE (1995). Molecular simulation of fluid adsorption in bucky tubes, *Langmuir*; **11**: 3988–3996.

Magrez A, Kasas S, Salicio V, Pasquier N, Seo JW, Celio M, Catsicas S, Schwaller B, Forró L (2006). Cellular toxicity of carbon-based nanomaterials, *Nano Lett*; **6**: 1121–1125.

Maiti A, Andzelm J, Tanpipat N, Von Allmen P (2001). Effect of adsorbates on field emission from carbon nanotubes, *Phys Rev Lett*; **87**: 155502.

Martin, CR, Kohli, P (2003). The emerging field of nanotube biotechnology, *Nat Rev Drug Discov*; **2**: 29–37.

McKenzie LC, Hutchison JE (2004). Green nanoscience. *Chim Oggi*; **22**: 30–33.

Mei X, Fan B, Sun K, Ouyang J (2009). High-performance dye-sensitized solar cells with nanomaterials as a counter electrode, nanoscale photonic and cell technologies for photovoltaics II, San Diego, CA, USA. **7411**: doi:10.1117/12.825858.

Minteer SD, Liaw BY, Cooney MJ (2007). Enzyme-based biofuel cells, *Curr Opin Biotech*; **18**: 228–234.

Mojumdar SC, Raki L (2005). Preparation and properties of calcium silicate hydrate-poly (vinyl alcohol) nanocomposite materials, *J Therm Anal Calorim*; **82**: 89–95.

Moon SI, Paek KK, Lee YH, Park HK, Kim JK, Kim SW, Ju BK (2008). Bias-heating recovery of MWCNT gas sensor, *Mater Lett*; **62**: 2422–2425.

Morones JR, Elechiguerra JL, Camacho A, Holt K, Kouri JB, Ramirez JT, Yacaman MJ (2005). The bactericidal effect of silver nanoparticles, *Nanotechnology*; **16**: 2346–2353.

Morozan A, Stamatin L, Nastase F, Dumitru A, Vulpe S, Nastase C, Stamatin I, Scott K (2007). The biocompatibility microorganisms-carbon nanostructures for applications in microbial fuel cells, *Phys Status Solidi (A)*; **204**: 1797–1803.

Muller J, Huaux F, Moreau N, Misson P, Heilier JF, Delos M, Arras M, Fonseca A, Nagy JB, Lison D (2005). Respiratory toxicity of multi-wall carbon nanotubes, *Toxicol Appl Pharmacol*; **207**: 221–231.

Narita M, Amano F, Kadoishi Y, Nishiumi H (2007). Phenol decomposition by visible-light catalysis with fullerene, *Kagaku Kogaku Ronbunshu*; **33**: 564–569.

Neimark AV, Ruetsch S, Kornev KG, Ravikovitch PI (2003). Hierarchical pore structure and wetting properties of single wall carbon nanotube fibers, *Nano Lett*; **3**: 419–423.

Nel A, Xia T, Madler L, Li N (2006). Toxic potential of materials at the nano level, *Science*; **311**: 622–627.

Nguyen HQ, Huh JS (2006). Behaviour of single-walled carbon nanotube-based gas sensors at various temperatures of treatment and operation, *Sens Actuator B Chem*; **117**: 426–430.

Nguyen LH, Phi TV, Phan PQ, Vu HN, Nguyen-Duc C, Fossard F (2007). Synthesis of multi-walled carbon nanotubes for NH_3 gas detection, *Phys E Low Dimens Syst Nanostruct*; **37**: 54–57.

Nuzzo JB (2006). The biological threat to U.S. water supplies: Toward a national water security policy, *Biosecur Bioterror Biodefense Strategy Pract Sci*; **4**: 147–159.

Papirer E, Brendle E, Ozil F, Balard H (1999). Comparison of the surface properties of graphite, carbon black and fullerene samples, measured by inverse gas chromatography, *Carbon*; **37**: 1265–1274.

Pejcic B, Eadington P, Ross A (2007). Environmental monitoring of hydrocarbons: Sensor perspective a chemical. *Environ Sci Technol*; **41**: 6333–6342.

Peng XJ, Li YH, Luan ZK, Di ZC, Wang HY, Tian BH, Jia ZP (2003). Adsorption of 1,2-dichlorobenzene from water to carbon nanotubes. *Chem Phys Lett*; **376**: 154–158.

Peng XJ, Luan ZK, Ding J, Di ZH, Li YH, Tian BH (2005). Ceria nanoparticles supported on carbon nanotubes for the removal of arsenate from water, *Mater Lett*; **59**: 399–403.

Penza M, Rossi R, Alvisi M, Cassano G, Serra E (2009a). Functional characterization of carbon nanotube networked films functionalized with tuned loading of Au nanoclusters for gas sensing applications, *Sens Actuator B Chem*; **140**: 176–184.

Penza M, Rossi R, Alvisi M, Signore MA, Cassano G, Dimaio D, Pentassuglia R, Piscopiello E, Serra E, Falconieri M (2009b). Characterization of metal-modified and vertically aligned carbon nanotube films for functionally enhanced gas sensor applications, *Thin Solid Films*, **517**: 6211–6216.

Pignatello JJ, Xing BS (1996). Mechanisms of slow sorption of organic chemicals to natural particles. *Environ Sci Technol*; **30**: 1–11.

Plata DL, Gschwend PM, Reddy CM (2008). Industrially synthesised single-walled carbon nanotubes: Compositional data for users, environmental risk assessments, and source apportionment, *Nanotechnology*, **19**: 185706.

Popov VN (2004). Carbon nanotubes: Properties and application, *Mater Sci Eng R*; **43**: 61–102.

Punyamurtula VK, Qiao Y (2007). Hysteresis of sorption isotherm of multiwall carbon nanotube in paraxylene, *Mater Res Innov*; **11**: 37–39.

Qi P, Vermesh O, Grecu M, Javey A, Wang Q, Dai H, Peng S, Cho KJ (2003). Toward large arrays of multiplex functionalized carbon nanotube sensors for highly sensitive and selective molecular detection, *Nano Lett*; **3**: 347–351.

Qiao Y, Li CM, Bao SJ, Bao QL (2007). Carbon nanotube/polyaniline composite as anode material for microbial fuel cells, *J Power Source*; **170**: 79–84.

Quang NH, Van Trinh M, Lee BH, Huh JS (2006). Effect of NH_3 gas on the electrical properties of single-walled carbon nanotube bundles, *Sens Actuator B Chem*; **113**: 341–346.

Ramasamy E, Lee WJ, Lee DY, Song JS (2008). Spray coated multiwall carbon nanotube counter electrode for triiodide reduction in dye-sensitized solar cells, *Electrochem Commun*; **10**: 1087–1089.

Rao GP, Lu C, Su F (2007). Sorption of divalent metal ions from aqueous solution by carbon nanotubes: A review, *Sep Purif Technol*; **58**: 224–231.

Ray SS, Vaudreuil S, Maazouz A, Bousmina M (2006). Dispersion of multi-walled carbon nanotubes in biodegradable poly(butylenes succinate) matrix, *J Nanosci Nanotechnol*; **6**: 2191–2195.

Reijnders L (2006). Cleaner nanotechnology and hazard reduction of manufactured nanoparticles, *J Clean Prod*; **14**: 124–133.

Rinzler AG, Liu J, Dai H, Nikolaev P, Huffman CB, Rodriguez-Macias FJ, Boul PJ, Lu AH, Heymann D, Colbert DT, Lee RS, Fischer JE, Rao AM, Eklund PC, Smalley RE (1998). Large-scale purification of single-wall carbon nanotubes: Process, product, and characterization, *Appl Phys A Mater Sci Process*; **67**: 29–37.

Saito R, Fujita M, Dresselhaus G, Dresselhaus MS (1992). Electronic-structure of chiral graphene tubules, *Appl Phys Lett*; **60**: 2204–2206.

Saito Y, Uemura S (2000). Field emission from carbon nanotubes and its application to electron sources, *Carbon*; **38**: 169–182.

Sawatsuk T, Chindaduang A, Sae-kung C, Pratontep S, Tumcharern G (2009). Dye-sensitized solar cells based on TiO_2-MWCNTs composite electrodes: Performance improvement and their mechanisms, *Diam Relat Mater*; **18**: 524–527.

Sayes CM, Liang F, Hudson JL, Mendez J, Guo WH, Beach JM, Moore VC, Doyle CD, West JL, Billups WE, Ausman KD, Colvin VL (2006). Functionalization density dependence of single-walled carbon nanotubes cytotoxicity in vitro, *Toxicol Lett*; **161**: 135–142.

Schwarzenbach RP, Westall J (1981). Transport of non-polar organic compounds from surface water to groundwater laboratory sorption studies, *Environ Sci Technol*; **15**: 1360–1367.

Scrivens WA, Tour JM, Creek KE, Pirisi L (1994). Synthesis of 14C-Labeled C60, its suspension in water, and its uptake by human keratinocytes, *J Am Chem Soc*; **116**: 4517–4518.

Sharma T, Reddy ALM, Chandra TS, Ramaprabhu S (2008). Development of carbon nanotubes and nanofluids based microbial fuel cell, *Int J Hydrogen Energy*; **33**: 6749–6754.

Sheldon RA (2007). Enzyme immobilisation: The quest for optimum performance, *Adv Synth Catal*; **349**: 1289–1307.

Shelimov KB, Esenaliev RO, Rinzler AG, Huffman CB, Smalley RE (1998). Purification of single-wall carbon nanotubes by ultrasonically assisted filtration, *Chem Phys Lett*; **282**: 429–434.

Shi ZJ, Lian YF, Liao FH, Zhou XH, Gu ZN, Zhang G, Iijima S (1999). Purification of single-wall carbon nanotube's, *Solid State Commun*; **112**: 35–37.

Shinohara H, Sato H, Saito Y, Takayama M, Izuoka A, Sugawara T (1991). Formation and extraction of very large all-carbon fullerenes, *J Phys Chem*; **95**: 8449–8451.

Sinha N, Ma JZ, Yeow JTW (2006). Carbon nanotube-based sensors, *J Nanosci Nanotechnol*; **6**: 573–590.

Song L, Qiu Z (2009). Crystallisation behaviour and thermal property of biodegradable poly(butylene succinate)/functional multi-walled carbon nanotubes nanocomposites, *Polym Degrad Stab*; **94**: 632–637.

Steinfeldt M, Gleich AV, Petschow U, Haum R (2007). *Nanotechnologies, Hazards and Resource Efficiency: A Three-Tiered Approach to Assessing the Implications of Nanotechnology and Influencing its Development*; Springer: Berlin.

Stone AJ, Wales DJ (1986). Theoretical studies of icosahedral C60 and some related species, *Chem Phys Lett*; **128**: 501–503.

Suehiro J, Zhou G, Hara M (2005). Detection of partial discharge in SF_6 gas using a carbon nanotube-based gas sensor, *Sens Actuator B Chem*; **105**: 164–169.

Sun YP, Fu KF, Lin Y, Huang WJ (2002). Functionalized carbon nanotubes: Properties and applications, *Acc Chem Res*; **35**: 1096–1104.

Tang YJJ, Ashcroft JM, Chen D, Min GW, Kim CH, Murkhejee B, Larabell C, Keasling JD, Chen FF (2007). Charge associated effects of fullerene derivatives on microbial structural integrity and central metabolism, *Nano Lett*; **7**: 754–760.

Tans SJ, Devoret MH, Dai H, Thess A, Smalley RE, Geerligs LJ, Dekker C (1997). Individual single-wall carbon nanotubes as quantum wires, *Nature*; **386**: 474–477.

Taylor R, Walton DRM (1993). The chemistry of fullerenes, *Nature*; **363**: 685–693.

Tenne R (2006). Inorganic nanotubes and fullerene-like nanoparticles, *Nat Nanotechnol*; **1**: 103–111.

Terashima M, Nagao S (2007). Solubilization of [60] fullerene in water by aquatic humic substances, *Chem Lett*; **36**: 302–303.

Thess A, Lee R, Nikolaev P, Dai HJ, Petit P, Robert J, Xu CH, Lee YH, Kim SG, Rinzler AG, Colbert DT, Scuseria GE, Tomanek D, Fischer JE, Smalley RE (1996). Crystalline ropes of metallic carbon nanotube's, *Science*; **273**: 483–487.

Tong ZH, Bischoff M, Nies L, Applegate B, Turco RF (2007). Impact of fullerene (C-60) on a soil microbial community, *Environ Sci Technol*; **41**: 2985–2991.

Tsai HY, Wu CC, Lee CY, Shih EP (2009). Microbial fuel cell performance of multiwall carbon nanotubes on carbon cloth as electrodes, *J Power Sources*; **194**: 199–205.

Ueda T, Bhuiyan MMH, Norimatsu H, Katsuki S, Ikegami T, Mitsugi, F (2008a). Development of carbon nanotube-based gas sensors for NO_x gas detection working at low temperature, *Phys E Low Dimens Syst Nanostruct*; **40**: 2272–2277.

Ueda T, Katsuki S, Takahashi, K., Narges, HA, Ikegami T, Mitsugi F (2008b). Fabrication and characterization of carbon nanotube based high sensitive gas sensors operable at room temperature, *Diam Relat Mater*; **17**: 1586–1589.

Ugarte D (1992). Curling and closure of graphitic networks under electron-beam irradiation, *Nature*; **359**: 707–709.

Upadhyayula VKK, Deng S, Mitchell MC, Smith GB, Nair VK, Ghoshroy S (2008a). Adsorption kinetics of Escherichia coli and Staphylococcus aureus on single-walled carbon nanotube aggregates, *Water Sci Technol*; **58**: 179–184.

Upadhyayula VKK, Ghoshroy S, Nair VS, Smith GB, Mitchell MC, Deng S (2008b). Single-walled carbon nanotubes as fluorescence biosensors for pathogen recognition in water systems, *Res Lett Nanotechnol*; **2008**: 156358.

Valentini L, Cantalini C, Armentano I, Kenny JM, Lozzi L, Santucci S (2004). Highly sensitive and selective sensors based on carbon nanotubes thin films for molecular detection, *Diam Relat Mater*; **13**: 1301–1305.

Van Hieu N, Thuy LTB, Chien ND (2008). Highly sensitive thin film NH_3 gas sensor operating at room temperature based on SnO_2/MWCNTs composite, *Sens Actuator B Chem*; **129**: 888–895.

Vaudreuil S, Labzour A, Sinha-Ray S, Mabrouk KE, Bousmina, M (2007). Dispersion characteristics and properties of poly(methyl methacrylate)/multiwalled carbon nanotubes nanocomposites, *J Nanosci Nanotechnol*; **7**: 2349–2355.

Walther JH, Jaffe RL, Kotsalis EM, Werder T, Halicioglu T, Koumoutsakos P (2004). Hydrophobic hydration of C-60 and carbon nanotubes in water, *Carbon*; **42**: 1185–1194.

Wang J (2005). Carbon-nanotube based electrochemical biosensors: A review, *Electroanalysis*; **17**: 7–14.

Wang XK, Chen CL, Hu WP, Ding AP, Xu D, Zhou X (2005). Sorption of 243 Am^{3+} to multiwall carbon nanotube's, *Environ Sci Technol*; **39**: 2856–2860.

Wang BC, Wang HW, Chang JC, Tso HC, Chou YM (2001). More spherical large fullerenes and multi-layer fullerene cages, *J Mol Struct Theochem*; **540**: 171–176.

Wang H, Xu P, Zhong W, Shen L, Du Q (2005). Transparent poly-(methyl methacrylate)/silica/zirconia nanocomposites with excellent thermal stabilities, *Polym Degrad Stab*; **87**: 319–327.

Watanabe K (2008). Recent developments in microbial fuel cell technologies for sustainable bioenergy, *J Biosci Bioeng*; **106**: 528–536.

Weber WJ, McGinley PM, Katz LE (1991). Sorption phenomena in subsurface systems concepts, models and effects on contaminant fate and transport, *Water Res*; **25**: 499–528.

Wei C, Dai L, Roy A, Tolle TB (2006). Multifunctional chemical vapour sensors of aligned carbon nanotube and polymer composites, *J Am Chem Soc*; **128**: 1412–1413.

Wei L, Shizhen H, Wenzhe C (2010). An MWCNT-doped SNO_2 thin film NO_2 gas sensor by RF reactive magnetron sputtering, *J Semicond*; **31**: 024006.

Wiesner MR (2006). Responsible development of nanotechnologies for water and wastewater treatment, *Water Sci Technol*; **53**: 45–51.

Wick P, Manser P, Limbach LK, Dettlaff-Weglikowska U, Krumeich F, Roth S, Stark WJ, Bruinink A (2007). The degree and kind of agglomeration affect carbon nanotube cytotoxicity, *Toxicol Lett*; **168**: 121–131.

Wijewardane S (2009). Potential applicability of CNT and CNT/composites to implement ASEC concept: A review article, *Sol Energy*; **83**: 1379–1389.

Worle-Knirsch JM, Pulskamp K, Krug HF (2006). Oops they did it again! Carbon nanotubes hoax scientists in viability assays, *Nano Lett*; **6**: 1261–1268.

Xia YN, Yang PD, Sun YG, Wu YY, Mayers B, Gates B, Yin YD, Kim F, Yan YQ (2003). One-dimensional nanostructures: Synthesis, characterization, and applications, *Adv Mater*; **15**: 353–389.

Yan H, Gong AJ, He HS, Zhou J, Wei YX, Lv L (2006). Adsorption of microcystins by carbon nanotubes, *Chemosphere*; **62**: 142–148.

Yang FH, Lachawiec AJ, Yang RT (2006). Adsorption of spill over hydrogen atoms on single-wall carbon nanotube's, *J Phys Chem B*; **110**: 6236–6244.

Yang S, Song H, Yi H, Liu W, Zhang H, Chen X (2009). Carbon nanotube capsules encapsulating SnO_2 nanoparticles as an anode material for lithium ion batteries, *Electrochim Acta*; **55**: 521–527.

Yang K, Wang XL, Zhu LZ, Xing BS (2006). Competitive sorption of pyrene, phenanthrene, and naphthalene on multiwalled carbon nanotubes, *Environ Sci Technol*; **40**: 5804–5810.

Yang K, Xing BS (2007). Desorption of polycyclic aromatic hydrocarbons from carbon nanomaterials in water, *Environ Pollut*; **145**: 529–537.

Yang K, Zhu LZ, Xing BS (2006). Adsorption of polycyclic aromatic hydrocarbons by carbon nanomaterials, *Environ Sci Technol*; **40**: 1855–1861.

Zhao HY, Zhou HM, Zhang JX, Zheng W, Zheng YF (2009). Carbon nanotube-hydroxyapatite nanocomposite: A novel platform for glucose/O_2 biofuel cell, *Biosens Bioelectron*; **25**: 463–468.

Zhang Q, Huang J, Wei F, Xu G, Wang Y, Qian W, Wang D (2007). Large scale production of carbon nanotube arrays on the sphere surface from liquefied petroleum gas at low cost, *Chin Sci Bull*; **52**: 2896–2902.

Zhang WD, Zhang WH (2009). Carbon nanotubes as active components for gas sensors, *J Sensors*; **2009**: 1–16.

Zhu SB, Xu TG, Fu HB, Zhao JC, Zhu YF (2007). Synergetic effect of Bi_2WO_6 photocatalyst with C-60 and enhanced photoactivity under visible irradiation, *Environ Sci Technol*; **41**: 6234–6239.

Chapter 9

Role of Computational Tools in Designing Enzymatic Biosensors for the Detection of Pesticides in Environment

Mohd. Shahbaaz, Suvardhan Kanchi, Myalowenkosi Sabela, and Krishna Bisetty

Department of Chemistry, Durban University of Technology, Durban 4000, South Africa

ksuvardhan@gmail.com; bisettyk@dut.ac.za

9.1 Introduction

"The most intelligent species on earth poisons its food before eating it." These startling words by Rishi Miranhshah [1] reveal the stark reality of destruction that we face today due to the relentless and injudicious use of pesticides. Pesticides are the chemical agents which prevent the proliferation of harmful pests (weeds, insects, mould or fungi) but simultaneously also interfere with the normal biological processes of other living organisms [2, 3]. Pesticides are indispensable to the modern agriculture and used extensively in agricultural fields to control the attack of insects, fungus and rodents on crops; reduce the

Nanocomposites for Pollution Control
Edited by Chaudhery Mustansar Hussain and Ajay Kumar Mishra

ISBN 978-981-4774-45-1 (Hardcover), 978-1-315-14368-2 (eBook)
www.panstanford.com

growth of weeds; boost agricultural productivity, thus increases crop yields; and reduce post-harvest losses [4–7]. A large amount of pesticides reaches the soil, air, water, food and other various systems through drift that takes the pesticides to places downwind the point of application. This may prove to be detrimental to the various life forms if it enters into the food chains [8, 9]. Some of these pesticides are very much toxic [10] mutagenic, carcinogenic and tumourogenic [11, 12].

Majority of the pesticides are neurotoxic compounds and irreversibly inhibit the enzyme acetylcholinesterase, an essential enzyme for the functioning of the central nervous system in humans and insects. This inhibition leads to the accumulation of the neurotransmitter acetylcholine in nerves which interferes with muscular activities and the functioning of vital organs, hence producing symptoms of serious infections which may even lead to death [13, 14]. Target site of other various pesticides are: human erythrocytes [15], reproduction and development system [16], endocrine system [17], liver, central nervous system and immune system [18, 19], etc. Thus, the analysis of pesticide residues is an important concern due to their bioaccumulation effect, high toxicity and their long-term damage risk to both the environment and lives even though their concentration is very low.

Detection of pesticides at the levels established by the Environmental Protection Agency (EPA) remains a challenge. The chromatographic methods coupled to selective detectors have been traditionally used for the analysis of pesticide due to their sensitivity, reliability and efficiency. Nevertheless, chromatographic techniques are time-consuming and laborious, and require expensive equipment and highly trained technicians [20, 21]. Over the past decade, considerable attention has been given to the development of biosensors for the detection of pesticides as a promising alternative. Some electrochemical transducers have been used in biosensors for pesticides detection due to their high sensitivity [22–24]. Additionally, their low cost, simple design and small size makes them excellent candidates for the development of portable biosensors [25–28]. Acetylcholinesterase (ACH) is a neurotransmitter and easily undergoes hydrolysis to form choline and acetic acid. It is found at neuromuscular junctions and cholinergic synapses in the central nervous system, where its activity/concentration terminates synaptic transmission [29]. Cytochrome P450 (CYP) is a heme enzyme which enhances the different chemical reactions including heteroatom oxidation, epoxidation and hydroxylation. CYP enzyme is mainly involved in the metabolism of several drugs and xenobiotics for bio-activation [30]. Glutathione S-transferase (GST) is a detoxification enzyme, catalyzing glutathione (GSH) conjugation reactions. Protein phosphorylation process is the predominant molecular mechanism that can be achieved with Protein kinase C (PKC) [31].

Several bio-recognition methods that have been reported for the detection of pesticides include enzymatic, whole cell, immunochemical, and DNA biosensors. Designing of facile, selective and cost-effective biosensor technology is a growing area for the detection of various classes of pesticides. The biosensor with these features can be achieved only with the various bio-components with different transducers [32]. In fabrication of biosensors for pesticides, mainly enzyme are used due to the pesticide inhibition capability on the enzyme by various neurotoxins [33, 34] and are very sensitive (able to detect 10^{-10} M), but they offer poor specificity [35]. The approach used in this work took advantage of the characteristic feature of ACE, CYP, GST and PKC molecules probing its ability as biosensors for the detection of different class of pesticides. A survey of literature revealed that the quite a few reports were available on the ACE performance as a biosensor for the detection of pesticides [36–41]. However, in case of biomolecules such as CYP, GST and PKC, there are no reports in the literature. Accordingly, this work reports on the interaction studies of ACE, CYP, GST and PKC biomolecules with fenobucarb, dichlorodiphenyltrichloroethane (DDT) and parathion which gives a novel insights to confirm these biomolecules as good promising candidates for future biosensors.

In this study, the molecular docking techniques were used to infer the specificity of the proteins to recognize the used pesticides. Furthermore, the conformational behaviour of the stable complexes were analysed using molecular dynamics (MD) simulations in the explicit water conditions each for 50 ns time scale. The outcomes of this extensive analysis provide an insight into the structural features of the studied proteins responsible for the recognition of the pesticides. Novel protein candidate for the detection of the pesticides was also identified in this study.

9.2 Materials and Methods

9.2.1 Molecular Modelling

The structure of acetylcholinesterase (ACE, PDB ID – 5HF9), cytochrome P450 (CYP, PDB ID – 4D7D), glutathione S-transferase (GST, PDB ID – 4MPF), and protein kinase C (PKC, PDB ID – 5F9E) were remodelled to remove the discontinuities in the structural coordinates deposited in the biological database. The protein structures were predicted by satisfying the spatial restraints using MODELLER module of the Discovery Studio 2016 (DS, http://accelrys.com/products/collaborative-science/biovia-discovery-studio/). The generated 3-D models were minimized using the optimization and side chain refinement modules of DS. The structure of fenobucarb, DDT and parathion pesticides were constructed by means of drawing utilities present in the DS and the geometries of the resulted structures were optimized by using the DFT method of Dmol3 module of the DS.

9.2.2 Molecular Docking

The refined structures of proteins, as well as pesticides, were docked by utilizing the AutoDock 4 [29] package which enables prediction of the bound conformation on the basis of the free energy-based empirical force field coupled with Lamarckian Genetic Algorithm [42]. The grid boxes of dimensions 50 × 70 × 60 Å along the XYZ directions with grid spacing of 0.375 Å were created using the AutoGrid module. In order to increase the efficiency, the parameters associated with Lamarckian genetic algorithm were set to the maximum efficiency value; for example, the number of individuals in the population was set to 250, while the maximum number of energy evaluations was set to "longer". As a result, around 100 docked conformations grouped according to the RMSD tolerance of 2.0 Å were obtained. The generated conformations were re-scored on the basis of the scoring function present in DrugScoreX server [43] and the conformation with the highest score was selected for the MD simulations. In order to validate the generated outcomes the CDOCKER module was used, which are a CHARMM [44] force field-based docking algorithm implemented in DS.

Furthermore, the pharmacophore modelling was performed in order to identify the essential features between receptors and pesticides which are responsible for the inhibition activity of the studied compounds. The features were computed using the "automatic pharmacophore generation" as well as "receptor-ligand pharmacophore generation" modules of DS. The pharmacophore features such as hydrophobic group (H), aromatic ring (R), hydrogen bond donor (D), hydrogen bond acceptor (A), negatively ionizable group (N), and positively ionizable group (P) were calculated. The generated pharmacophore models were analyzed and compared to the binding priorities of the ligands with the studied proteins.

9.2.3 Molecular Dynamic Simulations

The generated docked complexes were subjected to MD simulations using GROMACS [45] (version 5.1.2, installed on the Center for High Performance Computing (CHPC), Cape Town, which provides 10 nodes with 24 cores per node of space for computation). The topologies of pesticide structures were generated by the means of GROMOS96 53a6 force field [46]. Due to the unavailability of suitable force field parameters for drug-like molecules in the GROMACS package, the PRODRG server [47] was used for the generation of the piperine in topologies and coordinate files. The partial charges were corrected by using DFT method of Gaussian which utilized the B3LYP 6-31G (d,p) basis set and CHELPG program [48]. After the successful topology generation, all docked complexes were immersed in SPC/E water model [49] and the systems were neutralized by adding the counter ions. The neutralized systems were energetically minimized by steepest descent and conjugate gradient algorithms with a convergence criterion of 0.005 kcal mol^{-1}. In order to increase the reliability of the MD simulations the restraints were applied to the structure of the pesticides before the equilibration phase.

The equilibration phase was carried out separately for 100 ps time scale in NVT (constant volume) as well as NPT (constant pressure) ensemble conditions. The temperature of the system was maintained at 300 K by using Berendsen weak coupling method in both ensemble conditions along with pressure which was maintained at 1 bar by utilizing Parrinello–Rahman barostat in constant pressure ensemble. The final MD simulations were produced using LINCS algorithm for 50 ns time scale. The knowledge extracted from the trajectory files were utilized for the analyses of each complex behaviour in the explicit water environment. The variations in the distances, H-bonds, RMSD (root mean square deviations), and Radius of Gyration (Rg) between the proteins and pesticides complexes were analyzed.

9.3 Results and Discussion

9.3.1 Calibre of Acetylcholinesterase (ACH) as a Future Biosensor: Interaction Studies of ACH with Fenobucarb, DDT and Parathion

The ACH was characterized to be a serine hydrolase and involved in the hydrolysis of acetylcholine, a neurotransmitter involve in the conducting of nerve impulses in a wide range of animals including humans [50]. The pesticide inhibition of the ACH takes place in an irreversible manner by the phosphorylation of serine hydroxyl, which increases the stability of the enzyme [51]. The ACH is widely used in the formation of inhibition-based biosensor for the detection of the harmful pesticides [51]. Therefore, the ACH was selected in addition to other proteins as a possible candidate which can be used for designing an enzyme-based biosensor. All the three categories of the pesticides were docked in the active site of ACH using Autodock package. The parathion showed highest inhibition constant score of 448.09 μM and free energy binding calculated to be –4.57 kcal/mol as presented in Table 9.1.

This is followed by fenobucarb and then DDT, with inhibition constants of 138.15 and 21.41 μM, respectively. The highest binding energy of –6.37 kcal/mol was obtained for DDT and therefore, the ACH and parathion complex was selected for further analysis. Interestingly, the parathion was found to interact with Tyr121, Trp283, Ser290, Val291, and Tyr338 as depicted in Fig. 9.1A. The binding with Ser290 confirmed phosphorylation of it hydroxyl group as the two-step process of the inhibition of the Ach processes [50]. Furthermore, the pharmacophore analysis showed the presence of 24 features, while only two were observed to be common with the receptor ACH protein (Fig. 9.1B).

Table 9.1 Interaction parameters generated after molecular docking of ACH with fenobucarb, DDT and parathion

	Pesticide		
Generated parameters	**Fenobucarb**	**DDT**	**Parathion**
Selected pose	21	27	8
Free energy of Binding (kcal/mol)	-5.27	-6.37	-4.57
Inhibition Constant (μM)	138.15	21.41	448.09
vdW + Hbond + desolvEnergy (kcal/mol)	-6.46	-7.25	-7.02
Intermolecular energy (kcal/mol)	-6.46	-7.27	-6.66
Total internal (kcal/mol)	-0.73	-0.98	-0.92
Torsional energy (kcal/mol)	1.19	0.89	2.09

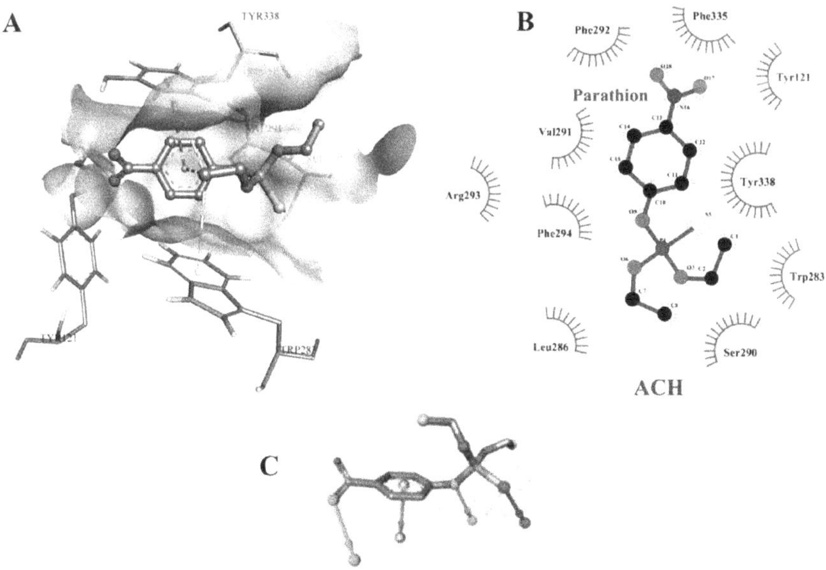

Figure 9.1 (A) The 3-D and 2-D representation of stable docked conformation generated for ACH-parathion complex. (B) The generated pharmacophore features for the docked complex (HB_ACCEPTOR: 13, HB_DONOR: 6, HYDROPHOBIC: 2, POS_IONIZABLE: 1, RING_AROMATIC: 2 and two features match the receptor-ligand interactions: AA).

Similar docked residues were observed in the case of fenobucarb and DDT (Fig. 9.2A,B). The docked complex of ACH-parathion was subjected to 50 ns MD simulations. The partial charges were corrected by using the DFT-based analysis (Fig. 9.3). The distance between the ACH and parathion was calculated using GROMACS utilities. The initial distance was observed to be around 0.4 nm and gradually decreases up to 10 ns time period,

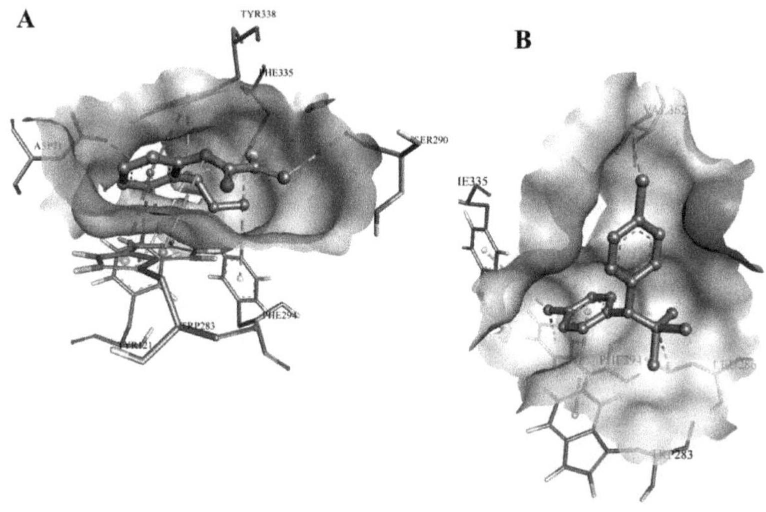

Figure 9.2 (A) The obtained docked complexes for ACH with fenobucarb. (B) The resulting complex of ACH with DDT.

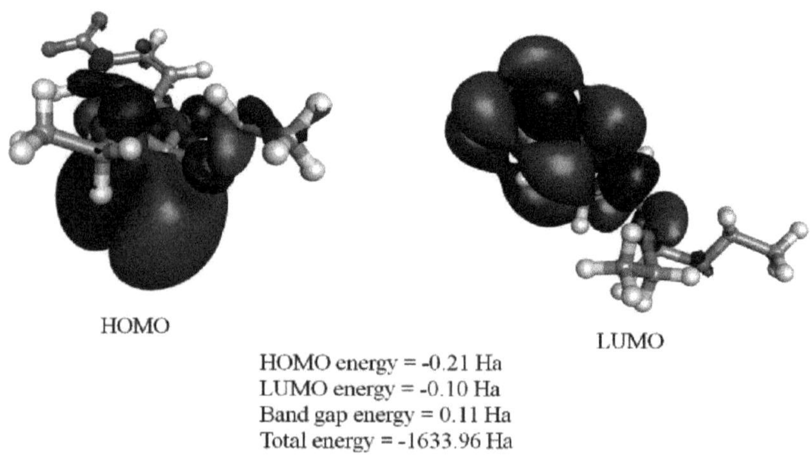

Figure 9.3 The DFT-generated parameters of ACH-parathion complex.

then continue fluctuating between 0.25–0.3 nm but showed sudden increase around 50 ns as illustrated in Fig. 9.4A. The hydrogen bond estimation between the receptor and pesticide may reach up to 8 during the course of simulations as shown in Fig. 9.4B. The continuous fluctuations were observed in the RMSD and Rg plots which indicated that the ACH is energetically active as depicted in Figs. 9.4C–D). The binding of parathion sifted the peaks of the curve indicated that the conformation of the ACH changed during the inhibition.

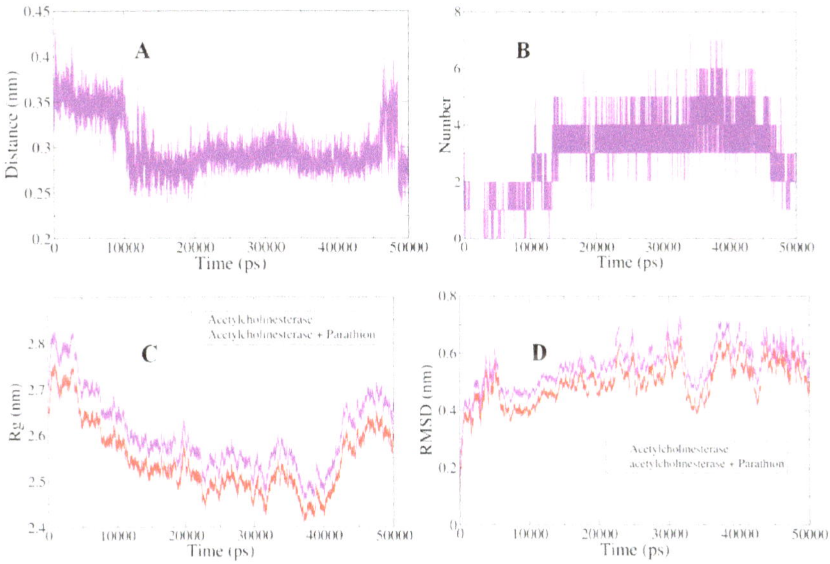

Figure 9.4 (A) The graph depicting the variation of the distance between ACH and parathion. (B) The curve illustrating the variations in the hydrogen bonds observed during 50 ns MD simulations. (C) and (D) The graphical representation of the fluctuations observed in the Rg and RMSD values.

9.3.2 Calibre of Cytochrome P450 (CYP) as a Future Biosensor: Interaction Studies of CYP with Fenobucarb, DDT and Parathion

The human hepatic CYP is known to catalyze the bio-transformation of pesticides, resulting in the detoxification of these compounds [52]. However, the metabolism of organophosphorus and carbamate insecticides leads to the formation of higher toxic metabolites [53, 54]. Since there is proven record on a variety of pesticides that induces expression of the CYP gene [55, 56], this protein is a promising candidate in the development of the pesticide specific biosensor. Accordingly, the molecular docking studies were performed and validated using the information available in the literature [53, 54]. The fenobucarb showed the most suitable combination of generated parameters and therefore selected for further analysis. The free energy of binding and inhibition constant for CYP-fenobucarb complex was calculated to be around –5.1 kcal/mol and 183.49 μM, respectively as tabulated in Table 9.2.

Table 9.2 Interaction parameters generated after molecular docking of CYP with fenobucarb, DDT and parathion

	Pesticide		
Generated parameters	**Fenobucarb**	**DDT**	**Parathion**
Selected pose	6	5	12
Free energy of Binding (kcal/mol)	–5.1	–6.54	–5.03
Inhibition Constant (μM)	183.49	16.15	206.93
vdW + Hbond + desolvEnergy (kcal/mol)	–6.11	–7.42	–6.24
Intermolecular energy (kcal/mol)	–6.29	–7.43	–7.11
Total internal (kcal/mol)	–0.72	–1.04	–0.77
Torsional energy (kcal/mol)	1.19	0.89	2.09

This is followed by fenobucarb and then DDT, with inhibition constants of 138.15 and 21.41 μM, respectively. The highest binding energy of –6.37 kcal/mol was obtained for DDT and therefore, the ACH and parathion complexes were selected for further analysis. Interestingly, the parathion was found to interact with Tyr121, Trp283, Ser290, Val291, and Tyr338 as depicted in Fig. 9.5A. The binding with Ser290 confirmed phosphorylation of its hydroxyl group as the two-step process of the inhibition of the Ach processes [50]. Furthermore, the pharmacophore analysis showed the presence of 24 features, while only two were observed to be common with the receptor ACH protein (Fig. 9.5B). Similar docked residues were observed in the case of fenobucarb and DDT (Fig. 9.6A,B).

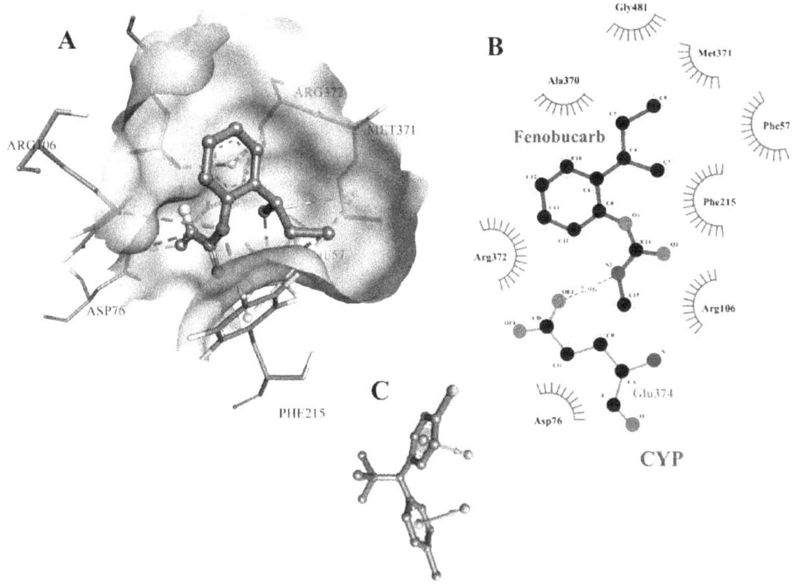

Figure 9.5 (A) The selected conformational pose of CYP-fenobucarb docked complex showing the interacting residues. (B) The generated pharmacophore features containing: HB_ACCEPTOR: 13, HB_DONOR: 6, HYDROPHOBIC: 2, POS_IONIZABLE: 1, RING_AROMATIC: 2 and two features match for receptor-ligand.

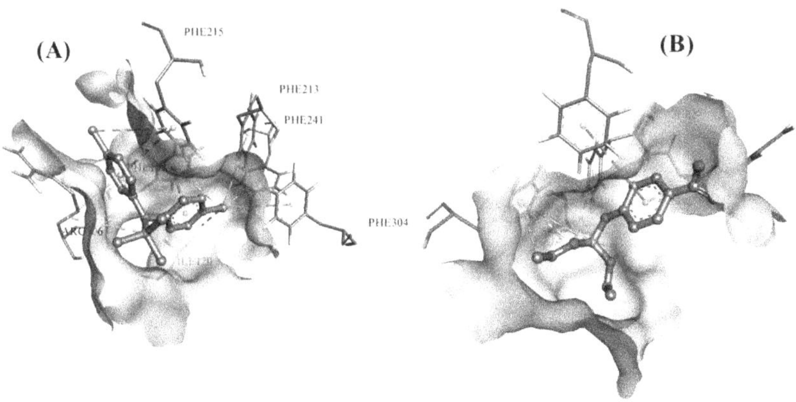

Figure 9.6 (A) The docked complexes showing interaction of CYP with DDT. (B) The resulted complex of CYP with parathion.

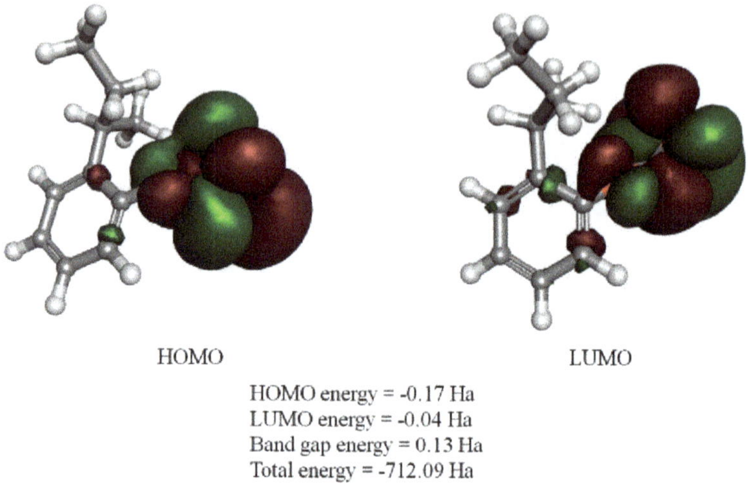

Figure 9.7 The DFT-generated parameters of CYP-fenobucarb complex.

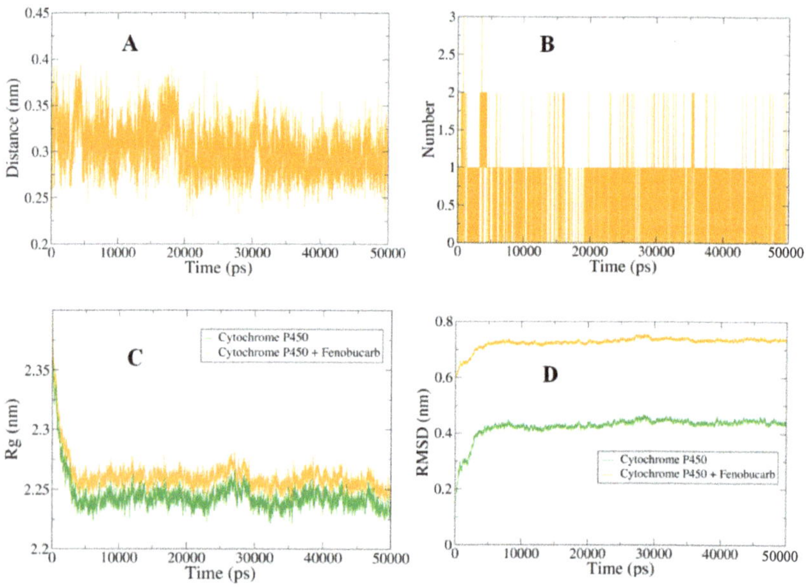

Figure 9.8 (A) and (B) The graphical view illustrating the dynamics observed in the bonding patterns between CYP and fenobucarb. (C) and (D) Rg and RMSD curves highlighting the changes in the conformational behaviour observed during 50 ns MD simulations.

Furthermore, the CYP-fenobucarb complex was subjected to 50 ns MD simulations in explicit solvent conditions with pesticide topology generated using DFT-based methods (Fig. 9.7). The dynamics of bonding between the CYP and fenobucarb was inferred from the calculated distance, which was fluctuating between 0.25–0.35 nm as shown in Fig. 9.8A. While a lesser number of hydrogen bonds were calculated in comparison with ACH-parathion complex (see Fig. 9.8B). The Rg plots signified the stable compactness of the complex as the variation in the values was observed around 2.25 nm (see Figs. 9.8C,D). However, the RMSD variations were significantly affected by the binding of fenobucarb with CYP as the values started fluctuating at 0.6 nm from 0.4 nm, which indicated that the increased energetic attributes of the protein. These observations highlighted the potential of CYP in the biosensing of pesticide compounds.

9.3.3 Calibre of Glutathione S-Transferase (GST) as a Future Biosensor: Interaction Studies of GST with Fenobucarb, DDT and Parathion

The GST is an enzyme involved in the formation of the defence against the insecticides such as decamethrin [57] and usually considered as the target in designing novel insecticides [58]. Therefore, the GST can also be considered as the potential candidate for designing a biosensor for the detection of pesticide level. The predicted 3-D structure of GST was used as receptor and docked with all the categories of the pesticides discussed in above sections. The complex with DDT, the most widely used insecticide, was used for the further analyses, which showed highest inhibition constant as compared to the other studied protein as presented in Table 9.3.

The parathion and fenobucarb also showed highest inhibition constant than it was calculated for the other proteins. DDT was found to interact with Gly14, Arg15, Leu72, and Glu162 residues of GST as shown in Fig. 9.9A while similar interaction behaviour was observed for other pesticides (see Figs. 9.10A,B). Furthermore, three pharmacophore features were calculated to be involved in the inhibition (see Fig. 9.9B), while similar distribution was observed in the orbitals after DFT-based calculations (see Fig. 9.11).

Table 9.3 Interaction parameters generated after molecular docking of GST with fenobucarb, DDT and parathion

	Pesticide		
Generated parameters	**Fenobucarb**	**DDT**	**Parathion**
Selected pose	26	13	4
Free energy of Binding (kcal/mol)	−4.5	−5.43	−4.11
Inhibition Constant (μM)	501.29	103.88	968.91
vdW + Hbond + desolvEnergy (kcal/mol)	−5.81	−6.3	−5.79
Intermolecular energy (kcal/mol)	−5.7	−6.33	−6.2
Total internal (kcal/mol)	−0.72	−0.99	−0.88
Torsional energy (kcal/mol)	1.19	0.89	2.09

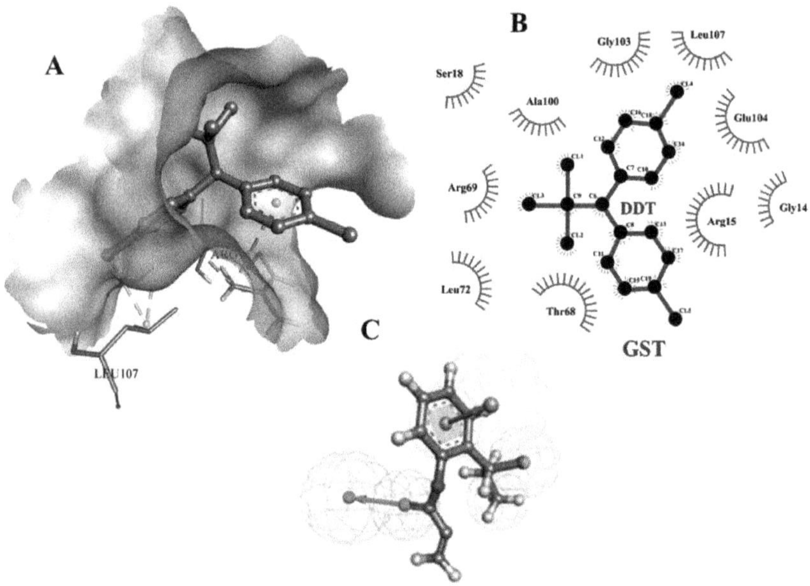

Figure 9.9 (A) The stable pose generated after molecular docking between the GST and DDT. (B) The pharmacophore features HYDROPHOBIC: 5 and RING_AROMATIC: 4 were detected and three features common to both receptor as well as ligand.

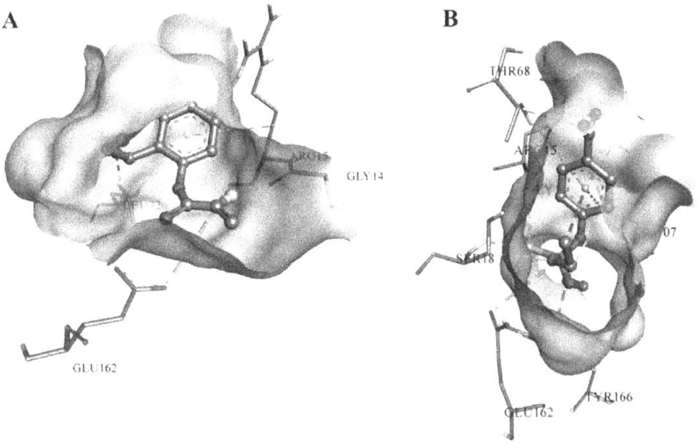

Figure 9.10 (A) The docking results showing the interacting residues of GST with fenobucarb. (B) The generated complex of GST with parathion.

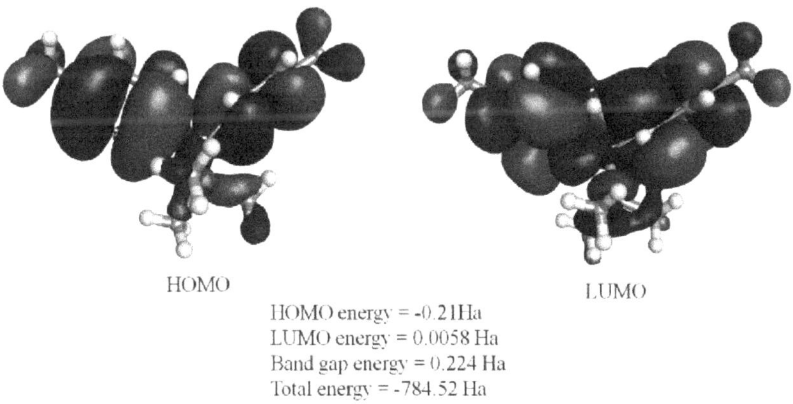

Figure 9.11 The parameters obtained after quantum calculations of GST-DDT complex.

Afterwards, the GST-DDT docked complex was centrally immersed in the solvation box containing water molecules and simulated for 50 ns time scale.

The analyses of the trajectories showed that the distance between GST and DDT fluctuated in the range of 0.25–0.3 nm as shown in Fig. 9.12A, while the hydrogen bond dynamics illustrated the presence of three bonds between the docked molecules (Fig. 9.12B). Similarly, the Rg and RMSD plots signified that binding of DDT resulted in little changes in the conformational preferences (Figs. 9.12C,D). Therefore, GST can be used for the detection of both insecticide and pesticide compounds.

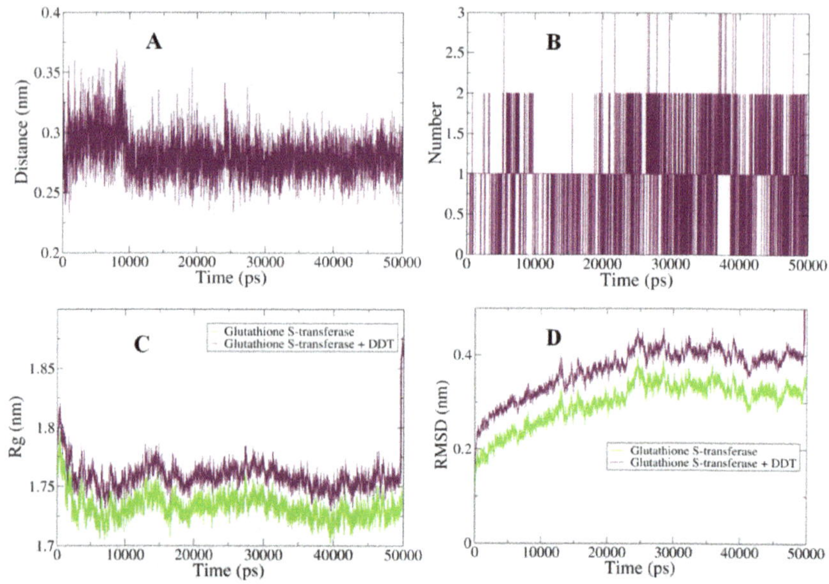

Figure 9.12 (A) and (B) The 2-D diagram describing the changes observed in the distance and hydrogen bonding patterns for GST and DDT complex. (C) and (D) The curves illustrating the changes in the Rg and RMSD values during 50 ns MD simulations.

9.3.4 Calibre of Protein Kinase C (PKC) as a Future Biosensor: Interaction Studies of PKC with Fenobucarb, DDT and Parathion

The PKC belongs to the family of serine/threonine kinases and catalyzed the phosphorylation of γ-aminobutyric acid receptors and regulates the effects of insecticides on neuronal channels and receptors present in insects [59]. However, the effect of PKC in the modulation of the insecticide sensitivity is still remains unclear [59]. Consequently, in order to understand the structural basis of PKC interaction with the pesticides, its structure was predicted and utilized for further analysis. The fenobucarb, DDT, and parathion were docked in the active site of PKC as depicted in Figs. 9.13A and 9.14. The highest score for the inhibition constant was obtained for PKC-fenobucarb complex (see Table 9.4), in addition to the four common pharmacophore features observed for studied receptor and pesticide (Fig. 9.13B). As the minimum features were detected in this case, the selectivity

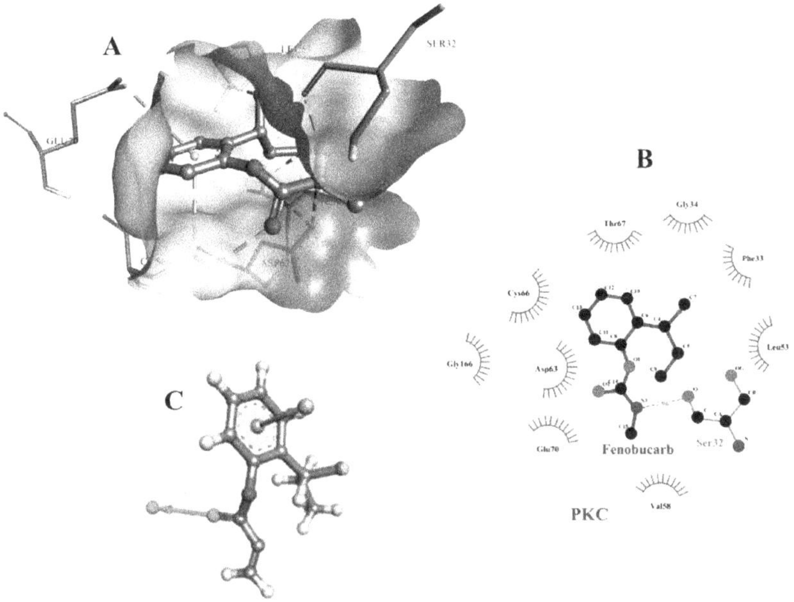

Figure 9.13 (A) 3-D and 2-D representation of the interacting residues observed in the docked complex of PKC and fenobucarb. (B) The nine generated pharmacophore features with four common for both PKC-fenobucarb complex with selectivity score of 5.6124 was generated.

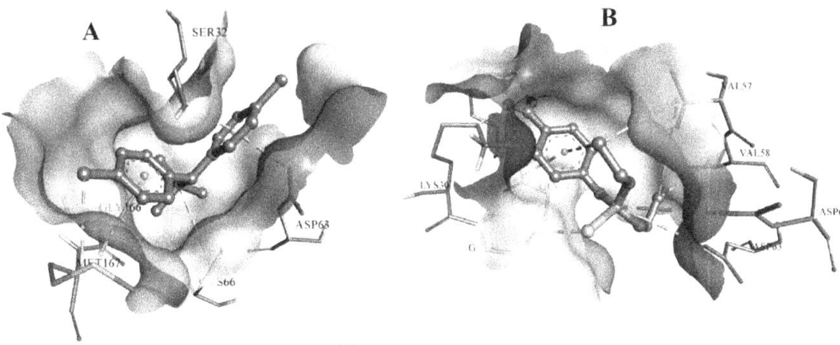

Figure 9.14 (A) The resulting interacting residues of PKC with DDT. (B) The generated complex of PKC with parathion.

score of 5.6124 was predicted by DS indicating the specificity of PKC. The fenobucarb was found to interact with Ser32, Leu53, Val58, Asp63, Cys66, and Glu70 as shown in Fig. 9.13.

Table 9.4 Interaction parameters generated after molecular docking of protein kinase C with fenobucarb, DDT and parathion

	Pesticide		
Generated parameters	**Fenobucarb**	**DDT**	**Parathion**
Selected pose	26	21	29
Free energy of Binding (kcal/mol)	−4.42	−6.11	−5.32
Inhibition Constant (μM)	575.53	33.22	125.1
vdW + Hbond + desolvEnergy (kcal/mol)	−5.63	−7.04	−5.12
Intermolecular energy (kcal/mol)	−5.61	−7.0	−7.41
Total internal (kcal/mol)	−0.73	−0.94	−0.96
Torsional energy (kcal/mol)	1.19	0.89	2.09

The in-depth knowledge about the conformational behaviour of PKC-fenobucarb complex was obtained through the principles of MD simulations and DFT studies (see Fig. 9.15).

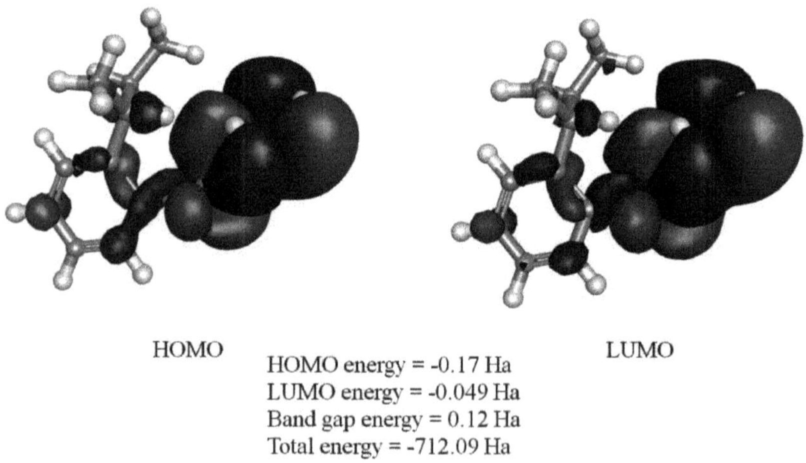

Figure 9.15 The quantum calculation-based generated parameters for PKC-fenobucarb complex.

The GROMACS utilities revealed that the distance between the PKC and fenobucarb was maintained in the range of 0.25–0.35 nm and hydrogen bonds also showed high fluctuations with maximum three bonds was present in the complex (see Figs. 9.16A,B). The Rg values showed a steady decrease up to 50 ns time scale indicated that the compactness of the structure was increased continuously throughout the MD simulations (see Fig. 9.16C). While the values in the RMSD curve showed a steady increase indicated that the PKC become energetically active with the advent of the MD simulations (see Fig. 9.16D). These observations indicated that the PKC is selectively inhibited by the pesticides and can be used for the specific detection.

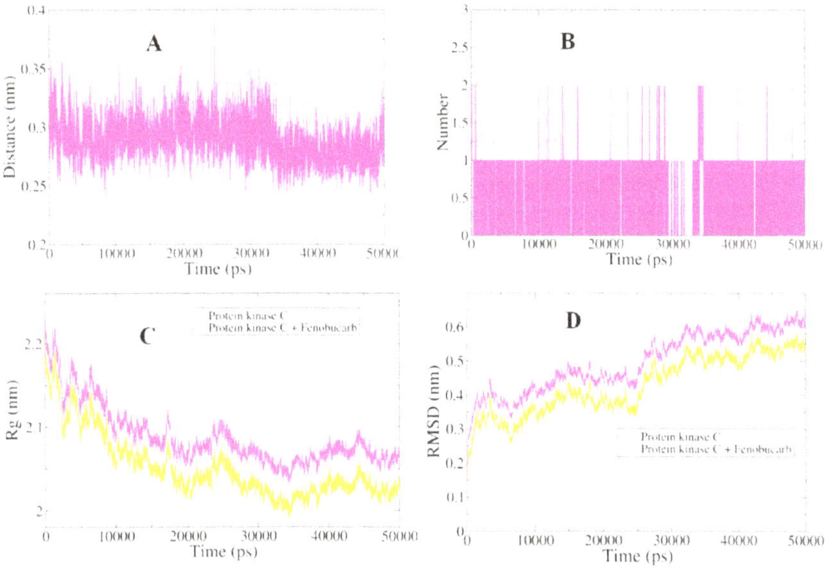

Figure 9.16 (A) and (B) The illustration of the dynamics observed between PKC and fenobucarb in terms of distance and Hydrogen bonding. (C) and (D) The curves illustrating the changes in the Rg and RMSD values during 50 ns MD simulations.

9.4 Conclusions

In this study the structural basis of the possible candidates for the biosensing of the pesticides were analyzed and their interaction patterns were compared. In silico studies provide a new insight into the interacting residues of the proteins involved in the p rocess of inhibition. Furthermore, MD simulation studies provided the understanding about the conformational behaviours of the studied proteins-pesticides complexes. In addition to the ACH, which was validated to be the primary target for organophosphorus and carbamates, their role in the inhibition of other metabolically significant proteins have been established in this study. The generated outcomes can be used in other experimental studies for assessing the toxic effects of the pesticides. The GST showed highest values of inhibition constants indicating its efficiency of binding with the pesticides, but the PKC showed highest specificity. Therefore, these proteins can further be utilized in designing a suitable biosensor for the detection of the concentration level of the pesticides in the water resources and food materials.

Acknowledgement

The authors thank the Center for High Performance Computing (CHPC), South Africa, for providing the supercomputing facility.

References

1. Revolutionary thought, *The Tribune*, 23 October, 2011.
2. P. Skladal, G. S. Nunes, H. Yamanaka, M. L. Ribero, Detection of carbamate pesticides in vegetable samples using cholinesterase-based biosensors, *Electroanalysis* 9(1997) 1083–1087.
3. A. F. S. Amaral, Pesticides and asthma: Challenges for epidemiology, *Front Pub. Heal.*, 2(6) (2014) 1–3.
4. C. Linan, *Vademecum de products fito sanitarios* (10th ed.) (1994), Madrid, Spain.
5. F. Hua, Y. Yunlong, C. Xiaoqiang, W. Xiuguo, Y. Xiaoe, Y. Jingquan, Degradation of chlorpyrifos in laboratory soil and its impact on soil microbial functional diversity, *J. Environ. Sci.*, 21 (2009) 380–386.
6. G. Odukkathil, N. Vasudevan, Toxicity and bioremediation of pesticides in agricultural soil, *Rev. Environ. Sci. Biotechnol.*, 12 (2013) 421–444.
7. X. W. Meng, J. F. Wei, X. L. Ren, J. Ren, F. Tang, A simple and sensitive fluorescence biosensor for detection of organophosphorus pesticides

using H_2O_2-sensitive quantum dots/bi-enzyme, *Biosens Bioelectron.*, 47 (2013) 402–407.

8. F. L. McEwen, G. R. Stephenson, *The Use and Significance of Pesticides in the Environment*, John Wiley and Sons (1979), New York, USA.
9. P. N. Viswanathan, Environmental toxicology in India, *Membr. Biol.*, 11 (1985) 88–97.
10. S. D. DeFlora, L. Vigano, F. Agoslini, A. Camorirano, M. Bagnasco, C. Bennicelli, F. Melodia, A. Arillo, Multiple genotoxicity biomarkers in fish exposed in situ to polluted river water, *Mutat. Res.*, 319 (1993) 167–177.
11. K. Kuroda, Y. Yamagachi, G. Endo, Mitotic toxicity, sister chromatid exchange, and rec assay of pesticides, *Arch. Environ. Contam. Toxicol.*, 23 (1992) 13–18.
12. Z. Rehana, A. Malik, M. Ahmad, Mutagenic activity of the Ganges water with special reference to the pesticide pollution in the river between Kachla to Kannauj (U.P.), India, *Mutat. Res.*, 343(2–3) (1995) 137–144.
13. W. J. Donarski, D. P. Dumas, D. H. Heitmeyer, V. E. Lewis, F. M. Raushel, Structure-activity relationships in the hydrolysis of substrates by the phosphotriesterase from Pseudomonas diminuta, *Biochemistry*, 28 (1989) 4650–4655.
14. S. Chapalamadugu, G. R. Chaudhry, Microbiological and biotechnological aspects of metabolism of carbamates and organophosphates, *Crit. Rev. Biotechnol.*, 12 (1992) 357–389.
15. M. Suwalsky, C. Rodriguez, F. Villena, C. P. Sotomayor, Human erythrocytes are affected by the organochloride insecticide chlordane, *Food Chem. Toxicol.*, 43 (2005) 647–54.
16. A. T. Farag, A. H. Radwan, F. Sorour, A. E. I. Okazy, E. I. S. E. I. Agamy, Chlorpyrifos induced reproductive toxicity in male mice, *Reprod. Toxicol.*, 29 (2010) 80–85.
17. N. C. Rawlings, S. J. Cook, D. Waldbillig, Effects of the pesticides carbofuran, chlorpyrifos, dimethoate, lindane, triallate, trifluralin, 2,4-D, and pentachlorophenol on the metabolic endocrine and reproductive endocrine system in Ewes, *J. Toxicol. Environ. Health A*, 54 (1998) 21–36.
18. L. S. Gold, T. H. Slone, N. B. Manley, Bernstein L. Target organs in chronic bioassays of 533 chemical carcinogens, *Environ. Health Persp.*, 93 (1991) 33–46.

19. E. A. Fendick, E. Mather-Mihaich, K. A. Houck, Ecological toxicology and human health effects of heptachlor, *Rev. Environ. Contam. Toxicol.*, 11 (1990) 161–142.

20. N. Jaffrezic-Renault, New trends in biosensors for organophosphorus pesticides, *Sensors*, 1 (2001) 60–74.

21. M. N. Velasco-Garcia, T. Mottram, Biosensor technology addressing agricultural problems, *Biosyst. Eng.*, 84 (2003) 1–12.

22. S. Andreescu, J. L. Marty, Twenty years research in Cholinesterase biosensors: From basic research to practical applications, *Biomol. Eng.*, 23(1) (2006) 1–15.

23. M. D. Luque de Castro, M. C. Herrera, Enzyme Inhibition-ased biosensors and biosensing systems: Questionable analytical devices, *Biosens. Bioelectron.*, 18 (2–3)(2003) 279–294.

24. D. Grieshaber, R. MacKenzie, J. Vöros, E. Reimhult, Electrochemical biosensors-sensor principles and architectures, *Sensors*, 8(3) (2008) 1400–1458.

25. R. S. Freire, C. A. Pessao, L. D. Mello, L. T. Kubota, Direct electron transfer: An approach for electro-chemical biosensors with higher selectivity and sensitivity, *J. Braz. Chem. Soc.*, 14(2) (2003) 230–243.

26. D. R. Thévenot, K. Toth, R. A. Durst, G. S. Wilson, Electrochemical biosensors: Recommended definitions and classification, *Pure Appl. Chem.*, 71(12) (1999) 2333–2348.

27. N. J. Ronkainen, H. B. Halsall, W. R. Heineman, Electrochemical biosensors, *Chem. Soc. Rev.*, 39(11) (2010) 1747–1763.

28. U. Yogeswaran, S. M. Chen, A review on the electrochemical sensors and biosensors composed of nanowires as sensing material, *Sensors*, 8(1) (2008) 290–313.

29. P. Norouzi, M. Pirali-Hamedani, M. R. Ganjali, F. A. Faridbod, A novel acetyl cholinesterase biosensor based on chitosan–gold nanoparticles film for determination of monocrotophos using FFT continuous cyclic voltammetry, *Int. J. Electrochem. Sci.*, 5 (2010) 1434–1446.

30. Ortiz de Montellano, P. R. (ed.), *Cytochrome P450: Structure, Mechanism and Biochemistry* 3rd ed. Kluwer Academic/Plenum Publishers, New York (2004).

31. C. A. Stanyon, O. Bernard, LIM-kinase1, *Int. J. Biochem. Cell Biol.*, 31 (1999) 389–394.

32. N. Verma, S. S. Dhillon, Biosensors for monitoring insecticides and herbicides—a survey, *Int. J. Environ. Stud.*, 60 (2003) 29–43.

33. N. Mionetto, J. L. Marty, I. Karube, Acetylcholinesterase in organic solvents for the detection of pesticides: Biosensor application, *Biosens. Bioelectron.*, 9 (1994) 463–470.

34. R. P. Deo, J. Wang, I. Block, A. Mulchandani, K. A. Joshi, M. Trojanowicz, F. Scholz, W. Chen, Y. Lin, Determination of organophosphate pesticides at a carbon nanotube/organophosphorus hydrolase electrochemical biosensor, *Anal. Chim. Acta*, 530 (2005) 185–189.

35. A. L. Simonian, E. I. Rainina, J. R. Wild, A new approach for discriminative detection of organophosphate neurotoxins in the presence of other Cholinesterase inhibitors, *Anal. Lett.*, 30 (1997) 2453–2468.

36. A. Krasinski, Z. Radic, R. Manetsch, J. Raushel, P. Taylor, K. B. Sharpless, H. C. Kolb, In situ selection of lead compounds by click chemistry: Target-guided optimization of acetylcholinesterase inhibitors, *J. Am. Chem. Soc.*, 127 (2005) 6686–6692.

37. R. Manetsch, A. Krasinski, Z. Radic, J. Raushel, P. Taylor, K. B. Sharpless, H. C. Kolb, In situ click chemistry: Enzyme inhibitors made to their own specifications, *J. Am. Chem. Soc.*, 126 (2004) 12809–12818.

38. G. Lin, Y. R. Lee, Y. C. Liu, Y. G. Wu, Ortho effect for inhibition mechanisms of butyrylcholinesterase by o-substitute phenyl L-butyl carbamates and comparison with acetylcholinesterase, cholesterol esterase, and phenol, *Chem. Res. Toxicol.*, 18 (2005) 1124–1131.

39. H. Schulze, S. B. Muench, F. Villatte, R. D. Schmid, T. T. Bachmann, Insecticide detection through protein engineering of Nippostrongylus brasiliensis acetylcholinesterase B, *Anal. Chem.*, 77 (2005) 5823–5830.

40. S. Andreescu, A. Avramescu, C. Bala, V. Magearu, J.-L. Marty, Detection of organophosphorus insecticides with immobilized acetylcholinesterase: Comparative study between two enzyme sensors, *Anal. Bioanal. Chem.*, 374 (2002) 39–45.

41. T. Montesinos, S. Pérez-Munguia, F. Valdez, J. L. Marty, Disposable cholinesterase biosensor for the detection of pesticides in water miscible- organic solvents, *Anal. Chim. Acta*, 431 (2001) 231–237.

42. G. M. Morris, R. Huey, W. Lindstrom, M. F. Sanner, R. K. Belew, D. S. Goodsell, A. J. Olson, AutoDock4 and AutoDockTools4: Automated docking with selective receptor flexibility, *J. Comput. Chem.*, 30(16) (2009) 2785–2791.

43. G. Neudert, G. Klebe, DSX: A knowledge-based scoring function for the assessment of protein-ligand complexes, *J. Chem. Inf. Model*, 51(10) (2011) 2731–2745.

44. K. Vanommeslaeghe, E. Hatcher, C. Acharya, S. Kundu, S. Zhong, J. Shim, E. Darian, O. Guvench, P. Lopes, I. Vorobyov, A. D. Mackerell, Jr., CHARMM general force field: A force field for drug-like molecules compatible with the CHARMM all-atom additive biological force fields, *J. Comput. Chem.*, 31(4) (2010) 671–690.

45. S. Pronk, S. Pall, R. Schulz, P. Larsson, P. Bjelkmar, R. Apostolov, M. R. Shirts, J. C. Smith, P. M. Kasson, D. van der Spoel, B. Hess, E. Lindahl, GROMACS 4.5: A high-throughput and highly parallel open source molecular simulation toolkit, *Bioinformatics*, 29(7) (2013) 845–854.

46. C. Oostenbrink, A. Villa, A. E. Mark, W. F. van Gunsteren, A biomolecular force field based on the free enthalpy of hydration and solvation: The GROMOS force-field parameter sets 53A5 and 53A6, *J. Comput. Chem.*, 25(13) (2004) 1656–1676.

47. A. W. Schuttelkopf, D. M. van Aalten, PRODRG: A tool for high-throughput crystallography of protein-ligand complexes, *Acta Crystallogr. D Biol. Crystallogr.*, 60(Pt 8) (2004) 1355–1363.

48. M. J. Frisch, G. W. Trucks, H. B. Schlegel, G. E. Scuseria, M. A. Robb, J. R. Cheeseman, G. Scalmani, V. Barone, B. Mennucci, G. A. Petersson, H. Nakatsuji, M. Caricato, X. Li, H. P. Hratchian, A. F. Izmaylov, J. Bloino, G. Zheng, J. L. Sonnenberg, M. Hada, M. Ehara, K. Toyota, R. Fukuda, J. Hasegawa, M. Ishida, T. Nakajima, Y. Honda, O. Kitao, H. Nakai, T. Vreven, J. A. Montgomery Jr., J. E. Peralta, F. Ogliaro, M. J. Bearpark, J. Heyd, E. N. Brothers, K. N. Kudin, V. N. Staroverov, R. Kobayashi, J. Normand, K. Raghavachari, A. P. Rendell, J. C. Burant, S. S. Iyengar, J. Tomasi, M. Cossi, N. Rega, N. J. Millam, M. Klene, J. E. Knox, J. B. Cross, V. Bakken, C. Adamo, J. Jaramillo, R. Gomperts, R. E. Stratmann, O. Yazyev, A. J. Austin, R. Cammi, C. Pomelli, J. W. Ochterski, R. L. Martin, K. Morokuma, V. G. Zakrzewski, G. A. Voth, P. Salvador, J. J. Dannenberg, S. Dapprich, A. D. Daniels, Ö. Farkas, J. B. Foresman, J. V. Ortiz, J. Cioslowski, D. J. Fox, Gaussian 09, Gaussian, Inc., Wallingford, CT, USA (2009).

49. J. Zielkiewicz, Structural properties of water: Comparison of the SPC, SPCE, TIP4P, and TIP5P models of water, *J. Chem. Phys.*, 123(10) (2005) 104501.

50. T. R. Fukuto, Mechanism of Action of organophosphorus and carbamate insecticides, *Environ. Health Perspectives*, 87 (1990) 245–254.

51. C. S. Pundir, N. Chauhan, Acetylcholinesterase inhibition-based biosensors for pesticide determination: A review, *Anal. Biochem.*, 429(1) (2012) 19–31.

52. K. Abass, V. Lämsä, P. Reponen, J. Küblbeck, P. Honkakoski, S. Mattila, O. Pelkonen, J. Hakkola, Characterization of human cytochrome P450 induction by pesticides, *Toxicology*, 294(1) (2012) 17–26.

53. T. S. Poet, H. Wu, A. A. Kousba, C. Timchalk, In vitro rat hepatic and intestinal metabolism of the organophosphate pesticides chlorpyrifos and diazinon, *Toxicol. Sci.*, 72(2) (2003) 193–200.

54. K. A. Usmani, E. Hodgson, R. L. Rose, In vitro metabolism of carbofuran by human, mouse, and rat cytochrome P450 and interactions with chlorpyrifos, testosterone, and estradiol, *Chem. Biol. Interact.*, 150(3) (2004) 221–232.

55. G. de Sousa, F. Fontaine, M. Pralavorio, D. Botta-Fridlund, Y. Letreut, R. Rahmani, Insecticide cytotoxicity and CYP1A1/2 induction in primary human and rat hepatocyte cultures, *Toxicol. In Vitro*, 11(5) (1997) 451–457.

56. N. Liu, M. Li, Y. Gong, F. Liu, T. Li, Cytochrome P450s–Their expression, regulation, and role in insecticide resistance, *Pest Biochem. Physiol.*, 120 (2015) 77–81.

57. I. Kostaropoulos, A. I. Papadopoulos, A. Metaxakis, E. Boukouvala, E. Papadopoulou-Mourkidou, Glutathione S-transferase in the defence against pyrethroids in insects, *Insect. Biochem. Mol. Biol.*, 31(4–5) (2001) 313–319.

58. Z. Wang, Z. Zhao, M. M. Abou-Zaid, J. T. Arnason, R. Liu, B. Walshe-Roussel, A. Waye, S. Liu, A. Saleem, L. A. Caceres, Q. Wei, I. M. Scott, Inhibition of insect glutathione S-transferase (GST) by conifer extracts, *Arch. Insect. Biochem. Physiol.*, 87(4) (2014) 234–249.

59. L. Murillo, A. Hamon, Z. Es-Salah-Lamoureux, V. Itier, S. Quinchard, B. Lapied, Inhibition of protein kinase C decreases sensitivity of GABA receptor subtype to fipronil insecticide in insect neurosecretory cells, *Neuro Toxicol.*, 32(6) (2011) 828–835.

Chapter 10

Core–Shell Quantum Dots: Sensing Applications

Suvardhan Kanchi,[a] Myalowenkosi Sabela,[a] Krishna Bisetty,[a] and Venkatasubba Naidu Nuthalapati[b]

[a]*Department of Chemistry, Durban University of Technology, Durban 4000, South Africa*
[b]*Department of Chemistry, Sri Venkateswara University, Tirupati 517502, Andhra Pradesh, India*

ksuvardhan@gmail.com

10.1 Introduction

Quantum dot (QD) research has attracted researchers' attention due to QDs' wide applications such as sensor and biomedical applications [1–4]. QDs are fluorescent semiconducting nanocrystals (NCs) comparable with the Bohr exciton in terms of radius of the material [5]. The principle behind the Bohr exciton is due to the excitation of electron from the valence band to conduction band in the semiconductors, resulted in the formation of an electron–hole pair. During this process, a weak columbic force of attraction exists between the hole and electron pair, and an exciton is created. The natural physical separation distance in the

Nanocomposites for Pollution Control
Edited by Chaudhery Mustansar Hussain and Ajay Kumar Mishra

ISBN 978-981-4774-45-1 (Hardcover), 978-1-315-14368-2 (eBook)
www.panstanford.com

excitons varies from material to material. In large quantities, the dimensions of semiconducting materials are much larger than the Bohr radius, and the energy levels are continuous. When the size of the QDs decreases and comparable to that of the Bohr exciton radius, the energies are discrete, this results in a large effective band gap leading to a blue shift in the optical transition compared to that of bulk materials. In a QD, the electrons and holes are in a confined space in all of the three directions that results in quantization of the electron and hole energy levels. This process is considered as a zero-dimensional confinement effect. Therefore, the electronic properties of QDs are between that of a bulk semiconductor and discrete molecules. In the early 1980s, Alexei Ekimov discovered QDs in a glass matrix [6]. The stable and sustainable QD confinement is done by the encapsulation of organic surfactant on the surface of nanomaterial. In the QD, the surface to volume ratio is high due to the high number of atoms present on the surface. Organic encapsulation acts as a surface trap state that aids the non-radiative de-excitation channels for the charge generated by the photon, which leads to an overall reduction in the fluorescent quantum yield. This could be overcome by growing epitaxial layers of inorganic material over the QD core material to obtain a core–shell quantum dot (CSQD) structure, which results in a substantial improvement in the photoluminescence efficiency of the QDs compared to their organically capped counterparts. This enhancement in the quantum efficiency is due to the increased confinement of the electron and hole to the pair in the core and the dangling bonds on the surface of the QDs. In a CSQD, it is observed that upon increasing the shell thickness there is a significant red shift in the threshold energy, also the nonlinear absorption coefficients enhance drastically and all the refractive index changes, independently on the dielectric environments. Similar trend has also been observed in most of the cases when an impurity in CSQD is displaced from the core centre to the shell centre. In contrast with a dielectrically homogeneous system, dispersing the systems into a matrix with lower dielectric constant blue shifts all the peak positions and the refractive index also changes. However, the corresponding magnitudes are substantially reduced [7, 8].

This chapter describes the recent advancements in the organic and inorganic surface modifications of the CSQDs. The surface modification plays an essential role in identifying the properties of the CSQDs. The functional groups present on the terminal end determine the function of the CSQD, which can permit for probable applications ranging from sensor to biomedical. In addition, applications of CSQDs in the field of sensing have also been reported.

10.2 Core–Shell QDs and Their Types

The different types of CSQDs mainly depend on the location of the valence and conduction band and the indispensable energy gap between them in the semiconductors. Type-A CSQDs could have a larger band gap than the outer shell or vice versa. The conduction band or valence band of the core could be located within the band gap of the shell categorized Type-B [9]. The QDs captured with a material that has a larger band gap can make the surface non-radiative recombination sites which can lead to the perfection in the quantum yield of photoluminescence. This modification using inorganic shell is more advantageous over its organic shell due to its high tolerance to CSQD processing conditions. The typical examples of CSQDs include (CdSe) CdS, (ZnS) CdSe [5] and (CdSe) InAs [10]. The (CdS) HgS [11], (CdS) CdSe [12] and (ZnSe) CdSe [13] are typical examples of Type-A CSQDs in which the band gap of the core is greater than the band gap of the shell. The charge carriers are partially delocalised within the shell, and the thickness of the shell controls the overall emission wavelength. In Type-B (CdTe) CdSe [14], (CdTe) CdSe [15], (CdSe) ZnTe and (CdSe) ZnTe CSQDs, one carrier is confined to the core and the other is primarily located in the shell. This separation acts as the chromophore and one of the charge carriers is injected into the matrix prior to recombination.

10.3 QD Surface Modification

Organometallic or aqueous phase synthesis is mainly used for the synthesis of CSQDs, however in the later stages; CSQDs are prepared under normal atmospheric conditions without special experimental set-up. The core–shell metal nanoparticles are generally synthesized by well-known high-temperature thermal decomposition of organometallic compound to metal in the absence of water and oxygen in a non-aqueous media. This method is carried out in the absence of water and oxygen, where the organometallic compounds are decomposed at high temperatures to metal in a non-aqueous media [16]. In this method, surfactants were used for the surface modification in which the surface acts as a mediator in controlling the particle size.

10.3.1 Surface Modification with Inorganic Materials

Bleuse and co-workers first reported an inorganic over coating of the core–shell system, where an intermediate shell was sandwiched between an outer shell and an inner QD to reduce the lattice mismatch [17]. For the CSQD modified with inorganic materials, results in a multilayer semiconductor heterogeneous system with relative valence band and conduction band. The principal benefit of such a system is that it provides higher quantum yield, extraordinary photoluminescence, enhanced optical properties, increased half-life time, improved stability towards photo-oxidation and finally better structural properties. If an inorganic semiconducting layer is provided over the CSQD such that its band gap is higher than that of the shell, then the particle is called a quantum dot quantum well (QDQW). The main aspect to be considered while developing such a material is the band gap and their respective lattice mismatch with CSQD. A typical example of QDQW is ((CdS) HgS) CdS [18, 19]. The photo-optical stability is comparable to that of (CdS) HgS CSQD because of over-coating a high band gap semiconducting material like CdS. For the same reason, quantum yield also increases. Therefore, a low band gap compound when sandwiched between the core and outer coating acts as a quantum well. Core–shell structured materials combine favourable properties of the magnetic core with a protective polymer, silica, gold, metal oxide or carbon shell. The silica coating on metal oxide core has many advantages. They mainly reduce the bulk conductivity and increase the suspension stability of the core particles. Being chemically inert, it can be used as an effective blocking agent that can cover the core preventing it from degradation. It is also optically transparent which is helpful in the spectroscopic investigation of core. Therefore, metals, inorganic composites, oxides and other combinations are used.

The common metal cores are from noble metals. Researchers are paying more attention to the coatings made using gold and silver coated with silica. Among the possible metal cores, noble metals such as Au and Ag coated with silica have been studied by a number of researchers [20, 21]. In addition to the noble metals, transition metals such as nickel, cobalt, iron and binary metal composites such as Fe-Ni have also been reported [22, 23]. The gold coating enhances the physico-chemical properties, protects the core material from corrosion. It also provides bioaffinity through functionalization of amine/thiol terminal groups. Other shell metals such as Ni, Co, Pd, Pt and Cu are also important for some specific application like catalysis, solar energy absorption and permanent magnetic properties.

There is also an inorganic modification applicable when a core is made of a polymer or different copolymers. The shell can also be made from different materials, such as metals, metal oxides, metal chalcogenides or silica [24–26]. These types of particles retain the properties of both the inorganic and organic materials. The inorganic shell over an organic core is beneficial in several respects, such as increased strength of the overall material, resistance to oxidation, thermal and colloidal stability and abrasion resistance [27, 28].

10.3.2 Surface Modification with Organic Materials

The modification with organic materials involves the suitable functional group on the surface of QDs. The functional moieties attached could be either a covalent bond or a mere physical adsorption on the surface of QDs maintains their stability. This capping provides the necessary chemical accessibility for the QDs by varying the terminal groups of the ligands pointing to the outside environment.

Chen and co-workers designed a thiolated ligand and fabricated the QD surface to produce CSQDs. In this study, hydrophobic ligands cannot be used directly in applications that require aqueous solubility or an effective charge transport property. In this method, Thioglycerol, mercaptoacetate, 1,4-dithiothreitol,

2-mercaptoethanol, cysteine, methionine and glutathione have been used as capping ligands [29]. Quantum dots can be also be encapsulated by a shell of material such as a polymer, micelle or bead that makes them more soluble in a particular media. Such encapsulation significantly increases the volume of the quantum dot-based material. This is not desirable in some applications, such as biosensors and live cell imaging. TOPO-capped CdSe QDs dispersed in poly [2-methoxy, 5-(2-ethyl-hexyloxy)-1,4-phenylene vinylene] (MEH-PPV) and poly(3-hexylthiophene) (P3HT) were used by Umesh and co-workers [30].

Recent efforts have concentrated on replacing TOPO as the solvent during the synthesis with a ligand that possesses functionality that allows for subsequent surface grafting without the need for ligand exchange. Phenyl bromide-functionalized dioctylphosphine oxide was used recently. The latter proved stable to the high-temperature reaction conditions of quantum dot growth. Subsequently, Heck-type coupling was used to grow poly- or oligo-(phenylene vinylene) (PPV or OPV) ligands from the functional quantum dots [31]. One-step DNA functionalization on core QDs or during core/shell QD synthesis in aqueous solution was reviewed by Samantaet and co-workers [32] (CdSe) ZnS-ssDNA-fluorescent dye conjugates were used as bioprobes by Huang et al. [33], to detect micrococcal nuclease with high specificity and sensitivity. The (CdSe) ZnS-2 mercaptoethyltrimethyl ammonium chloride conjugate was titrated with solutions of ATP and ADP. Guano sine and fluorescence quenching was observed. Water-soluble (CdTe)ZnS (after encapsulation) also served as a pH probe for enzyme kinetics [34] and tiopronin determination [35]. A polydentate phosphine coating on type II quantum dots have been employed in major cancer surgery [36] in large animals via imaging. In addition, capped (InP) ZnS have been used for cellular imaging [37, 38].

10.4 Applications

A novel electrochemical sensor for the detection of L-cysteine was proposed based on immobilizing poly(alizarin yellow R)/ carbon quantum dots film on glassy carbon electrode. In this report, hydrothermal treatment was used to prepare carbon quantum dots. Electrochemical impedance spectroscopy, cyclic voltammetry and amperommetry were utilized to confirm the successful stepwise assembly procedure of the sensor. The electrocatalytic behaviour of the sensor was also investigated by cyclic voltammetry and amperommetry. The results showed that poly(alizarin yellow R)/carbon dots exhibited a remarkable electrocatalytic activity for the oxidation of L-cysteine under optimal conditions. The electrocatalytic response of the sensor was proportional to the L-cysteine concentration in the range of (0.3 to 3.6 μM) and (3.9 to 7.2 μM) with a limit of detection and sensitivity of 90 nM and 0.482 μA/μM, respectively [39]. Wang and co-workers reported a 3-aminobenzeneboronic acid functionalized Mn^{2+}-doped ZnTe/ZnSe quantum dots (APBA-dQDs) for the biological applications. The APBA functional groups had strong binding ability with F^-, resulting in the quenchment of dQD photoluminescence (PL). Under the optimal condition, the fluorescence intensity of APBA-dQDs was related linearly to the concentration of F^- in the range of 0.25–1.5 mmol/L with a detection limit of 0.1 mmol/L. The selectivity of fluorescence quenching of APBA-dQDs for F^- was enhanced drastically. The results indicate that the APBA-dQDs are promising candidates for intracellular in MC3T3-E1 osteoblastic cells [40]. A new approach for developing a fluorimetric aptasensor has been described and applied for determination of a highly toxic cation, As(III). In this method, an aptamer was used to aggregate cationic cysteamine-stabilized CdTe/ZnS core/shell quantum dots; as a result, fluorescence quenching was accrued. In the presence of As(III), the aptamer and As(III) make a complex, which prevents aggregation of the quantum dots. Thus, the fluorescence intensity of the quantum dots was enhanced up on the de-aggregation, which depends on the concentration of As(III). The fluorimetric assay has a very low detection limit of 1.3 pmol L^{-1} As(III) with a dynamic range of 1.0×10^{-11} to 1.0×10^{-6} mol L^{-1}. The interference effect of a wide variety of cations and anions was investigated, and the obtained results confirm high selectivity of the aptasensor for As(III) detection. The present assay was successfully applied for the determination of As(III) in several water samples [41].

In this report, a simple, ultrasensitive and rapid cost-effective sensing technique based on mercaptosuccinic acid (MSA) capped

CdTe/ZnS core/shell (CS) QD for the detection of toxic Hg(II) in aqueous media. The present system demonstrated a remarkably low detection limit of 1 pico molar (pM). The CdTe/ZnS CS QD was synthesized by using a simple one pot aqueous method as shown in Fig. 10.1, and was characterized with the help of UV–vis, photoluminescence, XRD, TEM and FT-IR analysis. Due to the interaction between CS QDs and the heavy metal ions, the photoluminescence (PL) intensity of QDs quenches via excited state electron transfer mechanism, which was confirmed from cyclic voltammetry and zeta potential. The CdTe/ZnS QDs are found to be highly selective for Hg (II) ions, whereas the interference from other toxic metal ions is insignificant. The real-time analysis was further carried out with drinking water and tap water solutions and the QDs show remarkably good quenching in these solutions [42].

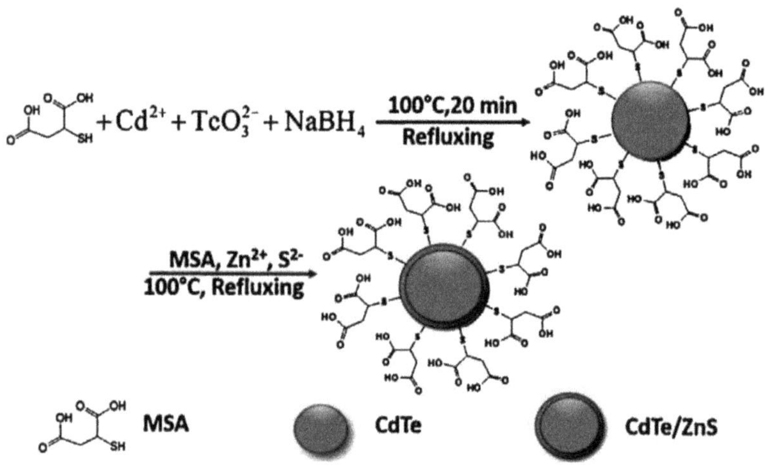

Figure 10.1 The synthesis procedures of MSA-coated CdTe/ZnS CS QD. Reproduced with permission from Ref. [42].

A facilely prepared fluorescent sensor was developed for dopamine (DA) detection with high sensitivity and selectivity based on polypyrrole/graphene quantum dots (PPy/GQDs) core/shell hybrids as illustrated in Fig. 10.2. The composites exhibit strong fluorescence emission, which is dramatically enhanced as high as three times that of pristine GQDs. The prepared sensor allows a highly sensitive determination of DA by fluorescent intensity decreasing with the addition of DA and presents a good linearity in range of 5–8000 nM with the detection limit of 10 pM (S/N = 3). Furthermore, the application of the proposed approach have been demonstrated in real samples and showed promise in diagnostic purposes [43].

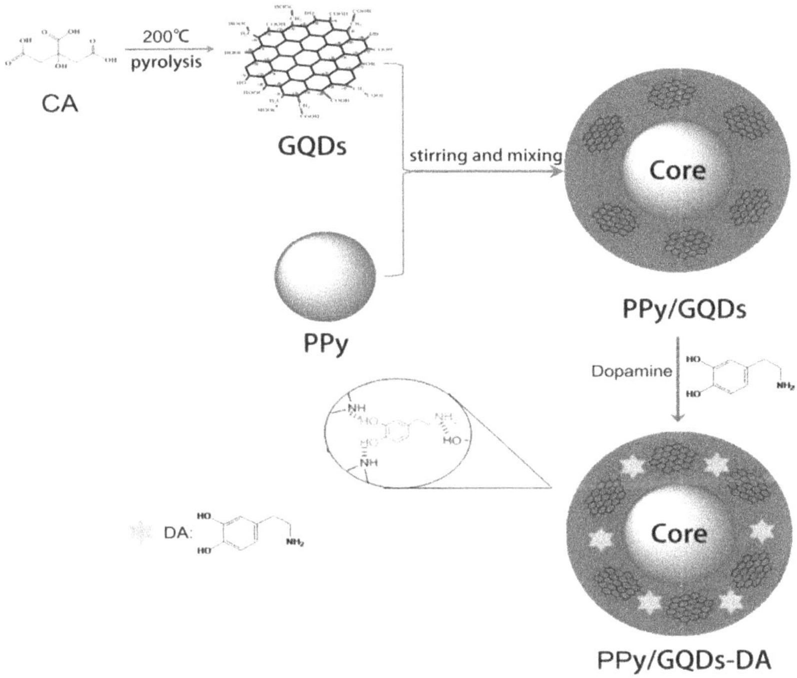

Figure 10.2 Illustration of the preparation of PPy/GQDs. Reproduced with permission from Ref. [43].

A facile, economical, and green method has been developed to prepare a new fluorescent sensor based on lignin sulphonate/graphene quantum dots (SL/GQDs) core/shell hybrids for label-free sensitive and selective detection of Fe^{3+} shown in Fig. 10.3. The combination of these two materials brings both excellent photoluminescent properties (the fluorescence intensity of SL/GQDs composites is four times higher than the free GQDs) and nice selectivity. In this composite, SL molecule plays two key roles in the system as a π-rich and sulphur-incorporated compound to enhance the fluorescence and as a chelator to detect Fe^{3+}. The obtained sensor presents excellent sensitivity and selectivity toward Fe^{3+} and exhibits a good linearity in the range of 0.005–500 μM with a detection limit as low as 0.5 nM [44].

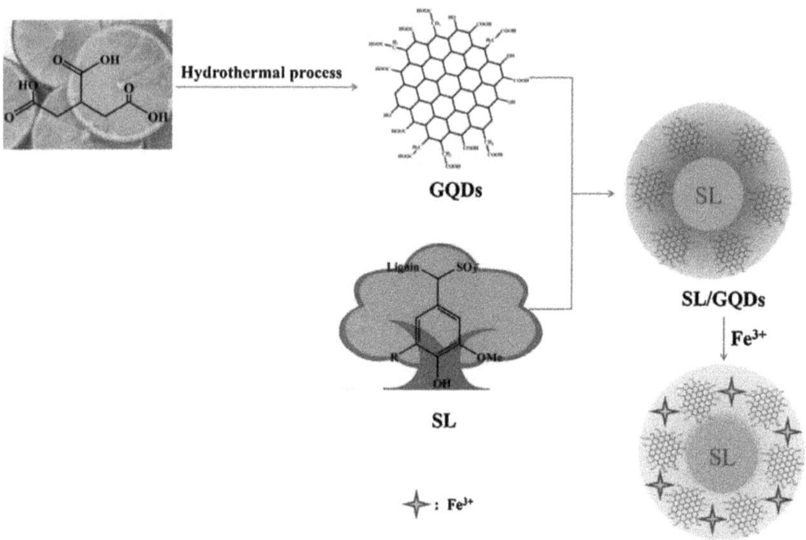

Figure 10.3 A low-cost, green route to construct a novel sensing platform for Fe^{3+} detection based on core/shell hybrids by using citric acid and sodium lignosulphonate through one-step treatment. Reproduced with permission from Ref. [44].

Uddin and co-workers developed a Pd@Pt core–shell nanocrystal-based resistivity-type sensor for fast-response hydrogen (H_2) detection. In this study, the effect of shell thickness on sensing properties were also reported. Pd@Pt core–shell nanocrystals as colloids were synthesized via a chemical route and were then carefully assembled on a SiO_2/Si substrate using the self-assembly monolayer (SAM) technique. Three different Pd@ Pt core–shell nanocrystals with varied Pt shell thicknesses were synthesized for the SAM process. The short diffusion length of Pd@Pt due to the thin Pt shell layer and the monolayer distribution of the Pd@Pt core–shell preferentially accelerated the adsorption–desorption of the H_2 molecules on the sensing interface, resulting in a very quick response to H_2 gas. At an optimum operating temperature of 150°C, the sensor showed a fast-response time of 6 s, a maximum response of 3.6% to 1 vol.% gas concentration, and a broad detection range of 0.001–4 vol.%

H_2 concentrations [45]. Stable and water-soluble cadmium telluride quantum dots (CdTe QDs) have been used to evaluate the ultra-sensing ability of Ag(I). In this study, the authors have designed and synthesized thiolactic acid (TLA) capped exquisite CdTe QDs having exclusive selectivity for silver. The limit of detection (LOD) of Ag(I) is 50 nM. The pendant methyl group of TLA effectively impedes the precursor aggregation by inhibiting the secondary coordination of carboxyl oxygen of mercapto acid with the surrounding Cd(II), and this results in highly luminescent TLA-CdTe QDs. This pendant methyl group of TLA also facilitates exclusive surface adsorption of QDs by incoming metal and renders sensing selectivity. Selective adsorption of silver on TLA-CdTe QD surface provides a turn-off photoluminescence-based assay for sensitive detection of Ag(I) without any interference of other commonly coexisting metal ions [46]. The fluorescence response of water dispersible CdSe/ZnS core–shell quantum dots capped with 2-mercaptopropionic acid (2MPA) towards different concentrations of heavy metal cations, namely mercury, lead and cadmium was studied by Vasudevan et al. [47]. Upon exposure to different concentrations of the various metal cations, a concentration-dependent decrease in the QDs' fluorescence emission was observed, which was decaying exponential in nature. The greatest degree of quenching was achieved in the presence of mercury. The resultant quantum dots were subsequently characterised by UV–vis spectroscopy, photo-luminescent spectroscopy, Raman Spectroscopy and X-ray diffraction. The quantitative detection of mercury, lead and cadmium cations by these capped quantum dots makes them exciting candidates for heavy metal sensing applications [47].

Highly luminescent, good stable and low toxic N-acetyl-l-cysteine (NAC) capped CdTe/CdS@ZnS–SiO_2 near-infrared (NIR)-emitting QDs were successfully fabricated in aqueous solution via microwave irradiation reduction route, in which thiol-capped CdTe/CdS QDs were employed as core templates and $ZnCl_2$, NAC and tetraethyl orthosilicate as shell precursors. This presented ZnS-like clusters filled hybrid SiO_2 model not only greatly improved the brightness and stability of original CdTe/CdS QDs, but also tremendously decreased the cytotoxicity towards HeLa cells. Furthermore, it was found that Hg^{2+} could effectively selective quench the QD NIR emission based on electron transfer process. On the basis of this fact, a simple, rapid and specific method for trace Hg^{2+} determination was designed. Under optimal conditions, the fluorescence intensity decreased linearly with the concentration of Hg^{2+} ranging from 5.0×10^{-9} to 1.0×10^{-6} M and the limit of detection for Hg^{2+} was 1.0×10^{-9}M (S/N = 3) [48].

10.5 Conclusion

This chapter provided a widespread report of a number of different tactics and approaches employed by various researchers for the generation of functional moieties on the surface of CSQDs including covalent and non-covalent methodologies as well as in situ synthesis procedures. The methods described in this chapter are targeted at reproducibility and enhanced productivity. The practical fabrication of CSQDs in terms of size, shape, morphology, crystallinity and chemical composition remains an active area of study. The properties of CSQDs are not just a simple sum of the layers present, and potential improvement in the constructive properties of these discrete systems could impact a wide range of areas including catalysis, sensing, optoelectronic devices, drug delivery and biolabelling. New constructive attachments on CSQD's could be performed with several nanomaterials, such as carbon ribbons, carbon nanotube and other metal and metal-oxide nanoparticles, that can create a new research dimension in a multidisciplinary science. We also believe that CSQD will be a key architecture in the on-going challenge of developing improved energy storage devices as well as alternative power sources. In addition, CSQDs could become a widely accepted biomedical and sensing tool in the near future.

Acknowledgements

The authors thank the Durban University of Technology, South Africa, for the financial support.

References

1. Zhang, C. Y., Yeh, H. C., Kuroki, M. T., and Wang, T. H. (2005). Single-quantum-dot-based DNA nanosensor, *Nat. Mater.* 4, 826–831.

2. Shi, L., De Paoli, V., Rosenzweig, N., and Rosenzweig, Z. (2006). Synthesis and application of quantum dots FRET-based protease sensors, *J. Am. Chem. Soc.* 128, 10378–10379.

3. Bagalkot, V., Zhang, L., Levy-Nissenbaum, E., Jon, S., Kantoff, P. W., Langer, R., and Farokhzad, O. C. (2007). Quantum dot-aptamer conjugates for synchronous cancer imaging, therapy, and sensing of drug delivery based on bi-fluorescence resonance energy transfer, *Nano Lett.* 7, 3065–3070.

4. Oh, E., Hong, M. Y., Lee, D., Nam, S. H., Yoon, H. C., and Kim, H. S. (2005). Inhibition assay of biomolecules based on fluorescence resonance energy transfer (FRET) between quantum dots and gold nanoparticles, *J. Am. Chem. Soc.* 127, 3270–3271.

5. Ziaudeen, S. A., Gaddam, R. R., Pallapothu, P. K., Sugumar, M. K., and Rangarajan, J. (2013). Supra gap excitation properties of differently confined PbS-nano structured materials studied with opto-impedance spectroscopy, *J. Nanophotonics* 7, 073075.

6. Dabbousi, B. O., Rodriguez-Viejo, J., Mikulec, F. V., Heine, J. R., Mattoussi, H., Ober, R., and Bawendi, M. G. (1997). (CdSe) ZnS core–shell quantum dots: Synthesis and characterization of a size series of highly luminescent nanocrystallites, *J. Phys. Chem. B* 101, 9463–9475.

7. Murray, C., Norris, D. J., and Bawendi, M. G. (1993). Synthesis and characterization of nearly monodisperse CdE (E = sulphur, selenium, tellurium) semiconductor nanocrystallites, *J. Am. Chem. Soc.* 115, 8706–8715.

8. Baskoutas, S., and Terzis, A. F. (2006). Size-dependent band gap of colloidal quantum dots, *J. Appl. Phys.* 99, 013708–013708.

9. Bera, D., Qian, L., Tseng, T. K., and Holloway, P. H. (2010). Quantum dots and their multimodal applications: A review, *Materials* 3, 2260–2345.

10. Reiss, P., Protiere, M., and Li, L. (2009). Core/shell semiconductor nanocrystals, *Small* 5, 154–168.

11. Cao, Y. W. U. (1999). Synthesis and characterization of InAs/InP and InAs/CdSe core/shell nanocrystals, *Angew. Chem. Int. Ed.* 8, 3692–3694.

12. Mews, A., Eychmüller, A., Giersig, M., Schooss, D., and Weller, H. (1994). Preparation, characterization and photophysics of the quantum dot quantum well system cadmium sulfide/mercury sulfide/cadmium sulfide, *J. Phys. Chem.* 98, 934–941.

13. Battaglia, D., Li, J. J., Wang, Y., and Peng, X. (2003). Colloidal two-dimensional systems: CdSe quantum shells and wells, *Angew. Chem. Int. Ed.* 42, 5035–5039.

14. Zhong, X., Xie, R., Zhang, Y., Basché, T., and Knoll, W. (2005). High-quality violet-to red emitting ZnSe/CdSe core/shell nanocrystals, *Chem. Mater.* 17, 4038–4042.

15. Kim, H., Achermann, M., Balet, L. P., Hollingsworth, J. A., and V. I. Klimov, V. I. (2005). Synthesis and characterization of Co/CdSe core/shell nanocomposites: Bifunctional magnetic–optical nanocrystals, *J. Am. Chem. Soc.* 127, 544–546.

16. Nadagouda, M. N., Varma, R. S. (2007). A greener synthesis of core (Fe, Cu)-shell (Au, Pt, Pd, and Ag) nanocrystals using aqueous vitamin C, *Cryst. Growth Des.* 7, 2582–2587.

17. Bleuse, J., Carayon, S., and Reiss, P. (2004). Optical properties of core/multishell CdSe/Zn (S,Se) nanocrystals, *Phys. E: Low-Dimensional Syst. Nanostruct.* 21, 331–335.

18. Lifshitz, E., Porteanu, H., Glozman, A., Weller, H., Pflughoefft, M., and Echymüller, A. (1999). Optically detected magnetic resonance study of CdS/HgS/CdS quantum dot quantum wells, *J. Phys. Chem. B* 103, 6870–6875.

19. Schooss, D., Mews, A., Eychmüller, A., and Weller, H. (1994). Quantum-dot quantum well CdS/HgS/CdS: Theory and experiment, *Phys. Rev. B* 49, 17072.

20. Wang, J., Ren, F., Yi, R., Yan, A., Qiu, G., and Liu, X. (2009). Solvothermal synthesis and magnetic properties of size-controlled nickel ferrite nanoparticles, *J. Alloys Comp.* 479, 791–796.

21. Lee, J., Lee, Y., Youn, J. K., Na, H. B., Yu, T., Kim, H., and Hyeon, T. (2008). Simple synthesis of functionalized superparamagnetic magnetite/silica core/shell nanoparticles and their application as magnetically separable high-performance biocatalysts, *Small* 4, 143–152.

22. Ung, T., Liz-Marzán, L. M., and Mulvaney, P. (1998). Controlled method for silica coating of silver colloids. Influence of coating on the rate of chemical reactions, *Langmuir* 14, 3740–3748.

23. Ammar, M., Mazaleyrat, F., Bonnet, J. P., Audebert, P., Brosseau, A., Wang, G., and Champion, Y. (2007). Synthesis and characterization of core–shell structure silica coated $Fe_{29.5}Ni_{70.5}$ nanoparticles, *Nanotechnology* 18, 285606.

24. Chen, Y. C., Zhou, S. X., Yang, H. H., and Wu, L. M. J. (2005). Sol–gel science structure and mechanical property of polyurethane/silica hybrid coatings, *Macromol. Mater. Eng.* 290, 1001.

25. Sun, Y., Zhou, B., Gao, P., Mu, H., and Chu, L. (2010). Single-crystalline Ag_2S hollow nanoparticles and their ordered arrays, *J. Alloys Comp.* 490, 48–51.

26. Kim, I. J., Kwon, O. S., Park, J. B. Joo, H. (2006). Synthesis and characterization of ABS/silica hybrid nanocomposites, *Curr. Appl. Phys.* 6, 43–47.

27. Feng, L., Wang, Y., Wang, N., and Ma, Y. (2009). Preparation of poly (ethylene glycol)-grafted silica nanoparticles using a facile esterification condensation method, *Polym. Bull.* 63, 313–327.

28. Zhou, S. X., Wu, L. M., Sun, J., and Shen, W. D. (2003). Effect of nanosilica on the properties of polyester-based polyurethane, *J. Appl. Pol. Sci.* 88, 189–193.

29. Chen, L. Y., Chou, H. L., Chen, C. H., and Tseng, C. H. (2012). Surface modification of CdSe and CdS quantum dots-experimental and density function theory investigation, *Nanocryst.–Synth., Charact. Appl.* 8, 149.

30. Kumar, U., Kumari, K., Sharma, S. N., Kumar, M., Vankar, V. D., Kakkar, R., and Kumar, V. (2010). Role of surface modification of colloidal CdSe quantum dots on the properties of hybrid organic–inorganic nanocomposites, *Colloids Poly. Sci.* 288, 841–849.

31. Hammer, N. I., Emrick, T., and Barnes, M. D. (2007). Quantum dots coordinated with conjugated organic ligands: New nanomaterials with novel photophysics, *Nanoscale Res. Lett.* 2, 282–290.

32. Samantha, A., Deng, Z., Liu, Y., and Yan, H. (2013). A perspective on functionalizing colloidal quantum dots with DNA, *Nano Res.* 6, 853–870.

33. Huang, S., Xiao, Q., He, Z. K., Liu, Y., Tinnefeld, P., Su, X. R., and Peng, X. N. (2008). A high sensitive and specific QDs FRET bioprobe for MNase, *Chem. Commun.* 45, 5990–5992.

34. Duong, H. D., and Rhee, J. I. (2007). Use of CdSe/ZnS core–shell quantum dots as energy transfer donors in sensing glucose, *Talanta* 73, 899–905.

35. Yu, D., Wang, Z., Liu, Y., Jin, L., Cheng, Y., Zhou, J., and Cao, S. (2007). Quantum dot-based pH probe for quick study of enzyme reaction kinetics, *Enzyme Microb. Technol.* 41, 127–132.

36. Kim, S., Lim, Y. T., Soltesz, E. G., De Grand, A. M., Lee, J., Nakayama, A., and Frangioni, J. V. (2004). Near-infrared fluorescent type II quantum dots for sentinel lymph node mapping, *Nat. Biotechnol.* 22, 93–97.

37. Kobayashi, H., Hama, Y., Koyama, Y., Barrett, T., Regino, C. A., Urano, Y., and Choyke, P. L. (2007). Simultaneous multicolor imaging of five different lymphatic basins using quantum dots, *Nano Lett.* 7, 1711–1716.

38. Stojanovic, M. N., and Stefanovic, D. (2003). A deoxyribozyme-based molecular automaton, *Nat. Biotechnol.* 21, 1069–1074.

39. Amini, N., Shamsipur, M., Gholivanda, M. B., and Barati, A. (2017). A glassy carbon electrode modified with carbon quantum dots and polyalizarin yellow R dyes for enhanced electrocatalytic oxidation and nanomolar detection of L-cysteine, *Microchem. J.* 131, 9–14.

40. Wang, H., Hu, T. Y., Zhao, Z. T., Zhang, X. Y., Wang, Y., Duan, X. Q., Liu, D. W., Jing, L., and Ma, Q. (2016). A novel Mn^{2+}-doped core/shell quantum dot-based intracellular probe for fluoride anions sensing in MC3T3-E1osteoblastic cells, *Talanta* 149, 285–289.

41. Ensafi, A. A., Kazemifard, N., and Rezaei, B. (2016). A simple and sensitive fluorimetric aptasensor for the ultrasensitive detection of arsenic(III) based on cysteamine stabilized CdTe/ZnS quantum dots aggregation, *Biosens. Bioelectronics* 77, 499–504.

42. Saikia, D., Dutta, P., Sarma, S. N., and Adhikary, N. C. (2016). CdTe/ZnS core/shell quantum dot-based ultrasensitive PET sensor for selective detection of Hg (II) in aqueous, *Sens. Actuators B* 230, 149–156.

43. Zhou, X., Ma, P., Wang, A., Yu, C., Qian, T., Wu, S., and Shen, J. (2015). Dopamine fluorescent sensors based on polypyrrole/graphene quantum dots core/shell hybrids, *Biosens. Bioelectronics* 64, 404–410.

44. Xua, L., Mao, W., Huang, J., Li, S., Huang, K., Li, M., Xia, J., and Chen, Q. (2016). Economical, green route to highly fluorescence intensity carbon materials based on lignin sulfonate/graphene quantum dots composites: Application as excellent fluorescent sensing platform for detection of Fe^{3+} ions, *Sens. Actuators B* 230, 54–60.

45. Uddin, A. S. M. I., Yaqoob, U., Hassan, K., and Chun, G. S. (2016). Effects of Pt shell thickness on self-assembly monolayer Pd@Pt core–shell nanocrystals based hydrogen sensing, *Int. J. Hydr. Energy* 41, 15399–15410.

46. Mahapatra, N., Mandal, A., Panja, S., and Halder, M. (2017). Emergence of the selective ultra-sensing of Ag(I): Thiolactic acid as

efficient capping agent for cadmium chalcogenide quantum dots in modulating photoluminescence and metal reception, *Sens. Actuators B* 240, 543–552.

47. Vasudevan, D., Trinchi, A., Hardin, S. G., and Cole, I. S. (2015). Fluorescent heavy metal cation sensing with water dispersible 2MPA capped CdSe/ZnS quantum dots, *J. Luminescence* 166, 88–92.

48. Wang, J., Li, N., Shao, F., and Han, H. (2015). Microwave-assisted synthesis of high-quality CdTe/CdS@ZnS–SiO_2 near-infrared-emitting quantum dots and their applications in Hg^{2+} sensing and imaging, *Sens. Actuators B* 207, 74–82.

Chapter 11

Use of Carbon Nanotubes as Sorbents for Heavy Metal Remediation from Wastewater

Akil Ahmad,[a,b] **David Lokhat,**[a] **Siti Hamidah Mohd Setapar,**[b] **Asma Khatoon,**[b] **Mohammad Shahadat,**[c] **and Mohd Rafatullah**[d]

[a]*Department of Chemical Engineering, College of Agriculture, Engineering and Science, University of KwaZulu-Natal, Durban 4041, South Africa*
[b]*Centre of Lipids Engineering and Applied Research, Ibnu Sina Institute for Industrial and Scientific Research, Universiti Teknologi Malaysia, 81310 UTM Skudai, Johor, Malaysia*
[c]*Department of Biochemical Engineering and Biotechnology, Indian Institute of Technology IIT Delhi, Hauz Khas, New Delhi 110016, India*
[d]*School of Industrial Technology, Universiti Sains Malaysia, 11800 Penang, Malaysia*

akilchem@yahoo.com, mohd_rafatullah@yahoo.co.in

Currently, contamination of water resources due to the discharge of heavy metals is a serious concern worldwide. The challenge to remedy toxic metal-containing waste streams from present or former mining operations, industrial sites, and groundwater is immense. Adsorption is a simple, economical and widely used method which plays an important role in wastewater treatment. It is based on the physical interaction between metal ions and

Nanocomposites for Pollution Control
Edited by Chaudhery Mustansar Hussain and Ajay Kumar Mishra

ISBN 978-981-4774-45-1 (Hardcover), 978-1-315-14368-2 (eBook)
www.panstanford.com

sorbents. Various types of natural and chemically modified sorbents have been applied over the years for the elimination of metal ions in industrial wastewaters. With the development of nanotechnology, nanomaterials are used as the sorbents in wastewater treatment; several researches have proved that nanomaterials are new and effective sorbents for the removal of heavy metal ions from wastewater due to their unique structure properties. The discovery of carbon nanotubes (CNTs) as sorbents and the anticipation of developing novel carbon-based nanomaterials have attracted researchers worldwide. Carbon nanotubes (CNTs) and modified CNTs with chemical treatment are relatively new sorbents due to their unique properties such as chemical stability, mechanical and thermal stability, and the high surface area that have proven very efficient for treating many kinds of trace pollutants specially heavy metal ions. CNTs show high selectivity, regeneration and adsorption capacity for heavy metal ions. This chapter presents a summary of the recent information obtained using batch and column studies (preconcentration factor and limits, sorption capacity, detection limit and kinetics model) and deals with the mechanisms involved during adsorption.

11.1 Introduction

Over the past few decades, contamination of water bodies (lakes, rivers and canals) due to the enormous discharge of heavy metal ions has become a serious concern worldwide. Nationally and internationally, the challenge to remedy toxic metal-containing waste streams from present or former mining operations, industrial sites, and groundwater is immense. Heavy metals are natural constituents of the earth's crust having the relatively high density and atomic mass compared with other elements. These metals are introduced into the environment mainly from natural and anthropogenic origins. Due to the growing urbanization and industrialization, human exposure has increased dramatically as a result of an exponential increase of their use in several industrial, geogenic, agricultural, domestic, pharmaceutical and technological applications (Hu, 2000; Hu, 2005). Nonetheless, the prime sources are industrial activities, which include metal processing refineries, coal burning in power plants, industrial wastes, pollution, metal manufacturing, incineration plants, electroplating, fuel combustion, plastics, textiles, microelectronics, wood preservation and paper-processing plants. In recent years, the development of analytical procedures with high sensitivity for the determination of metal ions in various real samples has been considered a very active research area to prevent and control the adverse effects of these metal ions on humans, animals and ecosystems (Chapman, 2008; Sarkar, 2002; Pesavento et al., 2009). Table 11.1 presents the maximum allowable limit of metal ions in water and its effects on human beings. Sources of water contamination and its removal are shown in Fig. 11.1.

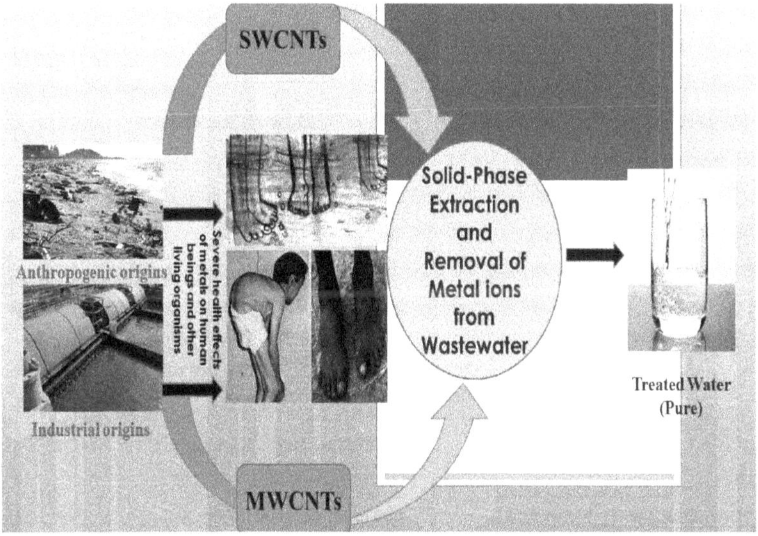

Figure 11.1 Sources of water contamination and its removal by CNTs.

Among several *methods for the determination of metal ions*, atomic spectrometric techniques *are commonly used due to its* rapid, reliable, most sensitive and selective characteristics. Today, the most common analytical techniques used for the detection of trace metal are flame atomic absorption spectroscopy (FAAS) (Islam et al., 2015; Islam et al., 2013; Ahmad et al., 2013; Islam et al., 2012a; Islam et al., 2012b; Islam et al., 2011a; Islam et al., 2011b; Islam et al., 2010a; Islam et al., 2010b), electrothermal atomic absorption spectrometry (ETAAS) (Asadollahi et al., 2010; Sardans et al., 2010; Naseri et al., 2008), inductively coupled plasma optical emission spectroscopy (ICP-OES) (Montaser and Golightly, 1992; Froes et al., 2009; Escudero et al., 2010), inductively coupled plasma mass spectrometry (ICP-MS) (Shih et al., 2011; Aydin and Soylak, 2010) and X-ray fluorescence (XRF) spectrometry (Devi et al., 1991; Rao and Chatt, 1993; Nomura et al., 2005). However, in certain cases these techniques have some limitations, especially in lower concentration of real samples and direct determination of metal ions is difficult due to the presence of contaminants and interferences from the sample matrix (Sung and Huang, 2003; Sabarudin et al., 2007).

Table 11.1 Maximum permissible limits of metal ions in water

Metals	Maximum permissible limit (mg L^{-1})	Potential health effects
Aluminium (Al)	0.2	Alzheimer's Disease and possibly other neurotoxic effects
Chromium (Cr)	0.1	Include shortness of breath, coughing, wheezing, ulcerations of the septum, bronchitis, decreased pulmonary function, and pneumonia. Asthma, nasal itching, and soreness, carcinogen
Nickel (Ni)	0.2	Nickel dermatitis, consisting of itching of the fingers, hands, and forearms, lung and nasal cancers
Cadmium (Cd)	0.01	Lung effects such as bronchial and pulmonary irritation
Cobalt (Co)	—	Decrease in ventilatory function, congestion, oedema, and haemorrhage of the lung, Cardiac effects
Copper (Cu)	3.0	Arthritis, fatigue, adrenal burnout, insomnia, scoliosis, osteoporosis, heart disease, cancer, migraine headaches, seizures, fungal and bacterial infections including yeast infection, gum disease, tooth decay, skin and hair problems

Metals	Maximum permissible limit (mg L^{-1})	Potential health effects
Lead (Pb)	0.05	Haematological damage, anaemia, kidney malfunctioning and brain damage
Arsenic (As)	0.05	Gastrointestinal effects (nausea, diarrhoea, abdominal pain) and central and peripheral nervous system disorders can occur from acute inorganic arsenic inhalation and ingestion
Iron (Fe)	0.3	Stomach and intestinal erosions and ulceration (i.e., hemorrhagic gastritis and enteritis with blood loss), induce oxidative stress and DNA damage
Zinc (Zn)	3.0	Papular-pustular skin eruptions in the axilla, inner thigh, inner arm, scrotum and pubic areas
Manganese (Mn)	0.1	Effects on the CNS, including slowed visual reaction time, hand steadiness, and eye-hand coordination, high levels result a syndrome called manganism
Mercury (Hg)	0.002	Effects on the CNS, including erethism (increased excitability), irritability, excessive shyness, and tremors, blurred vision, malaise, speech difficulties, and constriction of the visual field

Hence, in spite of the advances in analytical instrumentation, the preconcentration and separation procedures have been employed prior to the determination of metal ions, which decreases the detection limits. Solid phase extraction (SPE) is a most common preconcentration method, which designed for a rapid, selective sample preparation and purification prior to analysis that can be applied for both off-line and online systems. The basic principle of SPE is to transfer the analytes such as trace metal ions from the aqueous phase to the active sites of the solid sorbents. In SPE, solid sorbents are packed in column or microcolumn. After sorption of metal ions on active site of sorbents which can be desorbed with an appropriate eluting agent. Nowadays, various types of solid sorbents, including silica based sorbents, oxides of metals (e.g. aluminium, magnesium, zirconium, titanium and thorium), graphitized carbon or carbon based, polymeric sorbents (synthetic resins and its derivatives, styrene, acrylamide, methacrylic acid or methyl methacrylate) and biological substrates (bacteria, algae or fungus free or immobilized in a solid-support) are used in the solid-phase extraction (Pereira and Arruda, 2003; Camel, 2003).

In 1985, Kroto et al. was discovered the first carbon nanostructure, C60 buckminsterfullerene, which reported as an excellent solid material for metal extraction due to its unique properties, large surface area, and high inner volume (Kroto et al., 1985; Gallego et al., 1994). Different kinds of synthetic nanomaterials have been synthesized and applied as solid sorbents for the determination of metal ions in the past few years. In 1991, Iijima first noticed carbon nanotubes (CNTs) as a sorbent and later explained their use for different purposes in analytical science (Iijima, 1994). The discovery of CNTs as sorbents and the anticipation of developing novel carbon-based nanomaterials have attracted researchers worldwide. Carbon nanotubes and modified CNTs with chemical treatment are relatively new sorbents due to their unique properties such as chemical stability, chemical and thermal stability, unique morphology, relatively high reactivity, and the high surface area that have proven very efficient for treating many kinds of trace pollutants specially heavy metal ions (Kumar et al., 2008; Mauter and Elimelech, 2008; Smart et al., 2006; Tasis et al., 2006). CNTs show high selectivity, regeneration and adsorption capacity for heavy metal ions.

The objective of this chapter is to present an overview of the applications of single-walled carbon nanotubes (SWCNTs) and multiwalled carbon nanotubes (MWCNTs) for the solid-phase preconcentration of metals in diverse sample matrices and their subsequent determination by atomic spectrometric techniques. The mechanistic aspects of metal ions adsorption onto CNTs are also discussed for a better understanding of the process.

Table 11.2 Chemically modified SWCNTs and MWCNTs for determination of metal ions

Sorbents materials/chelate	Techniques coupled	Metals	Detection limit ($ng \cdot mL^{-1}$)	Ref.
Single-walled carbon nanotubes (SWCNTs)				
Bi/SWCNTs/GCE	cathodic stripping voltammetry	Cr(VI)	0.036 nM	Ouyang et al. (2013)
SWCNT-Schiff's Naphthol-2-(Pyridylazo-2)-1 (PAN)	FAAS	Cu(II)	0.10	Arbabi Rashid and Ali (2014)
SWCNT-HNO_3	ICP-MS	Cu(II), Co(II), Pb(II)	0.039, 0.0012, 0.0054	Chen et al. (2009)
l-cys/AuNPs/SWCNTs/GCE	stripping voltammetry	Cu(II)	0.02 nM	Cheng et al. (2013)
SWCNTs disk	FAAS	Fe(III), Cr(III)	2.12, 4.08	Soylak and Unsal (2010)
SWCNT-PhSH	stripping voltammetry	Hg(II)	3.0 nM	Wei et al. (2014)
SDS-coated SWCNTs	UV/vis/NIR spectrophotometer	Hg(II)	10 nM	Kim et al. (2009)
Nano alumina-SWCNTs	FAAS	Pb(II), Cu(II), Co(II)	2.1, 0.9, 1.1	Yalcinkaya et al. (2011)
SWCNTs	ICP-MS	Au, Pd	0.23, 0.06	Cheng et al. (2010)
Multiwalled carbon nanotubes (MWCNTs)				
MWCNT-dodecylsalicylaldoxime	FI-FAAS	Cu(II)	4.1	Tobiasz et al. (2012)

(*Continued*)

Table 11.2 (*Continued*)

Sorbents materials/chelate	Techniques coupled	Metals	Detection limit ($ng \cdot mL^{-1}$)	Ref.
MWCNT-HNO_3	X-ray fluorescence spectrometry	Cr(III), Mn(II), Fe(III), Co(II), Ni(II), Cu(II), Zn(II), Pb(II)	0.6, 0.6, 1.0, 0.7, 0.6, 0.5, 0.9, 1.9	Zawisza et al. (2012)
MWCNT-glutaric dihydrazide	AAS	Co(II), Cd(II), Pb(II), Pd(II)	0.16, 0.19, 0.17, 0.12	Tavallali et al. (2014)
GO-MWCNT-DETA	ICP-OES	Cr(III), Fe(III), Pb(II), Mn(II)	0.16, 0.50, 0.24, 0.38	Zhu et al. (2015)
MMWCNT	FAAS	Pb(II), Mn(II)	100, 60	Tarigh and Shemirani (2013)
MWCNT-pyridine	FAAS	Pb(II)	2.0	Torkian et al. (2014)
MWCNT-phosphonium IL	ETAAS	As(III), As(V)	0.0071	Grijalba et al. (2015)
MWCNT-TiO_2	anodic stripping voltammetry	Hg(II)	5.01	Mao et al. (2015)
MWCNT-HNO_3	FAAS	Pd(II)	0.3	Afzali et al. (2012)
MWCNT	ETA-AAS	Cd(II)	0.010	Méndez et al. (2015)
MWCNT-HNO_3	FAAS	Rh(III)	0.010	Ghaseminezhad et al. (2009)

Sorbents materials/chelate	Techniques coupled	Metals	Detection limit ($ng \cdot mL^{-1}$)	Ref.
MWCNT-HNO_3	ETAAS	Cd(II), Pb(II)	0.0097, 0.00013	Méndez et al. (2015)
MWCNT-N-(4-{4-[(anilinocarbothioyl) amino]benzyl}phenyl)-N-phenylthiourea	Voltammetric	Ag(I)	0.079	Cheraghi et al. (2015)
MWCNT-8-aminoquinoline	FAAS	Cd(II), Pb(II), Ni(II)	0.09, 0.72, 1.0	Taghizadeh et al. (2014)
MWCNT-polypropylene amine dendrimers	AAS	Au(III), Pd(II)	0.08, 0.12	Behbahani et al. (2014)
MWCNT-5-aminosalicylic acid	ICP-OES	Pb(II)	0.25	Soliman et al. (2013)
MWCNT-thiosemicarbazide	ICP-OES	Cd(II), Cu(II), Pb(II)	0.56, 0.22, 0.18	Zhang (2013)
MWCNT-3-hydroxy-4-((3-silylpropylimino) methyl) phenol	FAAS	Cu(II), Ni(II), Zn(II), Pb(II), Co(II), Fe(III)	1.73, 2.47, 2.46, 2.89, 2.77, 1.58	Ghaedi et al. (2013)
MWCNT-triphenylphosphine	anodic stripping voltammetry	Pb(II), Hg(II), Cd(II)	6.0×10^{-5}, 9.2×10^{-5} and 7.4×10^{-5} μM	Bagheria et al. (2013)
MWCNTs	ICP-AES	As, Bi, Cd, Pb, Hg and Ti	2.4, 4.08, 0.3, 2.1, 1.8, and 1.8	ALqadamio et al. (2013)

(*Continued*)

Table 11.2 (*Continued*)

Sorbents materials/chelate	Techniques coupled	Metals	Detection limit ($ng \cdot mL^{-1}$)	Ref.
MWCNT-3-(4-methoxybenzylideneamino)-2-thioxothiazolodin-4-one	anodic stripping voltammetry	Hg(II) and Pb(II)	0.18, 0.12	Afkhami et al. (2012)
MWCNT-Schiff base-chitosan	ICP-MS	V(V), Cr(VI), Cu(II), As(V), Pb(II)	0.0021, 0.0038, 0.0035, 0.0013, 0.0036	Dai et al. (2012)
MWCNT-Schiff base	FAAS	Cu(II), Pb(II)	0.80 and 1.80	Dalali et al. (2012)
MWCNT-poly(2-amino thiophenol)	AAS	Cd(II), Pb(II)	0.3, 1.0	Nabid et al. (2012)
MWCNT-phenyl-iminodiacetic acid	ICP-OES	Fe (III), Cu(II) and Pb(II)	0.26, 0.15, 0.18	Cui et al. (2012)
MWCNT-4-(4-isopropylbenzylideneamino) thiophenol	FAAS	Cd(II), Ni(II), Pb(II) and Zn(II)	1.6, 2.4, 5.6, 1.3	Mortazavi et al. (2011)
MWCNT-ethylenediamine	ICP-OES	Cr(III), Fe(III), Pb(II)	0.24, 0.19, 0.33	Zang et al. (2009)
MWCNT-ammonium pyrrolidine dithiocarbamate	FAAS	Cu(II), Cd(II), Pb(II), Zn(II), Ni(II), Co(II)	0.30, 0.45, 0.60, 0.35, 0.57, 0.40	Tuzen et al. (2008)

11.2 Carbon Nanotubes as Adsorbents

CNTs are allotropes of carbon, tube-shaped materials having the unique size, shape, thermal conductivity and remarkable physical properties: strength, stiffness, and toughness (Liang et al., 2004; Liang et al., 2005; Pyrzynska, 2008; Pyrzynska and Bystrzejewski, 2010). Currently, several methods are used for the synthesis of CNTs including arc discharge, laser ablation, electrolysis, sono-chemical or hydrothermal and chemical vapour deposition. CNTs acquire a few nanometres in diameter and several microns of length and have two basic forms: SWCNTs and MWCNTs. They have a very broad application, such as reinforcement composite, gas adsorption, as catalyst supports and in the electrochemistry, for instance, in the biosensors preparation as transducer, stabilizer and immobilization matrix. The various chemically modified SWCNTs and MWCNTs which have been used so far for the preconcentration and determination of metal ions are summarized in Table 11.2.

11.2.1 Single-Walled Carbon Nanotubes

Single-walled carbon nanotubes (SWCNTs) are one of the promising and important members in carbon family which have been anticipated and experimentally proved to retain excellent mechanical strength with superior heat and electric conductivity, high chemical stability and a large specific surface area, making them a better candidate for use as adsorbers (Eva et al., 2007). They are widely used as chemical sensors, nanobiomaterial, conductive nano-ink, nanodevices, conductive heating films, conductive transparent electrodes, and displays (backlight, flat lamp and field emitter). They have been produced by the arc-discharge process with the purity of about 95 to 98 wt% and include roughly 92 to 95 wt% of carbon nanoparticles.

A number of studies have been reported by authors on SWCNTs for the preconcentration and determination of metal ions. Table 11.2 presents some examples of application of chemically modified SWCNTs with organic moieties for determination of metal ions in various real samples. Bui et al. (2012) prepared a transparent and flexible SWCNT film electrode using the vacuum filtering methods. Cadmium and Lead were determined by square-wave stripping voltammetry. The finding reveals that the used flexible SWCNT film electrodes in reported work can be applied as simple, efficient, cost-effective, and/or disposable electrodes for simultaneous detection of heavy metal ions. Later, a highly selective and sensitive Bi film wrapped SWCNT-modified glassy carbon electrode (Bi/SWCNTs/GCE) was fabricated and used for the detection of ultratrace Cr(VI) (Ouyang et al., 2013). A catalytic adsorptive cathodic stripping voltammetry (AdCSV) was used for the analysis of Cr(VI) and detection limit was found to be 0.036 nM. Results suggest that the newly designed electrode exhibits better reproducibility and repeatability toward aqueous detection of trace Cr(VI) as comparison to previous reported work in terms of high selectivity and sensitivity. A Schiff's naphthol-2-(pyridylazo-2)-1 (PAN) complex on SWCNTs onto C18 cartridge has been reported by Arbabi Rashid and Ali (2014) and used as a selective material for the determination of Cu(II). The proposed method was successfully applied in spiked and unspiked water samples for Cu(II) determination with high preconcentration factor 125 for a 1000 mL sample volume and relative standard deviation (RSD) 1.25%. New and novel oxidized SWCNT adsorbents were prepared by Chen et al. (2009), who developed a preconcentration method for trace Cu, Co and Pb in biological and environmental samples prior to their determination by inductively coupled plasma mass spectrometry (ICP-MS). The authors studied the various physico-chemical factors pH, sample flow rate and volume, eluent concentration and interfering ions which greatly influenced the preconcentration and determination of the analytes. The proposed method was successfully applied for the determination of trace Cu, Co and Pb in real water sample with the recoveries of 96.0–109% and validation of method was checked by using the certified reference material of mussel. Cheng et al. (2013) developed an inorganic–organic hybrid electrochemical sensor, namely l-cysteine self-assembled gold nanoparticle/single-walled carbon nanotubes/glassy carbon electrode (l-cys/AuNPs/SWCNTs/GCE) and used it for the detection of trace Cu(II) by stripping voltammetry. The prepared sensor was successfully applied for the determination of Cu(II) in real environmental water samples and reported results were found to be satisfactory compared with the certified values. The detection limit of Cu was found to be 0.02 nM and interference experiments showed that Ag(I), Pb(II), Cd(II) and Hg(II) had little influence on Cu(II) signal. Soylak and Unsal (2010) prepared a SWCNT disk (5% (w/v) sodium dodecyl sulphate) and established a preconcentration–separation procedure for Fe(III) and Cr(III). Thirty milligram of nanotube was sufficient for the quantitative recovery of Fe(III) and Cr(III) at pH 8.0. The reported method was successfully applied in food and herbal plant samples collected from Turkey for the determination of iron and chromium and the method was validated by using various certified reference materials such as lichen (IAEA-336), CRM025-050 Metals on soil and BCR-032 Moroccan Phosphate rock. Wei et al. (2014) developed an excellent material by the immobilization of thiophenol functionalized (SWCNT-PhSH) onto the gold electrode surface. Various experimental parameters such as pH, preconcentration time, deposition potential, and deposition time have been investigated for the determination of Hg(II). Various interfering agents such as Cr(II), Mn(II), Fe(II), Co(II), Ni(II), Cu(II), and Zn(II) associated with Hg(II) have been studied with no significant effect and the low detection limit was found to be 3.0 nM (S/N = 3). Kim et al. (2009) developed a selective and sensitive SWCNT sensor and used it for the detection of mercury (Hg^{2+}). The developed sensor showed a detection limit of 10 nM for Hg^{2+} in water, which is proportionate with the maximum allowable limit of mercury ions in drinking water set by most government environmental protection agencies. Yalcinkaya et al. (2011) prepared a nano alumina–SWCNT hybrid sorbent and applied it for the determination of Pb(II), Cu(II) and Co(II) by FAAS in various real samples via solid-phase extraction technique. The detection limit and adsorption capacity of hybrid material for Pb(II), Co(II) and Cu(II) were found to be 2.1, 0.9, 1.1 μg L^{-1} and 101.2, 76.3, 53.2 mg g^{-1}, respectively. The method was applied for the determination of Pb(II), Co(II) and Cu(II) in tap water, dam water, soil and tea leaves and the method accuracy was validated by using various standard reference materials (BCR 150 Skim milk powder, CMI 7003 Silty clay) and spiked real samples. Cheng et al. (2010) developed a solid-phase extraction method for the determination of trace precious metals (Au and Pd). A potential and an excellent single-walled carbon nanotubes used as an adsorbent for the separation and preconcentration of Au and Pd and analysis performed by using inductively coupled plasma mass spectrometry (ICP-MS). The maximum adsorption capacity was achieved in the pH range 2.0–5.0 and the analytes were completely eluted with 2.0 mL of 3.0% thiourea in 0.5 mol L^{-1} HCl solution. The reported method was successfully applied in geological and water samples with recoveries of 93.6–103%.

11.2.2 Multi-Walled Carbon Nanotubes

Multiwalled carbon nanotubes (MWCNTs) have excellent mechanical strength with superior heat and electric conductivity and also have high specific surface area, high crystallinity, and high length-to-diameter ratio. Their novel properties make them potentially useful in a variety of applications in nanotechnology, optics, electronics, and other fields of materials science (ESD, heat exchanger, reinforced material and EMI shielding). Similar to SWCNTs, MWCNTs are relatively new and promising adsorbents that have been proven to possess great potential for removing several types of pollutants, especially heavy metal ions such as lead, chromium, copper, nickel, zinc and cadmium ions. Owing to their high surface area, chemical stability, hexagonal arrays of carbon atom and mechanical strength, MWCNTs are considered a new form of carbon, are fascinating as excellent functional materials, which may provide a novel solid-phase extractor for various inorganic and organic materials at trace levels.

Chemical modification and activation of MWCNTs makes them a promising adsorbent for the sorption of trace and precious metal ions and also improves the analytical detection limit by preconcentration and separation methods. Table 11.2 presents a few examples of the application of chemically modified MWCNTs in the determination of metal ions in various real samples. Some authors reported the work based on chemically modified MWCNTs for the separation and preconcentration of heavy metal ions. Tobiasz et al. (2012) synthesized MWCNTs using

5-dodecylsalicylaldoxime as a modifier and applied them for Cu(II) determination by flame atomic absorption spectrometric (FAAS). The maximum sorption capacity of 18.1 and 31.6 mg g^{-1} was achieved at pH 4.5–6.3 and pH 6.15–6.25, respectively, for both activated and non-activated CNTs. The sorption kinetics showed that the process followed the pseudo-second-order reaction model. The accuracy of the method was validated by two certified reference materials: wastewater (EU-H-3) and ground water (ES-H-2). Multiwalled carbon nanotubes were activated by concentrated nitric acid and used for the preconcentration of Cr(III), Mn(II), Fe(III), Co(II), Ni(II), Cu(II), Zn(II) and Pb(II) by wavelength-dispersive X-ray fluorescence analysis (WDXRF) (Zawisza et al., 2012). The proposed method was optimized by using 1 mg of adsorbents in 100 mL solution and satisfactory recoveries were obtained within 5 min. The preconcentration method coupled with WDXRF spectrometry was successfully applied in natural water samples for the determination of metal ions. Tavallali et al. (2014) chemically functionalized MWCNTs by using glutaric dihydrazide (GDH) and used them for the preconcentration of metal ions. The maximum sorption capacity was found to be 33.6, 29.2, 22.1, and 36.0 mg g^{-1} for Co(II), Cd(II), Pb(II), and Pd(II), respectively, at pH 4; 1.5 mol L^{-1} of HNO_3 was used as eluting agent for all the metals and the method was successfully used in soil, well water, and wastewater samples with satisfactory results for the determination of all analytes. Similarly, Zhu et al. (2016) synthesized MWCNTs dispersed in graphene oxide (GO) colloids and later functionalized them with diethylenetriamine (DETA) and finally prepared a novel adsorbent GO-MWCNTs-DETA nanocomposite. The prepared nanocomposite was successfully employed for the solid-phase extraction and determination of Cr(III), Fe(III), Pb(II), and Mn(II) ions at the trace levels in wastewater. The batch method was applied for the pH and sorption study and maximum sorption capacity was found to be 5.4, 13.8, 6.6 and 9.5 mg g^{-1} for Cr(III), Fe(III), Pb(II) and Mn(II) ions, respectively with the equilibrium time being 30 min. Tarigh and Shemirani (2013) developed a magnetic MWCNT nanocomposite and used it for the preconcentration and determination of lead(II) and manganese(II) by flame atomic absorption spectrometry (FAAS). This study suggested that the adsorbent was worthwhile over the conventional SPE in terms of operational simplicity and less time-consuming. The enhancement factors of the proposed method were 390 and 697 for Pb and Mn, respectively. The accuracy of the method was appraised by analysing the certified reference material Seronorm™ Urine LOTNO2525. Later, MWCNTs were modified by the pyridine group by using a silane agent (Torkian et al., 2014) and the adsorbent material characterized by infrared spectroscopy (IR), thermal analysis (TG/DTA), and elemental analysis (CHN) and scanning electron microscopy (SEM). Synthesized sorbents were successfully applied for the determination of Pb ions in aqueous samples using FAAS, and the accuracy was checked by using standard reference materials (NIST 1571 and NIST 1572). Afzali et al. (2012) prepared oxidized multiwalled carbon nanotubes with concentrated nitric acid and the method was applied for the preconcentration and determination of palladium at the trace level by FAAS. Various experimental factors, including sample flow rate, eluent flow rate, and eluent concentration, were studied. The maximum sorption of Pd(II) was reported in the pH range of 1.0–4.5 and quantitative desorption appeared with 3.0 mL 0.4 mol L^{-1} thiourea. The proposed method was successfully applied in water, road dust, and standard samples for the determination of Pd(II). Another important advantage of this material was studied by Méndez et al. (2015), who developed oxidized MWCNTs for the determination of Cd and Pb in urine samples with ETAAS. This procedure achieved the column preconcentration/separation resulting in an enhancement of sensitivity, simple operation, fast, efficient matrix elimination, as well as the utilization of low sample and reagent volumes. The recovery of Cd and Pb was achieved in the range of 96–102% and 97–101%, respectively.

11.3 Binding Mechanism of Metal Ions onto Modified CNTs

Carbon nanotubes (CNTs) chemically modified with organic moieties and treated with concentrated nitric acid containing significantly different chemical donor functionalities (-N, -P, -O, -S) find increasing and promising interest for the sorption of trace and precious metal ions. Several functional group atoms are capable of chelating trace elements. The atoms most frequently used are nitrogen (e.g. N present in amines, azo groups, amides and nitriles), oxygen (e.g. O present in carboxylic, hydroxyl, phenolic, ether, carbonyl, phosphoryl groups) and sulphur (e.g. S present in thiols, thiocarbamates and thioethers) (Ahmad et al., 2015; Latorre et al., 2012). The nature of the functional group will give an idea of the selectivity of the ligand towards trace elements. For soft metals, the following order of donor atom affinity is observed: O < N < S. A reversed order is observed for hard cations. For a bidentate ligand, affinity for a soft metal increases with the overall softness of the donor atoms: (O, O) < (O, N) < (N, N) < (N, S). The order is reversed for hard metals. In general, the competition for a given ligand essentially involves Group I and Group II metals for O sites, and metals of Group II and Group III for N and S sites. The competition between metals of Group I and Group III is weak.

Basically, metal ion adsorption may take place by both physisorption and chemisorption. Various factors such as solution conditions, nature of the adsorbent and the adsorbate, size of the adsorbent and the adsorbate, charge of the adsorbate and the adsorbent surface, pH and temperature of solution influence the adsorption onto CNTs. Various types of interaction and forces are involved in the sorption of metal ions onto modified CNTs, such as hydrogen bonding, electrostatic interaction, surface complexation and van der Waals forces, ion exchange, and so forth. The major groups such as carboxylic and phenolic participate in the sorption of metal ions by the proton exchange and chelation. Based on the electron-donating nature of the oxygen-containing groups in CNTs and the electron-accepting nature of metal ions, the ion exchange mechanism/chelation could be preferentially considered (Kumar et al., 2014; Liu et al., 2013). The mechanism of retention depends on the nature of the sorbent and may include simple adsorption, chelation or ion-exchange (Fig. 11.2). Also, for trace elements, ion-pair solid-phase extraction may be used.

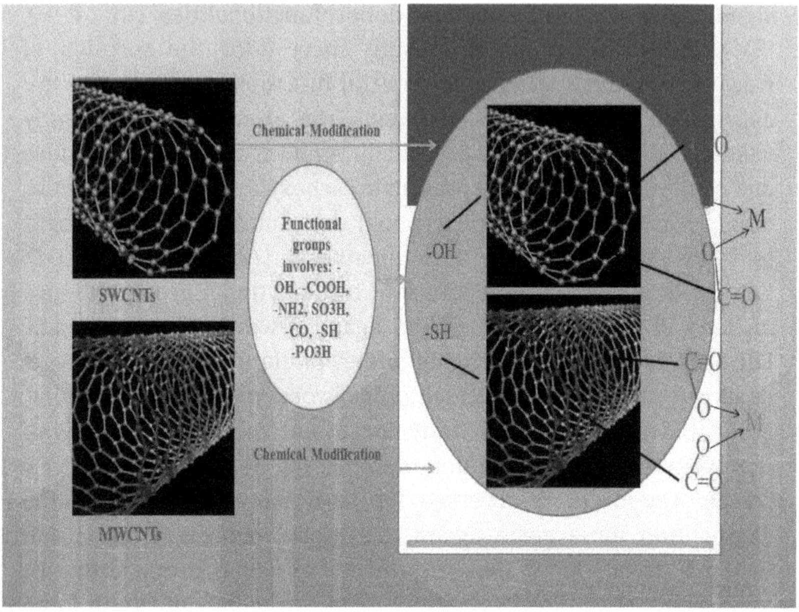

Figure 11.2 Co-ordination of metal ions with functionalized carbon nanotubes (CNTs).

11.4 Conclusion and Future Prospects

The present review focused on the application of CNTs for the remediation of heavy metals from wastewater. Carbon nanotubes are a revolutionary material and are expected to make a large impact on the environment in future. A number of synthesis route, treating agents and experimental conditions were used for the synthesis of CNT-based nanomaterials. The adsorption efficiency of these materials can be improved by their functionalization with acids or other chemicals. The existence of functional groups (e.g. hydroxyl, carboxylic, phenolic) on the surface of CNTs plays an important role for binding with metal ions. The adsorption efficiency of SWCNTs was found to be higher than MWCNTs, which may be due to the less number of active sites at the inner wall of nanotubes as well as lesser diffusion of metal ions in the multilayer structure. Therefore, the activation energy for the adsorption of metal ion onto SWCNTs was less than that of MWCNTs. In spite of the adsorption efficacy of CNTs, pH of the solution, surface complexation, hydrogen bond, electrostatic interaction, and ion exchange phenomena also play a vital role in the adsorption of metal ions.

Although, CNTs and their nanocomposites demonstrate excellent performance towards the adsorption for metallic pollutants, important investigation is required, such as the identification of active sites onto inner and outer wall of CNTs, modification of the CNT filter, sheet, and films. In addition, the mathematical modelling of the adsorption data needs more attention to understand the mechanism of adsorption. The regeneration of CNTs and the recovery of metal ions from saturated CNTs are necessary to produce cost-effective CNTs. Therefore, more work should be done to investigate the better desorbing agent having low cost and environmentally friendly nature (green eluent).

Despite the economical and toxicological drawbacks, CNTs are getting attention around the world. These nanomaterials not only remove metal pollutants but also exhibit vast application in diverse fields (e.g. modified electrodes, energy storage electrochemical capacitors, sensors, solar cells, antibiofouling agents, catalysts, fuel cells, semiconductors, adhesion agents, microwave filters and anticorrosion agents). Thus, on the basis of significant findings for CNTs, continued attention to this field remains obligatory.

Acknowledgments

The authors gratefully acknowledge the financial support given to this work by the National Research Foundation (Grant No. UID: 99623), University of KwaZulu-Natal, Durban, South Africa.

References

Afkhami A, Bagheri H, Khoshsafar H, Saber-Tehrani M, Tabatabaee M, Shirzadmehr A (2012). Simultaneous trace-levels determination of Hg(II) and Pb(II) ions in various samples using a modified carbon paste electrode based on multi-walled carbon nanotubes and a new synthesized Schiff base, *Anal Chim Acta*, **746**: 98–106.

Afzali D, Jamshidi R, Ghaseminezhad S, Afzali Z (2012). Preconcentration procedure trace amounts of palladium using modified multiwalled carbon nanotubes sorbent prior to flame atomic absorption spectrometry, *Arab J Chem,* **5**: 461–466.

Ahmad A, Khatoon A, Laskar MA, Islam A, Mohammed AW, Yong NL (2013). Use of 2-hydroxy-3-methoxybenzaldehyde functionalized Amberlite XAD-16 for preconcentration and determination of trace metal ions by Flame atomic absorption spectrometry, *Der Pharma Chem,* **5**: 12–23.

Ahmad A, Siddique JA, Laskar MA, Kumar R, Mohd-Setapar SH, Khatoon A, Shiekh RA (2015). New generation Amberlite XAD resin for the removal of metal ions: A review, *J Environ Sci,* **31**: 104–123.

ALqadami AA, Abdalla MA, ALOthman ZA, Omer K (2013). Application of solid phase extraction on multiwalled carbon nanotubes of some heavy metal ions to analysis of skin whitening cosmetics using ICP-AES, *Int J Environ Res Public Health,* **10**: 361–374.

Arbabi Rashid H, Ali M (2014). Extraction of trace amount Cu(II) by single walled carbon nanotubes (SWCNTs) in water samples and determination it by FAAS, *Orient J Chem,* **30**: 203–210.

Asadollahi T, Dadfarnia S, Shabani AMH (2010). Separation/preconcentration and determination of vanadium with dispersive liquid–liquid microextraction based on solidification of floating organic drop (DLLME-SFO) and electrothermal atomic absorption spectrometry, *Talanta,* **82**: 208.

Aydin FA, Soylak M (2010). Separation, preconcentration and inductively coupled plasma-mass spectrometric (ICP-MS) determination of thorium(IV), titanium(IV), iron(III), lead(II) and chromium(III) on 2-nitroso-1-naphthol impregnated MCI GEL CHP20P resin, *J Hazard Mater,* **173**: 669.

Bagheria H, Afkhami A, Khoshsafar H, Rezaei M, Shirzadmehr A (2013). Simultaneous electrochemical determination of heavy metals using a triphenylphosphine/MWCNTs composite carbon ionic liquid electrode, *Sens Actuators B,* **186**: 451–460.

Behbahani M, Gorji T, Mahyari M, Salarian M, Bagheri A, Shaabani A (2014). Application of polypropylene amine dendrimers (POPAM)-grafted MWCNTs hybrid materials as a new sorbent for solid-phase extraction and trace determination of gold(III) and palladium(II) in Food and Environmental Samples, *Food Anal Methods,* **7**: 957–966.

Bui MP, Li CA, Han KN, Pham XH, Seong GH (2012). Electrochemical determination of cadmium and lead on pristine single-walled carbon nanotube electrodes, *Anal Sci,* **28**: 699–704.

Camel V (2003). Solid phase extraction of trace elements, *Spectrochim Acta B*, **58**: 1177–1233.

Chapman TL (2008). *Genetic Heavy Metal Toxicity*, Universe, Lincoln, NE.

Chen S, Liu C, Yang M, Lu D, Zhu L, Wang Z (2009). Solid-phase extraction of Cu, Co and Pb on oxidized single-walled carbon nanotubes and their determination by inductively coupled plasma mass spectrometry, *J Hazard Mater*, **170**: 247–251.

Cheng X, Chen S, Wang X, Liu C (2010). Single-walled carbon nanotubes as solid-phase extraction adsorbent for the preconcentration and determination of precious metals by ICP-MS, *At Spectroscop*, **31**: 75–80.

Cheraghi S, Taher MA, Fazelirad H (2015). Voltammetric determination of silver with a new multi-walled carbon nanotube modified paste electrode, *Russian J Electrochem*, **51**: 271–277.

Cui Y, Hu Z-J, Yang J-X, Gao H-W (2012). Novel phenyl-iminodiacetic acid grafted multiwalled carbon nanotubes for solid phase extraction of iron, copper and lead ions from aqueous medium, *Microchim Acta*, **176**: 359–366.

Dai B, Cao M, Fang G, Liu B, Dong X, Pan M, Wang S (2012). Schiff base-chitosan grafted multiwalled carbon nanotubes as a novel solid-phase extraction adsorbent for determination of heavy metal by ICP-MS, *J Hazard Mater*, **219–220**: 103–110.

Dalali N, Ashouri M, Nakisa S (2012). Solid phase extraction based on modified multi-walled carbon nanotubes packed column for enrichment of copper and lead on-line incorporated with flame atomic absorption spectrometry, *J Iranian Chem Soc*, **9**: 181–188.

Devi PR, Gangaiah T, Naidu GRK (1991). Determination of trace metals in water by neutron activation analysis after preconcentration on a poly(acrylamidoxime) resin, *Anal Chim Acta*, **249**: 533.

Escudero LA, Martinez LD, Salonia JA, Gasquez JA (2010). Determination of Zn(II) in natural waters by ICP-OES with on-line preconcentration using a simple solid phase extraction system, *Microchem J*, **95**: 164.

Eva D, Saldador O, Aurelio V (2007). Adsorption of volatile organic compounds onto carbon nanotubes, carbon nanofibers, and high-surface-area graphites, *J Colloid Interf Sci*, **305**: 7–16.

Froes RES, Neto WB, Naveira RLP, Silva NC, Nascentes CC, da Silva JBB (2009). Exploratory analysis and inductively coupled plasma optical emission spectrometry (ICP-OES) applied in the determination of metals in soft drinks, *Microchem J*, **92**: 68.

Fu X-C, Wu J, Li J, Xie C-G, Liu Y-S, Zhong Y, Liu J-H (2013). Electrochemical determination of trace copper(II) with enhanced sensitivity and selectivity by gold nanoparticle/single-wall carbon nanotube hybrids containing three-dimensional l-cysteine molecular adapters, *Sens Actuators B,* **182**: 382–389.

Gallego M, Pena YP, Valcarcel M (1994). Fullerenes as sorbent materials for metal preconcentration, *Anal Chem,* **66**: 4074–4078.

Ghaedi M, Montazerozohori M, Rahimi N, Biysreh MN (2013). Chemically modified carbon nanotubes as efficient and selective sorbent for enrichment of trace amount of some metal ions, *J Ind Eng Chem,* **19**: 1477–1482.

Ghaseminezhad S, Afzali D, Taher MA (2009). Flame atomic absorption spectrometry for the determination of trace amount of rhodium after separation and preconcentration onto modified multiwalled carbon nanotubes as a new solid sorbent, *Talanta,* **80**: 168–172.

Grijalba AC, Escudero LB, Wuilloud RG (2015). Ionic liquid-assisted multiwalled carbon nanotube-dispersive micro-solid phase extraction for sensitive determination of inorganic As species in garlic samples by electrothermal atomic absorption spectrometry, *Spectrochim Acta Part B,* **110**: 118–123.

Hu H (2000). Exposure to metals, *Prim Care,* **27**: 983–996.

Hu H (2005). Heavy metal poisoning. In: Kasper, DL (ed), *Harrison's Principles of Internal Medicine,* 16th ed. Mc Graw-Hill, Medical Publishing Division, New York, vol. 2, pp. 2577–2580.

Iijima S (1991). Helical microtubules of graphitic carbon, *Nature,* **354**: 56–58.

Islam A, Ahmad A, Laskar MA (2011b). A newly developed salicylanilide functionalized Amberlite XAD-16 chelating resin for its use in preconcentration and determination of trace metal ions from environmental and biological samples, *Anal Methods,* **3**: 2041–2048.

Islam A, Ahmad A, Laskar MA (2012a). Preparation, characterization of a novel chelating resin functionalized with o-hydroxybenzamide and its application for preconcentration of trace metal ions, *CLEAN-Soil Air Water,* **40**: 54–65.

Islam A, Ahmad A, Laskar MA (2012b). Characterization of a chelating resin functionalized via azo spacer and its analytical applicability for the determination of trace metal ions in real matrices, *J Appl Polym Sci,* **123**: 3448–3458.

Islam A, Ahmad A, Laskar MA (2015). Flame atomic absorption spectrometric determination of trace metal ions in environmental and biological samples after preconcentration on a new chelating resin containing p-aminobenzene sulfonic acid, *J AOAC Int*, **98**: 165–175.

Islam A, Laskar MA, Ahmad A (2010a). Characterization and application of 1-(2-pyridylazo)-2-naphthol functionalized amberlite XAD-4 for preconcentration of trace metal ions in real matrices, *J Chem Eng Data*, **55**: 5553–5561.

Islam A, Laskar MA, Ahmad A (2010b). Characterization of a novel chelating resin of enhanced hydrophilicity and its analytical utility for preconcentration of trace metal ions, *Talanta*, **81**: 1772–1780.

Islam A, Laskar MA, Ahmad A (2011a). The efficiency of Amberlite XAD-4 resin loaded with 1-(2-pyridylazo)-2-naphthol in preconcentration and separation of some toxic metal ions by flame atomic absorption spectrometry, *Environ Monit Assess*, **175**: 201–212.

Islam A, Laskar MA, Ahmad A (2013). Preconcentration of metal ions through chelation on a newly synthesized resin containing O, O donor atoms for quantitative analysis of environmental and biological samples, *Environ Monit Assess*, **185**: 2691–2704.

Kim TH, Lee J, Hong S (2009). Highly selective environmental nanosensors based on anomalous response of carbon nanotube conductance to mercury ions, *J Phys Chem C*, **113**: 19393–19396.

Kroto HW, Heath JH, O'Brian SC, Curl RF, Smalley RE (1985). C_{60}: Buckminsterfullerene, *Nature*, **318**: 162–163.

Kumar R, Khan MA, Haq N (2014). Application of carbon nanotubes in heavy metals remediation, *Crit Rev Environ Sci Tech*, **44**: 1000–1035.

Kumar S, Kumar R, Jindal VK, Bharadwaj LM (2008). Immobilization of single walled carbon nanotubes on glass surface, *Mater Lett*, **62**: 731–734.

Latorre CH, Mendez JA, Garcia JB, Martin SG, Crecente RMP (2012). Carbon nanotubes as solid-phase extraction sorbents prior to atomic spectrometric determination of metal species: A review, *Anal Chim Acta*, **749**: 16–35.

Liang P, Ding Q, Song FJ (2005). Application of multiwalled carbon nanotubes as solid phase extraction sorbent for preconcentration of trace copper in water samples, *J Sep Sci*, **28**: 2339–2343.

Liang P, Liu Y, Guo L, Zeng J, Lu H (2004). Multiwalled carbon nanotubes as solid-phase extraction adsorbent for the preconcentration of trace metal ions and their determination by inductively coupled plasma atomic emission spectrometry, *J Anal Atom Spectrom*, **19**: 1489–1492.

Liu X, Wang M, Zhang S (2013). Application potential of carbon nanotubes in water treatment: A review, *J Environ Sci*, **25**: 1263–1280.

Mao A, Li H, Cai Z, Hu X (2015). Determination of mercury using a glassy carbon electrode modified with nano TiO_2 and multi-walled carbon nanotubes composites dispersed in a novel cationic surfactant, *J Electroanal Chem*, **751**: 23–29.

Mauter MS, Elimelech M (2008). Environmental applications of carbon-based nanomaterials, *Environ Sci Technol*, **42**: 5843–5859.

Méndez JÁ, García JB, Martín SG, Peña Crecente RM, Herrero Latorre C (2015). Determination of cadmium and lead in urine samples after dispersive solid–liquid extraction on multiwalled carbon nanotubes by slurry sampling electrothermal atomic absorption spectrometry, *Spectrochim Acta Part B*, **106**: 13–19.

Mendez JA, Garcia JB, Pena Crecente RM, Martin SG, Latorre CH (2011). A new flow injection preconcentration method based on multiwalled carbon nanotubes for the ETA-AAS determination of Cd in urine, *Talanta*, **85**: 2361–2367.

Montaser A, Golightly DW (1992). *Inductively Coupled Plasmas in Analytical Atomic Spectrometry*, 2nd ed, VCH, New York.

Mortazavi K, Ghaedi M, Montazerozohori M, Roosta M (2011). Multiwalled carbon nanotubes (MMWCNTS) modified with 4-(4-isopropyl benzylideneamino)thiophenol as efficient adsorbent for solid phase extraction and preconcentration of some metal ions in environmental and food samples, *Fresen Environ Bull*, **20**: 2847–2854.

Nabid MR, Sedghi R, Bagheri A, Behbahani M, Taghizadeh M, Oskooie HA, Heravi MM (2012). Preparation and application of poly(2-amino thiophenol)/MWCNTs nanocomposite for adsorption and separation of cadmium and lead ions via solid phase extraction, *J Hazard Mater*, **203–204**: 93–100.

Naseri MT, Hosseini MRM, Assadi Y, Kiani A (2008). Rapid determination of lead in water samples by dispersive liquid–liquid microextraction coupled with electrothermal atomic absorption spectrometry, *Talanta*, **75**: 56.

Nomura M, Sato M, Amakawa H, Oura Y, Ebihara M (2005). Precise determination of Sc in natural waters by neutron activation analysis coupled with preconcentration of Sc, *Anal Chim Acta*, **553**: 58.

Ouyang R, Zhang W, Zhou S, Xue Z-L, Xu L, Gu Y, Miao Y (2013). Improved Bi film wrapped single walled carbon nanotubes for ultrasensitive electrochemical detection of trace Cr(VI), *Electrochim Acta*, **113**: 686–693.

Pereira MG, Arruda MAZ (2003). Trends in preconcentration procedures for metal determination using atomic spectrometry techniques, *Microchim Acta,* **141**: 115–131.

Pesavento M, Alberti G, Biesuz R (2009). Analytical methods for determination of free metal ion concentration, labile species fraction and metal complexation capacity of environmental waters: A review, *Anal Chim Acta*, **631**: 129–141.

Pyrzynska K (2008). Carbon nanotubes as a new solid-phase extraction material for removal and enrichment of organic pollutants in water, *Sep Purif Rev*, **37**: 372–389.

Pyrzynska K, Bystrzejewski M (2010). Comparative study of heavy metal ions sorption onto activated carbon, carbon nanotubes, and carbon-encapsulated magnetic nanoparticles, *Colloids Surf A*, **362**: 102–109.

Rao RR, Chatt A (1993). Preconcentration neutron activation analysis of trace elements in seawater by coprecipitation with 1-(2-thiazolylazo)-2-naphthol, pyrrolidinedithio carbamate and N-nitroso-phenylhydroxylamine, *J Radioanal Nucl Chem*, **168**: 439.

Sabarudin A, Lenghor N, Oshima M, Hakim L, Takayanagi T, Gao YH, Motomizu S (2007). Sequential-injection on-line preconcentration using chitosan resin functionalized with 2-amino-5-hydroxy benzoic acid for the determination of trace elements in environmental water samples by inductively coupled plasma-atomic emission spectrometry, *Talanta*, **72**: 1609–1617.

Sardans J, Montes F, Penuelas J (2010). Determination of As, Cd, Cu, Hg and Pb in biological samples by modern electrothermal atomic absorption spectrometry, *Spectrochim Acta Part B*, **65**: 97.

Sarkar B (2002). *Heavy Metals in the Environment*, Marcel Dekker, New York.

Shih T-T, Chen W-Y, Sun Y-C (2011). Open-channel chip-based solid-phase extraction combined with inductively coupled plasma-mass spectrometry for online determination of trace elements in volume-limited saline samples, *J Chromatogr A*, **1218**: 2342.

Smart SK, Cassady AI, Lu GQ, Martin DJ (2006). The biocompatibility of carbon nanotubes, *Carbon*, **44**: 1034–1047.

Soliman EM, Marwani HM, Albishri HM (2013). Novel solid-phase extractor based on functionalization of multi-walled carbon nano tubes with 5-aminosalicylic acid for preconcentration of Pb(II) in water samples prior to determination by ICP-OES, *Environ Monit Assess*, **185**: 10269 10280.

Soylak M, Unsal YE (2010). Chromium and iron determinations in food and herbal plant samples by atomic absorption spectrometry after solid phase extraction on single-walled carbon nanotubes (SWCNTs) disk, *Food Chem Toxicol,* **48**: 1511–1515.

Sung H, Huang SD (2003). On-line preconcentration system coupled to electrothermal atomic absorption spectrometry for the simultaneous determination of bismuth, cadmium, and lead in urine, *Anal Chim Acta,* **495**: 165–176.

Taghizadeh M, Asgharinezhad AA, Samkhaniany N, Tadjarodi A, Abbaszadeh A, Pooladi M (2014). Solid phase extraction of heavy metal ions based on a novel functionalized magnetic multi-walled carbon nanotube composite with the aid of experimental design methodology, *Microchim Acta,* **181**: 597–605.

Tarigh GD, Shemirani F (2013). Magnetic multi-wall carbon nanotube nanocomposite as an adsorbent for preconcentration and determination of lead(II) and manganese(II) in various matrices, *Talanta,* **115**: 744–750.

Tasis D, Tagmatarchis N, Bianco A, Prato M (2006). Chemistry of carbon nanotubes, *Chem Rev,* **106**: 1105–1136.

Tavallali H, Malekzadeh H, Karimi MA, Payehghadr M, Deilamy-Rad G, Tabandeh M (2014). Chemically modified multiwalled carbon nanotubes as efficient and selective sorbent for separation and preconcentration of trace amount of Co(II), Cd(II), Pb(II), and Pd(II), *Arab J Chem,* http://dx.doi.org/10.1016/j.arabjc.2014.10.034.

Tobiasz A, Walas S, Hernández AS, Mrowiec H (2012). Application of multiwall carbon nanotubes impregnated with 5-dodecylsalicylaldoxime for on-line copper preconcentration and determination in water samples by flame atomic absorption spectrometry, *Talanta,* **96**: 89–95.

Torkian L, Amini MM, Gorji T, Sadeghi O (2014). A simple, rapid and sensitive method based on modified multiwalled carbon nanotube for preconcentration and determination of lead ions in aqueous media in natural pHs, *Arab J Chem,* http://dx.doi.org/10.1016/j.arabjc.2014.10.041.

Tuzen M, Saygi KO, Soylak M (2008). Solid phase extraction of heavy metal ions in environmental samples on multiwalled carbon nanotubes, *J Hazard Mater,* **152**: 632–639.

Wei J, Yang D, Chen H, Gao Y, Li H (2014). Stripping voltammetric determination of mercury(II) based on SWCNT-PhSH modified gold electrode, *Sens Actuators B,* **190**: 968–974.

Yalcinkaya O, Kalfa OM, Turker AR (2011). Preconcentration of trace copper, cobalt and lead from various samples by hybrid nano sorbent and determination by FAAS, *Curr Anal Chem,* **7**: 225–234.

Zang Z, Hu Z, Li Z, He Q, Chang X (2009). Synthesis, characterization and application of ethylenediamine-modified multiwalled carbon nanotubes for selective solid-phase extraction and preconcentration of metal ions, *J Hazard Mater,* **172**: 958–963.

Zawisza B, Skorek R, Stankiewicz G, Sitko R (2012). Carbon nanotubes as a solid sorbent for the preconcentration of Cr, Mn, Fe, Co, Ni, Cu, Zn and Pb prior to wavelength-dispersive X-ray fluorescence spectrometry, *Talanta,* **99**: 918–923.

Zhang J (2013). Preparation, characterization and application of thiosemicarbazide grafted multiwalled carbon nanotubes for solid-phase extraction of Cd(II), Cu(II) and Pb(II) in environmental samples, *J Environ Sci,* **25**: 2331–2337.

Zhu X, Cui Y, Chang X, Wang H (2016). Selective solid-phase extraction and analysis of trace-level Cr(III), Fe(III), Pb(II), and Mn(II) ions in wastewater using diethylenetriamine-functionalized carbonnanotubes dispersed in graphene oxide colloids, *Talanta,* **146**: 358–363.

Part 3

Nanocomposite Membranes for Pollution Control

Chapter 12

Nanocomposite Membranes for Heavy Metal Removal from Wastewater

Zulhairun Abdul Karim, Goh Pei Sean, and Ahmad Fauzi Ismail

Advanced Membrane Technology Research Centre (AMTEC), Universiti Teknologi Malaysia, 81310 UTM Skudai, Johor Darul Takzim, Malaysia

zulhairun@utm.my

The exposure to heavy metals such as lead, mercury and arsenic has imparted one of the alarming threats to human health and ecosystem. The conventionally used chemical precipitation, ion-exchange and electrochemical deposition processes suffer from inherent limitations such as energy intensive, excessive toxic sludge production which requires secondary treatment as well as the inefficiency to meet the desired treated water quality standard. Heavy metal removal using adsorbents has been proven technologically viable. In this context, nanomaterials hold immense potential to further improve the removal efficiency in order to remediate the heavy metal pollution issues. The unique features in terms of high surface area, fast kinetic, tunable porous structure and excellent heavy metal sorption capacity demonstrated by nanomaterials can be exploited for their

Nanocomposites for Pollution Control
Edited by Chaudhery Mustansar Hussain and Ajay Kumar Mishra

ISBN 978-981-4774-45-1 (Hardcover), 978-1-315-14368-2 (eBook)
www.panstanford.com

applications in this field. Nevertheless, the issue related to leaching of nanoparticles to the water body and the subsequent post treatment need to be addressed to eliminate the potential risks. In this regard, a new class of nanocomposite membrane has been developed by combining polymeric materials with nanoparticles to harness the advantages of both materials while addressing the limitation of nanoparticle adsorbents. Through the incorporation of nanomaterials within the polymer matrix, the leaching problem can be favorably resolved. On the other hand, this innovative approach also allows the generation of membranes hence is more cost effective. Furthermore, the advancement in nanomaterial science and engineering also enables the design of membrane architectures to render desired functionalities for effective heavy metal removal. This chapter presents the application of nanocomposite membranes for heavy metal removal from wastewater. The fabrications and performance evaluations of nanocomposite membranes incorporated with emerging nanomaterials such as carbon nanotubes (CNTs), graphene and its derivatives, metal oxides and zeolites are presented. The challenges and future direction in developing nanocomposite membranes for heavy metal removal are provided.

12.1 Introduction

Water is considered the most essential resource in all kind of industrial sectors where water is used for production, processing, washing and rinsing as well as cooling purposes. Industrial activities consume large amount of water hence producing tremendous amount of heavily polluted wastewater. Typical discharge from manufacturing industries (food products, textiles, paper and paper products, refined petroleum products, chemicals and chemical products, basic metals, automobiles, etc.) could exceed million liters of wastewater per day for large factories [1]. The characteristics of wastewater vary greatly from one industry to the others. Wastewater may contain insoluble and soluble substances. Insoluble or suspended pollutants can be easily separated from the wastewater. However, the industrial discharged water may be excessively alkaline or acidic, dyed, oiled, greased, bearing organic and inorganic toxic substances and perhaps contains bacterial pathogens with terrible odor [2]. Such effluent requires extensive treatments upon discharge. Purification of the wastewater to the acceptable level imposes considerably high cost corresponded to plant construction and operation, chemicals and maintenance. Some industries choose to release their untreated or undertreated wastewater into river bodies, lakes and coastline which would inflict serious threat to the environment and human's health [3]. One of the most hazardous pollutants that could possibly release from the industrial activities to the surface water and environment is heavy metal.

Heavy metal is characterized as the element whose atomic weight is between 63.5 and 200.6 and possesses specific gravity greater than 5. These metals could dissolve in water thus present in ionic state. Heavy metal contamination of surface and drinking water has been a global health concern [4]. The most common heavy metals found in water are arsenic, cadmium, chromium, copper, nickel, lead and mercury. Cadmium and nickel are carcinogens. They are extremely toxic even in low concentrations. Metal ions are non-biodegradable and will accumulate in the ecosystems and organisms. Environmental Protection Agency (EPA) of the United States has set a legal threshold limit on the amount of certain substances that public water could contain to ensure the safe use of water for the people as well as the environment. Table 12.1 lists some of the most hazardous heavy metals with their corresponding maximum contamination level (MCL) as dictated by EPA and recommended by World Health Organization (WHO).

Chemical intensive industries such as electroplating, metal product finishing, paint and batteries manufacturing generate significant quantities of contaminated wastewater containing chromium, cadmium, copper, lead, mercury, nickel, silver and zinc [7–12]. The concentrations of metals vary widely in different processes. Large electroplating industries might be producing wastewaters containing copper from 2,000 to as high as 15,000 mg/L depending on the chemicals and processes involved [10]. In printed circuit board manufacturing, the concentration of copper was reported to be around 500 mg/L [11]. Less extensive metal plating wastewater might contain less metal concentration (1.0–100 mg/L) [12]. A case study conducted in Dhaka, Bangladesh revealed that dissolved chromium in the water bodies exceeds 2000 mg/L resulted from tannery activities alone [3]. Other industries in Dhaka also found to be violating the permissible limit of metal ion concentration in their wastewater (see Table 12.2).

Table 12.1 Heavy metal toxicities and respective concentration limits [5–6]

Heavy metals	Heavy metal toxicities	EPA (mg/L)	WHO (mg/L)
Arsenic	Skin manifestations, visceral cancers, vascular disease	0.01	0.01
Cadmium	Kidney damage, renal disorder, human carcinogen	0.05	0.03
Chromium	Headache, diarrhea, nausea, vomiting, carcinogenic	0.10	0.05
Copper	Liver damage, Wilson disease, insomnia	1.3	2
Nickel	Dermatitis, nausea, chronic asthma, coughing, human carcinogen	0.20	0.07
Zinc	Depression, lethargy, neurological signs and increased thirst	0.80	3
Lead	Damage the fetal brain, diseases of the kidneys, circulatory system, bones, intestines, reproductive and nervous system	0.015	0.01
Mercury	Rheumatoid arthritis, and diseases of the kidneys, circulatory system, and nervous system	0.002	0.006

Table 12.2 Dissolved concentration of heavy metal (mg/L) in Dhaka wastewaters [3]

Industries	Cr	Ni	Cu	As	Cd	Pb
Tannery	1424–2485	0.28–6.5	0.68–9.0	1.7–10	0.08–2.1	0.14–2.9
Dye	2.1–5.3	0.53–1.6	0.47–2.2	2.2–4.4	0.51–2.8	0.08–0.94
Textile	1.3–2.6	0.51–3.7	1.3–5.2	0.81–1.7	0.01–0.07	0.06–0.07
Paper mill	0.76–1.3	0.04–1.4	1.3–3.4	0.29–0.81	0.90–2.0	0.84–1.4
Jute mill	0.96–1.2	0.90–1.1	1.7–2.5	0.53–0.54	0.53–1.0	0.34–0.44
Metal processing	1.7–2.5	1.4–5.5	4.3–8.0	0.77–1.8	0.52–2.6	2.2–5.2
Battery manufacturing	1.5–4.8	2.9–8.0	2.2–7.7	0.26–2.0	1.0–5.6	2.6–6.3

12.2 Conventional Heavy Metal Removal Treatment

Heavy metals can exist in particulate (either elemental or in some insoluble compound) or dissolved state. Among these two forms, it is much trickier to remove dissolved heavy metals compared to that in particulate form since they can be simply filtered out due to its bigger size. Metal contaminated wastewaters have been conventionally treated by chemical precipitation, flocculation, floatation, and chelation reaction [13]. Figure 12.1 provides an illustration of the different processes that are involved in order to remove the heavy metals using one of the more popular methods which is the combination of precipitation and flocculation. The issues with chemical precipitation methods include large quantities of chemicals and the generation of toxic sludge as a result of limestone precipitation. The produced sludge must be disposed in a sanctioned landfill. These techniques are not worth for treating high effluent volume with low heavy metal concentration.

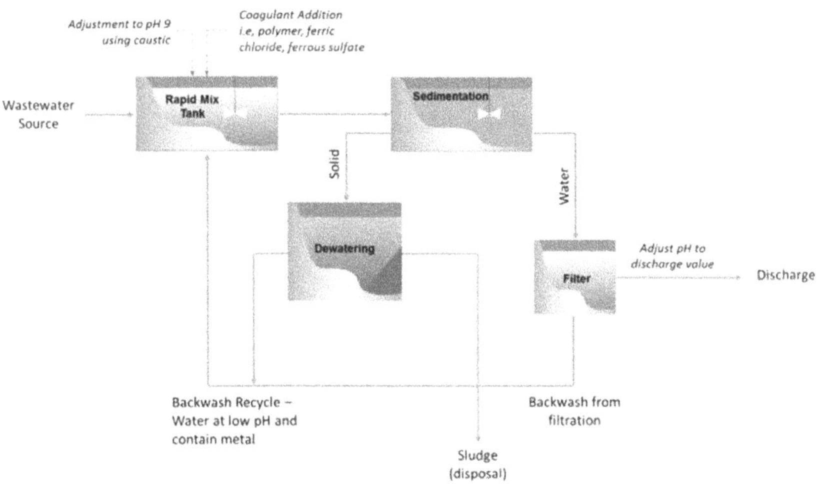

Figure 12.1 Schematic of precipitation and flocculation treatment system [13].

Other methods for removing heavy metals include fluidized bed granulation, ion exchange columns and electrocoagulation [12, 14, 15]. Using electrical methods although useful, the benefits are sometimes outweighed by few complications. Electrolytic recovery suffers from corrosion, especially at the anode. It also has high operational cost due to high energy consumption. While the ion exchange column foul very quickly and offers low volume solution. In addition, ion-exchange resins are not suitable for handling concentrated metal solutions, and the cost of the resins hinders their adoption by most industries [5, 16].

Highly efficient materials and physicochemical waste treatments to pull the toxins out of the water are necessary to reduce cost and environmental impact. Adsorption has been regarded as one of the most attractive options for heavy metal removal due to its simplicity, high efficiency, ease of operation, and its ability to remove multiple components simultaneously. Adsorbents with desirable properties (e.g., rapid sorption kinetics, high selectivity and capacity) are the key for the adsorption process [17].

Conventional adsorbents used for heavy metal adsorptions are natural minerals such as zeolite, clay, silica and activated carbons. These minerals were normally functionalized to increase the adsorption capacity towards metals [17–21]. In recent years, various novel adsorbents have been synthesized in the laboratory such as carbon based materials (carbon foam, graphene, CNTs) [22–25], modified biopolymers (derived from chitin, chitosan, starch) [26–29], magnetic nanoparticles [30–34], and hydrogels (cross-linked hydrophilic polymers) [35–36]. More innovative approaches in the development of novel adsorbent are by reclaiming agricultural wastes and biosorbents (nutshells, rice husks, algae) [37–42] and industrial by-products (fly ash, waste iron, iron slags, hydrous titanium oxide, etc.) [8, 17, 43, 44]. The adsorption process can be described in three major steps. First, the pollutant, in this case is heavy metals, are transported from bulk solution to the sorbent surface. It is then followed by the adsorption on the particle surface, and finally transported within the sorbent particle [13]. Therefore, efficient adsorbents should have numerous metal binding sites and high surface area. Reusability is another important criterion in the selection of the adsorbent [45]. Therefore, dispersing nano-adsorbent into the water as a mean to remove the heavy metals might be an unwise method since that could lead to unnecessary material loss [45] and risking the aquatic environment [46]. On the other hand, the uses of nano-adsorbents in fixed-bed columns are deemed unattractive due to extremely high pressure drop [47]. It is natural that nanoparticles tend to agglomerate due to their small size and high surface energy which results in lower surface area for adsorption [48]. They also lack of chemical and mechanical stability. Moreover, adsorption may not be the best option for small scale applications [49].

12.3 Membrane Technology

Membrane technology has been recognized as another viable option for treating metal contaminated wastewater. This technology offers considerable advantages over other established, conventional treatments due to small plant footprint, modularized design which make plant expansion and scale up simpler, high efficiency, low energy requirement and ease of operation with minimum supervision. In water purification applications, the applicability of the membranes to treat different types of water or wastewater is determined by its microstructural morphology which could be either dense, or porous. Depending on the separation goal, reverse osmosis (RO), nanofiltration (NF), ultrafiltration (UF) or microfiltration (MF) membranes could be used. These membranes are categorized based on their nominal pore size and molecular weight cut-off (MWCO) (Fig. 12.2).

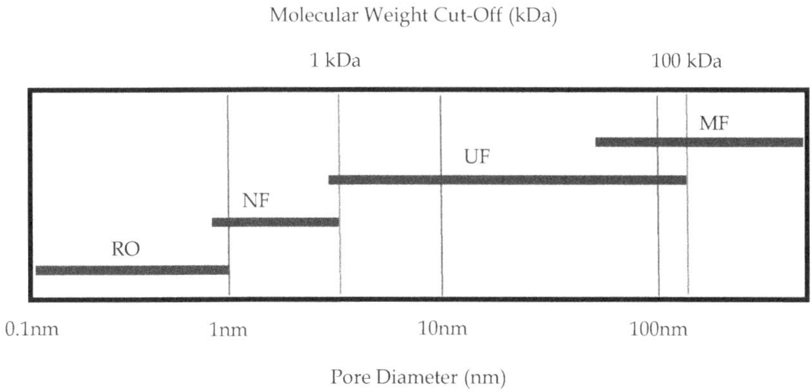

Figure 12.2 Membrane types based on pore diameter and molecular weight cut-off (MWCO).

Membrane filtration principally works based on size exclusion. This means only particles smaller than the size of the pores will pass through the membrane while the others are rejected. MF membranes only reject large particles including microorganisms due to its large pore size range. MF process could take place at very low pressure. Water flux is the highest for MF membranes. UF membranes are able to reject bacteria as well as proteins in addition to microorganisms given that the pore size is considerably smaller than MF membranes. Slightly higher pressure is needed to operate UF membranes compared to MF (up to 5 bar). The pores of an NF membrane are less than 1 nm. The separation capability could match the RO membranes but with higher fluxes. NF retains low molar mass species, organic compounds and inorganic salts by size exclusion as well as electrostatic repulsion (Donnan effect) between the ions and the charged membrane [50, 51]. Therefore, NF membranes pose higher selectivity for divalent ions over monovalent ions [52, 53]. Figure 12.3 illustrates the sieving capability of an NF membrane. Due to its very small pores, great driving force is needed thus requires high operating pressures (5–50 bar). RO membranes are essentially dense, thus rejecting everything but water. RO membrane is used for desalination of seawater, operating at very high pressures. RO seems inferior or less attractive to NF techniques due to its high energy requirement (higher operating pressures) and lower water flux.

These membranes, with different morphologies can be achieved by simply tuning the starting dope solution compositions, fabrication methods and its conditions. Membranes can be prepared from various materials ranging from polymers, ceramics, and metals, or the combinations of these materials. Since the birth of membrane technology, polymeric materials have been recognized as the most feasible, cost effective material for membrane fabrication. In fact, most commercial membranes are polymeric materials. The heavy metal removal performance of some commercial membranes such as NF90, NF270 (Dow Filmtec), AFC80 (PCI Membrane Systems), and ESNA1-K1 (Hydecanme) have been tested and reported in the literature [54–57]. To achieve better heavy metal removal efficiency using conventional membranes, operational conditions such as pH, transmembrane pressure and flow parameters have to be optimized [54].

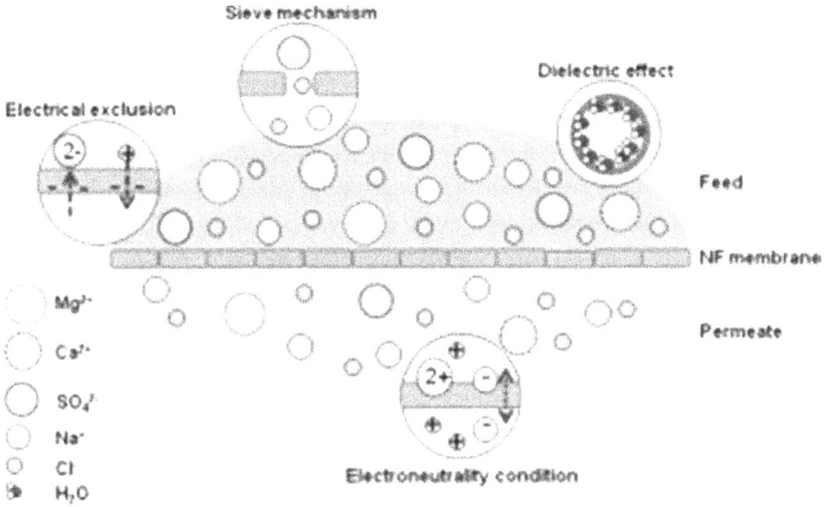

Figure 12.3 Separation mechanism through NF membranes [53].

Despite promising potential of RO and NF membranes in treating heavy metal containing wastewater [54, 58], they were found to be inefficient for treating low metal ions concentration effluents (<100 ppm) [59, 60]. Pressure-driven operation requires high energy, inflicting higher operational cost to operate RO and NF process, especially when subjected to large volume of wastewater. On the other hand, UF and MF membranes are not capable to retain heavy metals from the aqueous solutions since the ions are much smaller than the pores, unless the process takes place with the aid of complexing agent [60–62]. Complexing agent is added to the metal contaminated water to form high molecular-weight polymer–metal complexes which could be rejected by UF or even MF membranes. This strategy leads to better UF or MF heavy metal removal efficiency [63, 64]. However, the membranes will succumb to serious fouling problem due to the accumulation of polymer–metal complexes on the feed side of the membrane system. In order to overcome the limitations of the conventional polymeric membrane while retaining their advantages, recent efforts have been put on the development of nanocomposite membrane that incorporated with nanomaterials in the polymeric matrix.

12.4 Nanocomposite Mixed Matrix Membranes

Nanocomposite membranes can be fashioned from various materials into different morphologies. The simplest yet versatile approach in endowing heavy metal removal capability to polymeric membranes is by embedding adsorptive nanoparticles into the matrix of the membrane. Such membrane also called as the mixed matrix membrane (MMM). Nanocomposite mixed matrix membrane is prepared by blending a predetermined amount of nano-adsorbent with the polymer dope prior to membrane fabrication. Polysulfone have been an attractive choice of polymer for fabrication of micro-, ultra- and nanofiltration membranes due to its excellent processability and fair mechanical, thermal and chemical stability. It forms homogeneous blends with other polymers and various nanoparticles in the making of nanocomposite membranes. To obtain high flux membranes Loeb and Sourirajan phase inversion method [65] have been widely employed. Upon phase inversion, the polymer solidifies and the particles are immobilized throughout the membrane matrix.

In recent years, the preparation of nanocomposite heavy metal removal membranes incorporated with numerous types of nano-adsorbents has been increasingly reported. The combinations of materials with different properties yield a new functional material which could perform a task, or withstanding conditions which either of the material could not, individually. Nanocomposite membranes embedded with heavy metal adsorbent offers two-fold benefits: simultaneous water separation by size exclusion and selective adsorption of heavy metal ions onto the adsorbent. Throughout literature, the incorporation of nanoparticles into the matrix of polymeric membranes has often shown dramatic increase in water permeation fluxes due to increasing membrane porosity and hydrophilicity. In view of heavy metal removal capability, typical polymeric membranes have little to no ability to reject dissolved heavy metal ions since the pores of the membranes are relatively too big to sieve ionic molecules. With the presence of heavy metal ions selective sorbent in the membrane matrix, metal ions can be stripped out from the aqueous solution during filtration process simultaneously. The following sections discuss the fabrication and performance evaluations of nanocomposite membranes incorporated with emerging nanofillers, namely CNTs, graphene and derivatives, metal oxides, clays and zeolites.

12.4.1 Carbon Nanotubes Nanocomposite Membrane

Since its discovery, considerable attention has been paid to CNTs in wide range of applications due to its fascinating mechanical, physical, chemical and electronic properties. CNT is one-dimensional material made of rolled graphite sheets forming tubular structure with nanosize diameters and its length can be stretched up to a few microns. CNT can have single or multiple walls known as SWCNTs and MWCNTs, respectively. The tubular structure of CNTs is of great interest to be exploited in molecular transport and adsorption application [66]. The incorporation of various CNTs into the matrix of polymeric membranes have shown great advantages especially on improving the permeating flux of gases and liquids as well as rejection property even with very low loadings (less than 1 wt.% addition) [67–69]. Due to its strong modulus, the resultant nanocomposites have often showed improved mechanical properties as well. Despite its enormous potential in the development of high performance nanocomposite membranes, problems such as poor miscibility and phase separation between the CNTs and polymer matrix is kind of a setback [70]. To ensure optimum dispersion and good particle-polymer interaction, chemical modification on CNTs is recognized as one of the most effective and practical approaches. Throughout the literature, CNTs have been surface modified with various functional groups to boost the adsorption capacity of the nanoparticle and the adsorption kinetics [22, 24, 67, 68, 71].

Shah and Murthy [72] fabricated nanocomposite membranes with three different functionalized MWCNTs, i.e., oxidized (–COOH and –OH), amide (–CONH and $-CH_2NH_2$), and azide ($-CON_3$). All MWCNT-filled membranes demonstrated better hydrophilicity, thermal stability, and heavy metal rejections. Percentage rejection was found to be a function of the MWCNT loadings. The MWCNT nanocomposites with oxidized and amide functionalization exhibited greater than 90% removal of chromium, copper and lead, and greater than 80% removal of cadmium and arsenic, while the plain membrane only rejected around 10% of those metal ions. The introduction of the functional groups provides opportunity for complex formation with the heavy metal ions to facilitate effective removal.

In other works, Salehi et al. [73] functionalized MWCNTs with amine group via acid reaction which then incorporated into thin chitosan/poly(vinyl) alcohol adsorptive membrane for copper ion removal. Adsorption capacity for copper was found to be doubled for the membranes filled with amine-modified MWCNTs. They speculated that the enhancement of copper adsorption capacity was attributed by the increase in the number of reactive adsorption sites due to amine functionalization and the presence of free-volume pathways as a result of MWCNTs insertion in the membrane matrix. The nanochannel pathways enhanced the accessibility of the copper ions to the internal adsorption sites. They also hypothesized that magnetic anisotropy induced by the presence of benzene ring from the surface modification of the MWCNT enhanced the electrostatic attraction of copper ions, hence increase the adsorption capacity (Fig. 12.4).

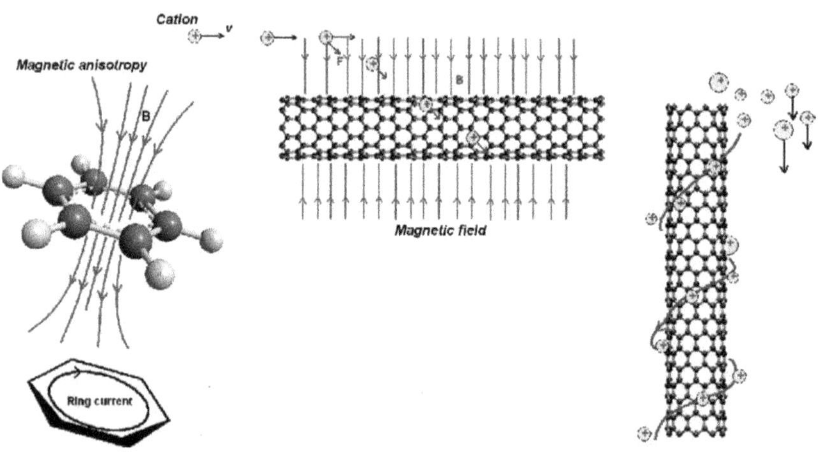

Figure 12.4 Schematic of the magnetic field around the MWCNTs [73].

12.4.2 Graphene Nanocomposite Membrane

Graphene oxide (GO), one of the graphene derivatives, is an oxidized graphene with abundant oxygen containing functional groups such as hydroxy, epoxy, carbonyl, and carboxylic acid to name a few [74, 75]. The available functional groups on GO surface can efficiently form metal ion complexes thus make it a promising next generation adsorbent for heavy metal pollution remediation. The adsorption capacity of modified GO was reported to be exceeding 500 mg/g for Pb^{2+} even with the presence of competing cations [74]. In most studies, GO has been synthesized in the laboratory involving chemical derivation from natural graphite using Hummers' method [23, 74, 75].

In a recent study, asymmetric PSF/GO nanocomposite mixed matrix membranes with different weight percentages were prepared by wet phase inversion technique [75]. The GO was measured to be negatively charged (−59 mV) thus rendering a net negative charge on the nanocomposite membranes (−12 to −29 mV). Due to the net negative charge, the GO nanocomposite membranes in turn showed high adsorption capacity for Pb^{2+} (79 mg/g), Cu^{2+} (75 mg/g), Cd^{2+} (68 mg/g) and Cr^{6+} (154 mg/g) at wide range of pH even with very small addition of GO (0.1–0.5%). Rejections ratio were found to be similar for all metal ions (90–96%) though significant difference in Cr^{6+} adsorption capacity was observed. Moderate Cr^{6+} ion rejection can be rationalized due to the fact that Cr ion forms hydroxide complex in basic solution hence it lowers the electrostatics interaction between complex and membrane surface due to the arising of similar charges. The membranes were easily regenerated via in-situ chemical regeneration by passing slightly acidic solution of pH 5.5 for one hour with flux recovery ratio over 0.9.

Chung's research group has successfully prepared novel nanofiltration membrane with ultrathin graphene oxide framework heavy metal selective layer via a layer-by-layer (LbL) approach [50]. The GO layer was deposited on the outer surface of hollow fiber membrane made of high performance polyamideimide polymer, Torlon®. The preparation of nanocomposite membranes involved cross-linking reaction between the hollow fiber substrate with polyethylenimine (HPEI), followed by repeated deposition cycle of GO and ethylenediamine (EDA) and amine-enrichment modification by HPEI (Fig. 5.5). The novel nanocomposite membrane showed remarkable rejection towards Pb^{2+}, Ni^{2+}, and Zn^{2+} (greater than 95% rejection) with a superior water permeability and excellent long-term performance stability during a 150-h NF test. In other similar research, an external electric field (EF) was applied to assist the preparation of graphene oxide (GO) and polyethyleneimine (PEI) LbL films on hydrolyzed polyacrylonitrile (H-PAN) nanofiltration substrate [76]. However, metal ions rejection studies were not performed using their newly synthesized membranes. Nevertheless, it shows the versatility of the LbL deposition approach and its potential to be extended to the fabrication of other nanomaterials in membrane fields.

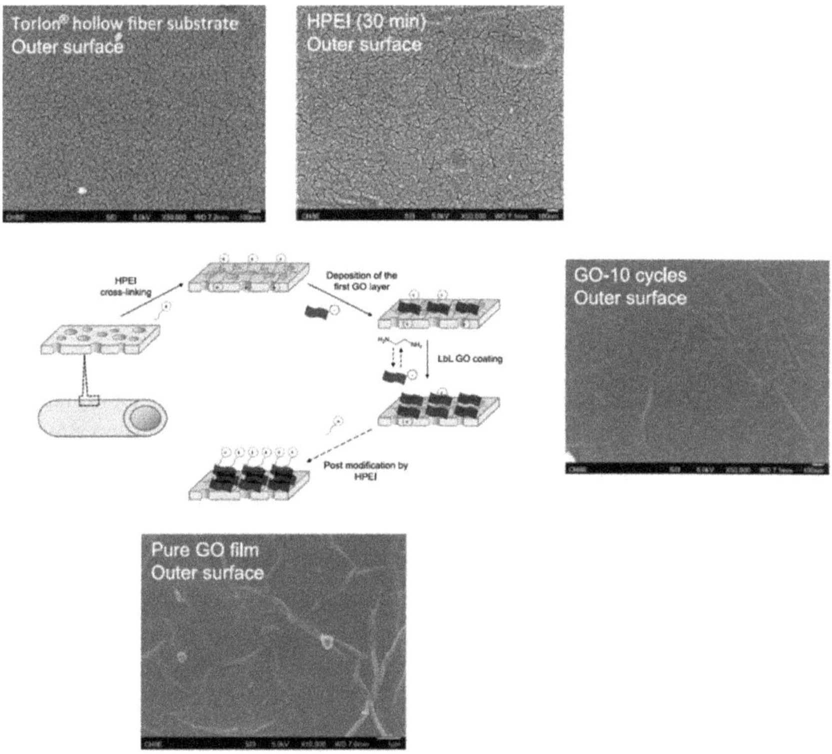

Figure 12.5 The construction of GO framework membrane via LbL approach with their respective surface morphology [50].

12.4.3 Metal Oxide Nanocomposite Membrane

Besides CNTs and graphene oxide, metal oxides nanoparticles have also attracted considerable attention in water treatment processes as attractive adsorbents due to their large specific surface area with surplus amount of active adsorption sites. Metal oxides such as ferric and manganese oxide nanoparticles have been specially synthesized in our laboratory for heavy metal adsorption purpose [77–79]. Abdullah et al. [77] synthesized hydrous ferric oxide (HFO) with particle size as low as 20 nm (Fig. 12.6). The HFO was incorporated into UF PSF membranes with several weight percentages for Pb^{2+} removal. Unlike usual MMMs, the membranes prepared were heavily loaded with the oxide nanoparticle, up to 1.5:1 (HFO:PSF) ratio. Due to the nature of nanoparticles, agglomeration was inevitable at such high loading. Though agglomeration was evident, the membranes exhibited drastic improvement in pure water flux permeability exceeding 900 $L/m^2 \cdot h \cdot bar$ which attributed to the improved membrane hydrophilicity, surface roughness and membrane porosity upon HFO incorporation. The incorporation of HFO has also resulted in excellent Pb^{2+} removal capability when 95% removal was achieved around neutral pH with Pb^{2+} adsorption capacity of 13.2 mg/g.

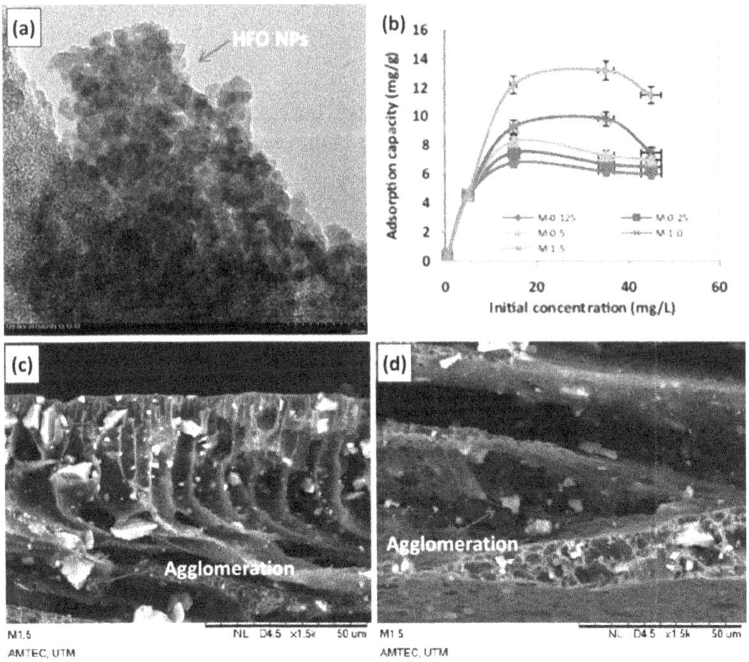

Figure 12.6 (a) TEM image of self-synthesized HFO NPs; (b) the effect of Pb^{2+} concentration on PSF/HFOs adsorption capacity; (c) HFO aggregates in the membranes microchannels; and (d) porous layer of membrane M-1.5 [77].

Jamshidi Gohari et al. [78] investigated the filtration and sorption of Pb^{2+} using hydrous manganese oxide (HMO) filled nanocomposite membranes. The addition of HMO resulted in smaller membrane pore size. Anyhow, overall porosity, surface roughness and hydrophilicity were markedly increased, thus high water flux was attainable. HMO loaded membrane demonstrated Pb^{2+} uptake as high as 204.1 mg/g around neutral pH. Subsequently, they synthesized Fe-Mn binary oxide (FMBO) and incorporated into UF PES membrane for adsorptive removal of As^{3+} from aqueous solution [79]. Similar to the previous work, FMBO affected the membrane morphology positively. Effective removal of arsenate in a broad pH range (from 2.0 to 8.0) was confirmed through batch experiment. The newly developed nanocomposite membranes can be effectively used for industrial wastewater as well as ground water where no pH adjustment is needed.

Magnetic magnetite (Fe_3O_4) is another potential metal oxide adsorbent for remediating heavy metal polluted water. It is non-toxic, hydrophilic and chemically stable. It exhibits electric and magnetic properties due to the electronic transfer in the octahedral sites between Fe^{2+} and Fe^{3+} [31]. High surface area, high surface energy, and magnetic property play very important role in metal ion adsorption process involving Fe_3O_4 [32]. The adsorption process is achieved mainly by ion exchange and complexation. Moradihamedani et al. [48] investigated the performance of PSF/talc and PSF/Fe_3O_4-talc nanocomposite membranes for Pb^{2+} and Ni^{2+} removal. The modification of talc powder with Fe_3O_4 nanoparticle (Fig. 12.7) led to the increase in the specific surface area of the adsorbent (from 6.675 to 37.079 m^2/g) thus more heavy metal adsorption sites are available. Therefore, PSF/Fe_3O_4-talc nanocomposite membranes exhibited more efficient rejections compared to PSf/talc membranes with 99.4% and 96.2% rejections for Pb^{2+} and Ni^{2+}, respectively. In other study, Fe_3O_4-MnO_2 nanoplates were synthesized by Zhao et al. [34] and astounding adsorption capacities for Pb^{2+}, Cu^{2+}, Cd^{2+}, Zn^{2+} and Ni^{2+} on Fe_3O_4-MnO_2 were measured to be 208.17, 111.90, 169.90, 100.24 and 55.63 mg/g, respectively.

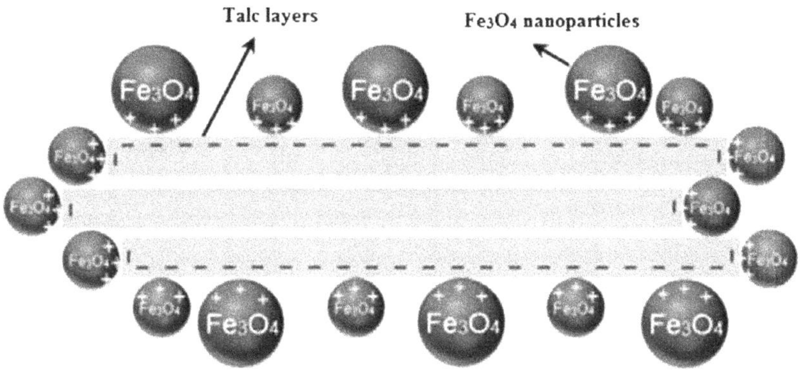

Figure 12.7 Schematic representation of Fe_3O_4-talc nanocomposite [48].

Ghaemi [80] explored the possibility to enhance the copper removal efficiency of polyethersulfone membrane by incorporating Al_2O_3 nanoparticle. Al_2O_3 has demonstrated good adsorption capacity towards metal ions though further studies are deemed necessary [81–83]. Due to the hydrophilicity rendered by Al_2O_3, PES/Al_2O_3 membranes showed better hydrophilicity, and higher water fluxes (three times higher than the plain PES membrane) even with very small addition of the nanoparticle. In the subsequent study [84], Al_2O_3 were surface modified with polypyrrole (PPy), a conducting polymer with conjugated backbone. Similar trends on pure water flux and water contact angle were observed after the PPy modification. In view of heavy metal removal efficiency, PES membrane only rejects 25% of copper from 20 mg/L of $Cu(NO_3)_2$ aqueous solution at pH 5.0, while PES/Al_2O_3 and PES/PPy-Al_2O_3 rejects up to 60% and 81%, respectively. The improvement seen for PES/PPy-Al_2O_3 was attributed to the auxiliary active sites from PPy which increase the chemical adsorption of heavy metals via complexation on the nanoparticle surface in the membrane matrix.

12.4.4 Clay Nanocomposite Membrane

Clay mineral is a form of natural layered silicate. There are five classes of clay minerals categorized based on their Al-Si atomic arrangements. They are kaolin (kaolinite, halloysite, dickite, and nacrite), smectite (montmorillonite, bentonite, nontronite, beidellite, and vermiculite), illite (clay-micas and glauconite), chlorite, and sepiolite family (sepiolite, attapulgite, and palygorskite) [85]. Among these clay families, montmorillonite has been extensively used for heavy metals adsorption due to its high specific surface area, intrinsic ion exchange behavior and complex formation capability [20, 56, 86–88]. In fact, its other geometrical and structural advantages also made it attractive in the development of polymer-clay nanocomposites in wide range applications (automobile, aerospace, packaging, insulators, paints, etc.) [89].

Luo et al. [56] prepared poly(m-phenylene isophthalamide)/organically modified montmorillonite (PMIA/OMMT) nanocomposite NF membranes with various OMMT contents (0.5%–5%) and 20 wt.% PMIA in the casting solution for the removal of perfluorooctane sulfonate (PFOS) in the presence of $PbCl_2$. Malandrino et al. [90] reported that, in comparison to Na-montmorillonite, vermiculite demonstrated higher adsorption capacity towards various metal ions which signify its great prospect to be employed as mixed matrix nanocomposite membrane fillers. Fe_3O_4/bentonite nanocomposite was prepared through solvothermal synthesis [91]. The nanocomposite adsorbent possess higher Pb^{2+}, Cd^{2+} and Cu^{2+} adsorption capacity due to higher surface area, pore volume and diameter than both adsorbent combined.

Besides montmorillonite, halloysite is another type of aluminosilicate clay mineral which is showing great potential in the membrane water treatment application. Halloysite from kaolin family exists in hollow, tubular form, with an internal diameter of 20–30 nm and an external diameter from 30 to 70 nm [92]. The tubular structure provides microchannel for water permeation while its surface is covered with hydroxyl groups, rendering high hydrophilicity [93]. Recent study by Zeng et al. [92] showed that by incorporating halloysite nanotubes (HNTs) into polyvinylidene fluoride (PVDF) NF membrane, the pure water flux was greatly enhanced due to increasing hydrophilicity as confirmed by the reduction of water contact angle from 83 to 58° with increasing amount of HNT modified with silane agent (3-aminopropyltriethoxysilane, APTES). The modification was performed to decorate the surface of HNT with amino groups to further increase the hydrophilicity of the clay as well as providing more adsorption sites for heavy metal chelation. The modification, as confirmed by many studies help to improve the dispersibility of nanomaterial in the polymer matrix and the compatibility between them [21, 60, 94]. HNT/PVDF nanocomposite membranes demonstrated higher rejection ratio and adsorption capacity towards Cu^{2+}, Cd^{2+} and Cr^{6+} with increasing HNTs loading (Fig. 12.8).

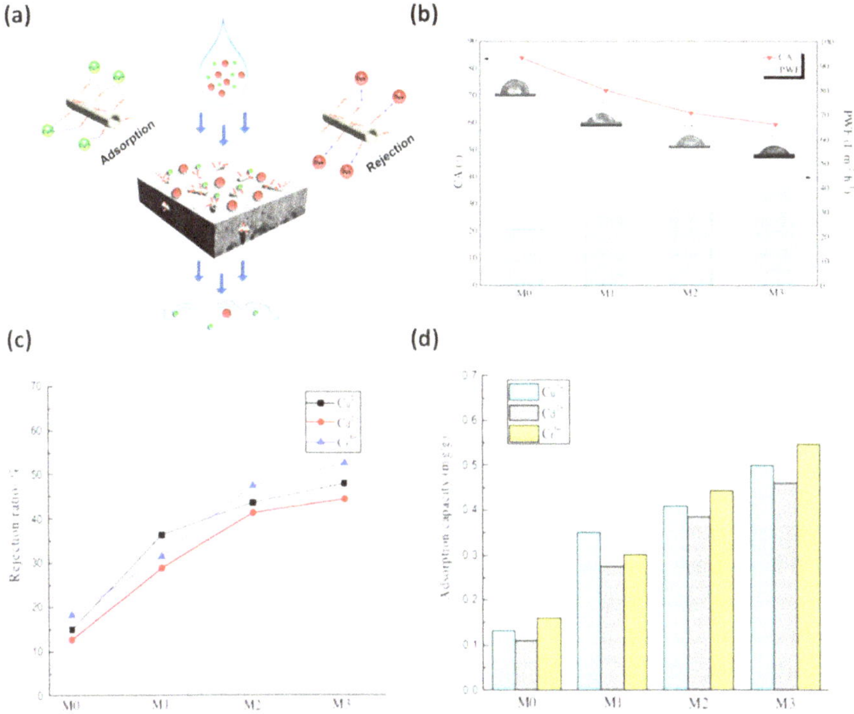

Figure 12.8 (a) Mechanism of dye and Cu^{2+} ions removal by A-HNT/PVDF membrane, (b) CA and PWF data of different membranes, (c) rejection ratio and (d) adsorption capacity of HNT/PDVF membranes for different heavy metals [92].

Hebbar et al. [93] also reported promising performance of HNT-filled nanocomposite membranes for Cd^{2+} and Pb^{2+} removal. Besides improving the filtration capability of the filled-membranes, the incorporation of HNT into the membrane matrix also enhanced the antifouling and biofouling properties as well as inhibiting microbial growth on the membrane surface.

12.4.5 Zeolite Nanocomposite Membrane

Zeolites are crystalline natural mineral, hydrated aluminosilicate with opened three-dimensional framework structures, regular intracrystalline cavities and channels of molecular dimension. Its porous structure makes it a fine choice of material to be used in the field of gas separation and storage. Zeolites are also considered as catalysts, ion-exchangers and absorbents due to their chemical structure [95]. In addition, its hydrophilic nature has also drawn significant interests to be exploited in the development of water filtration membranes [96]. UF membrane made of poly(phthalazinone ether sulfone ketone) (PPESK) was incorporated with NaA synthetic zeolite demonstrated improved hydrophilicity, water permeability, and antifouling property than some commercial membranes [95]. Similar positive effects were reported in the literature by other researchers using variety of zeolites incorporated in different polymer matrices [97, 98].

The net negative charge of zeolite renders its sorption property towards metal ions. In one of Nibou et al. [99] studies, NaA and NaX zeolites were used to treat an industrial effluent containing 80–100 mg/L of Zn^{2+} showed 70–80% percentage removal. Promising results were also reported for Fe^{3+} removal from aqueous solutions [100]. Recently, Yurekli [101] has incorporated NaX type zeolite to the matrix of asymmetrical polysulfone membrane for the removal of Pb^{2+} and Ni^{2+} ions from aqueous solutions (Fig. 12.9). Filtration and adsorption studies were performed at pH 6 revealed that 91% of lead was recovered after 60 minutes of operation when the initial concentration of metal ions was 100 mg/L. On the other hand, only 31% of nickel was recovered. Although the rejection seemed small for the nickel it was markedly improved over the unfilled membrane which showed no ability to reject nickel at all. Besides the adsorptive capability contributed by the addition of zeolite, the overall microstructural morphology of the membrane was also affected. With increasing amount of zeolite loaded into the casting solution, membrane with larger pore size was produced. This has in turn led to higher hydraulic water permeability. Water permeability was doubled after the addition of NaX zeolite (from 23 to 45 $L/m^2 \cdot h$).

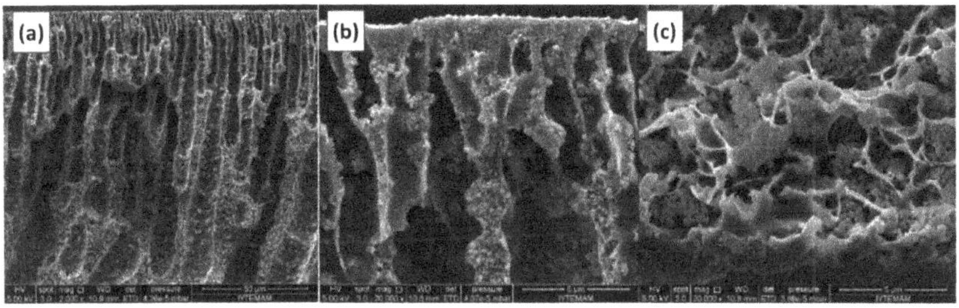

Figure 12.9 Polysulfone membrane incorporated with 10 wt% NaX zeolite. (a) Shows the overall morphology of the asymmetric membrane with evident distribution of zeolite through the whole cross section; (b) and (c) show the closed up views of zeolite embedded in the membrane matrix [101].

12.5 Electrospun Nanocomposite Membranes

Electrospinning is one of the techniques in which nanofibrous polymeric membranes are fabricated (Fig. 12.10). These membranes, often they are called nanofiber or electrospun membrane. They differ significantly from phase inversion membranes in terms of morphology. Among the attractive features of electrospun membranes are tuneable pore size, very high porosity and surface area per unit volume [102]. Such properties promise low pressure drop during operation thus better efficiency. Electrospun membrane has been found useful in widespread applications including environmental remediation, industrial processes, energy, electronics, and clinical purposes [103–105].

In the age of nanotechnology, advanced materials consist of the combinations of nanoparticles and nanofibers give rise to a new class high-performance nanocomposite fiber for even wider applications. Nanoparticles are employed to improve the surface chemistries of the nanofibers or to add special features which the nanofibers are initially do not possess, or to improve the mechanical and thermal properties. For instance, Bui and McCutcheon [106] synthesized PAN nanofiber as the substrate upon interfacial polymerization of polyamide layer. Mesoporous silica nanoparticle was blended with PAN polymeric solution prior to electrospinning process. The resultant nanocomposite fiber showed clusters morphology and become more evident with higher silica loadings (Fig. 12.11). Tensile strength and modulus of the resultant membranes increased two- to threefold. The water flux was also improved considerably. PVA/SiO_2 electrospun nanofiber membrane with surface areas up to 500 m^2/g was prepared by Wu and co-workers [107]. The mesoporous nanofiber was functionalized with abundant mercapto groups for Cu^{2+} ion sorption. High Cu^{2+} adsorption (approaching 500 mg/g) was recorded and it was ascribed to the ionic interaction between the metal ions and the fiber as well as chelation reaction due to the functionalization.

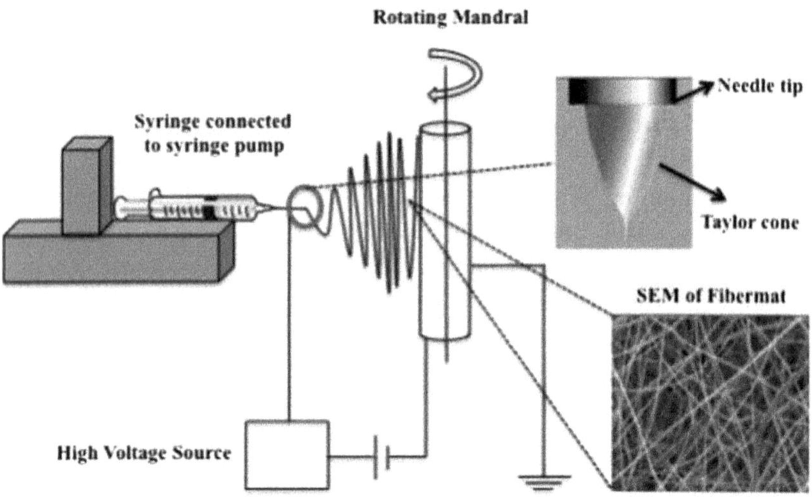

Figure 12.10 Schematic of Electrospinning apparatus set-up [103].

Moradi et al. [59] prepared organic-inorganic nanofiber membranes by electrospinning method. Titanium tetra isopropoxide (TTIP or $Ti(OiPr)_4$) was added to PVDF solution prior to electrospinning process. Hydroxylation and deprotonation reactions were performed on the nanofiber. The heavy metal removal performance of PVDF/TiO_2 nanofiber was evaluated through vacuum enhanced membrane distillation (VMD) process treating aqueous solutions containing various heavy metals (Co^{2+}, Zn^{2+}, Cu^{2+}, Ni^{2+}, Cd^{2+}, Pb^{2+}). The VMD process using PVDF/TiO_2 hybrid nanofiber led to total removal of trace heavy metal contaminants and producing ultra-pure water in one step.

In other studies, Hallaji et al. [108] prepared a novel electrospun PVA/ZnO nanofiber adsorbent for U^{6+}, Cu^{2+} and Ni^{2+}, while Rad et al. [109] incorporated NaX nanozeolite into PVA nanofiber prepared for Ni^{2+} and Cd^{2+} adsorption.

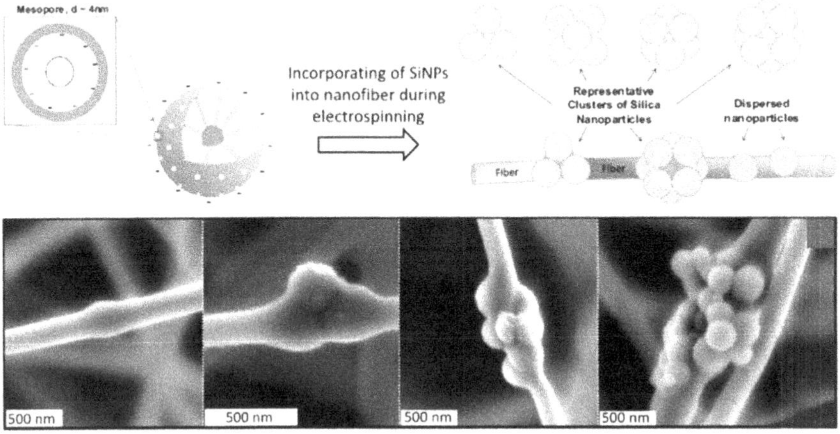

Figure 12.11 Schematic representation of silica nanoparticles impregnation on electrospun nanofiber membrane and the SEM photographs showing SiNP clusters [106].

Another interesting approach in the preparation of electrospun nanocomposite membrane was demonstrated by Sun et al. [110]. Electrospun PAN/$NaAlO_2$ composite was chemically modified (amination reaction), followed by hydrothermal reaction to form aminated PAN/γ-AlOOH nanofiber with hierarchical structure (Fig. 12.12). The growth of lamellae structure on the nanofiber resulted in significantly larger specific surface area. Adsorption studies on hierarchical aminated PAN/γ-AlOOH nanofiber revealed dramatic improvement over PAN/$NaAlO_2$. The maximum adsorption capacities for Pb(II), Cu(II) and Cd(II) were 180.83 mg/g, 48.68 mg/g and 114.94 mg/g, respectively. Superior adsorption capacity was attributed to the synergy of amine functional group and very large surface area for adsorption on the surface of flower-like structure γ-AlOOH crystals. Similar methodology was employed in another study for the preparation of hierarchical SiO_2@γ-AlOOH nanofiber membrane [111].

Figure 12.12 The morphology of (a) electrospun PAN/$NaAlO_2$ composite nanofibers, (b) aminated PAN/$NaAlO_2$ nanofibers, (c, d) the hierarchical aminated PAN/γ-AlOOH electrospun composite nanofiber membranes [110].

12.6 Thin Film Nanocomposite Membranes

Most commercial membranes are thin-film composites (TFC) e.g., NF70 and NF270 manufactured by Dow Filmtec, AFC80 by Paterson Candy International Ltd. TFC is composed of nanometer thick charged skin layer casted on thicker porous substrate made of different material. The thin skin layer is widely fabricated via interfacial polymerization (IP) process. The heavy metal removal performances of commercial TFC membranes have been proven attractive [55, 57, 58]. Plethora of innovative approaches on developing novel thin film nanocomposite (TFN) membranes for desalination and water treatment has been published in recent years [112–116]. Figure 12.13 illustrates the TFN membrane formation through interfacial polymerization with the incorporation of nanofiller at the aqueous or organic phase.

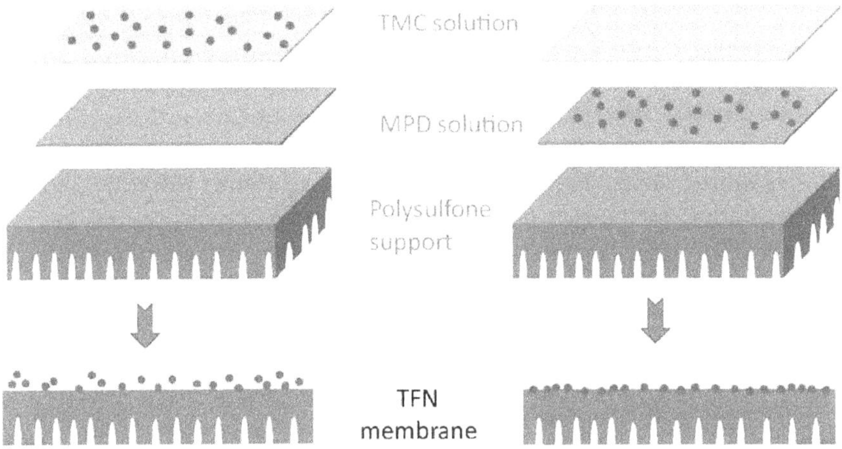

Figure 12.13 The formation of thin film nanocomposite membrane via interfacial polymerization technique [117].

Although limited attention has been placed on metal ions removal TFN membrane, there are a few studies done recently. Bao et al. [60] prepared TFN membrane via IP process of MPD-TMC on PVDF substrate. Amine functionalized mesoporous silica MCM-41 was grafted in the thin skin layer of the membrane surface. The addition of NH_2-MCM-41 to the polyamide layer increased the hydrophilicity of the membrane profoundly. The contact angle measured for unfilled membrane was 80°, while the nanocomposite was only 13°. The nanocomposite was tested for chromium and copper removal under continuous filtration process. The membrane exhibited fast adsorption (reaching equilibrium within 5 minutes of operation) with adsorption capacity of 2.8 mg/g for Cr^{6+} and 3.7 mg/g for Cu^{2+}. While, Zhang et al. [118] prepared a TFN with chitosan rejecting layer over PES porous support for Pb^{2+} and Ni^{2+} removal. Systematic investigation on the effects of chitosan concentration and drying duration on the membrane morphology and performance were conducted. Rejection studies at pH 5-6 demonstrated that high rejections towards Pb^{2+} and Ni^{2+} were achieved with high permeate flux. Chitosan was cross-linked with TMC, which resulted in the decrease in chitosan polymer crystallinity and the contraction of crystal size, hence increasing its selectivity towards metal ions.

Zhang et al. [119] fabricated metal ions selective NF membrane with nanometric GO framework layer of 70 nm thick (Fig. 12.14). GO was deposited on a commercial polycarbonate membrane (Whatman Cyclopore) via cross-linking process with ethylenediamine (EDA) followed by amine-enrichment post treatment with hyperbranched polyethylenimine (HPEI) and two other amine-rich modifiers. The post treatment ensured the integrity of GO framework on the membrane surface therefore eliminated the possibility of nanomaterial leaching into the water. More than 90% rejection was achieved by HPEI-modified GO membrane for metal ions in the order of $Cd^{2+} < Pb^{2+} < Ni^{2+} < Zn^{2+}$. Due to the highly positive charged membrane, efficient rejections can be achieved at neutral pH.

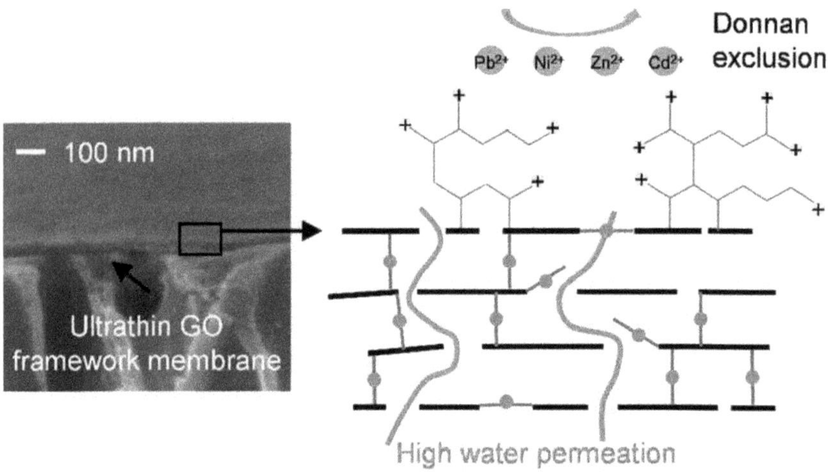

Figure 12.14 Morphology of GO framework layer and the proposed GO/EDA-HPEI 60K frameworks [119].

12.7 Future Direction and Concluding Remarks

Evidently, tremendous efforts have been made towards the progress in the development of nanocomposite membranes for water treatment, particularly heavy metal removal, in the past decade. The incorporation of nanomaterials has favorably tuned and optimized the membrane physicochemical properties of membranes in term of surface hydrophilicity, structural and mechanical properties. Additionally, the unique features of the nanomaterials also provide possibility to hierarchically design and induce desired functionality to further improve the properties of the resultant nanocomposite membrane for heavy metal removal. Table 12.3 lists the performance of recently developed nanocomposite membranes for heavy metal removal via various membrane fabrication approaches.

Despite the promising enhancement shown by this innovation, several key challenges still need to be tackled for practical applications, especially in large industrial scale. At the fundamental stage, more systematic studies and characterizations are needed to develop further understanding on the effects of nanomaterials on the membrane structure at the molecular level. This will allow the exceptional properties of the nanomaterials to be fully exploited and optimized. The predictions based on computational studies have undoubtedly shed some lights in this aspect; however, more experimental work should be conducted to provide verifications. Also, more breakthroughs and efforts are required to overcome the inherent limitations with the current synthesis/assembly approaches which are mainly related to process complexity, non-uniformity and agglomeration.

Another critical issue with the application of nanomaterials for wastewater treatment is the unforeseen environmental impacts. Currently, no clear regulations have been established for the monitoring of possible secondary water body pollution that caused by the leaching of nanomaterials. In fact, from the policy maker point of view, the nanomaterial leakage and its environmental toxicity need to be systematically evaluated. The possible negative effects rooted from direct release of nanomaterials into the water body may provoke public perceptions and present as a major hiccup to the deployment of this innovation for practical applications. To deal with this issue, it is important to improve the compatibility of the nanomaterials within the polymer matrix as it dictates the stability of the nanomaterials within the host polymer. Additionally, it is also important to minimize the loading concentration and improve the durability of the nanocomposite membranes. The long-term stability of the nanocomposite membranes should also be evaluated to ensure safe and reliable application.

Table 12.3 Performance summary of recently developed nanocomposite membranes for heavy metal removal

Nanocomposite membrane (configuration)	Metal ions	Metal concentration (mg/L)	Maximum adsorption capacity (g/mg)	Rejection (%)	pH	PWF ($L/m^2 \cdot h$)	Ref.
PSF/FAU NaX zeolite (MMM)	Pb(II)	100	—	91	6	45	[101]
		200		66			
		300		47			
		500		39			
	Ni(II)	50	—	42			
		100		31			
		200		13			
PSF/Fe_3O_2-Talc (MMM)	Pb(II)	50	—	99.4	5	30	[48]
		70		95			
		90		89			
	Ni(II)	50	—	96.2			
		70		91			
		90		87			
PSF/GO (MMM)	Pb(II)	50	79	95.5	6.7	30	[75]
	Cu(II)		75	91.4	9.5		
	Cd(II)		68	93.5	6.4		
	Cr(VI)		154	91.6	3.5		
PSF/HFO (MMM)	Pb(II)	35	13.2	95	7	942	[77]

Nanocomposite membrane (configuration)	Metal ions	Metal concentration (mg/L)	Maximum adsorption capacity (g/mg)	Rejection (%)	pH	PWF ($L/m^2 \cdot h$)	Ref.
PES/HMO (MMM)	Pb(II)	200	204.1	90	7	573.2	[79]
PES/FBMO (MMM)	As(III)	10	73.5	96	4	94.6	[78]
PES/Al_2O_3 (MMM)	Cu(II)	20	—	50	5	25.3	[80]
PVDF/APTES-HNT (MMM)	Cu(II) Cd(II) Cr(VI)	5 5 5	0.499	47.9 44.2 52.3	5.5	42.2	[92]
PEI-PDA/HNT (MMM)	Pb(II) Cd(II)	1000 1000	—	34 27	6	199.5	[93]
GO-EDA-HPEI60K-TORLON (LbL)	Pb(II) Ni(II) Zn(II)	1000	—	95.88 99.74 98.07		4.7	[50]
PEI/PDA-MWCNTs/TMC (TFN)	Cu(II) Zn(II)	1000	—	90.5 93.0	6	83.5	[120]
PAN/γ–AlOOH (ENM)	Pb(II) Cu(II) Cd(II)	200	180.83 48.68 114.94	—	—	—	[110]
CS-PVA/NH_2-MWCNT (MMM)	Cu(II)	30	18	70	5.5	—	[73]

Table 12.3 (*Continued*)

Nanocomposite membrane (configuration)	Metal ions	Metal concentration (mg/L)	Maximum adsorption capacity (g/mg)	Rejection (%)	pH	PWF ($L/m^2 \cdot h$)	Ref.
PVDF/NH_2-MCM-41 (TFNM)	Cr(VI)	0.5	2.8	—	3.5–6.5	58.4	[60]
	Cu(II)	1.0	3.7				
GO/EDA-HPEI60K (TFN)	Pb(II)	1000		95.7		5.01	[119]
	Ni(II)			96.0			
	Zn(II)			97.4			
	Cd(II)			90.5			
PES/ PPy-Al_2O_3 (MMM)	Cu(II)	20	—	81	5	20.6	[84]
PSF/Amide-MWCNT (MMM)	Cr(VI)	1000	—	94.2	2.6	—	[72]
	Cu(II)			93.1			
	Pb(II)			90.1			
	Cd(II)			78.2			
	As(III)			79.4			
PVA/SiO_2 (ENM)	Cu(II)		489.12	90	5	—	[107]
PVA/ZnO (ENM)	U(VI)	90	370.86		5	—	[108]
	Cu(II)		162.48				
	Ni(II)		94.43				
PVA/NaX (ENM)	Ni(II)	50	342.8		5	—	[109]
	Cd(II)		838.7				

Currently, the real application of nanocomposite membranes for heavy metal removal is still at its infancy stage. Many encouraging laboratory-based studies have been performed, but the large-scale production and applications of the nanomaterials and membranes for industrial scale has scarcely been reported. In term of commercialization, as the exploding benefits and potential of nanocomposite membranes can be retrofitted into the existing technologies without significant changes, it is believed that this state-of-the-art innovation will take relatively short time to headway from laboratory exploration to achieving operational maturity. It should be noted that, the complexity and economic measures related to the technological scale up should be taken into consideration from as early as in the research process until the product maturity period. The current approaches adopted for the synthesis and fabrication of nanomaterials or their devices may not be price competitive in comparison with existing materials, but looking at the broader sense, keeping the overall wastewater treatment investment and processing costs low enough through the improvement in energy and material efficiency could also prove a commercial success.

In conclusion, nanocomposite membranes have witnessed the improvements made in the field of wastewater treatment. In the context of this chapter, it is anticipated that with all the above-mentioned challenges and issues being addressed progressively through the advancement made in recent scientific and technological advances, nano-enabled membranes can undoubtedly provide more efficient, cost-effective and sustainable solutions for heavy metal removal.

Acknowledgments

The authors would like to acknowledge research management and financial support from Universiti Teknologi Malaysia and the Ministry of Higher Education under research grants 02G07.

References

1. Water use in industry—Statistics Explained, Europa.eu.

2. Shi, H. (2009). Industrial wastewater-types, amounts and effects, point sources of pollution: Local effects and its control, *Encyclopedia of Life Support Systems* (EOLSS), vol. 1.
3. Islam, S., Islam, S., Al-mamun H., Islam, S. A. (2016). Dennis Wayne Eaton, Total and dissolved metals in the industrial wastewater: A case study from Dhaka Metropolitan, Bangladesh, *Environ. Nanotech. Monitoring Manage.*, **5**, pp. 74–80.
4. Fernández-Luqueño, F., López-Valdez, F., Gamero-Melo, P., Luna-Suárez, S., Aguilera-González, E. N., Martínez, A. I., García-Guillermo, M. S., Hernández-Martínez, G., Herrera-Mendoza, R., Álvarez-Garza, M. A., Pérez-Velázquez, I. R. (2013). Heavy metal pollution in drinking water–a global risk for human health: A review, *Afr. J. Environ. Sci. Technol.*, 7(**7**), pp. 567–584.
5. Babel, S., Kurniawan, T. A. (2004). Cr(VI) removal from synthetic wastewater using coconut shell charcoal and commercial activated carbon modified with oxidizing agents and/or chitosan, *Chemosphere.*, **54**, pp. 951–967.
6. Ihsanullah, A. A., Adnan M. A., Tahar, L., Mohammed, J. A., Mustafa, S. N., Majeda, K., Muataz, A. A. (2016). Heavy metal removal from aqueous solution by advanced carbon nanotubes: Critical review of adsorption applications, *Sep. Purif. Technol.*, **157**, pp. 141–161.
7. Sahaa, B., Orvig, C. (2010). Biosorbents for hexavalent chromium elimination from industrial and municipal effluents, *Coord. Chem. Rev.*, **254**, pp. 2959–2972.
8. Hadi, P., Gao, P., Barford, J. P., McKay, G. (2013). Novel application of the nonmetallic fraction of the recycled printed circuit boards as a toxic heavy metal adsorbent, *J. Hazard. Mater.*, **252– 253**, pp. 166–170.
9. LaDou, J. (2006). Printed circuit board industry, *Int. J. Hyg. Environ. Health.*, **209**, pp. 211–219.
10. John, M., Heuss-Aßbichler, S., Ullrich, A., Rettenwander, D. (2016). Purification of heavy metal loaded wastewater from electroplating industry under synthesis of delafossite (ABO2) by "Lt-delafossite process", *Water Res.*, **100**, pp. 98–104.
11. Lou, J.-C., Huang, Y.-J., Han, J.-Y. (2009). Treatment of printed circuit board industrial wastewater by Ferrite process combined with Fenton method, *J. Hazard. Mater.*, **170**, pp. 620–626.
12. Al-Shannag, M., Al-Qodah, Z., Bani-Melhem, K., Qtaishat, M. R., Alkasrawi, M. (2015). Heavy metal ions removal from metal plating wastewater using electrocoagulation: Kinetic study and process performance, *Chem. Eng. J.*, **260**, pp. 749–756.

13. Barakat, M. A. (2011). New trends in removing heavy metals from industrial wastewater, *Arabian J. Chem.*, **4**, pp. 361–377.

14. Salcedo, A. F. M., Ballesteros, F. C., Vilando, A. C., Lu, M.-C. (2016). Nickel recovery from synthetic Watts bath electroplating wastewater by homogeneous fluidized bed granulation process, *Sep. Purif. Technol.*, **169**, pp. 128–136.

15. Akyol, A. (2012). Treatment of paint manufacturing wastewater by electrocoagulation, *Desalination*, **285**, pp. 91–99.

16. Zhao, M., Xu, Y., Zhang, C., Rong, H., Zeng, G. (2016). New trends in removing heavy metals from wastewater, *Appl. Microbiol. Biotechnol.*, DOI 10.1007/s00253-016-7646-x.

17. Reddy, D. H. K., Yun, Y. S. (2016). Spinel ferrite magnetic adsorbents: Alternative future materials for water purification?, *Coord. Chem. Rev.*, **315**, pp. 90–111.

18. Tounsadi, H., Abderrahim, K., Meryem, F., Mohamed, A., Noureddine, B. (2016). Experimental design for the optimization of preparation conditions of highly efficient activated carbon from Glebionis coronaria L. and heavy metals removal ability, *Process Saf. Environ. Prot.*, **102**, pp. 710–723.

19. Karnib, M., Ahmad, K., Hanafy, H., Zakia, O. (2014). Heavy metals removal using activated carbon, silica and silica activated carbon composite, *Energy Procedia.*, **50**, pp. 113–120.

20. Atta, A. M., Al-Lohedan, H. A., Al-Othman, Z. A., Abdel-Khalek, A. A., Tawfeek, A. M. (2015). Characterization of reactive amphiphilic montmorillonite nanogels and its application for removal of toxic cationic dye and heavy metals water pollutants, *J. Ind. Eng. Chem.*, **31**, pp. 374–384.

21. Huang, Q., Liu, M., Deng, F., Wang, K., Huang, H., Xu, D., Zeng, G., Zhang, X., Wei, Y. (2016). Mussel inspired preparation of amine-functionalized Kaolin for effective removal of heavy metal ions, *Mater. Chem. Phys.*, **181**, pp. 116–125.

22. Li, Q., Jingang, Y., Fang, Z., Xinyu, J. (2015). Synthesis and characterization of dithiocarbamate carbon nanotubes for the removal of heavy metal ions from aqueous solutions, *Colloids Surf. A: Physicochem. Eng. Aspects.*, **482**, pp. 306–314.

23. Dong, Z., Feng, Z., Dong, W., Xia, L., Jian, J. (2015). Polydopamine-mediated surface-functionalization of graphene oxide for heavy metal ions removal, *J. Solid State Chem.*, **224**, pp. 88–93.

24. Salam, M. A., Ghalia, A., Samia A. K. (2014). Removal of heavy metal ions from aqueous solution by multi-walled carbon nanotubes

modified with 8-hydroxyquinoline: Kinetic study, *J. Ind. Eng. Chem.*, **20**, pp. 572–580.

25. Lee, C., Mi-Kyung, S., Jae-Chun, R., Chanhyuk, P., Jae-Woo, C., Sang-Hyup, L. (2016). Application of carbon foam for heavy metal removal from industrial plating wastewater and toxicity evaluation of the adsorbent, *Chemosphere.*, **153**, pp. 1–9.
26. Salehi, E., Parisa, D., Arabi, S. A. (2016). A review on chitosan-based adsorptive membranes, *Carbohydr. Polym.*, **152**, pp. 419–432.
27. Zhou, K., Yaochi, L., Zhaoguang, Y., Ting, X., Houzhi, L., Chubin, Z. (2016). Adsorptive removal of heavy metals by a bio-based polymeric material PAO-CI from wastewater, *J. Taiwan Inst. Chem. Eng.*, **61**, pp. 342–350.
28. Vunain, E., Mishra, A. K., Mamba, B. B. (2016). Dendrimers, mesoporous silicas and chitosan-based nanosorbents for the removal of heavy-metal ions: A review, *Int. J. Biol. Macromol.*, **86**, pp. 570–586.
29. Zhang, L., Zeng, Y., Cheng, Z. (2016). Removal of heavy metal ions using chitosan and modified chitosan: A review, *J. Mol. Liq.*, **214**, pp. 175–191.
30. Ahmadi, A., Shahriar, H., Ahmad, R. M., Emaeel, D., Houshang, A. H. (2014). Optimization of heavy metal removal from aqueous solutions by maghemite (γ-Fe_2O_3) nanoparticles using response surface methodology, *J. Geochem. Explor.*, **147**, pp. 151–158.
31. Kumari, M., Charles, U. P. J., Dinesh, M. (2015). Heavy metals [chromium (VI) and lead (II)] removal from water using mesoporous magnetite (Fe_3O_4) nanospheres, *J. Colloid Interface Sci.*, **442**, pp. 120–132.
32. Jing, C., Zhong, Z., Xu, H., Yao, Z., Chen, R. (2013). Fabrication of poly(γ-glutamic acid)-coated Fe_3O_4 magnetic nanoparticles and their application in heavy metal removal, *Chin. J. Chem. Eng.*, **21**(11), pp. 1244–1250.
33. Barakat, M. A., Rajeev, K. (2015). Synthesis and characterization of porous magnetic silica composite for the removal of heavy metals from aqueous solution, *J. Ind. Eng. Chem.*, **23**, pp. 93–99.
34. Zhao, J., Jie, L., Ning, L., Wei, W., Jun, N., Zhiwei, Z., Fuyi, C. (2016). Highly efficient removal of bivalent heavy metals from aqueous systems by magnetic porous Fe_3O_4-MnO_2: Adsorption behavior and process study, *Chem. Eng. J.*, **304**, pp. 737–746.
35. Huang, Z., Qinglin, W., Shouxin, L., Tian, L., Bin, Z. (2013). A novel biodegradable-cyclodextrin-based hydrogel for the removal of heavy metal ions, *Carbohydr. Polym.*, **97**, pp. 496– 501.

36. Cavus, S., Yasar, G., Kaya, Y., Gönder, Z. B., Gürdaga, g., Vergili, I. (2016). Synthesis and characterization of gel beads based on ethyleneglycol dimethacrylate and 2-acrylamido-2-methyl-1-propane sulfonic acid: Removal of Fe(II), Cu(II), Zn(II), and Ni(II) from metal finishing wastewater, *Process Saf. Environ. Prot.*, **103**, pp. 227–236.

37. Nguyen, T. A. H., Ngo, H. H., Guo, W. S., Zhang, J., Liang, S., Yue, Q. Y., Li, Q., Nguyen, T. V. (2013). Applicability of agricultural waste and by-products for adsorptive removal of heavy metals from wastewater, *Bioresour. Technol.*, **148**, pp. 574–585.

38. Hegazi, H. A. (2013). Removal of heavy metals from wastewater using agricultural and industrial wastes as adsorbents, *HBRC J.*, **9**, pp. 276–282.

39. Feizi, M., Mohsen, J. (2015). Removal of heavy metals from aqueous solutions using sunflower, potato, canola and walnut shell residues, *J. Taiwan Inst. Chem. Eng.*, **54**, pp. 125–136.

40. Khandanlou, R., Hamid, R. F. M., Mansor, B. A., Kamyar, S., Mahiran, B., Katayoon, K. (2016). Enhancement of heavy metals sorption via nanocomposites of rice straw and Fe_3O_4 nanoparticles using artificial neural network (ANN), *Ecol. Eng.*, **91**, pp. 249–256.

41. Ali, R. M., Hesham, A. H., Mohamed, M. H., Gihan, F. M. (2016). Potential of using green adsorbent of heavy metal removal from aqueous solutions: Adsorption kinetics, isotherm, thermodynamic, mechanism and economic analysis, *Ecol. Eng.*, **91**, pp. 317–332.

42. Esmaeili, A., Beni, A. A. (2015). Novel membrane reactor design for heavy-metal removal by alginate nanoparticles, *J. Ind. Eng. Chem.*, **26**, pp. 122–128.

43. Tu, Y.-J., You, C.-F., Chang, C.-K., Wang, S.-L., Chan, T.-S. (2013). Adsorption behavior of As(III) onto a copper ferrite generated from printed circuit board industry, *Chem. Eng. J.*, **225**, pp. 433–439.

44. Visa, M. (2016). Synthesis and characterization of new zeolite materials obtained from fly ash for heavy metals removal in advanced wastewater treatment, *Powder Technol.*, **294**, pp. 338–347.

45. Vijwani, H., Nadagouda, M. N., Namboodiri, V., Mukhopadhyay, S. M. (2015). Hierarchical hybrid carbon nano-structures as robust and reusable adsorbents: Kinetic studies with model dye compound, *Chem. Eng. J.*, **268**, pp. 197–207.

46. Zhu, X., Zhu, L., Chen, Y., Tian, S. (2009). Acute toxicities of six manufactured nanomaterial suspensions to Daphnia magna, *J. Nanopart. Res.*, **11**, pp. 67–75.

47. Rivas, B. L., Sánchez, J., Urbano, B. F. (2016). Polymers and nanocomposites: Synthesis and metal ion pollutant uptake, *Polym. Int.*, **65**, pp. 255–267.
48. Moradihamedani, P., Katayoon, K., Abdul Halim, A., Noor Azian, M. (2016). High efficient removal of lead(II) and nickel(II) from aqueous solution by novel polysulfone/Fe_3O_4–talc nanocomposite mixed matrix membrane, *Desalin. Water Treat.*, DOI: 10.1080/19443994.2016.1193449.
49. Bae, J. Y., Lee, H.-J., Choi, W. S. (2016). Cube sugar-like sponge/polymer brush composites for portable and user-friendly heavy metal ion adsorbents, *J. Hazard. Mater.*, **320**, pp. 133–142.
50. Zhang, Y., Sui, Z., Jie, G., Tai-Shung, C. (2016). Layer-by-layer construction of graphene oxide (GO) framework composite membranes for highly efficient heavy metal removal, *J. Membr. Sci.*, **515**, pp. 230–237.
51. Zhao, K. (2014). Fabrication of a poly(piperazine-amide)/polysulfone hollow fiber membrane for effective removal of heavy metals in industrial wastewater, *7th Int. Conf. BioMed. Eng. Informatics* (BMEI 2014).
52. Izadpanah, A. A., Javidnia, A. (2012). The ability of a nanofiltration membrane to remove hardness and ions from diluted seawater, *Water*, **4**, pp. 283–294.
53. Nicolini, J. V., Borges, C. P., Ferraz, H. C. (2016). Selective rejection of ions and correlation with surface properties of nanofiltration membranes, *Sep. Purif. Technol.*, **171**, pp. 238–247.
54. Gherasim, C., Petr, M. (2014). Influence of operating variables on the removal of heavy metal ions from aqueous solutions by nanofiltration, *Desalination,* **343**, pp. 67–74.
55. Al-Rashdi, B. A. M., Johnson, D. J., Hilal, N. (2013). Removal of heavy metal ions by nanofiltration, *Desalination*, **315**, pp. 2–17.
56. Luo, Q., Liu, Y., Liu, G., Zhao, C. (2016). Preparation, characterization and performance of poly(m-phenylene isophthalamide)/organically modified montmorillonite nanocomposite membranes in removal of perfluorooctane sulfonate, *J. Environ. Sci.*, **46**, pp. 126–133.
57. Basaran, G., Kavak, D., Dizge, N., Asci, Y., Solener, M., Ozbey, B. (2015) Comparative study of the removal of nickel(II) and chromium(VI) heavy metals from metal plating wastewater by two nanofiltration membranes, *Desalin. Water Treat.*, DOI: 10.1080/19443994.2015.1127778.

58. Otero, J. A., Mazarrasa, O., Otero-Fernandez, A., Fernandez, M. D., Hernandez, A., Maroto-Valiente, A. (2012). Treatment of wastewater. Removal of heavy metals by nanofiltration. Case study: Use of TFC membranes to separate Cr(VI) in industrial pilot plant, *Procedia Eng.*, **44**, pp. 2020–2022.

59. Moradi, R., Saeed, M. M., Younes, A., Abolfazl, D. (2015). Vacuum enhanced membrane distillation for trace contaminant removal of heavy metals from water by electrospun PVDF/TiO_2 hybrid membranes, *Korean J. Chem. Eng.*, **32**(4), pp. 1–9.

60. Bao, Y., Xiaomin, Y., Wei, D., Xiaoni, X., Zhaoqi, P., Jialu, Z., Laisheng, L. (2015). Application of amine-functionalized MCM-41 modified ultrafiltration membrane to remove chromium (VI) and copper (II), *Chem. Eng. J.*, **281**, pp. 460–467.

61. Huang, Y., Dihua, W., Xiaodong, W., Wei, H., Darren, L., Xianshe, F. (2016). Removal of heavy metals from water using polyvinylamine by polymer-enhanced ultrafiltration and flocculation, *Sep. Purif. Technol.*, **158**, pp. 124–136.

62. Qiu, Y., Lian-Jun, M. (2013). Removal of heavy metal ions from aqueous solution by ultrafiltration assisted with copolymer of maleic acid and acrylic acid, *Desalination*, **329**, pp. 78–85.

63. Kumar, R., Arun, M. I., Ismail, A. F. (2014). Preparation and evaluation of heavy metal rejection properties of polysulfone/chitosan, polysulfone/N-succinyl chitosan and polysulfone/N-propylphosphonyl chitosan blend ultrafiltration membranes, *Desalination*, **350**, pp. 102–108.

64. Mbareck, C., Quang, T. N., Ouaf, T. A., Daniel, B. (2009). Elaboration, characterization and application of polysulfone and polyacrylic acid blends as ultrafiltration membranes for removal of some heavy metals from water, *J. Hazard Mater.*, **171**, pp. 93–101.

65. So, M. T., Eirich, F. R., Strathmann, H., Baker, R. W. (1973). Preparation of asymmetric Loeb-Sourirajan membranes, *J. Polym. Sci. B, Polym. Lett. Ed.*, **11**, pp. 201–205.

66. Ismail, A. F., Goh, P. S., Sanip, S. M., Aziz, M. (2009). Transport and separation properties of carbon nanotube-mixed matrix membrane, *Sep. Purif. Technol.*, **70**, pp. 12–26.

67. Wang, L., Song, X., Wang, T., Wang, S., Wang, Z., Gao, C. (2015). Fabrication and characterization of polyethersulfone/carbon nanotubes (PES/CNTs) based mixed matrix membranes (MMMs) for nanofiltration application, *Appl. Surf. Sci.*, **330**, pp. 118–125.

68. Ghaemi, N., Madaeni, S. S., Daraei, P., Rajabib, H., Shojaeimehr, T., Rahimpour, F., Shirvani, B. (2015). PES mixed matrix nanofiltration membrane embedded with polymer wrapped MWCNT: Fabrication and performance optimization in dye removal by RSM, *J. Hazard. Mater.,* **298,** pp. 111–121.

69. Yin, J., Zhu, G., Deng, B. (2013). Multi-walled carbon nanotubes (MWNTs)/polysulfone (PSU) mixed matrix hollow fiber membranes for enhanced water treatment, *J. Membr. Sci.*, **437**, pp. 237–248.

70. Park, C. H., Tocci, E., Fontananova, E., Bahattab, M. A., Aljlil, S. A., Drioli, E. (2016). Mixed matrix membranes containing functionalized multiwalled carbon nanotubes: Mesoscale simulation and experimental approach for optimizing dispersion, *J. Membr. Sci.*, **514**, pp. 195–209.

71. Gupta, A., Vidyarthi, S. R., Sankararamakrishnan, N. (2014). Enhanced sorption of mercury from compact fluorescent bulbs and contaminated water streams using functionalized multiwalled carbon nanotubes, *J. Hazard. Mater.*, **274**, pp. 132–144.

72. Shah, P., Murthy, C. N. (2013). Studies on the porosity control of MWCNT/polysulfone composite membrane and its effect on metal removal, *J. Membr. Sci.*, **437**, pp. 90–98.

73. Salehi, E., Madaeni, S. S., Rajabi, L., Vatanpour, V., Derakhshan, A. A., Zinadini, S., Ghorabi, Sh., Monfared, H. A. (2012). Novel chitosan/poly(vinyl) alcohol thin adsorptive membranes modified with amino functionalized multi-walled carbon nanotubes for Cu(II) removal from water: Preparation, characterization, adsorption kinetics and thermodynamics, *Sep. Purif. Technol.*, **89**, pp. 309–319.

74. Wan, S., Feng, H., Jiayu, W., Wubo, W., Yawei, G., Bin, G. (2016). Rapid and highly selective removal of lead from water using graphene oxide-hydrated manganese oxide nanocomposites, *J. Hazard. Mater.*, **314**, pp. 32–40.

75. Mukherjee, R., Prasenjit, B., Sirshendu, D. (2016). Impact of graphene oxide on removal of heavy metals using mixed matrix membrane, *Chem. Eng. J.*, **292**, pp. 284–297.

76. Wang, T., Lu, J., Mao, L., Wang, Z. (2016). Electric field assisted layer-by-layer assembly of graphene oxide containing nanofiltration membrane, *J. Membr. Sci.*, **515**, pp. 125–133.

77. Abdullah, N., Gohari, R. J., Yusof, N., Ismail, A. F., Juhana, J., Lau, W. J., Matsuura, T. (2016). Polysulfone/hydrous ferric oxide ultrafiltration mixed matrix membrane: Preparation, characterization and its

adsorptive removal of lead (II) from aqueous solution, *Chem. Eng. J.*, **289**, pp. 28–37.

78. Jamshidi, G. R., Lau, W. J., Matsuura, T., Halakoo, E., Ismail, A. F. (2013). Adsorptive removal of Pb(II) from aqueous solution by novel PES/HMO ultrafiltration mixed matrix membrane, *Sep. Purif. Technol.*, **120**, pp. 59–68.

79. Jamshidi, G. R., Lau, W. J., Matsuura, T., Ismail. A. F. (2013). Fabrication and characterization of novel PES/Fe–Mn binary oxide UF mixed matrix membrane for adsorptive removal of As(III) from contaminated water solution, *Sep. Purif. Technol.*, **118**, pp. 64–72.

80. Ghaemi, N. (2016). A new approach to copper ion removal from water by polymeric nanocomposite membrane embedded with γ-alumina nanoparticles, *Appl. Surf. Sci.*, **364**, pp. 221–228.

81. Fouladgar, M., Beheshti, M., Sabzyan, H. (2015). Single and binary adsorption of nickel and copper from aqueous solutions by γ-alumina nanoparticles: Equilibrium and kinetic modeling, *J. Mol. Liq.*, **211**, pp. 1060–1073.

82. Wang, X., Zhan, C., Kong, B., Zhu, X., Liu, J., Xu, W., Cai, W., Wang, H. (2015). Self-curled coral-like γ-Al_2O_3 nanoplates for use as an adsorbent, *J. Colloid Interface Sci.*, **453**, pp. 244–251.

83. Rahmani, H. M., Mousavi, H. Z., Fazli, M. (2010). Effect of nanostructure alumina on adsorption of heavy metals, *Desalination*, **253**, pp. 94–100.

84. Ghaemi, N., Daraei, P. (2016). Enhancement in copper ion removal by PPy@Al_2O_3 polymeric nanocomposite membrane, *J. Ind. Eng. Chem.*, **40**, pp. 26–33.

85. Pavlidou, S., Papaspyrides, C. D. (2008). A review on polymer–layered silicate nanocomposites, *Prog. Polym. Sci.*, **33**, pp. 1119–1198.

86. Akpomie, K. G., Dawodu, F. A. (2015). Treatment of an automobile effluent from heavy metals contamination by an eco-friendly montmorillonite, *J. Adv. Res.*, **6**, pp. 1003–1013.

87. Akpomie, K. G., Dawodu, F. A. (2016). Acid-modified montmorillonite for sorption of heavy metals from automobile effluent, *Beni-Suef Univ. J. Basic Appl. Sci.*, **5**, pp. 1–12.

88. Abollino, O., Aceto, M., Malandrino, M., Sarzanini, C., Mentasti, E. (2003). Adsorption of heavy metals on Na-montmorillonite. Effect of pH and organic substances, *Water Res.*, **37**, pp. 1619–1627.

89. Annabi-Bergaya, F. (2008). Layered clay minerals. Basic research and innovative composite applications, *Microporous Mesoporous Mater.*, **107**, pp. 141–148.

90. Malandrino, M., Abollino, O., Giacomino, A., Aceto, M., Mentasti, E. (2006). Adsorption of heavy metals on vermiculite: Influence of pH and organic ligands, *J. Colloid Interface Sci.*, **299**, pp. 537–546.

91. Yan, L., Li, S., Yu, H., Shan, R., Du, B., Liu, T. (2016). Facile solvothermal synthesis of Fe_3O_4/bentonite for efficient removal of heavy metals from aqueous solution, *Powder Technol.*, **301**, pp. 632–640.

92. Zeng, G., Yi, H., Yingqing, Z., Lei, Z., Yang, P., Chunli, Z., Zongxue, Y. (2016). Novel polyvinylidene fluoride nanofiltration membrane blended with functionalized halloysite nanotubes for dye and heavy metal ions removal, *J. Hazard. Mater.*, **317**, pp. 60–72.

93. Raghavendra, S. H., Isloor, M. A., Ananda, K., Ismail, A. F. (2015). Fabrication of antifouling, antimicrobial, well dispersed polydopamine functionalized halloysite nanotube-polyetherimide mixed matrix membranes for the heavy metal removal application, *J. Mater. Chem. A.*, DOI: 10.1039/C5TA09281G.

94. Sun, X., Jian, H. C., Zhenbo, S., Yihong, H., Xinfei, D. (2016). Highly effective removal of Cu(II) by a novel 3-aminopropyltriethoxysilane functionalized polyethyleneimine/sodium alginate porous membrane adsorbent, *Chem. Eng. J.*, **290**, pp. 1–11.

95. Han, R., Zhang, S., Liu, C., Wang Y., Jian, X. (2009). Effect of NaA zeolite particle addition on poly(phthalazinone ether sulfone ketone) composite ultrafiltration (UF) membrane performance, *J. Membr. Sci.*, **345**, pp. 5–12.

96. Wang, S., Peng, Y. (2010). Natural zeolites as effective adsorbents in water and wastewater treatment, *Chem. Eng. J.*, **156**, pp. 11–24.

97. Leo, C. P., Ahmad Kamil, N. H., Junaidi, M. U. M., Kamal, S. N. M., Ahmad, A. L. (2013). The potential of SAPO-44 zeolite filler in fouling mitigation of polysulfone ultrafiltration membrane, *Sep. Purif. Technol.*, **103**, pp. 84–91.

98. Liu, F., Ma, B.-R., Zhou, D., Xiang, Y., Xue, L. (2014). Breaking through tradeoff of Polysulfone ultrafiltration membranes by zeolite 4A, *Microporous Mesoporous Mater.*, **186**, pp. 113–120.

99. Nibou, D., Mekatel, H., Amokrane, S., Barkat, M., Trari, M. (2010). Adsorption of Zn^{2+} ions onto NaA and NaX zeolites: Kinetic, equilibrium and thermodynamic studies, *J. Hazard. Mater.*, **173**, pp. 637–646.

100. Nibou, D., Amokrane, S., Lebaili, N. (2010). Use of NaX porous materials in the recovery of iron ions, *Desalination*, **250**, pp. 459–462.

101. Yurekli, Y. (2016). Removal of heavy metals in wastewater using zeolite nano-particles impregnated Polysulfone membranes, *J. Hazard. Mater.*, **309**, pp. 53–64.

102. Feng, C., Khulbe, K. C., Matsuura, T., Tabe, S., Ismail, A. F. (2013). Preparation and characterization of electro-spun nanofiber membranes and their possible applications in water treatment, *Sep. Purif. Technol.*, **102**, pp. 118–135.

103. Goyal, R., Macri, L. K., Kaplan, H. M., Kohn, J. (2015). Nanoparticles and nanofibers for topical drug delivery, *J. Control. Release.*, http://dx.doi.org/10.1016/j.jconrel.2015.10.049.

104. Sundarrajan, S., Tan, K. L., Lim, S. H., Ramakrishna, S. (2014). electrospun nanofibers for air filtration applications, *Procedia Eng.*, **75**, pp. 159–163.

105. Ding, Y., Hou, H., Zhao, Y., Zhu, Z., Fong, H. (2016). Electrospun polyimide nanofibers and their applications, *Prog. Polym. Sci.*, **61**, pp. 67–103.

106. Bui, N.-N., McCutcheon, J. R. (2016). Nanoparticle-embedded nanofibers in highly permselective thin-film nanocomposite membranes for forward osmosis, *J. Membr. Sci.*, **518**, pp. 338–346.

107. Wu, S., Li, F., Wang, H., Fu, L., Zhang, B., Li, G. (2010). Effects of poly(vinyl alcohol) (PVA) content on preparation of novel thiol-functionalized mesoporous PVA/SiO_2 composite nanofiber membranes and their application for adsorption of heavy metal ions from aqueous solution, *Polymer*, **51**, pp. 6203–6211.

108. Hallaji, H., Keshtkar, A. R., Moosavian, M. A. (2015). A novel electrospun PVA/ZnO nanofiber adsorbent for U(VI), Cu(II) and Ni(II) removal from aqueous solution, *J. Taiwan Inst. Chem. Eng.*, **46**, pp. 109–118.

109. Rad, L. R., Momeni, A., Ghazani, B. F., Irani, M., Mahmoudi, M., Noghreh, B. (2014). Removal of Ni^{2+} and Cd^{2+} ions from aqueous solutions using electrospun PVA/zeolite nanofibrous adsorbent, *Chem. Eng. J.*, **256**, pp. 119–127.

110. Sun, B., Xiang, L., Rui, Z., Mingying, Y., Zongxin, W., Ziqiao, J., Ce, W. (2016). Hierarchical aminated PAN/γ–AlOOH electrospun composite nanofibers and their heavy metal ion adsorption performance, *J. Taiwan Inst. Chem. Eng.*, **62**, pp. 219–227.

111. Miao, Y.-E., Wang, R., Chen, D., Liu, Z., Liu, T. (2012). Electrospun self-standing membrane of hierarchical SiO_2@γ-AlOOH (boehmite) core/sheath fibers for water remediation, *ACS Appl. Mater. Interfaces.*, **4**, pp. 5353–5359.

112. Fathizadeh, M., Aroujalian, A., Raisi, A. (2011). Effect of added NaX nano-zeolite into polyamide as a top thin layer of membrane on water flux and salt rejection in a reverse osmosis process, *J. Membr. Sci.*, **375**, pp. 88–95.

113. Kim, E.-S., Deng, B. (2011). Fabrication of polyamide thin-film nano-composite (PA-TFN) membrane with hydrophilized ordered mesoporous carbon (H-OMC) for water purifications, *J. Membr. Sci.*, **375**, pp. 46–54.

114. Dong, L., Huang, X., Wang, Z., Yang, X., Wang, X., Tang, C. Y. (2016). A thin-film nanocomposite nanofiltration membrane prepared on a support with in situ embedded zeolite nanoparticles, *Sep. Purif. Technol.*, **166**, pp. 230–239.

115. Lai, G. S., Lau, W. J., Goh, P. S., Ismail, A. F., Yusof, N., Tan, Y. H. (2016). Graphene oxide incorporated thin film nanocomposite nanofiltration membrane for enhanced salt removal performance, *Desalination*, **387**, pp. 14–24.

116. Yin, J., Zhu, G., Deng, B. (2016). Graphene oxide (GO) enhanced polyamide (PA) thin-film nanocomposite (TFN) membrane for water purification, *Desalination*, **379**, pp. 93–101.

117. Liu, l., Zhu, G., Liu, Z., Gao, C. (2016). Effect of MCM-48 nanoparticles on the performance of thin film nanocomposite membranes for reverse osmosis application, *Desalination*, **394**, pp. 72–82.

118. Zhang, S., Peh, M. H., Thong, Z., Chung, T.-S. (2015). Thin film interfacial cross-linking approach to fabricate a chitosan rejecting layer over poly(ether sulfone) support for heavy metal removal, *Ind. Eng. Chem. Res.*, **54**, pp. 472–479.

119. Zhang, Y., Zhang, S., Chung, T.-S. (2015). Nanometric graphene oxide framework membranes with enhanced heavy metal removal via nanofiltration, *Environ. Sci. Technol.*, **49**, pp. 10235–10242.

120. Zhao, F.-Y., Ji, Y.-L., Weng, X.-D., Mi, Y.-F., Ye, C.-C., An, Q.-F., Gao, C.-J. (2016). High-flux positively charged nanocomposite nanofiltration membranes filled with poly(dopamine) modified multiwall carbon nanotubes, *ACS Appl. Mater. Interfaces.*, **8**, pp. 6693–6700.

Chapter 13

Nanocomposite Membrane-Based Photocatalytic Reactor for Degradation of Endocrine-Disrupting Compound in Water

Hazlini Dzinun, Mohd Hafiz Dzarfan Othman, A. F. Ismail, Mohd Hafiz Puteh, Mukhlis A. Rahman, and Juhana Jaafar

Advanced Membrane Technology Research Centre (AMTEC), Faculty of Chemical and Energy Engineering, Universiti Teknologi Malaysia, 81310 Skudai, Johor, Malaysia

hafiz@petroleum.utm.my

13.1 Introduction

The first attempt to fabricate polymeric dual-layer hollow-fiber (DLHF) membranes via phase inversion technique was conducted by DuPont in 1992. The membranes were formed by a co-extrusion process of two different polymers [1]. The morphology of the membrane inner and outer layers were formed based on the required properties in each layer by manipulating the extrusion parameters [2]. For example, to produce outer surface with useful opening pores, the amount of organic solvents such

Nanocomposites for Pollution Control
Edited by Chaudhery Mustansar Hussain and Ajay Kumar Mishra

ISBN 978-981-4774-45-1 (Hardcover), 978-1-315-14368-2 (eBook)
www.panstanford.com

as N-methylpyrrolidone (NMP) in the dope of outer layer can be manipulated, where the desired pore structure will be then induced during the phase inversion process [3].

To date, there has been vast research on the fabrication of DLHF membranes produced via the co-extrusion technique [3–8]. The main advantage of co-extrusion compared with other conventional techniques is it only involves a single-step process and thus leads to shorter fabrication time and lower cost. In spite of the single step, the produced membranes are able to deliver higher flux or equivalent to their counterpart produced by typical deposition techniques which might suffer from pore penetration, which is often a problem with dip-coating deposition.

Various studies clearly revealed the applicability of dual-layer fiber spinning technology to produce membranes for various gas [9–13] and liquid separations [9, 14–17]. Producing high performance DLHF membranes via co-extrusion process could be very challenging because it involves a sophisticated spinning mechanism where two different phase inversion pathways occur simultaneously. The main problem in DLHF membrane fabrication is the structure integrity and delamination between layers. Good adhesion between inner and outer polymer dopes can be achieved by optimizing the spinning parameters such as bore fluid composition, spinning process temperature, air gap height [2] and flow rate ratio of outer to inner layer [4].

Another important factor in fabricating DLHF membranes is the choice of outer and inner layer polymeric materials. Some polymer materials that have been chosen by the researchers to produce delamination-free DLHF membranes are polyvinylidene chloride (PVDC) with cellulose acetate (CA) [2], polyamide-imide (PAI) with polyethersulfone (PES) [13, 14] and poly(vinylidene fluoride) (PVDF) [15, 16, 19] for inner and outer dope.

In particular, PVDF has gained considerable attention as one of promising polymeric membrane materials due to its outstanding chemical and physical properties. The main advantages of PVDF include high hydrophobicity and excellent resistance to corrosive chemicals such as acids, bases, oxidants and halogens [20]. Compared to other hydrophobic materials such as polytetrafluoroethylene and propylene, PVDF possesses better process ability since it is easily soluble in common organic solvents. PVDF hollow fiber membranes with favorable asymmetric structure can be fabricated via dry-wet phase inversion process. With these superior properties, PVDF has been used in various membrane-based separations such as micro/nano/ ultrafiltration [21–25], membrane distillation [26], and membrane gas separation [13].

The incorporation of inorganic additives into polymeric membranes has been widely investigated for microfiltration, ultrafiltration, nanofiltration and membrane distillation. A variety of nanoparticles has been introduced to modify organic membranes, such as SiO_2, Al_2O_3, Fe_3O_4, ZrO_2 and TiO_2. Of these nanoparticles, TiO_2 has received the most attention because of its unique physical and chemical characteristics, high availability as well as great antifouling property [27]. Apart from adding anti-fouling property to membrane, TiO_2 particles can effectively degrade certain chemicals especially organic compounds with the presence of UV light.

However, TiO_2 nanoparticles has a tendency to agglomerate when it is blended with polymer casting solution due to their high surface area/particle size ratio that eventually reduces their overall efficiency [23]. Besides blending, in situ formation of TiO_2 within polymer matrix by sol-gel technique also stimulates such agglomeration [24, 25]. One method to minimize the agglomeration is by applying shear force by mixer/sonicator to physically disperse the TiO_2 particles in the solution. Another method is surface pre-treatment but it involves chemical modification that might change the TiO_2 properties itself [27]. Therefore, continuous studies are required to produce DLHF membranes with well-dispersed TiO_2 particles, especially at high loading.

The incorporation of TiO_2 nanoparticles in the membrane fabrication is to utilize their photocatalytic functionality. In the past few years, attention has been paid to immobilize TiO_2 nanoparticle on various materials. Since ultraviolet sources and TiO_2 are required for the photocatalytic reaction, thus, it is crucial to immobilize high concentration of TiO_2 on the outer surface of the polymeric membrane support. Therefore, by using co-extrusion approach, high loading of photocatalyst TiO_2 can be achieved, specifically on the outer layer of the hollow fiber membranes.

In this chapter, PVDF-TiO_2/PVDF DLHF membranes as illustrated in Fig. 13.1 were prepared by co-extrusion technique. The technique allows the nanoparticle distributed uniformly inside the membrane outer thin layer which encompasses most of the membrane activities and performances. The morphology and hydrophilicity of the membranes were characterized by scanning electron microscopy (SEM) and contact angle goniometer, respectively. Filtration experiments were carried out to examine the antifouling and fouling mitigation abilities of the membranes.

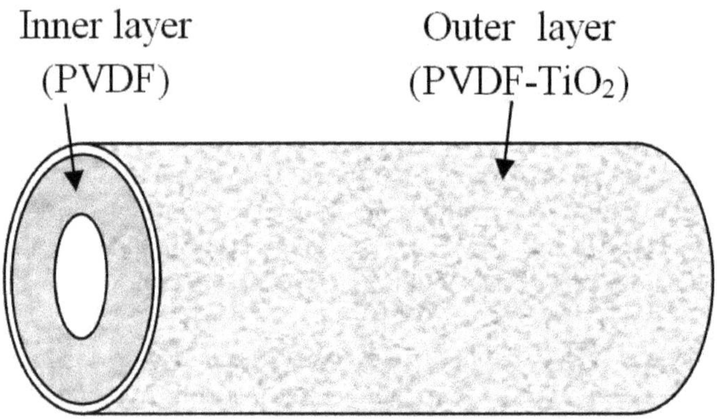

Figure 13.1 Dual-layer hollow fiber membrane [30].

13.2 Dual-Layer Hollow Fiber as Photocatalytic Membranes

The fabrication method of photocatalytic membranes is a key factor of their properties and performances. Although the unique properties of photocatalytic membranes provide a wide array of advantages when they are applied for water treatment compared to the conventional systems and processes, many technical issues still need to be resolved before this technique can be practically applied for water purification.

Photocatalytic membranes also encounter similar challenge as immobilized photocatalytic system, in which the contact area between photocatalysts, target pollutant and light intensity is lower than that of the suspended photocatalytic system. So, more efforts are required to improve the activity of photocatalytic membranes by using immobilized photocatalysts. Several methods have been proposed to immobilize TiO_2 onto/within solid supports using various techniques such as chemical vapor deposition, sol-gel dip coating, spinning or casting, atmospheric plasma spraying, electrospinning and hot press, and hydrothermal synthesis as stated. The primary aim of immobilized TiO_2 onto/within solid supports is to avoid the post separation associated with the powder form of the TiO_2 catalyst.

Generally, all the immobilization methods mainly focus on the modified TiO_2 in the membrane surface to enhance the photocatalytic activity. Among them, phase inversion method is more preferable in fabricating photocatalytic membrane in either flat sheet or hollow fiber configurations. This type of membrane is usually used in microfiltration or ultrafiltration processes due to its typical porous structure.

However, TiO_2 nanoparticles have a tendency to agglomerate when they are blended with polymer casting solution due to their high surface area/particle size ratio that eventually reduces their overall efficiency [23]. Besides blending, in situ formation of TiO_2 within polymer matrix by sol-gel technique also stimulates such agglomeration [28, 29]. One method to minimize the agglomeration is by applying shear force by mixer/sonicator to physically disperse the TiO_2 particles in the solution. Another method is surface pre-treatment but it involves chemical modification that might change the TiO_2 properties itself [27]. Therefore, continuous studies are required to produce membranes with well-dispersed TiO_2 particles, especially at high loading.

The crucial problem of the immobilization of TiO_2 within membrane is the limitation of TiO_2 nanoparticles exposed to UV light due to the polymer matrix hindered the TiO_2 nanoparticles in the membrane structure. Therefore, in 2009, Tahiri et al. investigated the effectiveness of TiO_2 nanoparticles immobilized within flat sheet membranes. They stated that the reaction kinetics were controlled by the kinetics of product transport through the porous structure that entraps the photocatalyst, which means that the porosity and pore size are the most important parameter for immobilized TiO_2 of composite membrane in photocatalytic process. In their studies, TiO_2/PVDF ratio of 0.5 was selected because of its higher mean pore diameter and porosity [31].

13.3 Co-Extrusion Approach in the Preparation of Dual-Layer Hollow Fiber Membranes

The co-extrusion approach is defined as a process to extrude two or more polymeric membrane solutions through a single die. Passing through the respective channels in die, the membrane solutions merge and join together at the orifices. The shape and structure of the orifices are specially designed to meet different requirements. The co-extrusion approach can be employed to produce multi layer-sheets, wire coating and other products. The advantages of the co-extrusion approach include:

(i) Each layer imparts a desired characteristic property, such as stiffness, permeability or resistance to some environments. It would be impossible for any single material to attain all of these properties.
(ii) It is able to make multi-layer, multi functional structures in a single process.
(iii) It is possible to lower down the cost because the amount of premium and expensive materials used for equal performance is optimized/minimized.
(iv) It provides the possibility to handle polymers that are of excellent permselectivity but could not form an integrated asymmetric membrane on their own.
(v) It provides the flexibility for fabrication process. A co-extrusion line could be used to produce single extrusion products without any modification but not vice versa.

The co-extrusion approach is not a new technology. It has been used the early 1950s to improve product quality and process efficiency. However, the application of the co-extrusion in the fabrication of composite membranes for separation was implemented in the 1970s. The applications of the co-extrusion approach in the preparation of hollow fiber membranes include dual bath approach for surface morphology control in the formation asymmetric hollow fiber membranes, preparation of ceramic composite hollow fiber membranes and fabrication of dual-layer asymmetric hollow fiber composite membranes as shown in Fig. 13.2. The dual-layer spinneret design of outer layer dope, inner layer dope and bore fluid are delivered to the orifice by passing through three independent channels as shown in Fig. 13.3.

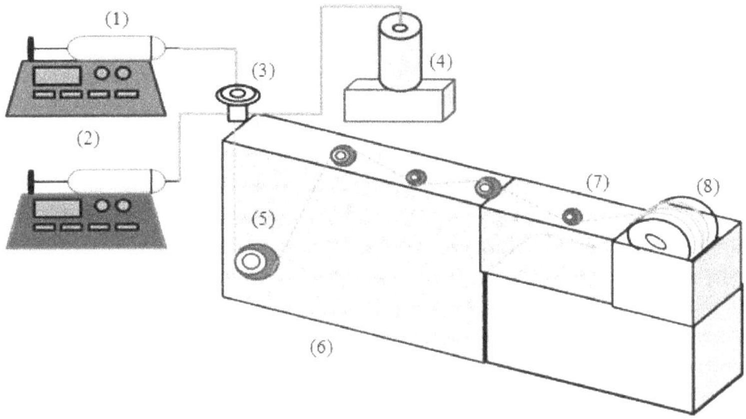

Figure 13.2 Schematic diagram of the spinning system: (1) syringe tube, (2) syringe pump, (3) triple orifice spinneret, (4) syringe pump, (5) roller, (6) first coagulation bath, (7) second coagulation bath and (8) take-up drum.

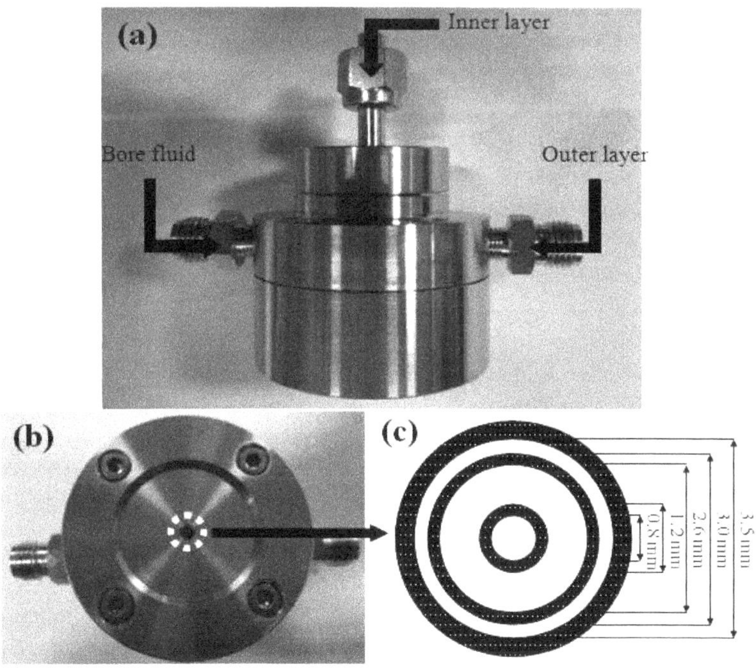

Figure 13.3 Photographic images of (a) triple orifice spinneret from side, (b) triple orifice spinneret from bottom, and (c) dimension of the triple orifice spinneret [32].

The dual-layer hollow fiber spinning is more complicated than the single layer one because the process involves two dopes being extruded and precipitating simultaneously. In addition to the inter-layer diffusion phenomenon between the two individual layers, the rapid external coagulation induced by water bath and the slow internal coagulation induced by the bore fluid may directly effect on phase inversion and the resultant morphology of the two skins. The dope flow rates for the inner and outer layers influence their layer thicknesses after precipitation, which play important roles on membrane performance [9].

13.3.1 Polymer Dope Preparation

The PVDF and TiO_2 powder were dried in a 50°C vacuum oven for 24 h to remove moisture prior to dope preparation. To prepare the dope solution, the TiO_2 and DMAc at desired amount were mixed as listed in Table 13.1 and stirred inside a Scott bottle with an overhead stirrer (IKA RW 20 digital) at 400 rpm for 24 h. After the TiO_2 mixture became a well-dispersed solution, different amounts of PVDF was gradually added. Ultrasonication was used for the dispersion of TiO_2 nanoparticles in dope solution. 20 min of ultrasonication (180W) is sufficient for the dispersion of the TiO_2 nanoparticles in dope solution, which can remain stable for more than 1 h as reported by Qi et al. in 2013 [33].

Table 13.1 Dope composition of the DLHF membranes

Membrane identification code	TiO_2/PVDF ratio	Dope composition (Outer layer)			Dope composition (Inner layer)	
		PVDF (wt%)	TiO_2 (wt%)	DMAc (wt%)	PVDF (wt%)	DMAc (wt%)
OL-0.0	0.0	15.0	0.0	85.0	18.0	82.0
OL-0.2	0.2	15.0	3.0	82.0	18.0	82.0
OL-0.5	0.5	15.0	7.5	77.5	18.0	82.0
OL-0.7	0.7	15.0	10.5	74.5	18.0	82.0
OL-1.0	1.0	15.0	15.0	70.0	18.0	82.0
OL-1.2	1.2	15.0	18.0	67.0	18.0	82.0

The prepared solutions were degassed in 1 h using an ultrasonic bath system at an ambient temperature before they were subjected to spinning process. Then, the spinning dope mixture was extruded using a triple orifice spinneret to form DLHF membranes.

Before spinning, both PVDF and TiO_2/PVDF dope solutions were degassed using an ultrasonic bath system at ambient temperature for 1 h. The outer dope solution flow rate was 1 mL/min, while the inner dope solution flow rate was 8 mL/min. In order to achieve delamination free, the ratio of the outer-layer dope flow rate to inner-layer dope flow rate should be higher or equal to 0.125 as reported by Li Dongfei in 2004 [8]. The bore fluid composition used was 100% distilled water with a flow rate of 8 mL/min. The nascent dual-layer hollow fiber membrane was passed through an air gap with a distance of 10 cm at ambient temperature and was then immediately drawn into a coagulation bath. Tap water at room temperature was used as the coagulant in this study. The as-spun dual-layer hollow fiber membranes were taken using take up drum with 0.18 m/s of speed.

During spinning process, the small loss of TiO_2 nanoparticles from membrane structure to coagulation bath cannot be observed with naked eyes in this work, which probably due to very small amount of TiO_2 nanoparticles loss as noted by Chong et al. in 2010 [34].

13.3.2 Post Treatment

The as-spun DLHF was immersed into water bath for 1 day to remove residual solvent. Prior to air drying, the fibers was post-treated using (ethanol:water, 50:50 wt.%) for 1 h and then 100% of ethanol for another 1 h in order to improve the membrane wettability and pore collapse [31]. Finally, the hollow fiber membranes were dried at room temperature for 3 days before module preparation.

13.4 Delamination

Delamination between the layers is often observed in the DLHF spinning process. The delamination may lead to poor membrane mechanical and separation properties. The mechanism of lamination/delamination at the interface of the outer and inner layers of dual-layer membranes is important to understand. The delamination is mainly controlled by the rates of the non-solvent (water) diffusion through the outer ($J_{NS\text{-}outer}$) and inner layer ($J_{NS\text{-}inner}$), which also has a direct impact on the growth of the macrovoids. There are two scenarios as illustrated in Fig. 13.4 in which the "up" and "down" arrows denote increase and decrease in the quantity, respectively:

(i) When the non-solvent (water from the external coagulant bath) has a higher diffusion rate in the outer layer as compared to that in the inner layer, the outer layer tends to expand to form large macrovoids and to hold more water at the interface. As a result, the accumulated water can impede the adhesion of the two layers, leading to a delamination.
(ii) If water has a slower penetration rate through the outer layer dope than the inner dope, good adhesion of these layers formed.

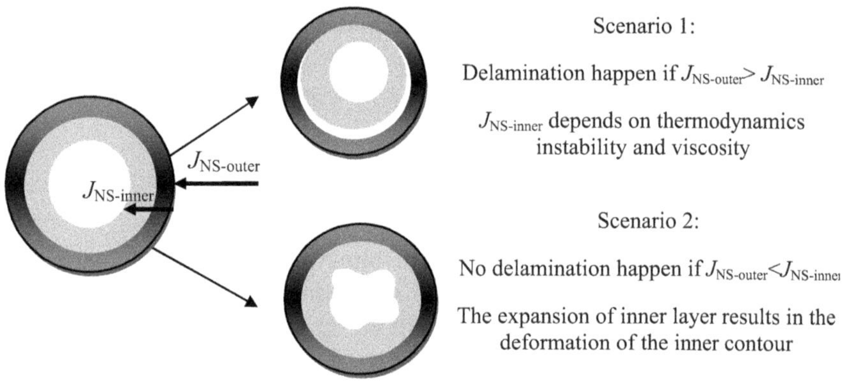

Figure 13.4 Mechanism of lamination/delamination at the interface of the layers and regular/irregular morphology of dual-layer hollow fiber membranes [18].

The causes for delamination may be due to the parameters of spinning process. Bhandari et al., in 2013 studied the effect of bore fluid composition, spinning process temperature and air-gap height. The result shows that bore fluid in the range of

70/30–60/40 wt% NMP/Water bore fluid composition gave a delamination free. The spinning process temperature (dope, transfer lines and spinneret) of 25 and 40°C both gave good layer adhesion. The inner and outer layers were well adhered for air gaps between 3 and 15 cm. Furthermore, Li et al., in 2004 stated that 100% water bore fluid composition and ratio of outer-layer dope flow rate to inner-layer dope flow rate of 0.125, the gap between two layers disappears [8].

Bonyadi and Chung in 2007 fabricated PVDF/PAN DLHF for membrane distillation. They found that the membrane morphology and structures without any delamination were obtained. However, the interface between the two layers was peeled off and no separation could be achieved when they were applied in the direct contact membrane distillation (DCMD) process. This might be due to different thermal expansion coefficients of these two polymers or the repulsive force acted on the interface between hydrophilic and hydrophobic layers in an aqueous environment.

Another important factor to produce delamination-free DLHF membranes is the choice of outer and inner layer polymeric materials. Some polymer materials that have been chosen by the researchers are polyvinylidene chloride (PVDC) with cellulose acetate (CA) [2], polyamide-imide (PAI) with polyethersulfone (PES) [7, 18] and PVDF [15, 16, 19] for inner and outer dope. As reported by Teoh et al. in 2011, both layers would have a good compatibility with each other and possess delamination-free structure when the same polymer was used for both inner and outer layer dopes. In fact, when mutual diffusion of polymers occurred between the inner and outer layers, the resultant combined layers would have seamless interface as shown in Fig. 13.5.

Since the PVDF concentration used for the outer layer dope solution (15 wt.%) was lower than its inner counterpart (18 wt.%), shrinkage rate of the outer layer is expectedly greater than the inner layer during the solvent/non-solvent exchange process [9]. As a result, the inner membrane layer is tightly enveloped by the outer layer, resulting in a seamless interface between the layers. No interlayer was found in the DLHF membranes.

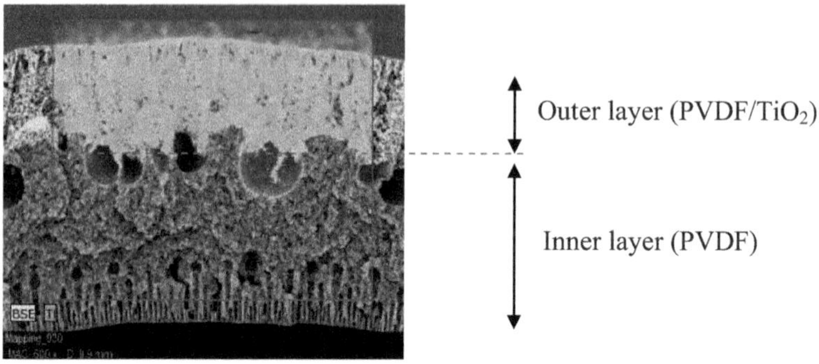

Figure 13.5 Seamless interface of outer and inner layer in dual-layer hollow fiber membranes [30].

13.5 Morphology and Physical Properties of Dual-Layer Hollow Fiber Membranes

The TiO_2 addition into the membrane matrixes has shown significant effects on their pore sizes and morphologies. Figure 13.6 shows SEM images of the DLHF membranes with different TiO_2 loadings in their outer layers. A finger-like structure was developed in both the outer and inner layers which was separated by a sponge-like structure in between as shown in Fig. 13.6 (a2-f2). This structure is known as sandwich-like structure. The formation of such sandwich-like structure might be due to suspension-coagulant interface instability occurred during the phase-inversion process. It can also be observed that different ratios of TiO_2/PVDF produced different membrane morphologies as indicated by their different finger-like void lengths formed in the outer layer.

Table 13.2 shows the effects of TiO_2 loadings on the average length of the finger-like voids, which was measured directly from the SEM images in Fig. 13.6. The effects on length can be explained by looking at the hydrophilicity and viscosity of the dope solution, as both factors could affect the solvent/non-solvent exchange rate during phase inversion process. As shown in Fig. 13.7, the formation of finger-like voids occurred in both layers. This formation is triggered when the nascent membranes are in contact with the coagulation bath, where the

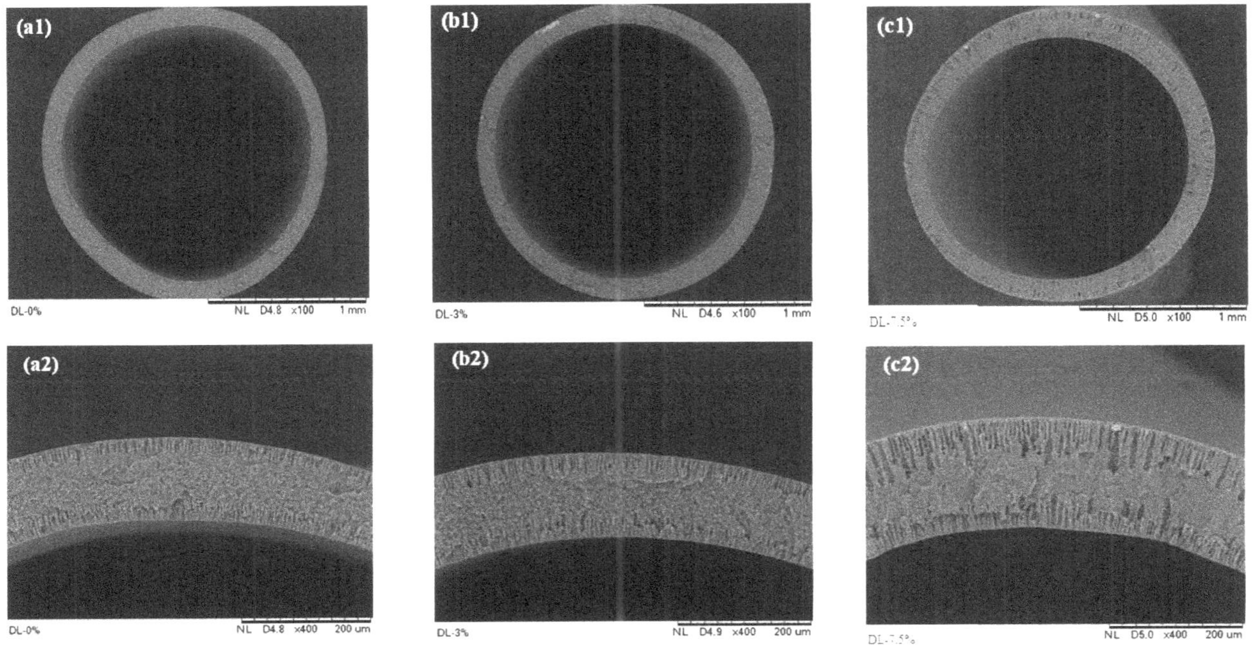

Figure 13.6 Overall and partial cross sectional images of the DLHF membranes with different TiO_2/PVDF ratios; (a) 0, (b) 0.2, (c) 0.5, (d) 0.7, (e) 1 and (f) 1.2 [30].

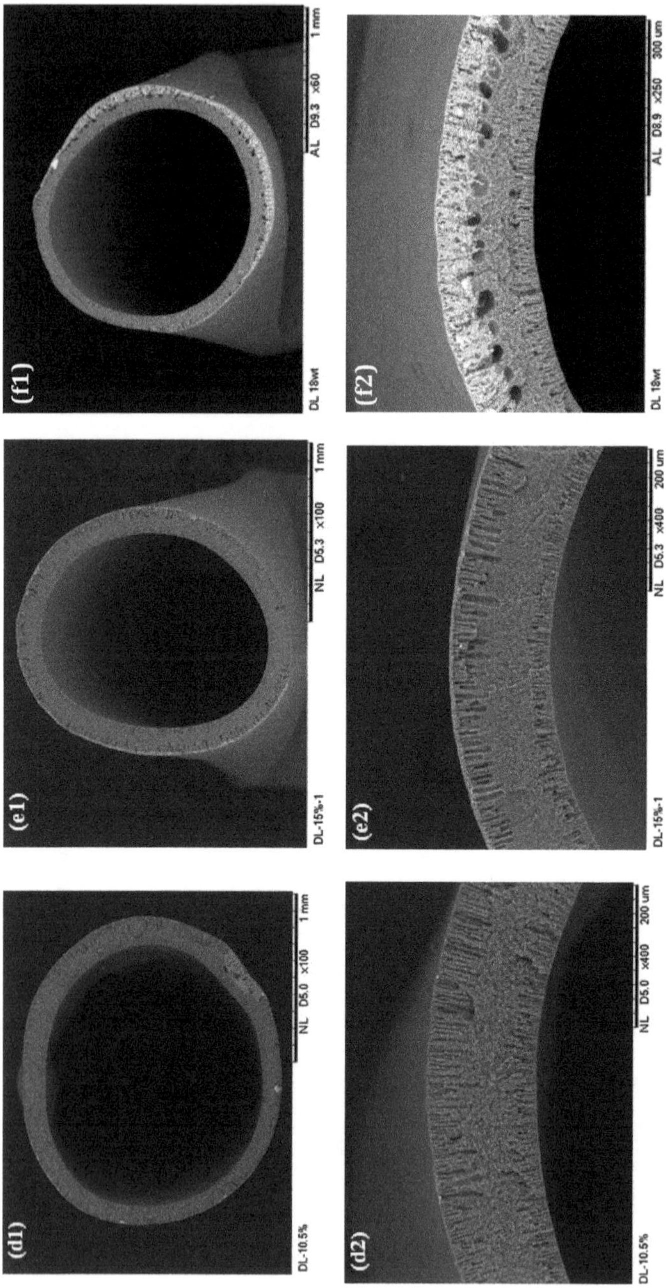

Figure 13.6 *(Continued)*

Table 13.2 Characteristics of DLHF membranes with different TiO_2 composition

Membrane identification code	Viscosity of outer layer (Pa·s)	Contact angle (°)	Mean pore size (nm)	Average length of finger-like void (μm)	
				Outer layer	inner layer
OL-0.0	1.836	91.1	263.99	22.4±2.13	21.5±2.10
OL-0.2	1.958	79.4	259.58	26.0±3.20	34.3±3.42
OL-0.5	3.345	77.6	238.66	68.8±2.12	50.4±2.20
OL-0.7	5.812	75.2	231.25	52.9±1.54	39.6±2.14
OL-1.0	7.144	67.3	163.95	50.9±2.50	29.6±2.34
OL-1.2	9.345	65.3	360.46	40.8±4.25	29.7±2.66

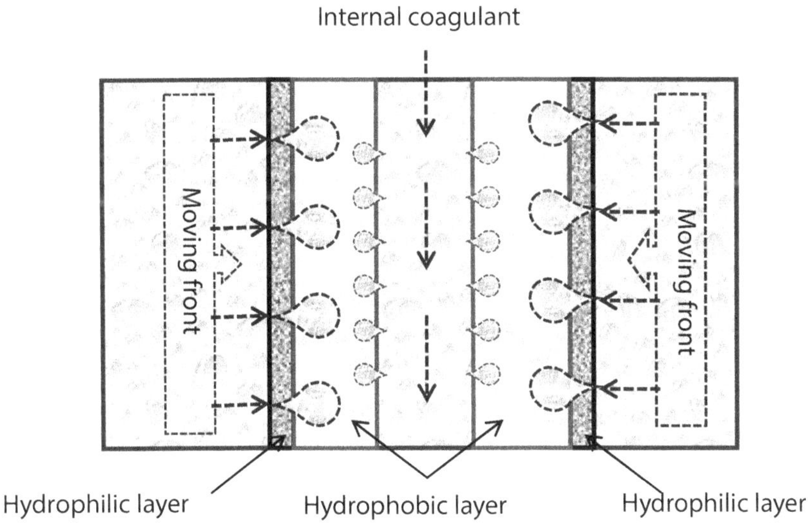

Figure 13.7 Schematic diagram showing the formation and growth of finger-like structure in the DLHF membranes [30].

non-solvent continuously diffuses into the membranes through the non-solvent/membrane interface. This formation and growth is further promoted when the nucleation is coupled with a rapidly moving front of the phase separation region. Rapid solvent exchanged across the wall of nucleus makes the composition inside the nucleus similar to the composition of the external coagulant. This rapid solvent exchange rate has a major impact on the osmotic pressure across the nucleus wall [8]. Due to the hydrophilicity properties of TiO_2 in the outer layer, it would attract large amount of the non-solvent (water) flow into the nucleus then enlarges the volume of the nucleus. Since the hydrophilicity of the dope solution increased with the increase of TiO_2/PVDF fraction as shown in Table 13.2, therefore, it can be seen that the finger-like length increased with the increase of TiO_2 loading up to 0.5.

However, it is also believed that the finger-like void formation does not solely depend on the hydrophilicity alone, but also the solution viscosity. As summarized in Table 13.2, the dope solution viscosity increased with the increase of TiO_2 loading. The viscosity of the dope would affect the rheological property during the solvent/non-solvent exchange in the phase inversion process, thus it would change the morphology of membranes. The viscosity increase would slow down the mass transfer rate during the membrane formation process and prevent the formation of macroporous structure. However, the effect of viscosity towards the formation of finger-like void is only dominant when the TiO_2/PVDF fraction is above 0.5, where the finger-like length became shorter for the membrane at TiO_2/PVDF fraction of 0.7 until 1.2. It is believed that the dope viscosity has exceeded its critical level (threshold) when the TiO_2/PVDF fraction reaches 0.7 and above, and as a result, the finger-like voids length decreased. The finger like length was measured by SEM images.

It is interesting to note that the length of the voids in the inner layer also increased with the addition of TiO_2 loading up to 0.5 ratio. It might be due to the hydrophilicity properties of TiO_2 in the outer layer, which would lead to the attraction of large amount of the non-solvent (water) flow from bore fluid (lumen) toward the outer layer via the inner surface of the nascent fiber. This has resulted to the enlargement of the finger-like void volume from the inner surface when more TiO_2 was added to the outer layer. However, as can be seen from Table 13.1, the finger-like length in the inner layer decreased when the TiO_2/PVDF ratio was more than 0.5 and the trend was found similar to the finger-like void length in the outer layer.

The SEM images depicting the outer surfaces of the DLHF membranes prepared without and with different ratios of TiO_2/

PVDF in the dope solutions are shown in Fig. 13.8. As can be seen, the TiO_2 nanoparticles were uniformly distributed on the membrane surfaces. Based on Fig. 13.8a, the surface of the neat membranes is composed of a microporous structure with relatively large pores, 263.99 nm was determined by AFM analysis. Pore sizes of more than 60 pores plotted against the median range were measured and listed in Table 13.2. Furthermore, with the increasing TiO_2 concentration, the amount of TiO_2 particles deposited on the surface was increased. The increasing number of TiO_2 particles on the membrane surfaces would provide additional hydrophilicity strength to the membranes, possibly increasing the permeate water flux and its antifouling properties. However, higher TiO_2 concentration (ratio of TiO_2/PVDF = 1 and 1.2) turned out to be promoting the aggregation of TiO_2 particles that eventually blocking the membrane pores.

As proved by FESEM images, TiO_2 nanoparticles distributed on the TiO_2/PVDF dual-layer hollow fiber membranes surface shows larger aggregation for ratio of 0.5 (Fig. 13.9a) and 1 (Fig. 13.9b) with 76 and 56 nm, respectively. These results were similar with what was observed by Teow et al. in 2012 who found that the higher TiO_2 concentration (0.1 g/L TiO_2) in casting solution might induce the aggregation of TiO_2 particles thus blocking the pores of membranes [35]. They stated that the major factor contribute to the particle aggregation is the higher surface tension between solvent (DMAc) and TiO_2. Besides, Mackay et al. in 2006 found that dispersion of nanoparticles into a polymeric liquid is thermodynamically stable for systems where the radius of gyration (R_g) of the linear polymer is greater than the radius of the nanoparticles [36].

Figure 13.10 presents three-dimensional AFM images and surface roughness (Ra) of DLHF membranes prepared at different TiO_2 loading at the outer layer dope. In these images, the brightest area represents the highest point of membrane surface and the dark regions indicate valley or membrane pores. The AFM images in Fig. 13.10 clearly show that the membrane surface roughness have significant different because the TiO_2 particles loading in the outer layer is varied. It might be considered that the increasing ratio of TiO_2/PVDF membrane led to smoother surface. It may due to the TiO_2 nanoparticles were filled up pores of outer membrane surfaces. By referring to the Rahimpour et al.'s result in 2011, they stated that the surface roughness parameters of membranes decreased with an increase the TiO_2 concentration in the dope solution. Since the roughness parameters depend on the Z-value, which is vertical distance that the piezoelectric scanner moves, this relationship is expected. When the surface includes deep depressions (pores) and high peaks (modules), the tip moves up and down over the wide range and the roughness parameter of the surface high.

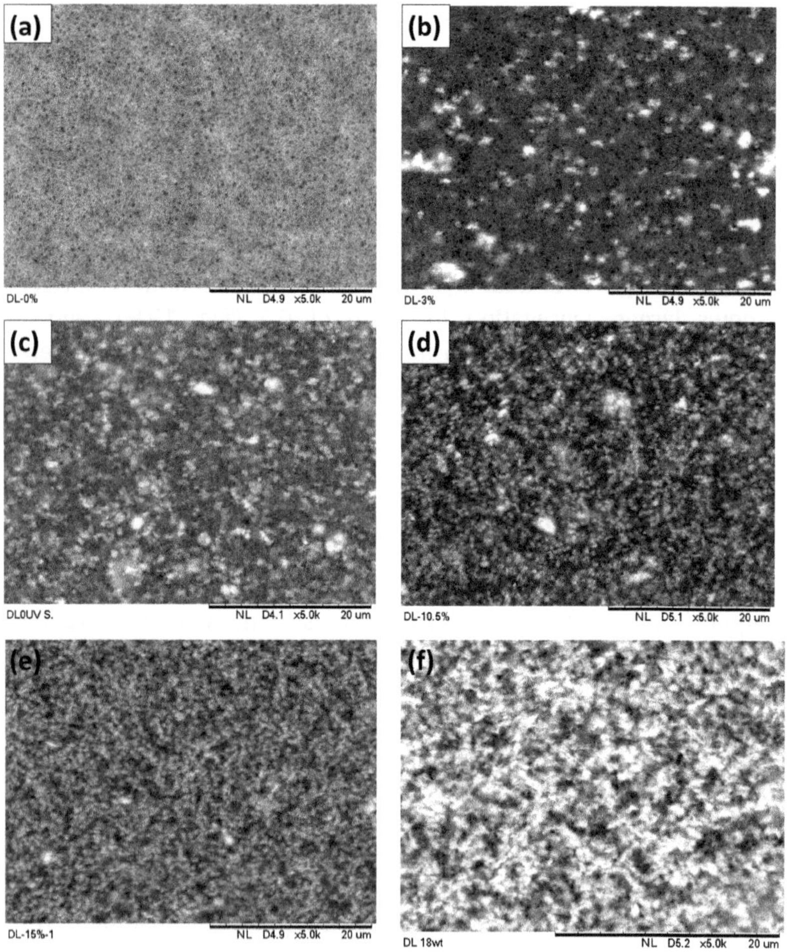

Figure 13.8 SEM images showing TiO_2 deposition on the surfaces of the DLHF membranes with different TiO_2/PVDF ratios: (a) 0, (b) 0.2, (c) 0.5, (d) 0.7, (e) 1 and (f) 1.2 [30].

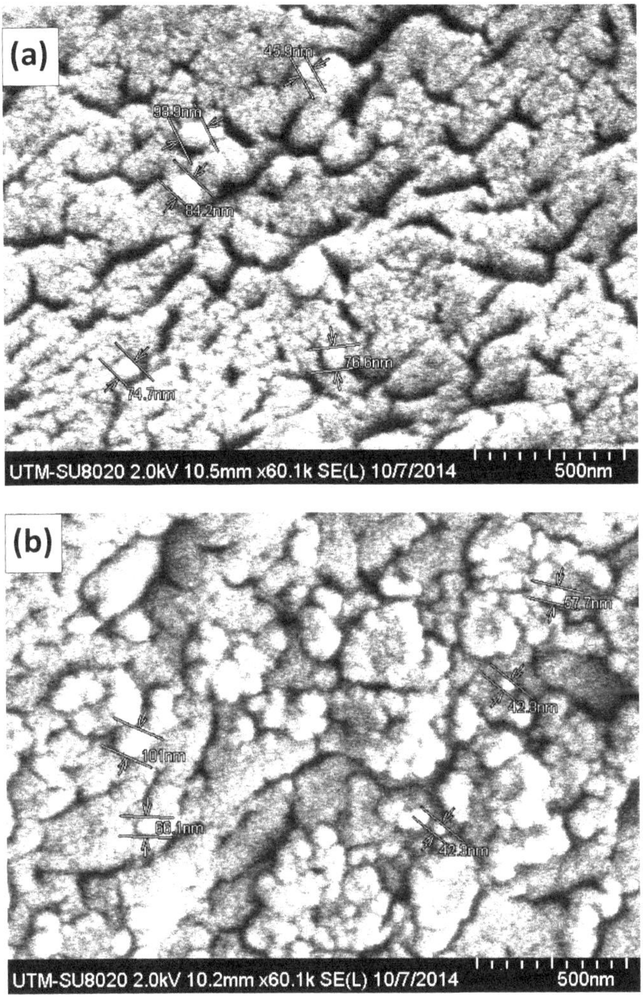

Figure 13.9 FESEM images reveal the details of the agglomeration of TiO_2 nanoparticles distribution on dual-layer hollow fiber membrane surfaces with ratio of (a) 0.5 and (b) 1 [30].

It is well known that the membrane with smoother surfaces has greater antifouling capability. Researchers in Yale University also considered that the membrane roughness was the most effective factor on membrane anti-fouling capability if the operating conditions were not taken into account [37].

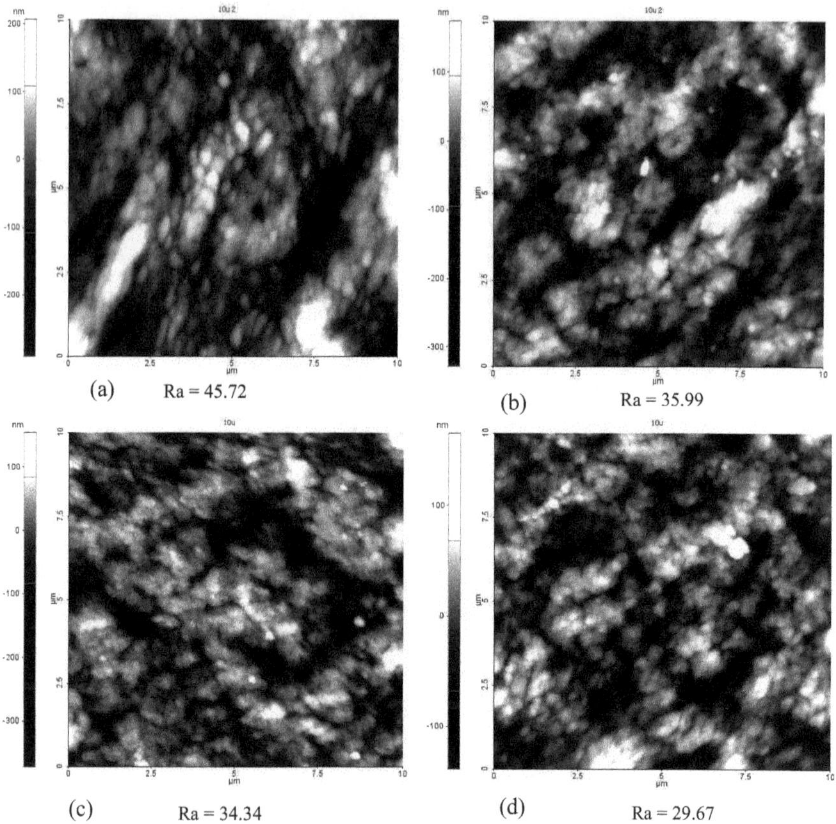

Figure 13.10 AFM topographic images at outer surface with different ratio of TiO_2/PVDF dual-layer hollow fiber membranes (a) 0, (b) 0.2, (c) 0.7 and (d) 1.2 [30].

As mentioned by Ong et al., in 2013, the membrane surface roughness was strongly dependent on the TiO_2 loading. The Ra value (measured by instruments) of dual-layer hollow fiber membranes decreased with the increase in the amount of TiO_2/PVDF ratio compared to the neat membrane. It may due to well distribution of TiO_2 on the outer membrane surface. It can be concluded that the outer surface containing TiO_2 particles is smoother than the neat membranes. It is also interesting to note that, the results obtained in this work is contrary to Ong et al.'s result, which reported that the Ra value of membrane increased from 10.4 to 31.7 nm with increasing concentration of TiO_2 from 0 to 4 wt%. Besides, the composition of dope solution that was employed by Ong and co-researchers was 18 wt% of PVDF with additional 5 wt% of polyvinylpyrrolidone (PVP) as pore forming agent. It should be noted that the addition of PVP has contributed greatly to high membrane porosity as reported by Yuan and Dan Li in 2008 and hence also affected to the surface roughness [38]. As mentioned by Khulbe et al. in 1997, the roughness of the outer surface could be affected by many things occur during the membrane fabrication process [39].

In this study, AFM analysis was used to measure the pore size distribution of the dual-layer hollow fiber membranes incorporated with different TiO_2 load as shown in Fig. 13.11. With the increase of TiO_2/PVDF fraction from 0 to 1.2, a wider pore size distribution and a larger pore size were observed. These pore characteristics are influenced both by the molecular characteristics of the polymer and the preparative method. Similar results were observed by Damodar et al. in which they showed that for composite TiO_2/PVDF flat sheet membranes with 1–4% TiO_2 content, the number of small pores increased compared to neat membrane, and higher addition of TiO_2 (4%) enhanced the formation of larger pore caused by the particle aggregation phenomena as shown in Fig. 13.11.

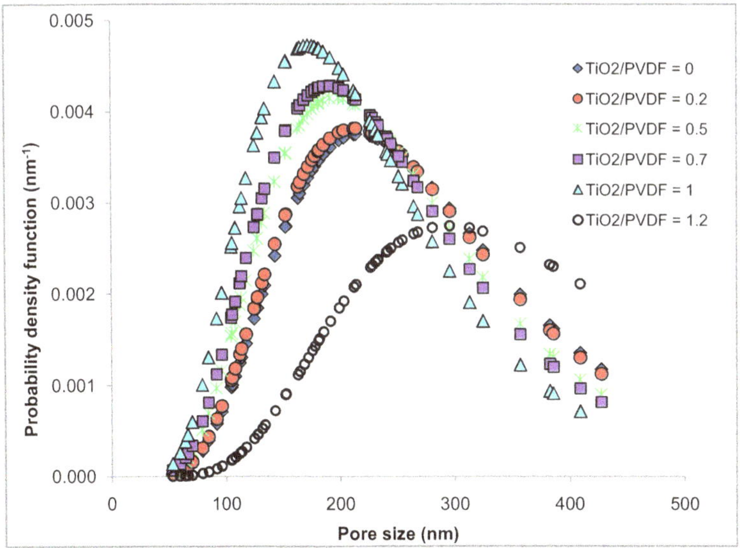

Figure 13.11 Probability density function curve generated from the pore sizes measured by AFM for dual-layer hollow fiber membranes prepared at different ratio of TiO_2/PVDF [40].

Since TiO_2 has higher affinity to water than polymer, the penetration velocity of water into the membrane may increase with TiO_2 content during the coagulation step. More porous were formed in membranes due to the increment of water solvent interdiffusion. During coagulation step, the DMAc which is soluble in water comes out of membranes, whereas TiO_2 particles remain inside the membrane matrix and plug some of the membrane pores. This process can also be affected by particle aggregation phenomena. Because of this, the addition of different amounts of TiO_2 yields different structures which are difficult to predict [25].

Estimation of the outer-layer thickness of the DLHF membranes was performed based on the following assumptions; (1) TiO_2 nanoparticles are distributed uniformly in the outer dope and (2) the diffusion of the TiO_2 nanoparticles into the inner-layer is negligible [19]. The estimated outer-layer thicknesses of the DLHF membranes with TiO_2/PVDF ratios of 0.5, 1.0 and 1.2 are shown in Fig. 13.12. Based on the EDX analysis, it clearly shows that the TiO_2 particles are distributed uniformly in the membrane surface. The thickness of the DLHF membranes was increased with increasing TiO_2 loading. The thickness of the DLHF membranes with TiO_2/PVDF ratio of 1 was two times higher than that of 0.5. Meanwhile, the thickness of the DLHF membranes with TiO_2/PVDF ratio of 1.2 was around 3 times higher than that of 1.0. It may be due to the excessive of TiO_2 nanoparticles over PVDF loadings. However, it also suggests that the single step co-extrusion technique applied in this work is able to produce a hollow fiber membrane with uniformly distributed TiO_2 particles at high loading specifically in the outer layer.

FTIR studied was carried out to ascertain the crystalline phase of PVDF and the relationship between the polymer molecule and the TiO_2 nanoparticles as shown in Fig. 13.15. It could be seen from Fig. 13.13 that the curves of TiO_2/PVDF and neat PVDF satisfactorily superimposed at the higher wavenumber and deviated at the lower wavenumber. The deviation was attributed to the addition of TiO_2. The analysis of the OH area stretched in the spectrum of immobilized TiO_2 in dual-layer hollow fiber membrane exhibited strong bands of asymmetric vibration at 1620 cm^{-1} and peaks below 800 cm^{-1}. The high intensity of these bands can be explained by the high affinity of TiO_2 nanoparticles entrapped in the surface structure of TiO_2/PVDF DLHF membrane to water. The strengthening of OH bonds in the TiO_2 immobilized in DLHF membranes is the main factor in the settlement of TiO_2 nanoparticles on the membrane surface structure.

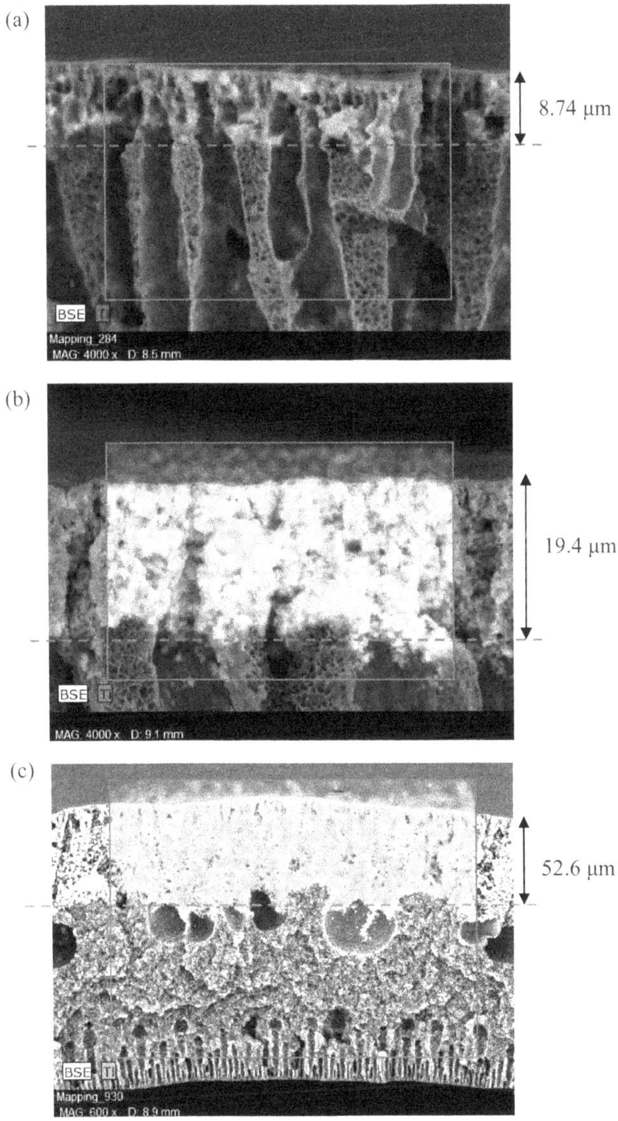

Figure 13.12 SEM-EDX images of the DLHF membranes with TiO_2/PVDF ratios of (a) 0.5, (b) 1 and (c) 1.2 [30].

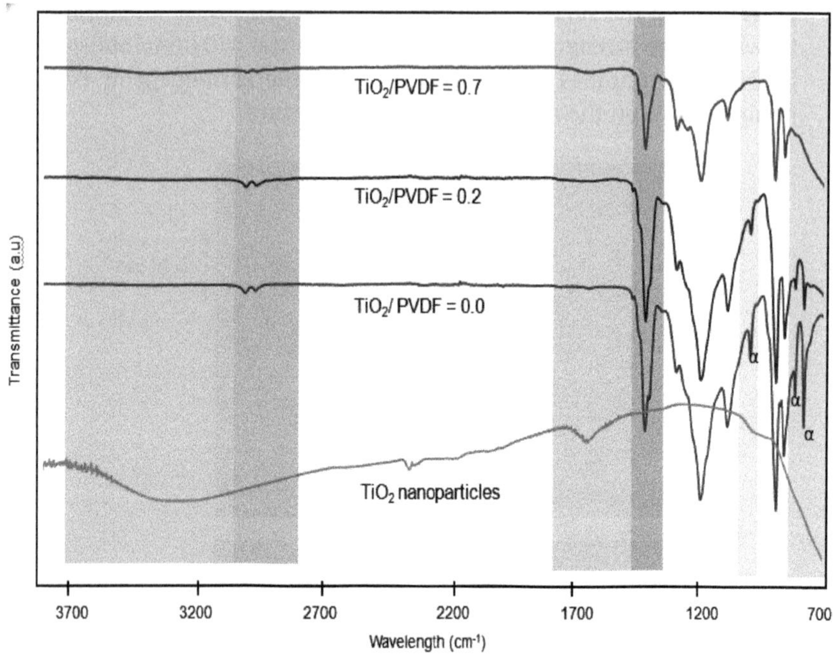

Figure 13.13 FTIR spectra of TiO_2, neat PVDF, 0.2 and 0.7 ratio of TiO_2/PVDF membrane.

The adsorption peak at 1180 cm^{-1} was assigned to the stretching vibration of CF_2 groups. The deformed vibration of CH_2 groups appeared at the frequency of 1404 cm^{-1}. The peaks of 765, 796, 976 and 1383 cm^{-1} could be assigned to the typical peaks of α-phase PVDF crystals [41]. As can be seen in Fig. 13.13, with the increasing of TiO_2 loading at the outer layer, the α crystalline of PVDF has been disappeared. It can be concluded that the α crystal phase of PVDF would shifted to β crystalline structure. This result was supported by XRD result as shown in Fig. 13.14.

XRD diffraction patterns of TiO_2 NPs, PVDF (powder) and PVDF/TiO_2 dual-layer hollow fiber membranes are shown in Fig. 13.14. It can be seen that the XRD pattern of PVDF/TiO_2 dual-layer hollow fiber membranes had three crystalline characteristic peaks at 2 of 20.395°, 25.405° and 38.712° that was analogous with the predominant characteristic peaks of PVDF membrane and TiO_2 crystal powder, respectively. The characteristic peaks of TiO_2 anatase crystalline at 25.4° and 38.7° were predominant in PVDF/TiO_2 dual-layer hollow fiber membranes. These characteristic peaks indicated that there are interactions between polymer and TiO_2 which influenced the PVDF crystal structure (transition of α to β phase) in the dual-layer hollow fiber membranes. The direct evidence of polymorph change of the PVDF membrane can also be seen in Fig. 13.14. The neat PVDF powder exhibited peaks at 2 of 18.69° and 20.11°; characteristic of α-polymorph, which was close to the literature data of 18.5° and 20.2° from Mago et al. in 2008 [42]. As proved by FTIR and XRD analysis, it can be concluded that the addition of TiO_2 in the DLHF membranes could affect the crystalline phase of PVDF in the co-extrusion process.

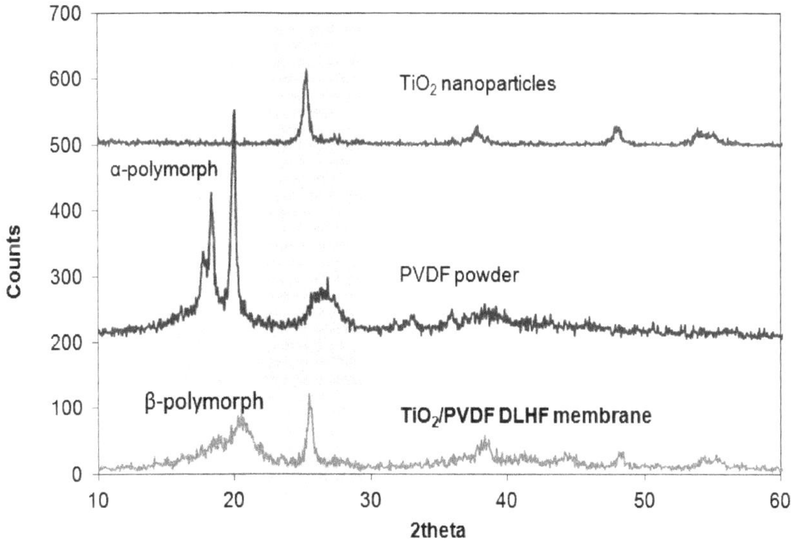

Figure 13.14 XRD pattern of TiO_2 nanoparticles, PVDF powder and TiO_2/PVDF DLHF membranes.

13.6 Filtration Performance

With the incorporation of a small proportion of TiO_2 particles in the outer dope solution, the hydrophilic nature of the TiO_2 particle has successfully improved the permeability of the membrane and achieved its maximum value when the TiO_2/PVDF ratio was 0.7, as shown in Fig. 13.15. The hydrophilic nature allows the TiO_2 to have higher affinity towards water. As the concentration of the TiO_2 was increased, the hydrophilicity of the membranes was improved as well (refer Table 13.2). Generally, water contact angle smaller than 90° indicates the solid surface is hydrophilic in nature, whereas more than 90° indicates the solid surface is hydrophobic [43]. However, when the TiO_2/PVDF ratio was 1 and 1.2, the permeabilities reduced since the finger-like void lengths for inner and outer layers decreased, therefore it hindered the water penetration rate. Another possibility is due to pore blockage by the excessive TiO_2 nanoparticles from the higher loading.

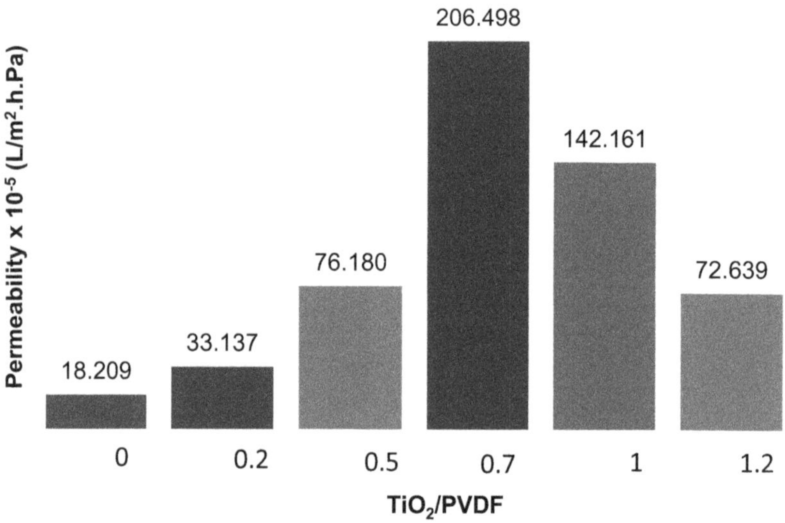

Figure 13.15 Permeabilities of the DLHF membranes with different ratios of TiO_2/PVDF [30].

13.7 Photocatalytic Performance in Hybrid Photocatalytic Membrane Reactor

The dispersion of immobilized TiO_2 particles is one of the most important factors in the fabrication of the hollow fiber membranes as it directly impacts the adsorption capacity and photocatalytic activity. In the photocatalytic degradation of employing dual-layer hollow fiber membranes, only the outer layer is responsible for the reaction to take place due to the presence of TiO_2 nanoparticles as a photocatalyst.

With the increase of TiO_2 concentration, the amount of TiO_2 particles deposited on the surface increased. This outer layer dope was developed in order to allow the pollutants to penetrate the outer layer so the degradation can take place. An increasing number of TiO_2 particles on the membrane surface will likely provide more surface area, thus accelerate the rate of photocatalytic degradation.

As mentioned by previous researchers, the challenge in immobilized photocatalytsts in the membrane is that the contact area between photocatalysts, target pollutant and light intensity is lower than in the suspended photocatalytic system [44–47].

Figure 13.16 shows the photocatalytic degradation of NP by using different ratio of TiO_2/PVDF in outer layer of DLHF membranes. As expected, no degradation occur under photolysis conditions (TiO_2/PVDF = 0). The photocatalytic degradation of NP did not occur since no catalyst was immobilized on the membrane. Nevertheless, the concentration of NP decreased from 10 to 1 and 0.2 mg/L in 330 min, for 0.2 and 0.7 ratio of TiO_2/PVDF respectively. When the ratio of TiO_2/PVDF was equal to 0.5 and 1, NP was not detected in 330 min and 150 min, respectively. This may due to the sufficient amount of catalyst for reaction completion and also the length of finger-like voids in the DLHF membranes in order to allow the NP to penetrate the outer layer to allow degradation.

For 1.2 ratio of TiO_2/PVDF, the highest degradation could be achieved in 50 min under UVA irradiation and after that slow degradation would take place. It may be due to the very high loading of TiO_2 immobilized on the membrane. At first 50 min under UVA irradiation, TiO_2 nanoparticles at the membrane surface were highly activated by UVA irradiation and formed many radicals and aggregated on the membrane surface. These radicals might form peroxide groups on the membrane surface and consequently reduced the photocatalytic performance [48].

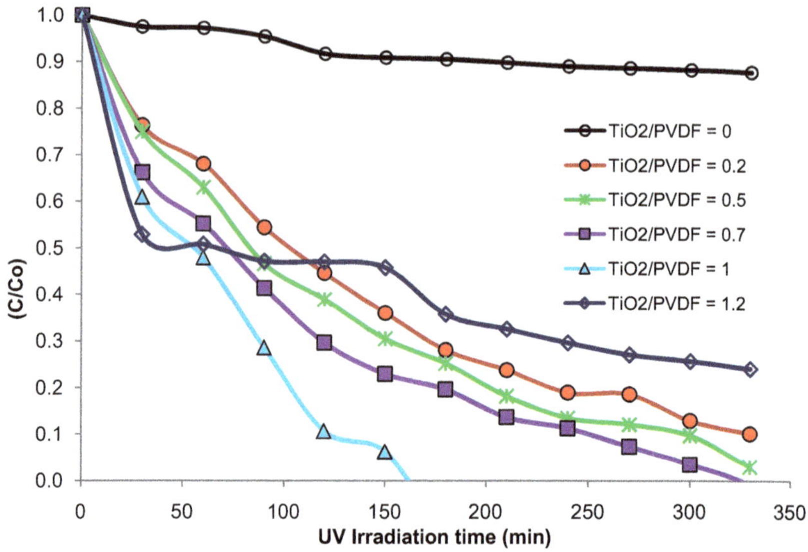

Figure 13.16 Normalized concentration of NP using different ratio of TiO_2/PVDF membrane.

Regarding the NP degradation rate, as reported in Fig. 13.17 and calculated using Langmuir-Hinshelwood model [49], it can be shown that these rates were enhanced by TiO_2 addition in the 0 to 1.0 of TiO_2/PVDF ratio. It is noteworthy that, the degradation rate of pollutant is influenced by the active site and the photoabsorption of the catalyst used. Higher loading of TiO_2 catalysts increases the generation rate of electron/hole pairs to enhance the degradation of NP pollutant.

However, further changes in the higher TiO_2 loadings up to 1.2 of TiO_2/PVDF ratio could affect the NP degradation rate due to the high density of these nanoparticles in the polymer matrix and increased outer thickness layer of DLHF membranes as well. As proved by SEM micrographs in Figure 13.6, the thickness of outer layer increasing with TiO_2 loading. Thus, this increase of the TiO_2 up to 1.0 of TiO_2/PVDF ratio, could achieve higher NP degradation rates. However, it appears that once the formed layer reaches a critical thickness, NP degradation rate decreased and the degradation rate becomes independent of the TiO_2 loading. As mentioned by Ibrahim and de Lasa (2002), the increase in TiO_2 loading produces more rapid rates of UV photon absorption and consequently, higher photoconversion rates up to a limit called "critical thickness" [50].

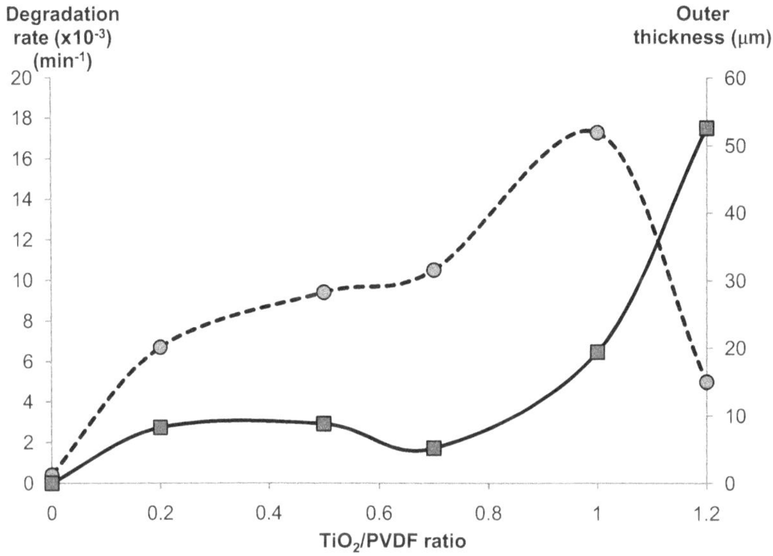

Figure 13.17 Relationship between TiO_2 loadings with outer thickness of DLHF membranes and NP degradation rate.

As shown in Fig. 13.17, the outer thickness was measured from SEM micrographs, allowed estimating the critical layer thickness to be in the 52.6 μm. This result could not be compared with literature due to the different method of immobilized TiO_2. Most of the researchers highlighted the critical thickness based on the TiO_2 film or deposited TiO_2 on the membrane support [51, 52].

13.8 Determination of Degradation Intermediates

The intermediates generated during the photocatalytic degradation process were analyzed by HPLC, according to the retention times and the UV spectra which were based on comparison with standards. HPLC analysis showed the disappearance of NP along with the formation of one product around 0.909 retention time as shown in Fig. 13.18. As reported by Neamţu and Frimmel in 2006, these peaks represent the formation of phenol [53]. These products also decreased with time due to the decomposition to CO_2 and H_2O. We were not able to detect benzoquinone because it converted to dihydroxybenzene during sampling.

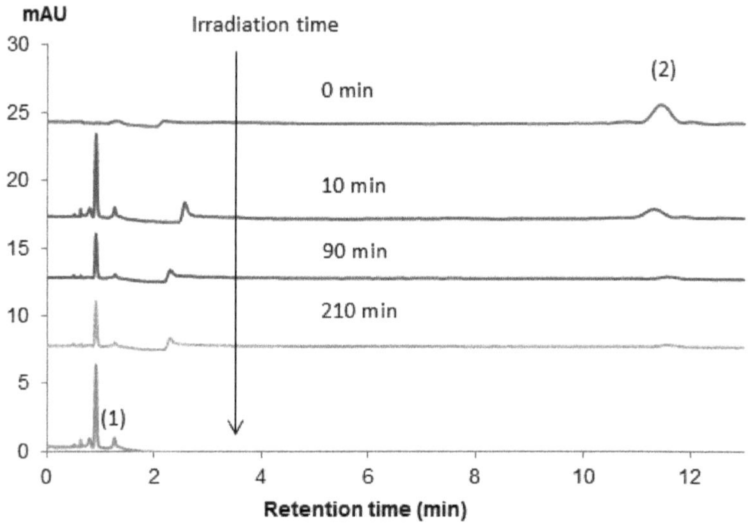

Figure 13.18 HPLC chromatogram of nonylphenol and its intermediate. Initial conditions: 10 mg/L of NP, 0.2 ratio of TiO_2/PVDF, 8W of UVA. Peak identities are as follows: (1) phenol; (2) nonylphenol [40].

As already proposed for other phenols, we assume that the degradation mechanism in the presence of TiO_2 involves the attack of •OH radicals [54–56]. Although available analytical information does not allow the prediction of •OH radical attack site, the considerations concerning previous literature and in analogy with octylphenol oxidation indicate that the primary step involves the hydrogen abstraction from the aliphatic chain [57].

As a membrane forms a barrier for mass transfer in space, one may raise the question whether the TiO_2/PVDF composite membrane also forms a barrier for the light in photocatalytic reaction. After 4 h of UV irradiation, DLHF membranes turned to yellowish color for immobilization of TiO_2 nanoparticles at the outer layer as shown in Fig. 13.19 (b2). On the contrary, the color of dual-layer hollow fiber membranes for the neat membrane does not change (Fig. 13.19 (a1-a2)). This could be explained by the fact that the light is scattered by the anatase particles and the PVDF ribbons can be used by the catalyst particles behind [31].

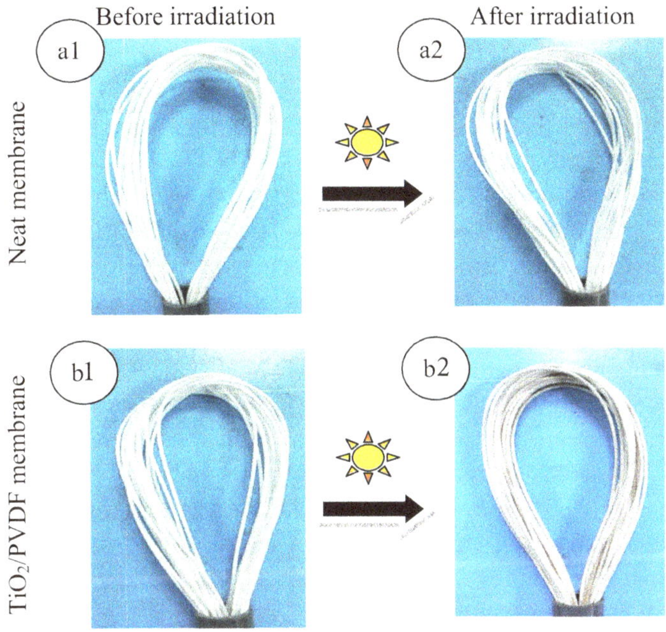

Figure 13.19 Dual-layer hollow fiber membranes before and after the UVA irradiation [40].

Identification of Ti in dual-layer hollow fiber membranes from feed solution and aliquot samples was analyzed by inductively coupled plasma atomic emission spectroscopy (ICPOES) as listed in Table 13.3. As could be seen, there are no Ti was leach out from dual-layer hollow fiber membranes in feed and aliquot solution.

Table 13.3 Ti element in TiO_2/PVDF dual-layer hollow fiber membranes of 1 ratio from ICPOES analysis

	Ti (mg/L)
Feed solution	0.03737 ± 0.000045
Aliquot solution (After 270 min UV irradiation)	0.03709 ± 0.000123

13.9 Conclusion

This chapter has discussed the fabrication of dual-layer hollow fiber membranes with different TiO_2 loadings in the outer layer by a single step co-extrusion technique. The catalytic efficiency of the immobilized TiO_2 with a different ratio of TiO_2/PVDF in the outer layer of the dual-layer hollow fiber membranes was evaluated and analyzed using a hybrid photocatalytic membrane reactor system.

The addition of TiO_2 nanoparticles to the membrane outer layers has improved the membrane properties by increasing their hydrophilicity, pore size and permeability. However, higher loading of TiO_2 has led to the decrease of membrane permeability due to pore blockage by the excessive presence of TiO_2 nanoparticles on the membrane surfaces.

In studying the effect of ratio TiO_2/PVDF on NP degradation, it can be seen that the NP degradation rate was increased with the increasing ratio of TiO_2/PVDF up to 1.0. Nevertheless, at 1.2 of TiO_2/PVDF ratio, NP degradation rate decreased due to the high density of these nanoparticles in the polymer matrix and increased outer thickness layer of DLHF membranes as well. At this ratio, the outer layer of DLHF membranes reaches a critical thickness, therefore, the degradation rate becomes independent of the TiO_2 loading. Overall, it can be concluded that the presence of TiO_2 nanoparticles on the outer layer of DLHF membranes greatly promotes the photocatalytic degradation of NP pollutant. Furthermore, the particles were well distributed in the membrane, thus significantly increased the efficiency of photodegradation. The use of DLHF membranes enables an efficient photocatalytic degradation of NP without additional separation steps for the catalyst particles.

References

1. S. P. Sun, K. Y. Wang, N. Peng, T. A. Hatton, and T.-S. Chung, Novel polyamide-imide/cellulose acetate dual-layer hollow fiber membranes for nanofiltration, *J. Memb. Sci.*, vol. 363, no. 1–2, pp. 232–242, Nov. 2010.

2. D. Bhandari, K. O. Olanrewaju, N. Bessho, V. Breedveld, and W. J. Koros, Dual layer hollow fiber sorbents: Concept, fabrication and characterization, *Sep. Purif. Technol.*, vol. 104, pp. 68–80, Feb. 2013.
3. T. He, M. H. V Mulder and M. Wessling, Preparation of Porous Hollow Fiber Membranes with a Triple-Orifice Spinneret. *J. Appl. Poly. Sci.*, vol. 87, pp. 2151–2157, 2003.
4. D. Li, T.-S. Chung, and R. Wang, Morphological aspects and structure control of dual-layer asymmetric hollow fiber membranes formed by a simultaneous co-extrusion approach, *J. Memb. Sci.*, vol. 243, no. 1–2, pp. 155–175, Nov. 2004.
5. C. Pereira, Hollow fiber membranes obtained by simultaneous spinning of two polymer solutions: A morphological study, *J. Memb. Sci.*, vol. 226, no. 1–2, pp. 35–50, Dec. 2003.
6. T. He, M. H. Mulder, H. Strathmann, and M. Wessling, Preparation of composite hollow fiber membranes: Co-extrusion of hydrophilic coatings onto porous hydrophobic support structures, *J. Memb. Sci.*, vol. 207, no. 2, pp. 143–156, Sep. 2002.
7. L. Setiawan, R. Wang, L. Shi, K. Li, and A. G. Fane, Novel dual-layer hollow fiber membranes applied for forward osmosis process, *J. Memb. Sci.*, vol. 421–422, pp. 238–246, Dec. 2012.
8. Li Dongfei, Dual layer hollow fiber membranes for gas separation. National University of Singapore, 2004.
9. L. Jiang, T. Chung, D. Fei, C. Cao, and S. Kulprathipanja, Fabrication of Matrimid/polyethersulfone dual-layer hollow fiber membranes for gas separation, vol. 240, pp. 91–103, 2004.
10. S. Husain, Mixed matrix dual layer hollow fiber membranes for natural gas separation, Georgia Institute of Technology, 2006.
11. N. Widjojo, T. Chung, and S. Kulprathipanja, The fabrication of hollow fiber membranes with double-layer mixed-matrix materials for gas separation, vol. 325, pp. 326–335, 2008.
12. D. Fei, T. Chung, R. Wang, and Y. Liu, Fabrication of fluoropolyimide/polyethersulfone (PES) dual-layer asymmetric hollow fiber membranes for gas separation, vol. 198, pp. 211–223, 2002.
13. A. L. Ahmad and W. K. W. Ramli, Hydrophobic PVDF membrane via two-stage soft coagulation bath system for membrane gas absorption of CO_2, *Sep. Purif. Technol.*, vol. 103, pp. 230–240, Jan. 2013.
14. S. P. Sun, T. A. Hatton, S. Y. Chan, and T.-S. Chung, Novel thin-film composite nanofiltration hollow fiber membranes with double

repulsion for effective removal of emerging organic matters from water, *J. Memb. Sci.*, vol. 401–402, pp. 152–162, May 2012.

15. P. Wang, M. M. Teoh, and T. Chung, Morphological architecture of dual-layer hollow fiber for membrane distillation with higher desalination performance, *Water Res.*, vol. 45, no. 17, pp. 5489–5500, 2011.
16. S. Bonyadi and T. S. Chung, Flux enhancement in membrane distillation by fabrication of dual layer hydrophilic–hydrophobic hollow fiber membranes, vol. 306, pp. 134–146, 2007.
17. Y. K. Ong and T. Chung, High performance dual-layer hollow fiber fabricated via novel immiscibility induced phase separation (I2PS) process for dehydration of ethanol, *J. Memb. Sci.*, vol. 421–422, pp. 271–282, 2012.
18. L. Setiawan, L. Shi, W. B. Krantz, and R. Wang, Explorations of delamination and irregular structure in poly (amide-imide)-polyethersulfone dual layer hollow fiber membranes, *J. Memb. Sci.*, vol. 423–424, pp. 73–84, 2012.
19. M. M. Teoh, T. Chung, and Y. S. Yeo, Dual-layer PVDF/PTFE composite hollow fibers with a thin macrovoid-free selective layer for water production via membrane distillation, *Chem. Eng. J.*, vol. 171, no. 2, pp. 684–691, 2011.
20. N. Peng, N. Widjojo, P. Sukitpaneenit, M. M. Teoh, G. G. Lipscomb, T.-S. Chung, and J.-Y. Lai, Evolution of polymeric hollow fibers as sustainable technologies: Past, present, and future, *Prog. Polym. Sci.*, vol. 37, no. 10, pp. 1401–1424, Oct. 2012.
21. H. Tseng, G. Zhuang, and Y. Su, The effect of blending ratio on the compatibility, morphology, thermal behavior and pure water permeation of asymmetric CAP/PVDF membranes, *DES*, vol. 284, pp. 269–278, 2012.
22. L. Yan, S. Hong, M. L. Li, and Y. S. Li, Application of the Al_2O_3–PVDF nanocomposite tubular ultrafiltration (UF) membrane for oily wastewater treatment and its antifouling research, *Sep. Purif. Technol.*, vol. 66, no. 2, pp. 347–352, Apr. 2009.
23. L. Yu, H. Shen, and Z. Xu, PVDF–TiO_2 composite hollow fiber ultrafiltration membranes prepared by TiO_2 sol–gel method and blending method, *J. Appl. Polymer Sci.*, vol. 113, no 35 1763–1772, 2009.
24. H. Song, J. Shao, Y. He, B. Liu, and X. Zhong, Natural organic matter removal and flux decline with PEG–TiO_2-doped PVDF membranes

by integration of ultrafiltration with photocatalysis, *J. Memb. Sci.*, vol. 405–406, pp. 48–56, Jul. 2012.

25. T.-H. Bae and T.-M. Tak, Effect of TiO_2 nanoparticles on fouling mitigation of ultrafiltration membranes for activated sludge filtration, *J. Memb. Sci.*, vol. 249, no. 1–2, pp. 1–8, Mar. 2005.
26. K. Yu, T. Chung, and M. Gryta, Hydrophobic PVDF hollow fiber membranes with narrow pore size distribution and ultra-thin skin for the fresh water production through membrane distillation, vol. 63, pp. 2587–2594, 2008.
27. A. Razmjou, J. Mansouri, and V. Chen, The effects of mechanical and chemical modification of TiO_2 nanoparticles on the surface chemistry, structure and fouling performance of PES ultrafiltration membranes, *J. Memb. Sci.*, vol. 378, no. 1–2, pp. 73–84, Aug. 2011.
28. G. Wu, S. Gan, L. Cui, and Y. Xu, Preparation and characterization of PES/TiO_2 composite membranes, *Appl. Surf. Sci.*, vol. 254, no. 21, pp. 7080–7086, Aug. 2008.
29. N. K. O. Cruz, G. U. Semblante, D. B. Senoro, S.-J. You, and S.-C. Lu, Dye degradation and antifouling properties of polyvinylidene fluoride/titanium oxide membrane prepared by sol–gel method, *J. Taiwan Inst. Chem. Eng.*, vol. 45, no. 1, pp. 192–201, Jan. 2014.
30. H. Dzinun, M. H. D. Othman, A. F. Ismail, M. Hafiz Puteh, M. A. Rahman, and J. Jaafar, Morphological study of co-extruded dual-layer hollow fibre membranes incorporated with different TiO_2 loadings, *J. Memb. Sci.*, DOI: 10.1016/j.memsci.2014.12.052 Jan. 2015.
31. O. Tahiri, Q. Trong, C. Mbareck, and T. Rhlalou, Elaboration and study of poly (vinylidene fluoride)-anatase TiO_2 composite membranes in photocatalytic degradation of dyes, *Appl. Catal. A: General*, vol. 358, pp. 13–20, 2009.
32. M. H. D. Othman, Z. Wu, N. Droushiotis, U. Doraswami, G. Kelsall, and K. Li, Single-step fabrication and characterisations of electrolyte/ anode dual-layer hollow fibres for micro-tubular solid oxide fuel cells, *J. Memb. Sci.*, vol. 351, no. 1–2, pp. 196–204, Apr. 2010.
33. J. Qi, Y. Y. Ye, J. J. Wu, H. T. Wang, and F. T. Li, Dispersion and stability of titanium dioxide nanoparticles in aqueous suspension: Effects of ultrasonication and concentration, *Water Sci. Technol.*, vol. 67, no. 1, p. 147–151, 2013.
34. M. N. Chong, B. Jin, C. W. K. Chow, and C. Saint, Recent developments in photocatalytic water treatment technology: A review., *Water Res.*, vol. 44, no. 10, pp. 2997–3027, May 2010.

35. Y. H. Teow, A. L. Ahmad, J. K. Lim, and B. S. Ooi, Preparation and characterization of PVDF/TiO_2 mixed matrix membrane via in situ colloidal precipitation method, *Desalination*, vol. 295, pp. 61–69, Jun. 2012.
36. M. E. Mackay, A. Tuteja, P. M. Duxbury, C. J. Hawker, B. van Horn, Z. Guan, and R. S. Krishnan, General strategies for nanoparticle dispersion, *Science*, no. 311, pp. 1740–1743, 2006.
37. X. Cao, J. Ma, X. Shi, and Z. Ren, Effect of TiO_2 nanoparticle size on the performance of PVDF membrane, *Appl. Surf. Sci.*, vol. 253, no. 4, pp. 2003–2010, Dec. 2006.
38. Z. Yuan and X. Dan-Li, Porous PVDF/TPU blends asymmetric hollow fiber membranes prepared with the use of hydrophilic additive PVP (K30), *Desalination*, vol. 223, no. 1–3, pp. 438–447, Mar. 2008.
39. K. C. Khulbe, T. Matsuura, G. Lamarche, and H. J. Kim, The morphology characterisation and performance of dense PPO membranes for gas separation, *J. Membr. Sci.*, vol. 135, pp. 211–223, 1997.
40. H. Dzinun, M. H. D. Othman, A. F. Ismail, M. H. Puteh, M. A. Rahman, and J. Jaafar, Photocatalytic degradation of nonylphenol using co-extruded dual-layer hollow fibre membranes incorporated with a different ratio of TiO_2/PVDF, *React. Funct. Polym.*, vol. 99, pp. 80–87, Feb. 2016.
41. F. Shi, Y. Ma, J. Ma, P. Wang, and W. Sun, Preparation and characterization of PVDF/TiO_2 hybrid membranes with different dosage of nano-TiO_2, *J. Memb. Sci.*, vol. 389, pp. 522–531, Feb. 2012.
42. G. Mago, D. M. Kalyon, and F. T. Fisher, Membranes of polyvinylidene fluoride and PVDF nanocomposites with carbon nanotubes via immersion precipitation, *J. Nanomater.*, vol. 2008, pp. 1–8, 2008.
43. Y. Yuan and T. R. Lee, *Surface Science Techniques*, vol. 51. Berlin, Heidelberg: Springer Berlin Heidelberg, 2013.
44. R. Molinari, M. Mungari, E. Drioli, A Di Paola, V. Loddo, L. Palmisano, and M. Schiavello, Study on a photocatalytic membrane reactor for water purification, *Catal. Today*, vol. 55, no. 1–2, pp. 71–78, Jan. 2000.
45. S. Kertèsz, J. Cakl, and H. Jiránková, Submerged hollow fiber microfiltration as a part of hybrid photocatalytic process for dye wastewater treatment, *Desalination*, vol. 343, 106–112, 2013.
46. G. Mascolo, R. Comparelli, M. L. Curri, G. Lovecchio, A Lopez, and A Agostiano, Photocatalytic degradation of methyl red by TiO_2: Comparison of the efficiency of immobilized nanoparticles versus conventional suspended catalyst., *J. Hazard. Mater.*, vol. 142, no. 1–2, pp. 130–7, Apr. 2007.

47. M. Umar and H. A. Aziz, Photocatalytic Degradation of Organic Pollutants in Water. *Organic Pollutants-Monitoring, Risk and Treatment*, pp. 195-208. 2013.

48. A. Rahimpour, S. S. Madaeni, A. H. Taheri, and Y. Mansourpanah, Coupling TiO_2 nanoparticles with UV irradiation for modification of polyethersulfone ultrafiltration membranes, *J. Memb. Sci.*, vol. 313, no. 1–2, pp. 158–169, Apr. 2008.

49. X. Chang, J. Huang, C. Cheng, W. Sha, X. Li, G. Ji, S. Deng, and G. Yu, Photocatalytic decomposition of 4-t-octylphenol over $NaBiO_3$ driven by visible light: catalytic kinetics and corrosion products characterization, *J. Hazard. Mater.*, vol. 173, no. 1–3, pp. 765–72, Jan. 2010.

50. H. Ibrahim and H. De Lasa, Photo-catalytic conversion of air borne pollutants effect of catalyst type and catalyst loading in a novel photo-CREC-air unit, *Appl. Catal. B Environ.*, vol. 38, pp. 201–213, 2002.

51. A. Danion, J. Disdier, C. Guillard, F. Abdelmalek, and N. Jaffrezic-Renault, Characterization and study of a single-TiO_2-coated optical fiber reactor, *Appl. Catal. B Environ.*, vol. 52, no. 3, pp. 213–223, Sep. 2004.

52. W. Choi, J. Chung, C. Cho, and J. Kim, Fabrication and photocatalytic activity of a novel nanostructured TiO_2 metal membrane, *Desalination*, vol. 279, no. 1–3, pp. 359–366, 2011.

53. M. Neamţu and F. H. Frimmel, Photodegradation of endocrine disrupting chemical nonylphenol by simulated solar UV-irradiation, *Sci. Total Environ.*, vol. 369, pp. 295–306, 2006.

54. K. Inumaru, M. Murashima, T. Kasahara, and S. Yamanaka, Enhanced photocatalytic decomposition of 4-nonylphenol by surface-organografted TiO_2: A combination of molecular selective adsorption and photocatalysis, *Appl. Catal. B: Environ.*, vol. 52, pp. 275–280, 2004.

55. Y. Li, X. Duan, X. Li, and D. Zhang, Photodegradation of nonylphenol by simulated sunlight, *Mar. Pollut. Bull.*, vol. 66, no. 1–2, pp. 47–52, 2013.

56. S. Naya, T. Nikawa, K. Kimura, and H. Tada, Rapid and complete removal of nonylphenol by gold nanoparticle/rutile titanium(IV) oxide plasmon photocatalyst, *ACS Catal.*, vol. 3, no. 5, pp. 903–907, May 2013.

57. N. Brand, G. Mailhot, and Å. L. E. Bolte, Degradation photoinduced by Fe(III): method of alkylphenol ethoxylates removal in water, *Environ. Sci. Technol.*, vol. 33, no. 18, pp. 2715–2720, Dec 1998.

Chapter 14

Frontiers of Application of Nanocomposites and the Wide Vision of Membrane Science: A Critical Overview and a Vision for the Future

Sukanchan Palit[a] **and Chaudhery Mustansar Hussain**[b]

[a]*Department of Chemical Engineering, University of Petroleum and Energy Studies, Post-Office Bidholi via Premnagar, Dehradun 248007, Uttarakhand, India*
[b]*Department of Chemistry and Environmental Science, NJIT, University Heights, Newark, New Jersey 07102, USA*

sukanchan68@gmail.com, sukanchan92@gmail.com, chaudhery.m.hussain@njit.edu

14.1 Introduction

The visionary world of environmental engineering science and chemical process engineering is undergoing rapid changes. Human scientific endeavours in the field of environmental engineering are focused towards the new-generation nanotechnology and the vast and wide applications of membrane science. Global water crisis, environmental engineering catastrophes and the imminent concern of ecological imbalance are ushering in a new era in scientific endeavour in membrane science. This chapter brings forward to the scientific domain the immense applications

Nanocomposites for Pollution Control
Edited by Chaudhery Mustansar Hussain and Ajay Kumar Mishra

ISBN 978-981-4774-45-1 (Hardcover), 978-1-315-14368-2 (eBook)
www.panstanford.com

of nanocomposites and the wide and versatile world of membrane science. The author delineates with deep comprehension and cogent insight the challenges, the scientific vision and the scientific astuteness in the future scientific endeavour in membrane science. In today's world, the science of membranes and environmental engineering science are linked by an unsevered umbilical cord. The author carefully observes the success of the application of nanocomposites in environmental engineering science and water treatment issues. Water and wastewater treatment issues are challenging the wide scientific horizon. This area of science and engineering needs to be revisited and re-envisioned with every step of human civilization. The author pointedly reviews both the domains of membrane science and nanocomposites with minute details. This scientific challenge and the wide scientific foresight will surely open up new windows of innovation and instinct in the decades to come. The world of environmental engineering science and chemical process engineering is moving from one paradigm to another. Science and engineering are immense colossus with a definite vision. Scientific rigour and scientific vision are changing the landscape of human scientific endeavour. Nanocomposites and their application are the new avenues of scientific innovation and deep scientific introspection. Water and wastewater pollution control are invariably in today's world linked with research and development initiatives in nanocomposites. Nanoadsorbents are another area of scientific vision. This chapter opens up a new chapter and a newer visionary future in the field of nanocomposites and their application in industrial pollution control. The challenge and vision of science are surpassing wide frontiers. Scientific validation and scientific scale-up are changing the face of chemical process design. The history of human scientific pursuit, the wide scientific progress and the immense scientific rigour will all lead a long way in the true emancipation of nano-science and nano-engineering. Environmental engineering applications of nanocomposites are another facet of this rigorous scientific endeavour. A new wave of science and engineering has ushered in the scientific horizon. The coin word of nanotechnology has revolutionized the scientific landscape. The author pointedly focuses on the wide and versatile domain of nanotechnology primarily on the application of nanocomposites. The vision of science is truly groundbreaking as human civilization moves from one visionary decade over another. Nanocomposites are the next great challenging area of nanotechnology. The application of nanocomposites in environmental protection is the major backbone of this chapter.

14.2 The Vision of the Chapter

The objective of this chapter is to unfold the wide domain of the application of nanocomposites in environmental engineering science and its relevant application in membrane science. Environmental engineering science today stands in the midst of deep scientific understanding and scientific introspection. Grave environmental engineering concerns, unmitigated environmental calamities and industrial pollution have urged the scientific community to devise new technologies. The author lucidly discusses the vision of the application of nanocomposites in membrane science and environmental protection. The vicious and arduous challenges facing nanotechnology today are immense and unimaginable. The author repeatedly lays stress on the future research and development initiatives, the futuristic trends of environmental engineering and the immense futuristic potential of nanocomposites and their applications.

14.3 The Need and the Rationale of the Study

The rationale of the study is immense, arduous and far-reaching. Human civilization and human scientific endeavour are in the path of immense re-envisioning. Environmental sustainability and sustainable development are in the avenue of scientific grit and scientific forbearance. Environmental disasters and grave ecological concerns stand as major imperatives to such a widely informed study. Nanocomposites and vast domains of nanotechnology are witnessing one paradigmatic change over another [25, 26]. The scientific urge to excel, the progress of technology and the futuristic potential of nanocomposites and membrane science will lead a long and visionary way in the true emancipation of both environmental and energy sustainability. Due to defunct sustainable development globally, the need for and the rationale of such a vast and versatile area is highly imperative. Membrane separation processes and novel separation phenomenon are witnessing global arduous challenges. The author, with cogent insight, discusses the success of application of nanocomposites in membrane science and the immense scientific furtherance of nano-science and nanotechnology [25, 26].

14.4 The Scope of the Study

The scope of the study of membrane science and nanocomposites is vast, versatile and arduous. Scientific vision and scientific candour are veritably changing the landscape of scientific innovation. Membrane science in today's world is linked by an umbilical cord with environmental engineering science. The scientific vision and the scientific journey of nanotechnology and membrane science are far-reaching. The imminent need and the veritable challenge of science today are the wide world of global water initiatives. Global water crisis and the innovations of nanotechnology are the backbones of this well-observed study. The vision of technology and engineering are wide and replete with scientific acuity. Membrane science is a wondrous branch of environmental engineering science today. The author repeatedly lays stress on the veritable facts about global water calamities. The history of science and engineering is witnessing drastic challenges as global water initiatives are being re-envisioned [25, 26].

14.5 What Are Composites?

Composites are combinations of two materials in which one of the materials, called the reinforcing phase, is in the form of fibres, sheets, or particles, and is embedded in the other material called the matrix phase. Typically, reinforcing materials are strong with low densities while the matrix is usually a ductile or tough material. If the composite is designed and fabricated correctly, it combines the strength of the reinforcement with the toughness of the matrix to achieve a combination of desirable properties not available in any single conventional material. The vision of science of composites is unimaginable and the relevant scientific endeavour is crossing vast scientific frontiers. Materials science and metallurgical engineering are the visionary hallmark of our times [25, 26].

Natural composites are wood and bamboo. Wood is a good example of a natural composite, combination of cellulose fibre and lignin. The cellulose fibre provides strength and the lignin is the "glue" that bonds and stabilizes the fibre. Bamboo is a very efficient wood composite structure. The components are cellulose and lignin, as in all other wood; however, bamboo is hollow. This results in a very light yet stiff structure [25, 26].

14.5.1 Fibre-Reinforced Composites

Fibre phase: The requirements of the fibre are that the small diameter fibre must be much stronger than the bulk material. It should also have higher strength (whiskers, fibres, wires) [25, 26].

Matrix Phase: It (1) binds fibres together, (2) acts as a medium through which externally applied stress is transmitted and distributed to the fibres, (3) protects fibres from surface damage, and (4) separates fibres and prevents a crack from one fibre propagating through another [25, 26].

14.5.2 Engineering Applications

Composite materials have been used in aerospace, automobile and marine applications. Recently, composite materials have been increasingly considered in civil engineering structures. The latter applications include seismic retrofit of bridge columns, replacements of deteriorated bridge decks and new bridge structures [25, 26].

14.5.3 Medical Applications

Stents are made with steel and more recently with composite polymers with shape memory effects. The material is deformed within a temperature range of glass transition temperature (T_g) of amorphous phase and melting temperature (T_m) of crystalline phase, then was cooled below T_g. After the material was reheated between T_g and T_m, the original structural shape was recovered. High dosage (up to 35% by weight) and at a high rate of release of medication are noted in various scientific research pursuits [25, 26].

14.5.4 Limitations of Composites

Properties of materials are highly anisotropic due to orientation of fibres. The strength perpendicular to the direction of alignment is considerably less. There is a loss of optical/electrical/chemical (barrier) properties. Although there are limitations of composites applications, the challenge and vision remain truly exemplary and groundbreaking. Composite science is undergoing one paradigmatic shift over another. The wide vision of science of composites are opening up one chapter over another in the scientific domain of materials science [25, 26].

14.6 What Do You Mean by Nanocomposites?

A nanocomposite is a multiphase solid material where one of the phases has one, two or three dimensions of less than 100 nanometres (nm), or structures having nano scale repeat distances between the different phases that make up the material. In the broadest sense, this definition can include porous media, colloids, gels and copolymers, but is more usually taken to mean the solid combination of a bulk matrix and nano-dimensional phases differing in properties due to dissimilarities in structure and chemistry. The mechanical, electrical, thermal, optical, electrochemical, catalytic properties of the nanocomposite will differ markedly from that of the component materials. Size limits for these effects have been proposed, less than 5 nm for catalytic activity, less than 20 nm for making a hard magnetic material soft, less than 50 nm for refractive index changes, and less than 100 nm for achieving superparamagnetism, mechanical strengthening or restricting matrix dislocation movement [25, 26].

Nanocomposites are found in nature, for example in the structure of the abalone shell and bone. The use of nanoparticle-rich materials long predates the understanding of the physical and chemical nature of the materials. From the mid-1950s nanoscale organo-clays have been used to control the flow of polymer solutions (e.g. as paint viscosifiers) or the constitution of gels (e.g. as a thickening substance in cosmetics, keeping the preparations in homogeneous form). By the 1970s polymer/clay composites were the topic of textbooks, although the term "nanocomposites" was not in common use [25, 26].

In mechanical terms, nanocomposites differ from conventional composite materials due to the exceptionally high surface to volume ratio of the reinforcing phase and/or its exceptionally high aspect ratio. The reinforcing material can be made of particles (e.g. minerals), sheets (e.g. exfoliated clay stacks) or fibres (e.g. carbon nanotubes or electrospun fibres). The area of the interface between the matrix and reinforcement phases is typically an order of magnitude greater than for conventional composite materials [25, 26]. The matrix material properties are significantly affected in the vicinity of the reinforcement.

14.7 Nanocomposites-Scientific Doctrine and Deep Scientific Vision

Nanocomposites, a high-performance material, exhibit unusual property combinations and unique design possibilities [25, 26]. With an estimated annual growth rate of about 25% and the fastest demand to be in engineering plastics and elastomers, their potential is so invigorating that they are useful in several areas ranging from packaging to biomedical applications. Scientific and technological vision is enshrined with the emergence of nanocomposites and with the forays of innovative avenue of materials science. Technology is a huge colossus with high vision. Human scientific research pursuit in nanocomposites is veritably opening new chapters in materials science. Chemical process engineering and nanotechnology are the burgeoning areas of science today [25, 26].

Nanocomposites are composites in which at least one of the phases shows dimensions in the nanometer range (1 nm = 10^{-9} m). Nanocomposite materials have emerged as suitable alternatives to overcome limitations of microcomposites and monolithic, while posing preparation challenges related to the control of elemental composition and stoichiometry in the nanocluster phase. The science and engineering of nanocomposites are surpassing wide and vast visionary frontiers. The challenge of scientific endeavour is truly unending. Nanocomposites are the materials of the 21st century in the view of possessing design uniqueness and property combinations that are not found in traditional composites. The general understanding of the science of nanocomposites and their properties is yet to be reached, even though the first inference on them was reported early as in 1992 [25, 26].

Nanocomposites are veritably a newer and visionary domain of science. The scientific evolution and the vast application domain of nanocomposites are far-reaching. The number of published papers containing words such as nanoscience, nanotechnology and nanomaterials doubled in 1.6 years in the late 1990s. It has been reported that changes in particle properties can be observed when the particle size is less than a particular level, called 'the critical size'. Additionally, as diameters reach the nanometer level, interactions at phase interfaces become largely improved, and this is important to enhance material properties. In this context, the surface area/volume ratio of reinforcement materials employed in the preparation of nanocomposites is crucial to the understanding of their structure–property relationships. Further, discovery of carbon nanotubes (CNT) in 1991 and the subsequent use to fabricate composites exhibiting some of the unique CNT-related mechanical, thermal and electrical properties added a new and interesting dimension to this far-reaching area [25, 26].

14.8 A Survey of the Applications of Nanocomposites

The following survey of nanocomposite applications introduces readers to the many avenues being explored, including the following:

Producing batteries with greater power output: Researchers have developed a method to make anodes for lithium ion batteries from a composite formed with silicon nanospheres and carbon nanoparticles. The anodes made of silicon–carbon nanocomposite make closer contact with the lithium electrolyte, which allows faster charging or discharging of power [25, 26].

Speeding up the healing process for broken bones: Researchers have shown that the growth of replacement bone is speeded up when a nanotube polymer nanocomposite is placed as a kind of scaffold which guides growth of replacement bone. They are conducting studies to better understand how this nanocomposite increases bone growth. The challenge of science is veritably beyond scientific imagination. In today's world, scientific vision and scientific sagacity are of utmost importance. Human scientific rigour, the advancements of science and the progress of engineering are the hallmarks towards a visionary application domain of nanocomposites and nanomaterials today [25, 26].

Producing structural components with high strength-to-weight ratio: For example, an epoxy containing carbon nanotubes can be used to produce nanotube-polymer composite windmill blades. This results in a strong but lightweight blade, which makes longer windmill blades practical. These longer blades increase the amount of electricity generated by each windmill [25, 26].

Using graphene to make composites with even higher strength to weight ratios: Researchers have found that adding graphene to epoxy composites may result in stronger/stiffer components than epoxy composites using a similar weight of carbon nanotubes. Graphene appears to bond better to the polymers in the epoxy, allowing a more effective coupling of the graphene into the structure of the composite. This property could result in the manufacture of components with higher strength-to-weight ratios for such uses as windmill blades or aircraft components. The vision and challenge of nanoscience is slowly unfolding with each step of scientific research pursuit [25, 26].

Making lightweight sensors with nanocomposites: A polymer nanotube nanocomposite conducts electricity; how well it conducts depends upon the spacing of the nanotubes. This property allows patches of polymer-nanotube nanocomposite to act as stress sensors on windmill blades [25, 26].

Using nanocomposites to make flexible batteries: A nanocomposite of cellulose materials and nanotubes could be used to make a conductive paper [25, 26].

Making it easier to see and remove tumours: Researchers are attempting to join magnetic nanoparticles and fluorescent nanoparticles in a nanocomposite particle that is both magnetic and fluorescent. The magnetic property of the nanocomposite particle makes the tumour more visible during an MRI procedure done prior to surgery [25, 26].

14.9 Scientific and Technological Objectives in the Application Domain of Nanocomposites

The objectives of science and technology are vast and varied. In a similar vein, nanocomposites and their application are surpassing wide frontiers of science. Nanocomposites and application of materials science are the forerunners towards a greater vision of nanoscience and nanotechnology. Human civilisation's progress, the immense academic rigour and the world of scientific notion all will lead a long way in the true realization of nanotechnology. Technological and scientific rigour in the field of application of nanocomposites is surpassing visionary frontiers. The challenge of the science of nanotechnology needs to be rebuilt and re-enshrined. The science of nanocomposites has a deep scientific vision of its own. The target of nanocomposites application should be towards innovative membrane fabrication and successful membrane separation phenomenon. True science of nanocomposites today is slowly unfolding.

Nanocomposites and their visionary applications are moving towards newer directions today. Human scientific endeavour and research pursuit are witnessing drastic challenges. The field of materials science and the engineering of composites are today crossing visionary boundaries [25, 26].

14.10 Nanocomposites and Environmental Protection

Nanocomposites and their application in environmental protection are of utmost importance in the success of membrane science endeavour today. The challenge, the scientific vision and the deep scientific fortitude are the pallbearers of vision and motivation today. The author deeply comprehends the present status and the future trends of the application of nanocomposites and the related areas of membrane science in the furtherance of environmental engineering science. Chemical process design is another facet of human scientific endeavour. Water treatment and industrial wastewater treatment are moving towards a globally new phase with the passage of human history and time. Nanotechnology re-envisions itself towards a newer scientific eon [25, 26].

14.11 Recent Scientific Research Pursuit in the Field of Composites

Campbell (2010) [1] dealt lucidly in a review paper on composite materials. The scientific vision and scientific understanding in the field of composites are widely expanding. This chapter dealt with both continuous and discontinuous polymer, metal, and ceramic matrix composites with an emphasis on continuous—fiber, high-performance polymer composites [1].

A composite material can be defined as a combination of two or more materials that results in better properties than those of the individual components alone. The challenge of science and its motivation and objectives are well defined at each step of scientific research pursuit. In contrast to metal alloys, each material retains its separate chemical, physical and mechanical properties. The two components are a reinforcement and a matrix [1].

Harris (1999) [2] comprehended with deep comprehension the science of engineering composite materials. Composite materials are extending the horizons of designers in all branches of engineering, and yet the degree to which this is happening can easily pass unperceived. The widening family of synthesized materials offers the possibility of exciting new solutions and new innovations to difficult engineering problems. This chapter goes deep into the science of composites—its scientific truth and scientific conscience. Conventional materials have lot of limitations. It is exceedingly difficult to draw up a table of material characteristics in order to assess the relative strengths and weaknesses of metals, plastics and ceramics because each of these terms covers whole family of materials within which the range of properties is often as broad as the differences between these three classes [2]. At a simplistic level, then:

- Plastics are of low density. They have good short-term chemical resistance but they lack thermal stability and have only moderate resistance to environmental degradation [2].
- Ceramics may be of low density (although some are very dense). They have great thermal stability and resistant to most forms of attack (abrasion, wear and corrosion) [2].
- Metals are mostly of medium to high density—only magnesium, aluminium and beryllium can compete with plastics in this respect. They have useful mechanical properties and high toughness [2].

Composite science is moving towards a definite direction. Future trends in research are extremely appealing and far-reaching. Technological motivation and objectives in the field of composite science research are drastically widening. The engineer who uses materials for structural or load-bearing purposes is quickly aware of an important feature of engineering solids, which is that they are never as strong as we would expect them to be from our knowledge of the strengths of the chemical bonds which hold them together.

14.11.1 Recent Scientific Endeavour in the Field of Nanocomposites

Nanocomposites are the next-generation materials. Materials science and chemical process engineering are gearing towards newer innovations and newer challenges. The aim of this chapter is to target nanocomposites as a next generation environmental friendly membrane material. Scientific endeavour in the field of nanocomposites are unfolding a new era of vision and immense scientific forbearance. Nanocomposites and materials science are today revolutionary areas of science and technology. The immense futuristic vision in the field of nanocomposites are witnessing drastic challenges.

Ma et al. (2010) [3]. Lucidly discussed in a well-researched review the dispersion and functionalization of carbon nanotubes for polymer-based nanocomposites. Carbon nanotubes (CNTs) hold the promise of delivering exceptional mechanical properties and multifunctional characteristics. The vision and the challenge of technology and engineering of nanocomposites are emboldened at each step of scientific rigour and scientific research pursuit. Ever-increasing interest in applying CNTs in many different fields has led to continuous efforts to develop dispersion and functionalization techniques. The vision of the application of CNTs lies in the fact that to employ CNTs as effective reinforcement in polymer nanocomposites, proper dispersion and appropriate interfacial adhesion between the CNTs and polymer matrix have been guaranteed. This review paper deeply comprehends the current understanding of CNTs and CNT/polymer nanocomposites with focus on two points: (1) the principles and techniques, (2) the effects of CNT dispersion and functionalization on the properties of CNT/polymer nanocomposites. The science and engineering of nanocomposites are changing the face of the domain of nanotechnology today. The fabrication techniques of the mentioned nanocomposites are also highlighted in details [3].

Liang et al. (2009) [4] with deep and cogent insight discussed molecular-level dispersion of graphene into poly (vinyl alcohol) and effective reinforcement of their nanocomposites. The scientific scenario is such that despite great and groundbreaking recent progress with carbon nanotubes and other nanoscale fillers, the development of strong, durable, and cost-effective multifunctional nanocomposite materials has yet to be achieved. The vision and the challenges are to achieve molecular level dispersion and maximum interfacial interaction between the nanofiller and the matrix at low loading. Nanocomposites, used as nanoscale fillers at low loading, have the potential to match or exceed the performance of significant quantities of conventional composite filler. The authors deeply comprehend the materials science portion of nanocomposites. Maximum mechanical enhancement can only be achieved when the "nanofiller" is dispersed homogeneously, or best at the molecular level, in the matrix and the external load is efficiently transferred via a strong interaction at the interface between the filler and the matrix. At present, although nanocomposites employing carbon-based reinforcement materials are dominated by carbon nanotubes, their intrinsic impurities from catalysts and their high cost have been hampering their application. Theoretical and experimental results show that single-layered two-dimensional (2D) graphene sheets are the strongest materials developed thus far. The challenge and vision of science are unfolding at every step of research pursuit in materials science and technology [4].

Gash et al. (2002) [5]. discussed deeply energetic nanocomposites with sol-gel with its synthesis, safety and characterization. The preparation and characterization of energetic composite materials containing nanometer-sized constituents is currently a very active and invigorating area of research. Some of these efforts have produced materials that have shown unique and important properties relative to traditional energetic materials. Progress of science, the immense academic rigour and the deep scientific introspection are the forerunners towards a greater emancipation and true realization of materials science application. The science of nanocomposites is gearing towards a newer challenge and a newer visionary eon [5].

Jancar et al. (2010) [6] lucidly discussed with deep and cogent insight the current issues in research on structure–property relationships in polymer nanocomposites. The understanding of the basic physical relationships between nano-scale structural variables and the macroscale properties of polymer nanocomposites today remains in deep infancy. The primary vision of this article is to ascertain the state of the art regarding the understanding and prediction of the macroscale properties of polymers reinforced with nanometer-sized solid inclusions over a wide temperature range. Polymer nanocomposites offer the possibility of substantial improvements in material properties—such as shear and bulk modulus, yield strength, toughness, film scratch resistance, optical properties, electrical conductivity, gas and solvent transport, among many—with only very small nanoparticles (NPs) dispersed in the polymer matrix. It has been known for about a century that the addition of nanofillers such as carbon black to rubbery polymers has a strong impact on the properties of materials [6].

Samzadeh-Kermani et al. (2016) [7] investigated on the synthesis and characterization of new biodegradable chitosan/polyvinyl alcohol/cellulose nanocomposite. In this report the authors lucidly discusses the synthesis of a new nanocomposite based on chitosan/polyvinyl alcohol/nanocrystalline cellulose. The morphology and particle size of NCC and nanocomposites were studied by scanning electron microscopy (SEM), X-ray diffraction analysis (XRD) and Fourier transform infrared (FT-IR) spectroscopy [7].

Irimpan et al. (2008) [8] deeply comprehend spectral and non-linear optical characteristics of nanocomposites of ZnO-CdS. The visionary challenge of science and engineering are slowly unfolding at each step of this scientific research pursuit. In this article, the authors present the spectral and non-linear optical properties of ZnO-CdS nanocomposites prepared by colloidal chemical synthesis. The optical band gap (E_g) of the material is tunable between 2.64 and 3.84 eV. The emission peaks of ZnO-CdS nanocomposites change from 385 to 520 nm almost in proportion to changes in E_g. It is possible to obtain a desired luminescence colour from UV to green by simply adjusting the composition [8].

Azeredo et al. (2009) [9] discussed lucidly in a review nanocomposites for food packaging applications. Most materials currently used for food packaging are non-degradable, generating environmental issues. Several biopolymers have been exploited to develop materials for eco-friendly food packaging. The challenge and scientific truth lies in the fact that the nanocomposites and degradable. The use of biopolymers has been limited because of their poor mechanical and barrier properties, which may be improved by adding reinforcing compounds (fillers), forming composites. Science and engineering of nanocomposites and biopolymers are reaching immense scientific heights. Biodegradability is the burning as well as a vexing question for the future. The scientific truth and the innovative scientific challenge behind biodegradability are opening up new avenues of research in years to come. A scientist's widening and unending vision are the precursors towards a newer scientific realm of biodegradability. Future of composite science lies in the hands of material scientists with a clear vision towards biodegradability [9].

Pedrazzoli et al. (2014) [10] in a recent scientific research pursuit discussed thermal, viscoelastic and mechanical behaviour of polypropylene with synthetic boehmite alumina nanoparticles. Effects of nanofiller concentration and surface treatment on the morphology, thermal, viscoelastic and mechanical behaviour of polypropylene copolymer (PP)/boehmite alumina (BA) nanocomposites were investigated with immense scientific innovation. Scientific truth, deep scientific comprehension and the success of research pursuit are the torchbearers towards a greater emancipation of composite technology today. Increasing scientific efforts are devoted to the research of thermoplastic nanocomposites exhibiting improved and novel properties. The challenge of the science of nanocomposites is slowly unfolding with immense scientific determination in decades to come [10].

Liu et al. (2012) [11] discussed lucidly in a far-reaching review recent advances in thermoelectric nanocomposites. The challenge, the scientific vision and the academic rigour in the field of thermoelectric nanocomposites are enhanced at each step of this scientific research pursuit. Thermoelectric power generation represents a class of energy conversion technology, which has been used in power supply of aeronautic and astronautic exploring missions, now so relevant to harvest the widely distributed waste heat and convert the abundant solar energy into electricity at lower cost. Recent years have seen remarkable and ever-growing interests in thermoelectric nanocomposite for energy conversion application. The challenge and scientific truth in nanocomposites is growing at a rapid pace. One decisive factor driving the current interest in nanocomposite thermoelectric study is the need for safe, clean and sustainable energy source. Solar energy will be the next possible energy alternative to fill the gap left by nuclear energy. Nuclear disasters and unviable nuclear technology has urged the scientific domain throughout the developed and developing economies to remove nuclear plants and gear towards alternate energy such as solar energy. The thermoelectric technology has been employed in Radioactive Thermoelectric generator (RTG) for satellite, deep space exploration missions, such as SNAP-19 for Pioneer 10 mission, which has been working for nearly 40 years. In today's world, thermoelectrics provide an alternative route to convert the solar energy into electric power, besides the technology of photovoltaic and thermionic energy conversion [11].

Veprek et al. (2002) [12] elucidated with deep comprehension the thorough understanding of mechanical properties of super and ultrahard nanocomposites. This chapter elucidates and explains in a simple way, understandable to a wide spectrum of readers, the unusual combination of the mechanical properties of the recently developed new class of superhard nanocomposites, such as high hardness, enhancement of the elastic modulus, very high elastic recovery and the absence of crack formation. Future experimental as been suggested in this work [12].

Ebrahimi et al. (2012) [13] dealt lucidly with new trends and development in the area of nanocomposites. The scientific vision and the scientific sagacity are deeply comprehended in this chapter. Nanoscience and nanotechnology have veritably become the central point of scientific and technical activity. Human scientific vision is ever-widening and surpassing notable and difficult boundaries. Over the last few years, the interest in nanostructures and their applications in various electronic devices, effective optoelectronic devices, biosensors, photodetectors, solar cells, nanodevices, plasmonic structures has been increasing tremendously. Thus, this chapter pointedly focuses on the challenges in research pursuit in the field of nanocomposites [13].

14.11.2 Recent Scientific Research Pursuit in the Field of Membrane Science and the Vision to Move Forward

The challenge of human civilization is inspiring and exceedingly groundbreaking. The targets of science and engineering in today's world are well defined with respect to forays into membrane science. Human scientific instinct, deep scientific rigour and the academic progress in the field of membrane science are drastically enhanced in today's scientific world. Drinking water crisis, global water challenges and the industrial wastewater imbroglio are the forerunners towards a greater emancipation of scientific endeavour and scientific realization in the field of membrane science and the broad domain of water pollution control.

Cheryan (1998) [14] deeply discusses with cogent details the success and future of membrane separation processes mainly ultrafiltration and microfiltration. Scientific research pursuit, scientific rigour and scientific forbearance are at its helm as science of membranes enters a newer eon. The development of Sourirajan–Loeb synthetic membrane in 1960 revolutionized the scientific horizon. However, it faced immense challenges in its earlier days. The situation is different today: Membranes are more robust, modules and equipment are better designed. Developments in nanofiltration, gas separations, pervaporation and bipolar membrane electrodialysis have widened the applicability of membranes. This well-informed and well-researched chapter discusses filtration, membrane science and classifications of membrane separation phenomenon. The author pointedly focuses on ultrafiltration and nanofiltration. Cheryan (1998) [14] goes deep into membrane chemistry, structure and function. The wide facets of this chapter are performance and engineering models, fouling and cleaning, process design and finally the large application domain of membrane science [14].

Fouling stands as a major impediment to membrane separation phenomenon. Scientific vision and scientific truth behind application of membrane science are surpassing wide and versatile boundaries. Zhu et al. (1997) [15] discussed in a well-informed chapter the colloidal fouling of reverse osmosis membranes with measurements and fouling mechanisms. The effect of chemical and physical interactions on the fouling rate of cellulose acetate and aromatic polyamide composite reverse osmosis (RO) membranes by silica colloids is investigated in deep details. The vision of science and technological objectives and motivation needs to be re-enshrined with each step of research pursuit in membrane science. Results of fouling experiments using a laboratory-scale unit demonstrate that colloidal fouling rate increases with increasing solution ionic strength, feed colloid concentration, and permeate water flux through the membrane. The challenge of science and engineering are observed in this detailed study in membrane fouling. It is observed that the rate of colloidal fouling is controlled by a unique interplay between permeation drag and electric double layer repulsion; that is, colloidal fouling of RO membranes involves inter-relationship (coupling) between physical and chemical interactions [15].

Jain et al. (2006) [16] discussed treatment of leather plant effluent by membrane separation processes. A detailed scheme is proposed for the treatment of leather plant effluent using membrane-based separation processes. The effluent coming out from the various upstream units of the leather plant (except chrome tanning) are combined and a two-step pressure driven membrane processes involving nanofiltration (NF) and reverse osmosis (RO) are adopted after a pre-treatment consisting of gravity settling, coagulation and cloth filtration. Science of membranes in today's world is crossing wide and versatile visionary boundaries.

This chapter goes beyond deep scientific rigour and unfolds new chapters in the field of membrane separation processes [16].

Razdan et al. (2003) [17] lucidly discussed novel membrane processes for the separation of organics. The article presents membrane processes for separation of organics listing different types of solvent resistant membranes, their applications, performance, advantages over conventional separation techniques and theoretical aspects dealing with solvent transport mechanisms. This chapter deeply transforms the scientific scenario and the scientific path leads towards newer innovations [17].

Ravanchi et al. (2009) [18] discussed with cogent insight in a review application of membrane separation processes in petrochemical processes. Scientific challenges and scientific sagacity are at its helm at each step of research pursuit in the field of membrane separation processes. In this paper, a general review on different membrane processes and membrane reactors are done. Processes such as olefin/paraffin separation, light solvent separation, solvent dewaxing, phenol and aromatic recovery, dehydrogenation, oxidative coupling of methane and steam reforming of methane were discussed in details [18].

14.12 Advancements in Nanotechnology, the Forays into Research in Nanocomposites and the Vision for the Future

Nanotechnology is a novel area of scientific endeavour and deep scientific rigour. The vision for the future is wide and bright. The challenge of human civilization is entering a newer eon as mankind gears up towards a newer innovation. Progress of nanotechnology, the grave concerns of environment and the visionary global water challenge are the forerunners towards a newer realization of environmental sustainability. The forays into the world of nanocomposites are latent yet not immature. Human scientific research pursuit needs to be re-envisaged at each step of frontiers of science. In the similar vein, nanotechnology advancements need to be re-envisioned at all scholarly and academic levels. Composites as well as nanocomposites are known for their mechanical properties. The success of application, the technological rigour and the futuristic vision are the forerunners towards a newer scientific imagination in the field of nanocomposites.

Nanotechnology is the visionary domain of science today. The vision for the future for the science and engineering of membranes is transforming vast and wide technological frontiers. Nanocomposites and materials science are the avenues of science which needs to be revisited and reframed. The vision for the future in the field of nanocomposites application is wide and far-reaching. The deep scientific vision and scientific cognizance are surpassing wide and visionary frontiers.

14.13 The Scientific Doctrine of Membrane Science

Membrane science and membrane separation processes are moving from one visionary avenue over another. Since the innovation of Loeb–Sourirajan model, membrane science has witnessed immense difficulties and re-envisioning [14, 19]. Scientific history, mankind's prowess and the immense scientific rigour are the pallbearers towards a newer visionary eon in membrane separation phenomenon. The classifications of membrane separation processes are presented in Table 14.1.

Table 14.1 Classification of membrane separation processes [14, 19]

Process	Driving force	Retentate	Permeate
Osmosis [1]	Chemical potential	Solutes/Water	Water
Dialysis	Concentration difference	Large molecules/ water	Small molecules/ water
Microfiltration	Pressure	Suspended particles/ water	Dissolved solutes/ water
Ultrafiltration	Pressure	Large molecules/ water	Small molecules/ water
Nanofiltration	Pressure	Small molecules/ divalent salts/ dissociated acids/ water	Monovalent ions/ undissociated acids/ water
Reverse Osmosis	Pressure	All solutes/water	Water
Electrodialysis	Voltage/current	Non-ionic solutes/ water	Ionized solutes/water
Pervaporation [1]	Pressure	Non-volatile molecules/water	Volatile small molecules/water

14.14 Definition of Membrane Separation Processes

Membrane separation phenomenon, with its scientific vision, is extremely complex and requires much scientific introspection. The challenge and the objective of scientific endeavour in the field of membrane science needs to be re-envisioned and rebuilt at each step of global scientific challenge. Global water initiatives in today's world are faced with unimaginable difficulties and unmitigated disasters. The scientific urge today should be targeted towards chemical process design and chemical process modeling. Separation phenomenon is at the heart of chemical process engineering. Membrane process design is, in a similar manner, at the heart of environmental engineering science. In this chapter, the author repeatedly stresses upon the world of challenges and the avenues of difficulties in the success of separation phenomenon. Effectiveness of separation phenomenon stands as of primary importance in the success of membrane science. In today's scientific paradigm, membrane science and membrane separation phenomenon are the heart of chemical process engineering. The challenge of science and engineering needs to be re-addressed and rebuilt with each step of scientific research pursuit [14, 19].

14.15 Scientific Forays in the Domain of Membrane Science and Recent Scientific Research Pursuit

Membrane science and environmental engineering science in today's world are linked by an unsevered umbilical cord. Membrane technology is the next generation revolutionary technology. Scientific horizon and scientific cognizance in the field of global water initiatives today stands in the midst of deep crisis. Vision of science, technological motivation and the immense scientific rigour in the field of membrane science and groundwater remediation are surpassing vast and versatile frontiers. Scientific forays in the domain of membrane science are challenged and needs to be re-envisioned at each step of human academic rigour. The author goes to the depth of membrane technology and pointedly focuses on the success of separation phenomenon attached to it. Today the challenge of science and engineering are unimaginable. Mankind's prowess, a scientist's deep insight and the rigours of environmental engineering are gearing towards a newer scientific emancipation. The author rigorously comprehends the unknown world of membrane separation phenomenon with delineation of membrane fouling and other intricacies of membrane separation. Scientific research pursuit discussion also leads to desalination science which stands as a major issue in many developed and developing economies. Groundwater metal contamination is a major burning and vexing issue. The author widens the review with special emphasis on barriers and difficulties in membrane technology. The challenge of science needs to be reviewed as science moves towards third decade of this visionary century.

14.16 Global Water Issues and Membrane Science

Human civilization is moving towards a newer eon. Groundwater metal contamination and drinking water issues are changing the face of scientific research pursuit. Arsenic groundwater contamination in developing and developed economies are a devastation to the fabric of science and technology. In such a critical juncture of human history and time, novel separation processes and non-traditional environmental engineering techniques are a boon to humankind. Membrane science is an apt classification of novel separation processes. Heavy metal groundwater contamination is an unsolved area of science. The challenge of technology needs to be re-envisioned and restructured. Global water initiatives are the imminent need of the hour. Groundwater quality management system is another burning and vexing issue today. Human civilisation's prowess, mankind's progress and the immense technological rigour will all lead a long way in the true emancipation and effective realization of integrated groundwater quality management system. Human scientific challenges today are unimaginable. Membrane separation phenomenon is the only decisive answer to the growing global water crisis. Desalination science and the wide visionary area of groundwater remediation are the other two feasible drinking water solutions. The scientific world today stands in the midst of immense scientific adjudication and scientific fortitude. Global water hiatus and global water catastrophe needs to be enshrined and envisioned at each step of this century. The rigours of science, the advancement of human intuition and scientific notion all will definitely lead a long way in true realization of the fundamentals of successful environmental sustainability.

14.17 Nanotechnology for Water Pollution Control

Water pollution control studies needs to be restructured with the inroads into scientific endeavour in membrane science and other novel separation processes. The application of nanotechnology in environmental pollution control is wide and versatile in spite of being latent. Drinking water treatment and industrial wastewater treatment today is linked by an umbilical cord with the vast and versatile world of nanotechnology. Nanocomposites are another wide facet of nanotechnology and environmental protection. The challenge of human scientific society is unimaginable. The vision of chemical technology and chemical process engineering are entering a new phase with the passage of human scientific history. Nanocomposites are showing human scientific generation a new phase.

Environmental engineering paradigm in today's world is witnessing drastic challenges. Provision of clean drinking water is in the state of immense disaster and the challenges of global water initiatives needs to be re-envisioned and re-built with every step of scientific endeavour. Nanotechnology is a burgeoning area of scientific pursuit. This revolutionary field needs to be restructured with respect to the application of nanotechnology in drinking water treatment and industrial wastewater treatment. Nanocomposites and nanomaterials are another burgeoning area of research and development endeavour. In this chapter, the author repeatedly stresses upon the immense potential of nanocomposites and nanomaterials in the wide and visionary domain of membrane science.

14.18 Membrane Science and Technology for Wastewater Reclamation

Human scientific endeavour is passing through drastic phases. Membrane science and technology are witnessing dramatic challenges. Global water initiatives are the imminent need of the hour. Wastewater is widely recognized as one of significant, ever-growing and reliable water resources. Wastewater production is the only potential water source which will increase as human population grows and the demand for fresh water grows. The management of wastewater reclamation has basic scope of balancing water availability and water demand, at reasonable cost and with acceptable environmental impacts. Wastewater reclamation technology needs to be appropriate and sustainable, operated and maintained thoroughly and very efficient in removing organic matter and pathogens present in wastewater. Scientific vision and scientific sagacity are immense in today's scientific horizon. The challenge and effectiveness of a nano-traditional environmental engineering technique are unimaginable. Scientific understanding and deep scientific hindsight are burgeoning in the avenues and aisles of human history [20, 21]. Global water issues should have links with wastewater reclamation technologies today [14, 19].

Although membrane technologies became commercially available more than 30 years ago, it is veritably experiencing rapid developments and fast-track improvements. A wide variety of membrane processes can be classified according to driving force, membrane material, membrane type and configuration, removal capabilities and mechanism, and membrane fouling and cleaning. For example pressure driven, membrane processes include microfiltration (MF), ultrafiltration (UF), nanofiltration (NF), and reverse osmosis (RO) [20, 21]. Among these membrane processes, membrane bioreactor (MBR) and reverse osmosis (RO) processes are exceedingly suited for application of wastewater reclamation process due to their effective energy requirement, compact configuration and high quality product. Science is a huge colossus with visionary potential. Global water concerns, global water imbroglio and the immense scientific rigour are the forerunners of technological validation today. Validation of science, technological prowess and the futuristic vision are the necessities of scientific progress today. Membrane science and water technology are facing the challenges of this century. Environmental sustainability, social sustainability and holistic development are the pallbearers of a greater visionary future [14, 19–21].

The existing conventional wastewater reclamation system effluent after secondary and tertiary treatment need high energy requirement due to supply water for removal of biodegradable organic matter (in solution or suspension) and suspended solids. It is costly and has a large footprint in installation of reclaimed water distribution system. The futuristic challenges of scientific research pursuit are unfolding visionary scientific chapters. Water reclamation is the imminent need of the hour for alleviating global water issues. Man's immense vision, technological advancements and scientific prowess are the torchbearers towards a visionary road to future of membrane technology.

14.19 Recent Advances in Membrane Science and Technology in Seawater Desalination and Its Application in Drinking Water Treatment

The success of human civilization and the progress of scientific research pursuit are opening up new chapters in the history of science and technology. Global water issues are re-invigorating the scientific arena. The development of seawater desalination membranes is continuously progressing and has effectively demonstrated highly remarkable achievement in performance. Indeed, there appears to be no limit to the performance of these membranes in sight. The first historic discovery of Loeb–Sourirajan asymmetric reverse osmosis (RO) membrane enabled seawater desalination on an industrial scale. Later, the membrane performance was improved by the development of thin-film-composite (TFC) membranes by Cadotte [20, 21]. The fluxes of currently available commercial membranes are an order of magnitude higher than the RO membranes of 1960s with salt rejection as high as 99.8%. Global water issues are changing the worldwide economic scenario. Scientific and academic rigour are in today's world in a state of immense catastrophe with respect to global water imbroglio. Membrane science, seawater desalination and the global water fiasco are the pallbearers towards a greater visionary tomorrow. Human civilization is moving towards a newer scientific urge—the vision of provision of basic human needs. In such a crucial juxtaposition, the challenge of science needs to be re-enshrined [14, 19–21].

Regarding membrane performance, attempts are being made to increase the flux and to improve membrane stability by incorporating nanoparticles and by modifying the membrane surface. Moreover, fabrication of membranes that can reject boron, arsenic, and organic compounds of low molecular weights to higher degrees than currently available membranes is required to produce less hazardous drinking water. Membranes that can withstand ultra-high pressure for seawater desalination and membranes that are operable at ultra-low pressures for brackish water desalination were achieved for drinking water production. Typically, membrane distillation, forward osmosis, and carbon nanotube membranes are considered potentially applicable for seawater desalination. The current status and future directions of membrane desalination in the Middle East and Singapore have been discussed in details in the following paragraphs [20, 21]. The author repeatedly points out the success, the scientific imagination and instinctive scientific cognizance in the field of membrane science and its linkages with water science and technology [14, 19–21].

14.20 Groundwater Remediation, the Success of Science and Technology and the Visionary Road towards Future

Groundwater remediation is the heart of environmental engineering science today. Technology and science needs to be rebuilt with immediate effect. Heavy metal contamination of groundwater is a vexing and a burning issue and stands in the midst of immense catastrophe. This chapter mainly focuses on the issues of global water challenges and the relevant applications of membrane science in alleviating global water shortage. Vision of science, the immense scientific and academic rigour and the progress of scientific research pursuit are the precursors towards a greater visionary era of environmental sustainability. Arsenic groundwater contamination is another area of scientific vision. The challenge of human civilization is immense and thoroughly inspiring. Today's success of science and technology lies in the hands of scientific domain. Mankind's immense prowess, human civilisation's scientific rigour and the progress of science and technology will all lead a long and visionary way in the true emancipation of global water challenges [20, 21].

14.21 Nanocomposites: The Visionary Domain of Future

Nanocomposites and their wide applications are veritably changing the scientific landscape. Materials science and chemical process engineering are being re-envisioned as humankind and human scientific mind targets for environmental sustainability. Holistic development of human civilization and the deep forays into research and development endeavour are the pallbearers towards a greater and effective realization of environmental sustainability. Composites science is not new yet immature. Nanocomposites are the next-generation material. The success of human scientific research pursuit is retrogressive today. Global water catastrophes, the environmental concerns and the deep scientific challenges will all lead a long and visionary way in the true realization of nanocomposites and membrane science applications.

14.22 Applications of Nanocomposites

Applications of nanocomposites in environmental protection will lead a long and visionary way in the true emancipation of the science and engineering of nanotechnology. Scientific rigour and scientific candour are at its helm at each step of scientific endeavour. The vision and prowess of nanotechnology and nanocomposites are changing the face of science and engineering. Mankind's vision, man's scientific prowess and the world of arduous challenges are the torchbearers towards a greater visionary future in the field of nanotechnology.

14.23 Environmental Sustainability and the Wide Vision for the Future

Environmental and energy sustainability are the pivotal point for the future of engineering science. Environmental engineering science today stands in the midst of immense scientific introspection. Ecological imbalance, deadly environmental disasters and the grave concerns of sustainable development are the torchbearers towards a newer visionary future in the field of environmental engineering applications. Membrane science and technology today is witnessing a new dawn of human scientific endeavour. Destruction of ecological diversity stands today in the midst of immense scientific introspection and deep crisis. The challenge of holistic sustainable development needs to be re-enshrined and restructured with the passage of human scientific endeavour.

Environmental sustainability today stands in the face of a major catastrophe. The progress and scientific rigour of human civilization is challenged today in the face of immense scientific candour. Man's vision, human civilisation's immense prowess and the march of technology are the forerunners towards a greater visionary future.

14.24 The Challenge of Scientific Endeavour in the Field of Nanocomposites

The challenge and the vision of research pursuit in the field of nanocomposites are vast, versatile and far-reaching. Environmental concerns, grave future of environmental sustainability and the immense potential of the science of nanotechnology will lead a long and visionary way in the true realization of sustainable development. The challenge, the scientific imagination and the immense vision in the field of application of nanocomposites are unimaginable and far-reaching. Technological motivation and immense scientific objectives are gearing towards a new path of change. Global water issues and the surmounting ground water challenges are veritably transforming the scientific landscape of research pursuit.

Nanocomposites and relevant areas of materials science are changing the avenues and paths of scientific research pursuit. The greatness of science, the unending rigour of technology and the progress of human civilization are all leading a long and visionary way in the true realization of application domain of nanocomposites and nanomaterials. The visionary domain of materials science and nanocomposites are changing the face of engineering science today.

14.25 The Scientific Challenges, the Scientific Sagacity and the Visionary World of Membrane Science

Scientific challenges and the subsequent scientific sagacity are changing the face of the vast and versatile world of membrane science. Technological vision and technological motivation are the ultimate drivers towards a greater emancipation of science and engineering today. In such a similar vein, membrane science and application of nanocomposites are witnessing drastic challenges. The scientific sagacity and the deep scientific insight are the proponents towards a greater visionary future in the field of membrane technology. In today's scientific genre, technology and engineering are moving at an immense and rapid pace. Membrane technology, desalination science and the future of global water shortage are all linked amongst themselves by an unsevered umbilical cord. In today's scientific genre and scientific candour, membrane separation processes and other non-traditional environmental engineering techniques are gaining immense ground in the futuristic success of environmental sustainability. The author repeatedly points out the success of novel separation processes and nanocomposites application in the furtherance of global science and technology. Nanocomposites application and innovative world of materials science are in the road towards greater scientific vision and scientific forbearance. Mankind's prowess, man's vision and the rigour of human civilization are entering into a new phase of advancement of science [22–24].

14.26 Nanotechnology, the Immense Scientific Barriers and the Road towards Future

Nanotechnology and nanoscience are leading a long way in the furtherance of humankind and human scientific rigour. Chemical process engineering, mathematical science and environmental engineering science are delving deep into the wide world of scientific vision. In a similar vein, membrane science and the wide world of nanocomposites are changing the immense face of chemical process engineering. Global water shortage is the other side of the scientific coin. Groundwater quality management is a burning issue today. Membrane separation phenomenon needs to be re-enshrined and revamped at each step of scientific thrust and technological foray. Nanotechnology today is challenged in the face of deep scientific hindsight. Barriers, difficulties and solutions are immense yet uncovered. Human scientific endeavour, the challenge of human scientific research pursuit and the immense academic rigour behind application of nanocomposite in membrane science are the scientific thrust of the day. The road towards the future is wide and visionary. Scientific ingenuity and scientific candour are today changing the face of academic rigour. The author repeatedly points out the success of application of nanocomposites in the scientific horizon of membrane technology and the subsequent furtherance of engineering science [22–24].

14.27 The Scientific Vision, the World of Chemical Process Engineering and Materials Science and Groundwater Remediation Technologies

Chemical process engineering and materials science are gearing forward towards newer challenges. Human scientific research pursuit today is a revolutionary avenue of scientific and academic rigour. Groundwater heavy metal contamination in today's human scientific ardour is facing immense challenges. Today's world of chemical process engineering and nanocomposites/materials science are surpassing visionary boundaries. The grave environmental concerns, the immense scientific rigour and the greater need of clean drinking water are the precursors towards a greater visionary emancipation in the field of heavy metal groundwater remediation.

14.28 Future of Nanoscience and Nanotechnology

Technological vision and cogent scientific objectives are the pallbearers of a greater visionary future of nanoscience and nanotechnology. The author repeatedly points out the intricacies in the application of nanocomposites and membrane science in environmental protection. Today the challenges of the application of nanotechnology in different areas of science are veritably widening. The future of nanotechnology is boundless and far-reaching. Man's immense vision, the deep scientific rigour and mankind's road towards future are the pallbearers towards a greater realization of environmental pollution control. Drivers of technological vision are the immense scientific effort and deep comprehension. Human scientific challenges and global environmental initiatives are changing the face of scientific progress. This chapter strongly comprehends the success of application of nanocomposites in the furtherance of science and engineering.

14.29 Futuristic Trends in the Application Areas of Nanocomposites in Membrane Science

Membrane science today stands in the midst of immense comprehension. Global water challenges are veritably changing the face of human civilization and innovative scientific pursuit. The hurdles, difficulties and barriers of membrane separation phenomenon needs to be re-envisioned and rebuilt. Heavy metal groundwater contamination are in the midst of deep catastrophe. Science and technology has few answers to this ravaging issue in developed and developing countries of the world. The success of applications of nanocomposites in membrane science are reframing the scientific and technological frontier. Materials science today is facing diverse challenges such as membrane fouling and subsequent prevention.

14.30 Future of Membrane Science Applications in Environmental Pollution Control

The future of membrane science applications is in the avenue of newer regeneration. Grave environmental scenario, global water crisis and loss of biodiversity are the pallbearers towards a greater realization of environmental pollution control. Technology is moving fast in today's scientific scenario. The fundamental applications of nanocomposites are in environmental pollution control. Global water imbroglio has urged scientists and environmental engineers to device new technologies innovative water policies. Arsenic groundwater remediation has challenged the scientific scenario. Technology and engineering are moving very fast in this century. Growing concerns for the environment has transformed the distant scientific horizon. Environmental pollution control is the necessity of the day. The world of chemical process engineering, the scientific horizon of materials science and the visionary avenues of membrane science are the pallbearers towards a greater emancipation of engineering science today. Nanocomposites is a revolutionary area of materials science. Composite science itself is not a new domain yet immature. The application domain needs to be re-envisioned and rebuilt at each step of scientific research forays. Water pollution control and global water initiatives are today linked by an unsevered umbilical cord. The challenge forward is awesome and veritably insightful. The future of environmental engineering science should target global water fiasco and provision of clean drinking water [24–26].

The science of environmental protection today stands in the midst of tremendous introspection and deep challenges. Materials science and chemical process engineering are witnessing a new dawn of human civilization. This chapter widely observes the scientific success, the scientific progress and the visionary scientific rigour in the field of membrane and nanocomposites application [24–26].

14.31 Vision of Science, the Road Forward and Success of Nanotechnology Applications in Industrial Pollution Control

The science and engineering of nanotechnology are crossing wide and visionary frontiers. In today's scientific world, nanotechnology has innumerable applications in environmental protection and industrial pollution control. Global water crisis today stands in the midst of devil and the deep sea. The challenge and vision need to be re-enshrined with each step of scientific research pursuit. Nanotechnology is opening up new visionary frontiers. Composite science is in the path of newer scientific rejuvenation. Nanoscience and nanotechnology are moving towards a new frontier of industrial wastewater treatment and drinking water treatment. Membrane science especially nanofiltration are the visionary domain of nanoscience today. This chapter ushers in a new era of environmental protection with the success of nanocomposites application in mind [24–26].

14.32 Conclusion

The science of nanocomposites is the newer revolutionary domain. Scientific and academic rigour needs to be re-envisioned and application of nanocomposites and membrane science needs to be rebuilt with step of research pursuit. Composites as well as nanocomposites are not new areas yet immature. Environmental protection and the area of renewable energy are the visionary domains of nanocomposite application. Provision of basic human needs such as water is the challenging area of nanocomposites and membrane science today. Engineering science needs to be restructured and revamped as the human civilization marches towards a newer global vision. The decisive areas of nanocomposites and membrane science are elucidated in details in this chapter. The global environmental engineering vision is today in a state of immense comprehension and deep scientific introspection.

Mankind needs to move forward as water challenge is changing itself from one paradigm towards another. This chapter focuses on the success of application domain of nanocomposites and membrane science. Diverse areas of research are focused in this well observed chapter. Global environmental concerns, industrial water pollution control and drinking water treatment are the forerunners of a greater vision today. A new dawn of human civilization is not far as scientific research pursuit in nanocomposites and membrane science enters into a new era.

References

1. F. C. Campbell, *Structural Composite Materials* (Chapter 1), ASM International (2010).
2. B. Harris, *Engineering composite materials*, Institute of Materials (1999).
3. P.-C. Ma, N. A. Siddiqui, G. Marom, J.-K. Kim, Dispersion and functionalization of carbon nanotubes for polymer based nanocomposites—a review, *Composites: Part-A* (2010), 41, 1345–1367.
4. J. Liang, Y. Huang, L. Zhang, Y. Wang, Y. Ma, T. Guo, Y. Chen, Molecular level dispersion of graphene into poly vinyl alcohol and effective reinforcement of their nanocomposites, *Advanced Functional Materials* (2009), 19, 1–6.
5. A. E. Gash, R. L. Simpson, J. H. Satcher, Jr, Energetic nanocomposites with sol-gel chemistry: Synthesis, safety and characterization, *29th International Pyrotechnic Seminar*, Westminster, USA (2002), July 14–19.
6. J. Jancar, J. F. Douglas, F. W. Starr, S. K. Kumar, P. Cassagnau, A. J. Lesser, S. S. Sternstein, M. J. Buehler, Current issues in research on structure–property relationships in polymer nanocomposites, *Polymer* (2010), 51, 3321–3343.
7. A. Samzadeh-Kermani, N. Esfandiary, Synthesis and characterization of new biodegradable chitosan/polyvinyl alcohol/cellulose nanocomposite, *Advances in Nanoparticles* (2016), 5, 18–26.
8. L. Irimpan, V. P. N. Nampoori, P. Radhakrishnan, Spectral and nonlinear optical characteristics of nanocomposites of ZnO-CdS, *Journal of Applied Physics* (2008), 103, 094914.

9. H. M. C. de Azeredo, Nanocomposites for food packaging applications, *Food Research International* (2009), 42, 1240–1253.

10. D. Pedrazzoli, V. M. Khumalo, J. Karger-Kocsis, A. Pegoretti, Thermal, viscoelastic and mechanical behavior of propylene with synthetic boehmite alumina nanoparticles, *Polymer Testing* (2014), 35, 92–100.

11. W. Liu, X. Yan, G. Chen, Z. Ren, Recent advances in thermoelectric nanocomposites, *Nanoenergy* (2012), 1, 42–56.

12. S. Veprek, A. S. Argon, Towards the understanding of mechanical properties of super and ultrahard nanocomposites, *Journal of Vacuum Science Technology* (2002), 20(2), 650–664.

13. F. Ebrahimi, *Nanocomposites: New Trends and Developments*, Intech Publishers (2012).

14. M. Cheryan, *Ultrafiltration and Microfiltration Handbook*, Technomic Publishing Co. Inc., USA (1998).

15. X. Zhu, M. Elimelech, Colloidal fouling of reverse osmosis membranes: measurements and fouling mechanisms, *Environmental Science and Technology* (1997), 31, 3654–3662.

16. S. K. Jain, M. K. Purkait, S. De, Treatment of leather plant effluent by membrane separation processes, *Separation Science and Technology* (2006), 41, 3329–3348.

17. U. Razdan, S. V. Joshi, V. J. Shah, Novel membrane processes for separation of organics, *Current Science* (2003), 85(6), 761–771.

18. M. T. Ravanchi, T. Kaghazchi, A. Kargari, Application of membrane separation processes in petrochemical industry: A review, *Desalination* (2009), 235, 199–244.

19. S. Palit, Filtration: Frontiers of the engineering and science of nanofiltration—a far-reaching review, in: *CRC Concise Encyclopedia of Nanotechnology* (Taylor and Francis), eds.: U. O.-Mendez, O. V. Kharissova, B. I. Kharisov (2016), pp. 205–214.

20. S. Palit, Advanced oxidation processes, nanofiltration, and application of bubble column reactor, in: *Nanomaterials for Environmental Protection*, eds.: B. I. Kharisov, O. V. Kharissova, H. V. Rasika Dias (Wiley, USA) (2015), pp. 207–215.

21. S. Palit, Microfiltration, groundwater remediation and environmental engineering science: A scientific perspective and a far-reaching review, *Nature, Environment and Pollution Technology* (2015), 14(4), 817–825.

22. S. Palit, Nanofiltration and ultrafiltration: The next generation environmental engineering tool and a vision for the future, *International Journal of Chem Tech Research* (2016), 9(5), 848–856.
23. S. Palit, Frontiers of nano-electrochemistry and application of nanotechnology: A vision for the future, in: *Handbook of Nanoelectrochemistry*, Springer International Publishing, Switzerland (2015).
24. S. Palit, Dependence of order of reaction on pH and oxidation-reduction potential in the ozone-oxidation of textile dyes in a bubble column reactor, *International Journal of Environmental Pollution Control and Management* (2011), 3(4), 69–78.
25. L. Singh, G. Goga, M. K. Rathi, Latest developments in composite materials, *IOSR Journal of Engineering* (2012), 2(8), 152–158.
26. G. Mago, D. M. Kalyon, S. C. Jana, Polymer nanocomposite-processing, characterization and applications, *Journal of Nanomaterials*, (Special Issue), (2011), Hindawi Publishing Corporation, USA.

Chapter 15

Polymer Nanocomposite Membranes Prepared by Electrospinning for Water Remediation

Pengchao Liu,[a] Xiangyang Shi,[b,c] and Chen Peng[d]

[a]*Department of Oral and Cranio-maxillofacial Science, Shanghai Ninth People's Hospital, Shanghai Jiao Tong University School of Medicine, Shanghai Key Laboratory of Stomatology, No. 639 Zhizaoju Road, Shanghai 200011, People's Republic of China*
[b]*State Key Laboratory for Modification of Chemical Fibers and Polymer Materials, College of Chemistry, Chemical Engineering and Biotechnology, Donghua University, Shanghai 201620, People's Republic of China*
[c]*CQM-Centro de Química da Madeira, Universidade da Madeira, Campus da Penteada, 9000-390 Funchal, Portugal*
[d]*Department of Radiology, Shanghai Tenth People's Hospital School of Medicine, Tongji University, 301 Yanchang Road, Shanghai 200072, People's Republic of China*

lpc08072006@126.com

15.1 Introduction

With the expansion of industrialization, the discharge of industrial wastewater and domestic sewage leads to the deterioration of water quality. Clean water shortage has become one of the biggest global crises to food security, human health, and natural ecology.

Nanocomposites for Pollution Control
Edited by Chaudhery Mustansar Hussain and Ajay Kumar Mishra

ISBN 978-981-4774-45-1 (Hardcover), 978-1-315-14368-2 (eBook)
www.panstanford.com

By 2025, more than 50% of the countries around the world will experience water scarcity [1]. In the wastewater, the most major pollutants include organic compounds (dyes, pesticide) and heavy metal ions (Cr(VI), Pb(II), Fe(III), As(V), Cu(II), Zn(II), Hg(II), and Ni(II)). Both of them are undesirable for water bodies as they are toxic, carcinogenic and mutagenic products [1]. They can remain in the wastewater for a long time. Therefore, great attention must be paid to addressing them. To remove these contaminants in water, the most commonly methods are chemical treatments, biochemical process, and physicochemical treatments. Physicochemical methods include (i) precipitation as hydroxides, carbonates or sulfides and subsequent liquid–solid separation by gravity settling and flotation of filtration; (ii) sorption (e.g., adsorption and ion exchange); (iii) membrane processes; (iv) electrolytic recovery; and (v) liquid–liquid extraction [2].

Among these methods, membrane technology has been identified as one of the most robust and flexible technologies to improve water quality, by removing undesirable compounds and contaminants. This is due to its permselectivity, high efficiency, greatly reduced energy consumption, pollution, and emission of carbon dioxide. A major breakthrough was achieved in 1960, when Loeb and Sourirajan developed cellulose diacetate membrane [3]. Since then, membrane technology developed from mere laboratory tool to a well-developed industrial process where potable water was produced for industries [4]. And now, membrane has been applied in the field of nanofiltration, microfiltration, ultrafiltration, and reverse osmosis for water remediation.

15.2 Electrospinning

Electrospinning has been used as a facile and efficient technique to fabricate membranes consisting of many nanofibers. This method was developed in 1934 by Formhals [5], who designed the first set of preparation device of polymer nanofibers under electric field. During the preparation process, spinning solution (melt, emulsion, or suspension) was drawn under the effect of electric field force. Taylor cone forms at the needle when the electric field force exceeds the surface tension of spinning solution, and then nanofibers could be collected accompanying by a solvent evaporation. The diameter of fibers ranges from tens of nanometers to a few micrometers. A typical electrospinning device [6] involves a high voltage supplier, an injection device, and a collector (Fig. 15.1).

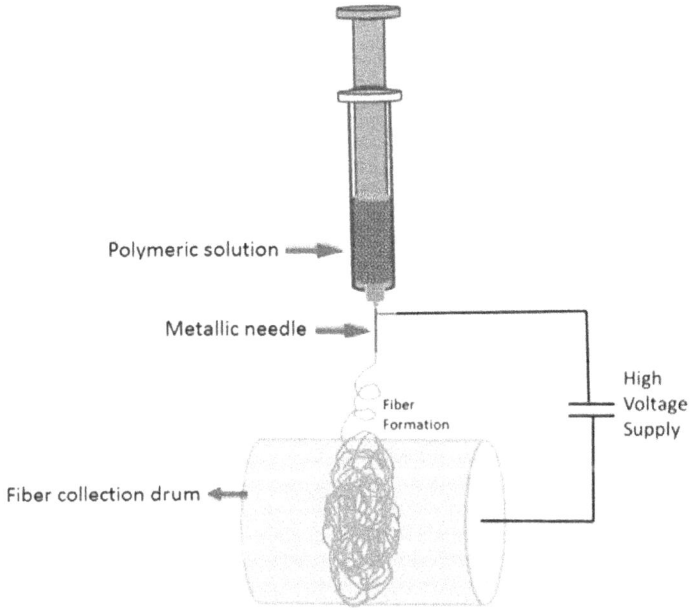

Figure 15.1 Schematic diagram of an electrospinning apparatus. Reproduced with permission from Ref. 6, Copyright 2015, Elsevier.

In the conventional electrospinning device, there is only one spraying needle, namely single capillary electrospinning [7]. Since 2003, coaxial electrospinning [8] was developed to produce fibers. It is similar to the conventional one except for the use of a spinneret containing two coaxial capillaries. In a typical procedure, two viscous liquids are simultaneously fed through the inner (core) and outer (shell) capillaries, respectively. In 2004, Li et al. [9] used coaxial electrospinning to produce core–shell nanofibers and hollow fibers. The complete immiscibility of core and shell solutions enabled the formation of a regular core–shell structure along the fibers.

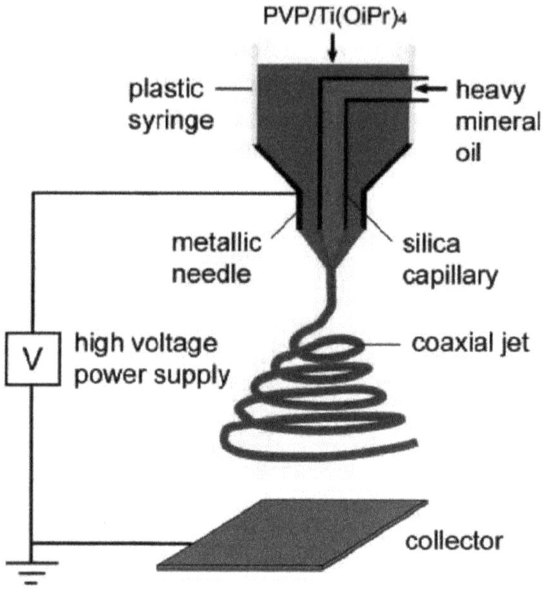

Figure 15.2 Schematic illustration of coaxial electrospinning. Reproduced with permission from Ref. 8, Copyright 2004, American Chemical Society.

By electrospinning method, polymer, inorganic, and composite nanofiber membranes can be fabricated. Nanofiber membranes have high porosity and interconnected pore structure, which endow the materials with high permeability for fluid. In addition, nanofiber membranes have good adsorption and selective properties due to their small pore size, large specific surface area, flexibility in surface functionalities and designability. These advantages make them excellent membrane candidates for water remediation [10].

15.3 Polymer Nanocomposite Electrospun Membranes (PNEMs)

Polymer nanocomposite membrane could be fabricated by electrospinning with fibers in random, aligned, cross-aligned, hollow, yarns, and core/shell morphologies. PNEMs are a kind of composite materials, in which nanofillers are added to a polymer to reinforce and provide novel functions. The possibility of combining a variety of polymers and particulate nanofillers through electrospinning leads to development of nanocomposite membranes with a much broader domain of environmental applications than their neat counterparts [11].

15.3.1 Methods of Fabricating PNEMs

Various techniques have been developed and applied for preparation of PNEMs based on electrospinning approach. According to the formation processes, they could be generally divided into three categories as follows:

15.3.1.1 Direct compounding electrospinning [12]

The most straightforward methodology for incorporating nanoparticles into polymer fibers is to directly disperse them in the polymer solution before electrospinning. To obtain a homogeneous suspension/mixture, stirring or sonication is also needed. In such a manner, the nanoparticles are encapsulated in the solid nanofibers. However, the simple direct compounding strategy is quite problematic when a high loading of nanoparticles is desired. This is due to a high viscosity of the resulting dispersion prevents fiber formation eventually and a high tendency of the nanoparticles to form larger aggregates during blending. To overcome this issue, a surfactant is sometimes used.

15.3.1.2 In situ synthesis

The second methodology used to obtain PNEMs derives from the in situ synthesis of nanoparticles by using suitable precursors. This approach is widely used to prepare PNEMs by preloading metal ions within polymer matrix to serve as nanoparticle precursors first. Then, metal ions/polymer composite nanofiber membranes are obtained by electrospinning. At last, the composite nanofibers are exposed to corresponding liquid or gas containing S^{2-}, OH^-, or Se^{2-} to in situ synthesize the target nanoparticles [13].

15.3.1.3 Post-treatment of electrospun fibers

PNEMs can also be prepared by post-treatment after electrospinning of polymer nanofibers. Generally, polymer nanofiber membranes are electrospun first, and then nanoparticles are doped onto the materials by surface modification or functionalization.

Liquid Phase Deposition is used to generate inorganic films on the polymer nanofibers' substrate through their immersion in the reactant solution. The first report on the production dates back to 2003 [14]. PAN electrospun fibers were covered with tin oxide or titanium oxide.

Atomic Layer Deposition enables the deposition of inorganic films with a precise control of thickness at the atomic level on the polymer nanofibers. Oldham et al. [15] prepared nylon-6 fibers coated with either ZnO or Al_2O_3, which could control the fiber wetting properties and chemical resistance.

Hydrothermal Synthesis could be used to produce crystals of oxides and hydroxides by the chemical reaction of precursors in water solution under controlled conditions of temperature and pressure [16]. This method is limited to electrospun semicrystalline fibers with high melting temperatures or amorphous fibers with high glass transition temperatures, because they can resist high temperature conditions. He et al. [17] prepared PVDF fibers covered with TiO_2 by hydrothermal method. Patel et al. [18] prepared PAN nanofibers covered with iron oxide, as shown in Fig. 15.3.

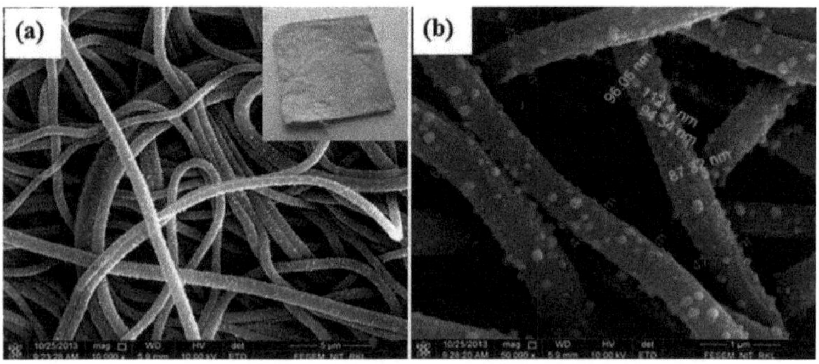

Figure 15.3 Low and high magnification FE-SEM images (a and b) of PAN/iron composite nanofibers (inset: photograph). Reproduced with permission from Ref. 17, Copyright 2016, Royal Society of Chemistry.

Reduction of Metal Ions: Metallic nanoparticles could be obtained by introducing metal ions into polymer fibers followed by chemical reduction treatment. Two different approaches have been described in the literature. One is using polymer fibers, on which functional groups able to coordinate the metal ions when the fibers are immersed in a solution of their salt. And then, the ions attached to the fibers are reduced by a chemical agent. The other approach is incorporating the metal ions in the polymer fibers during electrospinning process, followed by reduction to gain metallic particles. Shi et al. [19] reported a facile approach to assembling low generation poly (amidoamine) (PAMAM) dendrimer-stabilized gold nanoparticles (Au DSNPs) onto electrospun polyacrylic acid (PAA)/polyvinyl alcohol (PVA) nanofibrous mats for catalytic applications. In another study, Shi et al. [20] prepared multiwalled carbon nanotubes (MWCNTs) enhanced polyacrylic acid (PAA)/polyvinyl alcohol (PVA) nanofiber membrane, and then immobilized zero-valent iron nanoparticles (ZVI NPs) into electrospun polymer nanofibers.

15.3.2 Functions of Nanoparticles

In the PNEMs, a combination of polymer and nanoparticles offers the benefits of individual components together. Polymers provide lightweight, flexibility, and moldability, whereas nanoparticles endow the PNEMs with high strength, chemical resistance, and thermal stability. Most importantly, the nanoparticles also have some other special functions such as filtration, adsorption, catalysis, and antimicrobial/fouling.

15.3.2.1 Nanoparticles as adsorbents

Adsorption effect is found between nanomaterials and pollutants. Currently, many different porous materials have been developed such as carbon nanomaterials, clays, zeolites, mesoporous oxides [21]. Carbon nanomaterials (activated carbon, nanotubes, and graphene) have been chosen as adsorbents due to their high surface area, non-corrosive property, presence of oxygen-containing functional groups, tunable surface chemistry, and scalable production [22]. They can be also chemically regenerated after being exhausted [23].

Carbon-based nanoparticles

Activated carbon from various sources, such as coconut coir, jute stick, rice husk, etc., is the most popular adsorbent. Carbon nanotubes (CNTs) are composed of cylindrical shape rolled up in a tube like structure. CNTs include single walled carbon nanotubes and multiwalled carbon nanotubes. The adsorption capacity of CNTs is mainly determined by the chemical nature of CNTs, the surface area, and the number of oxygen functional groups [24]. The mechanism of metal ion sorption on CNT surface has been related to electrostatic interactions and sorption–precipitation between metal ions and the oxygen-containing groups [25].

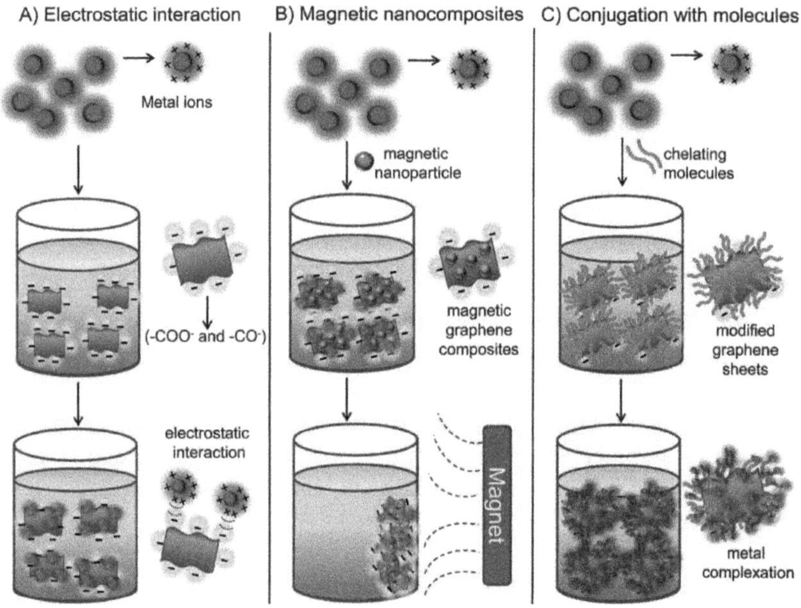

Figure 15.4 Main strategies to apply graphene-based materials as adsorbents for the removal of metal ions from aqueous solutions. (A) The sorption process can be performed using non-modified graphene oxide (GO), graphene, or reduced graphene oxide (rGO); the mechanism of adsorption is mostly due to electrostatic interaction between the negatively charged GO sheets and the positively charged metal ions. (B) Graphene sheets can be functionalized with magnetic nanoparticles to improve adsorption capacity; since the GO nanocomposites possess magnetic properties, metal ions can be removed from water by magnetic attraction. (C) Modification of graphene sheets with organic molecules can be used to prepare graphene-based adsorbents with improved effectiveness; the mechanism of adsorption is attributed to a synergetic effect between the chelating properties of the organic molecules and adsorption capacity of the graphene sheets. Reproduced with permission from Ref. 26, Copyright 2015, Royal Society of Chemistry.

Graphene-based materials are potential adsorbents for metal ions complexation through both electrostatic and coordinate approaches [26]. They serve as efficient adsorbents due to their large specific surface area and electron rich environment. GO has better performance than pristine graphene in metal ion adsorption due to GO's high content of oxygen groups available to interact with metal ions. Of particular interest is the conjugation of graphene with magnetic nanoparticles (e.g., iron or iron oxide), which has been the most common approach to prepare graphene-based composites for the removal of metal ions [27]. Besides metal ion removal, grapheme-based adsorbents have also been applied to remove various organic pollutants from water [21]. In addition, Graphene-magnetic nanoparticle composites have improved adsorption performance, which may be attributed to a combined effect of metal complexation on the nanoparticles and on the adsorption sites of graphene aromatic layer [27a]. Figure 15.4 illustrates the different methods of applying grapheme-based materials as adsorbents to remove metallic contaminants from aqueous solutions [28].

Metal oxide based nanoparticles

Nano metal oxides, including nano-sized aluminum oxide, titanium oxide, iron oxide, and magnesium oxide are promising adsorbents for removal of pollutants from aqueous system [2, 29]. This is due to their large surface areas and high activities caused by the size quantization effect [30]. These pollutants include organic dyes and heavy metals.

15.3.2.2 Nanoparticles as photocatalysts

TiO_2 nanoparticles are the most widely studied photocatalyst for photo-degradation of aqueous pollutants. They have been extensively used for the oxidative and reductive transformation of organic and inorganic contaminants in water [31]. TiO_2 has three metastable phases, known as rutile, anatase, and brookite. The anatase phase is used as the most powerful photocatalyst under UV light in industry [32]. The photocatalysis efficiency of TiO_2 is dependent on the initial concentration of the contaminants, the amount of light penetration to the surface of TiO_2, and the duration time of light irradiation. The surface area of the photocatalyst is a key factor in catalytic reactions. Besides TiO_2, other nanoparticles were also investigated as photocatalysts to purify the water, including ZnO [33], Bi_2O_3/BiOCl [34], WO_3 [35], Fe_2O_3 [36], and CuS [37]. These photocatalysts can be used to remove dye and heavy metals.

15.3.2.3 Nanoparticles as antibacterial agents

Besides organic dyes and heavy metals in the wastewater, there are also microorganisms and bacteria. They can cause infectious diseases to human. For preventing such type of pollutants, nanoparticles could be applied as antibacterial agents.

Silver nanoparticles are known for their strong antibacterial effects against a wide array of organisms (e.g., viruses, bacteria, fungi). Therefore, silver nanoparticles are widely used for the disinfection of water [38]. Ag^+ ions interact with thiol groups in proteins, which leads to inactivation of respiratory enzymes and the production of reactive oxygen species [39]. It was also shown that Ag^+ ions can prevent DNA replication and affect the structure and permeability of the cell membrane [40]. Silver ions are also photoactive in the presence of UV irradiation, causing an improvement in the UV inactivation of bacteria and viruses [41]. Nano-sized TiO_2 was also reported to kill viruses, including poliovirus 1 [42], the hepatitis B virus [43], the Herpes simplex virus [43] and MS2 bacteriophage [44]. The concentration of TiO_2 required to kill bacteria varies between 100 and 1000 ppm depending on the size of the particles and the intensity and wavelength of the light used [45]. CuO nanoparticles was also reported to impart antibacterial functions to textile materials [46]. ZnO nanoparticles exhibit strong antibacterial activities on a broad spectrum of bacteria [47].

15.3.2.4 Nanoparticles as reductants

Nanoscale zero-valent metals (NZVMs) (Fe^0, Zn^0, Ti^0, Ni^0, Pd^0, Mg^0, Al^0, Au^0, etc.) with significantly strong chemical reducibility, high efficiency, and large specific surface have the most promising applications in water remediation [48]. They have been applied for the efficient removal of different metals, such as Cr(VI) [49], Hg(II) [50], U(IV) and U(VI) [51], and Co(II) [52]. The removal processes of contaminants include both a 'physical' removal from solution to an immobile phase and a 'chemical' removal by reaction to form less hazardous products.

Bimetallic particles are composed of two types of zero-valent metals. The bimetals' structures include cluster-in-cluster and core–shell [53]. Fe^0/Ni^0 has better degradation rates for the dehalogenation of trichloroethylene [54]. Pd^0/Fe^0 has rapid and complete dechlorination of all the chlorinated pollutants for water remediation. Shi et al. [54c] proved that co-reduced Fe^0/Ni^0 nanoparticles-containing nanofibrous mats have much higher decoloration efficiency than the single iron NP-immobilized polymer mats. In addition, Fe/Al was used for the removal of chlorinated organic solvents, nitrate, and heavy metals, such as Cr(VI) and Cu(II), as well as perchlorate ions from wastewaters [53, 55]. Cu/Al particles have shown a better degradation of CCl_4 and CH_2Cl_2 [56].

Although these nanoparticles show great potential for water remediation, one major drawback has to be solved for their practical use. These nanomaterials are not stable due to small size and high reactivity. Polymer nanocomposite electrospun membranes can solve this issue by introducing nanoparticles into the polymer nanofibers' matrix. So they can be used for water remediation.

15.4 Polymer Nanocomposite Membrane for Water Remediation

15.4.1 Water Filtration

Polymer nanocomposite electrospun membranes are good at water filtration. Water filtration membranes are divided into sieving and affinity membranes, including micro-, ultra-, and nanofiltration ones for remediation of wastewater. Pant et al. [57] prepared a composite membrane of nylon-6/TiO_2 nanoparticles. TiO_2 induced a higher hydrophilicity, mechanical strength, antimicrobial, and UV protecting ability to the material. This nanostructured membrane with an improved wettability, also antifouling property, can be a potential candidate for future water filtration applications. Bui et al. [58] prepared mesoporous silica nanoparticles-embedded nanofiber membranes for forward osmosis. They found that the membranes showed a remarkable 7-fold and 3.5-fold enhancements in osmotic water permeability and water/sodium chloride selectivity, respectively, compared to standard commercial forward osmosis membranes. To reinforce an electrospun nanofibrous microfiltration membrane made of polyether sulfone (PES) mechanically and chemically (to improve wettability), zirconia nanoparticles [59] and TiO_2 nanoparticles [60] were added to the nanofibers.

15.4.2 Oil/Water Separation

Oil/water separation is very meaningful in the field of water remediation, especially as to oil leakage in the sea. Ding et al. successfully prepared a series of superwetting hierarchical electrospun nanofiber membranes that could separate water-in-oil [61] as well as oil-in-water [62] micro-emulsions (as shown in Fig. 15.5). In particular, porous silica-based electrospun nanofiber membrane system having superhydrophilicity and underwater superoleophobicity was demonstrated using a facile combination of electrospun silica nanofibers and in situ polymerized polybenzoxazine (from an aldehyde benzoxazine monomer) functional layer containing SiO_2 nanoparticles.

15.4.3 Adsorption of Heavy Metal Ions

By introducing some functional nanoparticles to nanofibers, PNEMs are able to remove heavy metal. The selective sorption of these PNEMs toward heavy metals can be explained on the basis of their specific structure including: (i) the negatively charged host material, and (ii) the dispersion of the nanoparticles onto the inner surface of the polymers. Such sorption preference is mainly attributed to two mechanisms: (i) Donnan's membrane effect caused by the negatively charged supporting material, and (ii) specific affinity between nanoparticles and heavy metals.

Several nanoparticles were used to fabricate this kind of PNEMs, in most cases by in situ synthesis. Iron minerals have been recognized as an effective media to remove various heavy metals such as As(III)/As(V) [63], Cr(VI) [64] and Pd(II) [65]. The hierarchical aminated PAN/γ-AlOOH composite nanofibers were proven to be good adsorbents toward Pb(II), Cu(II) and

Cd(II) ions in aqueous solution. The maximum adsorption capacities for Pb(II), Cu(II) and Cd(II) were 180.83, 48.68 and 114.94 mg/g, respectively [66]. In a research by Hota et al. [67], a PNEM was developed by incorporation of nano-boehmite into nylon-6 and polycaprolactone (PCL) electrospun nanofibers. This membrane showed a Cd^{2+} removal ability of 0.2 μg/mg. In another study by Xiao et al. [68], a PNEM, based on polyacrylic acid (PAA)/polyvinyl alcohol (PVA) nanofibers containing multiwalled carbon nanotubes (MWNTs), was fabricated. The nanocomposite nanofibers were further immobilized with zero-valent iron (ZVI) nanoparticles. While MWNTs were incorporated into the nanofibers to optimize their mechanical stability, ZVI nanoparticles were added to interact and remove Cu^{2+} from water.

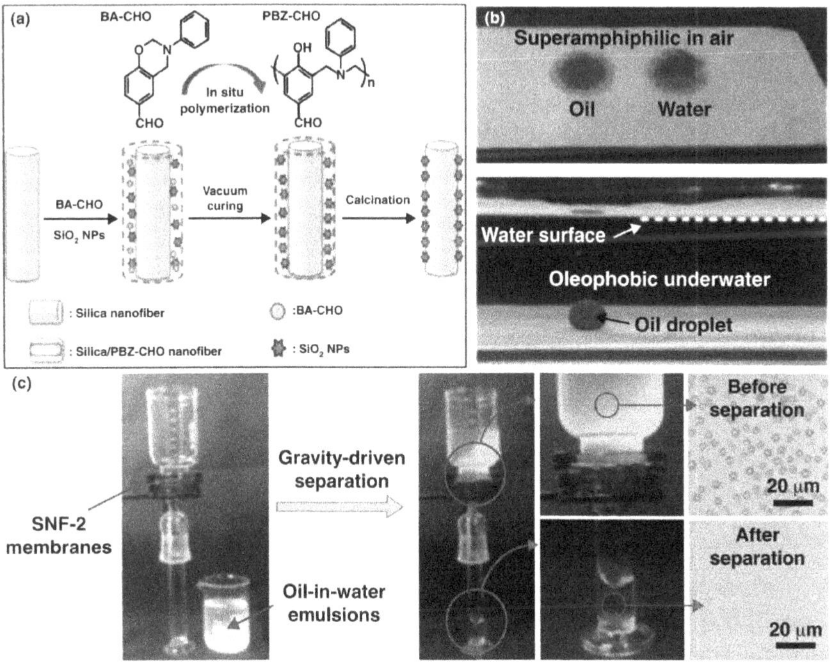

Figure 15.5 (a) Illustration of the synthesis of hierarchical porous silica nanofibrous (SNF) membranes by combining the nanofibers with in situ polymerization. (b) Photograph of oil and water on the SNF-0 membranes in air, and an underwater oil droplet on the SNF-0 membranes. (c) Photographs showed the facile gravity-driven separation of oil-in-water micro-emulsions using SNF-2 membranes. Reproduced with permission from Ref. 61, Copyright 2014, Royal Society of Chemistry.

15.4.4 Removal of Organic Compounds (Adsorption and Degradation)

Water quality is highly sensitive to some organic pollutants, such as oil, protein, and humic acid. Such organics, even at a very low concentration, e.g., less than 1% of the pollution in a river, are able to use up the dissolved oxygen, making the water free of any life [69]. Organic pollutants, such as dyes and pesticides, are readily decomposed in aqueous solutions by photocatalysis in the presence of metal oxides such as titanium oxide (TiO_2), zinc oxide (ZnO), tin oxide (SnO_2), and copper oxide (CuO).

Textile industry which is one of the most important water consumers produces highly colored and complex wastewater. More than 100,000 commercial dyes with over 7.105 t were produced yearly [70]. Pathania et al. [70] found that Guar gum/Al_2O_3 (GG/AO) nanocomposite could decompose efficiently malachite green dye. In another study by this group, TiO_2 particles were uniformly dispersed on PAN nanofibers to create a PNEM for the UV light induced photo-decomposition of RhB dye [71]. Ye et al. [72] made a PNEM through incorporation of ZnO nanocrystals into cellulose nanofibers. This PNEM benefits from not only the photocatalytic ability of ZnO nanoparticles but also the thermal and mechanical stability and solvent resistance of the cellulose nanofibers. A strong photocatalytic efficiency towards the degradation of RhB was observed.

15.4.5 Removal of Microorganisms

The pathogenic microorganisms such as *Cryptosporidium parvum* and *Giardia lamblia* are so hazardous that their presence in water can cause serious illnesses. For this reason, in many countries, removal of them from water is compulsory. By virtue of available high surface area, antimicrobial agents incorporated PNEMs can offer a very promising efficiency in removal of such pollutants. Lala et al. [73] fabricated several silver-impregnated polymeric nanofibrous membranes and evaluated their antimicrobial capability using two gram negative bacterial groups: *E. coli* and *P. aeruginosa*. The results were quite promising in terms of antimicrobial activity of the membranes when incubated with bacteria [74]. In addition to incorporating antimicrobial properties, silver nanoparticles can minimize the formation of biofilm on the surface of membranes hence inducing an antibiofouling effect [67a]. Park et al. [75] developed nylon-6 nanofibers containing silver nanoparticles. Such a PNEM showed an optimized antibacterial activity as compared to its neat counterpart.

15.5 Conclusions and Perspectives

Polymer nanocomposite electrospun membranes (PNEMs) are currently studied extensively for water remediation. Different methods for producing PNEMs were described: (1) direct compounding electrospinning, (2) in situ synthesis, and (3) post-treatment of electrospun fibers, including liquid phase deposition, atomic layer deposition, hydrothermal synthesis, reduction of metal ions. By doping inorganic fillers with different functions, PNEMs possess both the advantages of polymers and nanoparticles. With regard to the water remediation, addition of inorganic fillers can confer not only a higher mechanical stability, but also oil/water separation ability, removal ability for heavy metal ions, organic compounds (dyes, pesticides), and microorganisms. On the whole, the future membranes for water remediation would be based on cost effective and energy saving PNEMs. Undoubtedly, PNEMs would play a significant role when considering their tunable selectivity, designability, and versatility.

References

1. Homaeigohar, S., Elbahri, M., Nanocomposite electrospun nanofiber membranes for environmental remediation. *Materials* 2014, 7(2), 1017–1045.

2. Peng, C., Zhang, J., Xiong, Z., Zhao, B., Liu, P., Fabrication of porous hollow γ-Al_2O_3 nanofibers by facile electrospinning and its application for water remediation. *Microporous Mesoporous Mater.* 2015, 215, 133–142.

3. Sidney, L., Srinivasa, S., High flow porous membranes for separating water from saline solutions. US: 1964.

4. Shenvi, S. S., Isloor, A. M., Ismail, A. F., A review on RO membrane technology: Developments and challenges. *Desalination* 2015, 368, 10–26.
5. Anton, F., Process and apparatus for preparing artificial threads. US: 1934.
6. Ahmed, F. E., Lalia, B. S., Hashaikeh, R., A review on electrospinning for membrane fabrication: Challenges and applications. *Desalination* 2015, 356, 15–30.
7. Liu, P., Zhu, Y., Ma, J., Yang, S., Gong, J., Xu, J., Preparation of continuous porous alumina nanofibers with hollow structure by single capillary electrospinning. *Colloids Surf. A Physicochem. Eng. Asp.* 2013, 436, 489–494.
8. Larsen, G., Velarde-Ortiz, R., Minchow, K., Barrero, A., Loscertales, I. G., A method for making inorganic and hybrid (organic/inorganic) fibers and vesicles with diameters in the submicrometer and micrometer range via sol–gel chemistry and electrically forced liquid jets. *J. Am. Chem. Soc.* 2003, 125(5), 1154–1155.
9. D. L., Xia, Y., Direct fabrication of composite and ceramic hollow nanofibers by electrospinning. *Nano Lett.* 2004, 4(5), 933–938.
10. Miao, Y. E., Wang, R., Chen, D., Liu, Z., Liu, T., Electrospun self-standing membrane of hierarchical SiO_2@γ-AlOOH (boehmite) core/sheath fibers for water remediation. *ACS Appl. Mater. Interfaces* 2012, 4(10), 5353–5359.
11. Homaeigohar, S., Elbahri, M., Nanocomposite electrospun nanofiber membranes for environmental remediation. *Materials* 2014, 7(2), 1017.
12. (a) Sahay, R., Kumar, P. S., Sridhar, R., Sundaramurthy, J., Venugopal, J., Mhaisalkar, S. G., Ramakrishna, S., Electrospun composite nanofibers and their multifaceted applications. *J. Mater. Chem.* 2012, 22(26), 12953–12971; (b) Ayutsede, J., Gandhi, M., Sukigara, S., Ye, H., Hsu, C., Gogotsi, Y., Ko, F., Carbon nanotube reinforced bombyx mori silk nanofibers by the electrospinning process. *Biomacromolecules* 2006, 7(1), 208–214.
13. (a) Dong, Y., Jie, L., Jiang, Z., Lu, L., Xue, C., Chitosan/TiO_2 nanocomposite pervaporation membranes for ethanol dehydration. *Chem. Eng. Sci.* 2009, 64(13), 3130–3137; (b) Fang, X., Ma, H., Xiao, S., Shen, M., Guo, R., Cao, X., Shi, X., Facile immobilization of gold nanoparticles into electrospun polyethyleneimine/polyvinyl alcohol nanofibers for catalytic applications. *J. Mater. Chem.* 2011, 21(21), 4493–4501.

14. Drew, C., Liu, X., Ziegler, D., Wang, X., Bruno, F. F., Whitten, J., Samuelson, L. A., Kumar, J., Metal oxide-coated polymer nanofibers. *Nano Lett.* 2003, 3(2), 143–147.

15. Oldham, C. J., Senecal, K. J., Godfrey, T. A., Bo, G., Spagnola, J. C., Jur, J. S., Parsons, G. N., Encapsulation and chemical resistance of electrospun nylon nanofibers coated using integrated atomic and molecular layer deposition. *J. Electrochem. Soc.* 2011, 158(9), D549–D556.

16. Shi, W., Song, S., Zhang, H., Hydrothermal synthetic strategies of inorganic semiconducting nanostructures. *Chem. Soc. Rev.* 2013, 42(13), 5714–5743.

17. He, T., Zhou, Z., Xu, W., Ren, F., Ma, H., Wang, J., Preparation and photocatalysis of TiO_2-fluoropolymer electrospun fiber nanocomposites. *Polymer* 2009, 50(13), 3031–3036.

18. Patel, S., Hota, G., Iron oxide nanoparticle-immobilized PAN nanofibers: Synthesis and adsorption studies. *Rsc Adv.* 2016, 6(19), 15402–15414.

19. Hu, D., Huang, Y., Liu, H., Wang, H., Wang, S., Shen, M., Zhu, M., Shi, X., The assembly of dendrimer-stabilized gold nanoparticles onto electrospun polymer nanofibers for catalytic applications. *J. Mater. Chem. A* 2014, 2(7), 2323–2332.

20. Xiao, S., Shen, M., Guo, R., Huang, Q., Wang, S., Shi, X., Fabrication of multiwalled carbon nanotube-reinforced electrospun polymer nanofibers containing zero-valent iron nanoparticles for environmental applications. *J. Mater. Chem* 2010, 20(27), 5700–5708.

21. Santhosh, C., Velmurugan, V., Jacob, G., Jeong, S. K., Grace, A. N., Bhatnagar, A., Role of nanomaterials in water treatment applications: A review. *Chem. Eng. J.* 2016, 306, 1116–1137.

22. Sitko, R., Zawisza, B., Malicka, E., Modification of carbon nanotubes for preconcentration, separation and determination of trace-metal ions. *Trac Trends Anal. Chem.* 2012, 37(37), 22–31.

23. Yang, K., Xing, B., Desorption of polycyclic aromatic hydrocarbons from carbon nanomaterials in water. *Environ. Pollut.* 2007, 145(2), 529–537.

24. Stafiej, A., Pyrzynska, K., Adsorption of heavy metal ions with carbon nanotubes. *Sep. Purif. Technol.* 2007, 58(1), 49–52.

25. Rao, G. P., Lu, C., Su, F., Sorption of divalent metal ions from aqueous solution by carbon nanotubes: A review. *Sep. Purif. Technol.* 2007, 58(1), 224–231.

26. (a) Wu, W., Yang, Y., Zhou, H., Ye, T., Huang, Z., Liu, R., Kuang, Y., Highly efficient removal of Cu(II) from aqueous solution by using graphene oxide. *Water Air Soil Pollut.* 2013, 224(1), 1–8; (b) Sitko, R., Turek, E., Zawisza, B., Malicka, E., Talik, E., Heimann, J., Gagor, A., Feista, B., and Wrzalik, R., Adsorption of divalent metal ions from aqueous solutions using graphene oxide. *Dalton Trans.* 2013, 42(16), 5682–5689.

27. (a) Sumesh, E., Bootharaju, M. S., Anshup, Pradeep, T., A practical silver nanoparticle-based adsorbent for the removal of Hg 2+ from water. *J. Hazard. Mater.* 2011, 189(1–2), 450–457, (b) Pradeep, T., Anshup, Noble metal nanoparticles for water purification: A critical review. *Thin Solid Films* 2009, 517(24), 6441–6478.

28. Perreault, F., Faria, A. F. D., Elimelech, M., ChemInform abstract: Environmental applications of graphene-based nanomaterials. *Chem. Soc. Rev.* 2015, 44(16), 5861–5896.

29. (a) Visa, M., Andronic, L., Enesca, A., Behavior of the new composites obtained from fly ash and titanium dioxide in removing of the pollutants from wastewater. *Appl. Surf. Sci.* 2015, 388, 359–369; (b) Saharan, P., Chaudhary, G. R., Mehta, S. K., Umar, A., Removal of water contaminants by iron oxide nanomaterials. *J. Nanosci. Nanotechnol.* 2014, 14(1), 627–643; (c) Rafiq, Z., Nazir, R., Durr-e-Shahwar, Shah, M. R., Ali, S., Utilization of magnesium and zinc oxide nano-adsorbents as potential materials for treatment of copper electroplating industry wastewater. *J. Environ. Chem. Eng.* 2014, 2(1), 642–651.

30. El-Sayed, M. A., Some interesting properties of metals confined in time and nanometer space of different shapes. *ChemInform* 2001, 34(26), 257–264.

31. Kamat, P. V., Dan, M., Nanoscience opportunities in environmental remediation. *Comptes Rendus Chim.* 2003, 6(8), 999–1007.

32. Bet-Moushoul, E., Mansourpanah, Y., Farhadi, K., Tabatabaei, M., TiO_2 nanocomposite based polymeric membranes: A review on performance improvement for various applications in chemical engineering processes. *Chem. Eng. J.* 2016, 283, 29–46.

33. Choina, J., Bagabas, A., Fischer, C., Flechsig, G. U., Kosslick, H., Alshammari, A., Schulz, A., The influence of the textural properties of ZnO nanoparticles on adsorption and photocatalytic remediation of water from pharmaceuticals. *Catal. Today* 2015, 241, 47–54.

34. Mushtaq, F., Guerrero, M., Sakar, M. S., Hoop, M., Lindo, A. M., Sort, J., Chen, X., Nelson, B. J., Pellicer, E., Pané, S., Magnetically driven

Bi_2O_3/BiOCl-based hybrid microrobots for photocatalytic water remediation. *J. Mater. Chem. A* 2015, 3(47), 23670–23676.

35. Lee, H., Choi, J., Lee, S., Yun, S. T., Lee, C., Lee, J., Kinetic enhancement in photocatalytic oxidation of organic compounds by WO_3 in the presence of Fenton-like reagent. *Appl. Catal. B Environ.* 2013, 138–139(11), 311–317.
36. Miao, Z., Tao, S., Wang, Y., Yu, Y., Meng, C., An, Y., Hierarchically porous silica as an efficient catalyst carrier for high performance vis-light assisted Fenton degradation. *Microporous Mesoporous Mater.* 2013, 176(176), 178–185.
37. Gupta, V. K., Pathania, D., Agarwal, S., Singh, P., Adsorptional photocatalytic degradation of methylene blue onto pectin–CuS nanocomposite under solar light. *J. Hazard. Mater.* 2012, 243(4), 179–186.
38. Chou, K. S., Lu, Y. C., Lee, H. H., Effect of alkaline ion on the mechanism and kinetics of chemical reduction of silver. *Mater. Chem. Phys.* 2005, 94(2–3), 429–433.
39. Matsumura, Y., Yoshikata, K., Kunisaki, S., Tsuchido, T., Mode of bactericidal action of silver zeolite and its comparison with that of silver nitrate. *Appl. Environ. Microbiol.* 2003, 69(7), 4278–4281.
40. Feng, Q. L., Wu, J., Chen, G. Q., Cui, F. Z., Kim, T. N., Kim, J. O., A mechanistic study of the antibacterial effect of silver ions on *Escherichia coli* and *Staphylococcus aureus*. *J. Biomed. Mater. Res.* 2000, 52(4), 662–668.
41. Kim, J. S., Kuk, E., Yu, K. N., Kim, J. H., Park, S. J., Hu, J. L., Kim, S. H., Park, Y. K., Yong, H. P., Hwang, C. Y., Antimicrobial effects of silver nanoparticles. Nanomed: NBM 3: 95–101. *Nanomed. Nanotechnol. Biol. Med.* 2007, 3(1), 95–101.
42. Watts, R. J., Kong, S., Orr, M. P., Miller, G. C., Henry, B. E., Photocatalytic inactivation of coliform bacteria and viruses in secondary wastewater effluent. *Water Res.* 1995, 29(1), 95–100.
43. Zan, L., Fa, W., Peng, T., Gong, Z. K., Photocatalysis effect of nanometer TiO_2 and TiO_2-coated ceramic plate on Hepatitis B virus. *J. Photochem. Photobiol. B Biol.* 2007, 86(2), 165–169.
44. Cho, M., Chung, H., Choi, W., Yoon, J., Different inactivation behaviors of MS-2 phage and Escherichia coli in TiO_2 photocatalytic disinfection. *Appl. Environ. Microbiol.* 2005, 71(1), 270–275.
45. Wei, C., Lin, W. Y., Zainal, Z., Williams, N. E., Zhu, K., Kruzic, A. P., Smith, R. L., Rajeshwar, K., Bactericidal activity of TiO_2 photocatalyst in

aqueous media: Toward a solar-assisted water disinfection system. *Environ. Sci. Technol.* 1994, 28(5), 934–938.

46. Lubasova, D., Yalcinkaya, F., Komarek, M., Sanetrnik, F., Maryska, J. In *Producing Antibacterial Textile Material by Weaving PVB/CuO Nanocomposite Fiber Covered Yarn*, Nanocon, 2014.
47. Adams, L. K., Lyon, D. Y., Alvarez, P. J. J., Comparative eco-toxicity of nanoscale TiO_2, SiO_2, and ZnO water suspensions. *Water Res.* 2006, 40(19), 3527–3532.
48. Li, L., Hu, J., Shi, X., Fan, M., Jin, L., Wei, X., Nanoscale zero-valent metals: A review of synthesis, characterization, and applications to environmental remediation. *Environ. Sci. Pollut. Res.* 2016, 23(18), 17880–17900.
49. Xu, Y., Zhao, D., Reductive immobilization of chromate in water and soil using stabilized iron nanoparticles. *Water Res.* 2007, 41(10), 2101–2108.
50. Lisha, K. P., Anshup, Pradeep, T., Towards a practical solution for removing inorganic mercury from drinking water using gold nanoparticles. *Gold Bull.* 2009, 42(2), 144–152.
51. Dickinson, M., Scott, T. B., The application of zero-valent iron nanoparticles for the remediation of a uranium-contaminated waste effluent. *J. Hazard. Mater.* 2010, 178(1–3), 171–179.
52. Üzüm, Ç., Shahwan, T., Eroðlu, A. E., Lieberwirth, I., Scott, T. B., Hallam, K. R., Application of zero-valent iron nanoparticles for the removal of aqueous Co^{2+} ions under various experimental conditions. *Chem. Eng. J.* 2008, 144(2), 213–220.
53. Khin, M. M., A review on nanomaterials for environmental remediation. *Energy Environ. Sci.* 2012, 5(8), 8075–8109.
54. (a) Schrick, B., Blough, J. L., Jones, A. D., Mallouk, T. E., Hydrodechlorination of trichloroethylene to hydrocarbons using bimetallic nickel–iron nanoparticles. *Chem. Mater.* 2002, 14(12), 5140–5147; (b) Feng, J., Lim, T. T., Pathways and kinetics of carbon tetrachloride and chloroform reductions by nano-scale Fe and Fe/Ni particles: Comparison with commercial micro-scale Fe and Zn. *Chemosphere* 2005, 59(9), 1267–1277; (c) Ma, H., Huang, Y., Shen, M., Hu, D., Yang, H., Zhu, M., Yang, S., Shi, X., Enhanced decoloration efficacy of electrospun polymer nanofibers immobilized with Fe/Ni bimetallic nanoparticles. *Rsc Adv.* 2013, 3(3), 6455–6465.
55. And, X. L., Zhang, W., Sequestration of metal cations with zerovalent iron nanoparticles a study with high resolution X-ray

photoelectron spectroscopy (HR-XPS). *J. Phys. Chem. C* 2007, 111(19), 6939–6946.

56. Lien, H. L., Zhang, W., Enhanced dehalogenation of halogenated methanes by bimetallic Cu/Al. *Chemosphere* 2002, 49(4), 371–378.
57. Pant, H. R., Bajgai, M. P., Nam, K. T., Yun, A. S., Pandeya, D. R., Hong, S. T., Kim, H. Y., Electrospun nylon-6 spider-net like nanofiber mat containing TiO_2 nanoparticles: A multifunctional nanocomposite textile material. *J. Hazard. Mater.* 2010, 185(1), 124–130.
58. Bui, N. N., Mccutcheon, J. R., Nanoparticle-embedded nanofibers in highly permselective thin-film nanocomposite membranes for forward osmosis. *J. Membr. Sci.* 2016, 518, 338–346.
59. Homaeigohar, S. S., Elbahri, M., Novel compaction resistant and ductile nanocomposite nanofibrous microfiltration membranes. *J. Colloid Interface Sci.* 2012, 372(1), 6–15.
60. Lijo, F., Marsano, E., Vijila, C., Barhate, R. S., Vijay, V. K., Ramakrishna, S., Thavasi, V., Electrospun polyimide/titanium dioxide composite nanofibrous membrane by electrospinning and electrospraying. *J. Nanosci. Nanotechnol.* 2011, 11(2), 1154–1159.
61. Huang, M., Si, Y., Tang, X., Zhu, Z., Ding, B., Liu, L., Zheng, G., Luo, W., Yu, J., Gravity driven separation of emulsified oil-water mixtures utilizing in situ polymerized superhydrophobic and superoleophilic nanofibrous membranes. *J. Mater. Chem. A* 2013, 1(45), 14071–14074.
62. Yang, S., Si, Y., Fu, Q., Hong, F., Yu, J., Aldeyab, S. S., Elnewehy, M., Ding, B., Superwetting hierarchical porous silica nanofibrous membranes for oil/water microemulsion separation. *Nanoscale* 2014, 6(21), 12445–12449.
63. Chandra, V., Park, J., Chun, Y., Lee, J. W., Hwang, I.-C., Kim, K. S., Water-dispersible magnetite-reduced graphene oxide composites for arsenic removal. *ACS Nano* 2010, 4(7), 3979–3986.
64. Hu, J., Chen, G., Lo, I. M. C., Removal and recovery of Cr(VI) from wastewater by maghemite nanoparticles. *Water Res.* 2005, 39(18), 4528–4536.
65. Shin, S., Jang, J., Thiol containing polymer encapsulated magnetic nanoparticles as reusable and efficiently separable adsorbent for heavy metal ions. *Chem. Commun.* 2007, 41(41), 4230–4232.
66. Sun, B., Li, X., Zhao, R., Yin, M., Wang, Z., Jiang, Z., Wang, C., Hierarchical aminated PAN/γ–AlOOH electrospun composite nanofibers and their heavy metal ion adsorption performance. *J. Taiwan Inst. Chem. Eng.* 2016, 62, 219–227.

67. (a) Teo, W. E., Ramakrishna, S., Electrospun nanofibers as a platform for multifunctional, hierarchically organized nanocomposite. *Composites Sci. Technol.* 2009, 69(11–12), 1804–1817; (b) Hota, G., Kumar, B. R., Ng, W. J., Ramakrishna, S., Fabrication and characterization of a boehmite nanoparticle impregnated electrospun fiber membrane for removal of metal ions. *J. Mater. Sci.* 2008, 43(1), 212–217.

68. Xiao, S., Ma, H., Shen, M., Wang, S., Huang, Q., Shi, X., Excellent copper(II) removal using zero-valent iron nanoparticle-immobilized hybrid electrospun polymer nanofibrous mats. *Colloids Surf. A Physicochem. Eng. Aspects* 2011, 381(1), 48–54.

69. Ramakrishna, S., Fujihara, K., Teo, W. E., Yong, T., Ma, Z., Ramaseshan, R., Electrospun nanofibers: Solving global issues. *Mater. Today* 2006, 9(3), 40–50.

70. Pathania, D., Katwal, R., Sharma, G., Naushad, M., Khan, M. R., Al-Muhtaseb, A. A. H., Novel guar gum/Al_2O_3 nanocomposite as an effective photocatalyst for the degradation of malachite green dye. *Int. J. Biol. Macromol.* 2016, 87, 366–374.

71. Ji, S. I., Min, I. K., Lee, Y. S., Preparation of PAN-based electrospun nanofiber webs containing TiO_2 for photocatalytic degradation. *Mater. Lett.* 2008, 62(21), 3652–3655.

72. Ye, S., Zhang, D., Liu, H., Zhou, J., ZnO nanocrystallites/cellulose hybrid nanofibers fabricated by electrospinning and solvothermal techniques and their photocatalytic activity. *J. Appl. Polymer Sci.* 2011, 121(3), 1757–1764.

73. Lala, N. L., Ramaseshan, R., Bojun, L., Sundarrajan, S., Barhate, R. S., Ying-Jun, L., Ramakrishna, S., Fabrication of nanofibers with antimicrobial functionality used as filters: Protection against bacterial contaminants. *Biotechnol. Bioeng.* 2007, 97(6), 1357–1365.

74. Chonkaew, W., Electrospun nanofibers in energy and environmental applications. *Energy Environ. Sci.* 2008, 1(2), 205–221.

75. Park, S. W., Bae, H. S., Xing, Z. C., Kwon, O. H., Huh, M. W., Kang, I. K., Preparation and properties of silver-containing nylon 6 nanofibers formed by electrospinning. *J. Appl. Polymer Sci.* 2009, 112(4), 2320–2326.

Part 4

Bio-Nanocomposites for Pollution Control

Chapter 16

Biosynthesized and Bio-Inspired Functional Nanocomposites for Pollution Control

Akeem Adeyemi Oladipo

Faculty of Engineering, Cyprus Science University, Ozankoy, Girne, TRNC via Mersin 10, Turkey

akeem.oladipo@kiu.edu.tr

High-performance bio-inspired functional nanocomposites fuse the benefits of both functionalized nanomaterials and hierarchical structures of bio-based inorganic particles for effective pollution control. These functional bio-inspired nanocomposites possess excellent abilities to decontaminate and degrade a range of pollutants from aqueous solutions and, have attracted concerted interests in academia and industry across the globe. This chapter provides a broad spectrum of recent techniques of preparation of bio-inspired functional nanocomposites, their structural properties and wide coverage in practical pollution control. Representative samples are discussed to illustrate the efficiency of bio-inspired

Nanocomposites for Pollution Control
Edited by Chaudhery Mustansar Hussain and Ajay Kumar Mishra

ISBN 978-981-4774-45-1 (Hardcover), 978-1-315-14368-2 (eBook)
www.panstanford.com

nanocomposites in the field of pollution control. This chapter intends to provide an update on advances in the fabrication of novel bio-inspired functional nanocomposites for environmental remediation applications.

16.1 Introduction

Increasing threat from environmental pollution has become a global issue and received special attention worldwide. Environmental pollutants are the major component of the pollution process and can be referred to as the actual "*executing agents*" of environmental pollution. The threat of non-biodegradable pollutants is alarming as they persist in the biosphere for acutely long periods of time as compared with the biodegradable pollutants [1]. Dyes, heavy metals, organophosphorus compounds (insecticides and pesticides) and phenolic wastewaters are major non-biodegradable environmental pollutants generated from the peoples' activities and industries [1–3].

These pollutants cause **harm** and **discomfort** to humans and other living organisms and ultimately damage the environment. For instance, it is reported that ionic dyes can accumulate in human bodies if exposed to trace amounts of dye-contaminated wastewater for a long time and can induce cancer-related diseases [2–5]. The organophosphorus-based pesticides exhibit toxic effects on humans and create poisoning which may result in death, indicating that their toxicity threshold is of lethal dose [6]. Long-term exposure to a trace amount of heavy metals can cause serious health effects, including organ damage, reduced growth, and nervous system damage. An exposure to lead and mercury may cause a person's immune system to attack its cells (autoimmunity). This condition can lead to rheumatoid arthritis and damage of the foetal brain [7, 8].

There are many types of environmental pollution, but this chapter will focus on water pollution. Water pollution is one of the most important types of environmental pollution caused by various factors, including an oil spill, leakage of fertilizers, industrial waste disposal, herbicides, and extraction of fossil fuels, etc. Water pollution has become a serious issue globally; thus, it is highly necessary to control and decontaminate pollutant-laden wastewaters before releasing into the environment due to their adverse effects.

Recently, there is increasing interest in the development of bio-inspired or biosynthesized composite materials for pollution control. The unique properties of bio-inspired nanoscale materials have given rise to immense research activity directed towards the applications of bio-inspired nanocomposites for pollution control. The properties of bio-inspired materials may differ from materials prepared by other traditional methods. The potential range of compositions, shapes, and sizes of biosynthesized materials translates into a broad domain of new and existing nanocomposite applications. Although the use of bio-inspired nanocomposites has extended to almost all fields, this chapter considers biosynthesized inorganic semiconductors being utilized with other biological and polymer components for driving an environmental remediation process.

16.2 Treatment Techniques for Pollution Control

A number of technologies exist for the remediation of non-biodegradable environmental pollutants, and thus have definite outcomes such as: (a) extraction of pollutants for disposal or further treatment (b) substantial or complete degradation/destruction of the pollutants, (c) stabilization of pollutants to forms less toxic or mobile, (d) containment of the polluted material to limit exposure of the entire environment and (e) separation and recycling of non-contaminated materials from polluted materials [9, 10].

These treatment technologies are classified into three categories viz. physico-chemical, chemical, and biological/ biochemical technologies. Although the mentioned techniques can be utilized for the treatment of pollutant laden-wastewater, it is paramount to state that the selection of the most suitable treatment methods depends on the content of the wastewater, plant flexibility, initial pollutant concentration, environmental impact, reliability, capital investment and operational cost, etc. [11, 12]. Here, adsorption and degradation techniques have been comprehensively discussed with specific examples due to their efficiencies.

16.2.1 Adsorption Process

Adsorption is the process of accumulating a substance (adsorbate) on the surface of another substance (adsorbent) and becomes bound by chemical and/or physical interactions [13]. The adsorption mechanism can be a single or a combination of multiple phenomena, including electrical attraction, chemical complex formation and exclusion of the adsorbate from the bulk solution. Adsorption results from stronger intermolecular bonding between molecules of solvent than those existed between the solute and molecules of solvent (water) [14].

Adsorption process differs from absorption mechanism, in which the *absorbate* is dissolved by the *absorbent*. Absorption process involves the entire bulk of the material while adsorption is a surface-based process as illustrated in Fig. 16.1. The term sorption encompasses both absorption and adsorption processes, while desorption is the reverse of it. The adsorption process is often illustrated through isotherms (Langmuir, Freundlich, etc.) that is, the amount of adsorbate on the adsorbent as a function of its concentration at constant temperature [1, 14]. Figure 16.2 depicts the typical adsorption process.

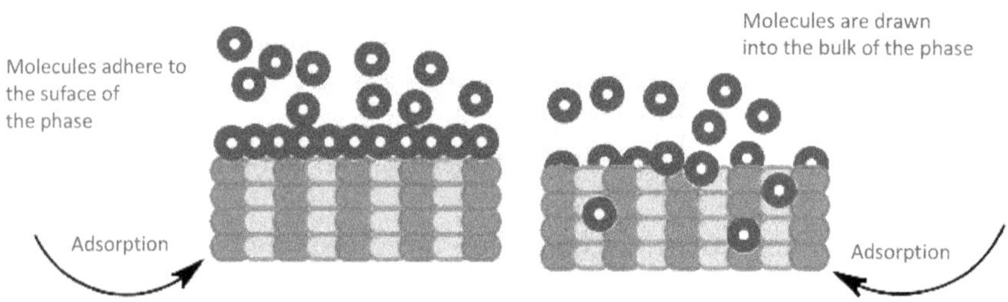

Figure 16.1 Illustrative difference between adsorption and absorption processes.

The properties of adsorbents and adsorbates depend on their constituents and are quite specific. Physisorption occurs if the interaction between the adsorbed molecules and the solid surface has a physical nature. In this case, the interactions are van der Waals forces and reversible due to the weakness of the forces. Also, physisorption occurs close or lower to the critical temperature of the adsorbed substance and accompanied by a decrease in free energy and entropy of the adsorption system and, thus, it is exothermic [15]. On the other hand, chemisorption process occurs if the attraction forces between the solid surface and the adsorbed molecules are due to chemical bonding. Contrary to physisorption, chemisorption occurs only via a monolayer mechanism and, substances chemisorbed on the solid surface are hardly removed due to stronger forces involved. Under favourable conditions, both processes can occur simultaneously or alternatively.

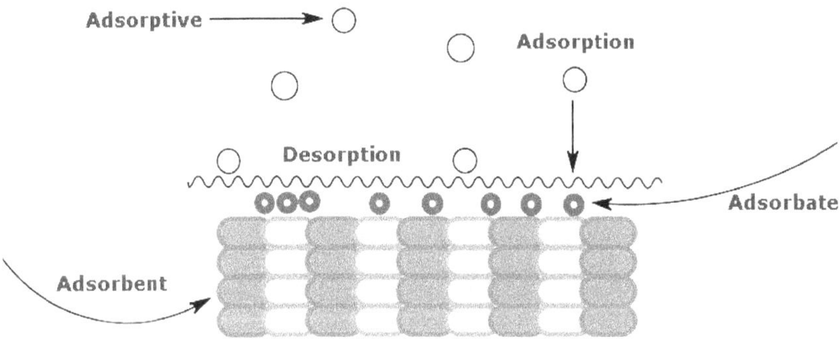

Figure 16.2 Typical adsorption process.

The adsorption process is widely considered due to its convenience, low operational cost, the simplicity of design and ease of operation. Recently, the search for high-performance and low-cost adsorbents has intensified. Although, locally available materials such as agricultural wastes, bioadsorbents and industrial wastes have been utilized as low-cost adsorbents as shown in Fig. 16.3, low efficiency and poor recoverability limit their industrial utilization. Also, various polymer-based materials have been utilized for the treatment of pollutant-laden wastewaters due to their bioavailability, modifiable functional groups and ease of processing [16, 17]. In recent years, concerted efforts have been directed to developing bio-inspired nanocomposites for wastewater treatment due to their strength, insensitivity to various toxic pollutants and higher activity than the pure components.

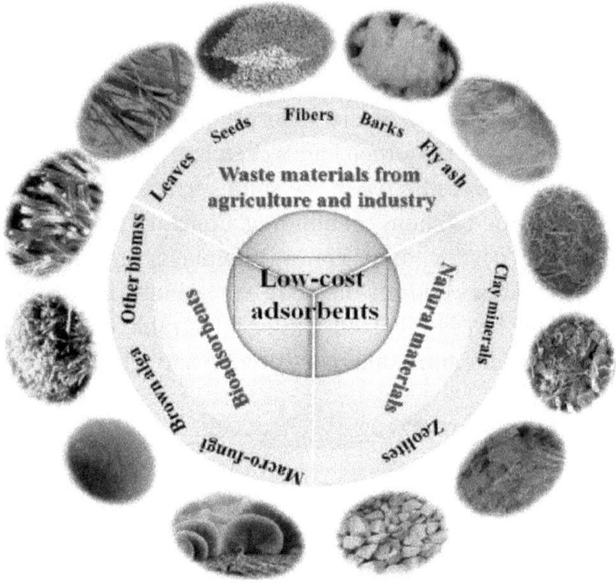

Figure 16.3 Low-cost adsorbents for pollutant removal from aqueous solution. Reproduced with permission from Elsevier from Mu and Wang [18].

16.2.2 Advanced Oxidation

A growing number of research works have been directed to study the treatment of pollutant-laden wastewater by advanced oxidation processes (AOPs). The AOPs are series of oxidation methods (photocatalysis, ozonation, electrochemical oxidation, Fenton and Fenton-like processes) based on the use of highly reactive oxidizing agents such as hydroxyl radicals (OH) resulting in the degradation of the target pollutants. To activate the AOPs, catalysts (TiO_2, ZnO), visible light or ultraviolet, and different oxidants (O_2, O_3, H_2O_2) are necessary.

For instance, photocatalysis generates holes and free radicals for degradation of recalcitrant organic contaminants using semiconducting materials as shown in Fig. 16.4. The advantages of the AOPs include non-selective oxidation of multiple pollutants at the same time and rapid reaction rates [19]. However, there are numerous factors that limit the efficacy of the AOPs such as reaction parameters, high cost of hydrogen peroxide and excess consumption of chemicals [1]. This chapter reviews the bio-inspired nanocomposites utilized via the AOPs for degradation of various pollutants in wastewater.

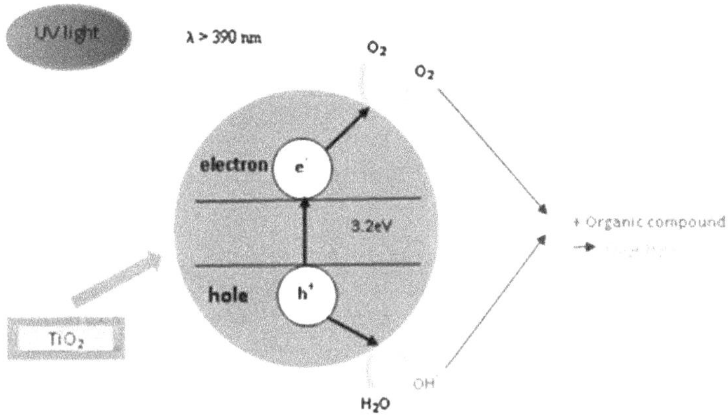

Figure 16.4 Schematic illustration of oxidative species production and photocatalysis based on TiO_2. Reproduced with permission from Elsevier from Asghar et al. [20].

16.3 Bio-Inspired Fabrication

Nature is a source of inspiration for the development of various types of high-performance bio-inspired materials [21]. **Bio-inspired fabrication** is a scientific discipline that utilizes biological principles to fabricate novel solutions for the environment, medicine, and many other fields. The emergence of Bio-inspired fabrication is the unification of the physical sciences and engineering with life sciences. Bio-inspired fabrication offers the opportunity to control shape, particle distribution, size, orientation, composition, crystal structure of nanoscale assemblies, which is difficult to attain using conventional routes [22]. In recent years, many hybrid inorganic/organic-based nanocomposites with special hierarchies have been fabricated via bio-inspired techniques with the assistance of various templates, such as proteins, synthetic polymers, low mass surfactant molecules and self-assembling peptides, etc. [22]. For instance, the nanostructure of bone provided the inspiration for the development of the amyloid–graphene nanocomposites with useful properties, such as shape-memory properties and enzyme activity as illustrated in Fig. 16.5.

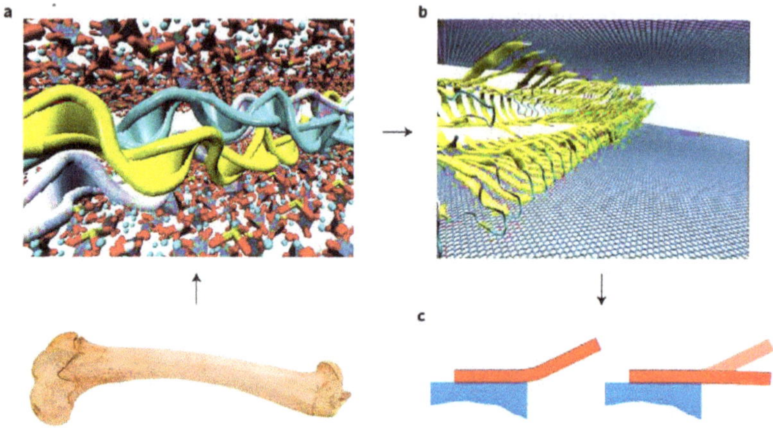

Figure 16.5 Schematic illustration of bio-inspired nanocomposite with enzyme activity and shape-memory properties. Adapted from *Nature Nanotechnology* 7, 417 and reproduced with permission from Springer Nature (2012).

16.3.1 Benefits of Using Biomolecules for Fabrication of Bio-Inspired Nanocomposites

The application of biomolecules during the syntheses of nanocomposites is beneficial and technically promising. The common biomolecules used in bio-inspired synthesis can be grouped into four classes including polysaccharides, proteins, nucleic acids and peptides. The first potential benefit of using biomolecules during the bio-inspired synthesis is the production of advanced materials under mild reaction conditions contrary to most conventional techniques that require severe reaction conditions. Bio-inspired syntheses are inherently "**green**" techniques due to the avoidance of harmful solvents and reduction of energy input. The second main benefit is the effective control on the chemistry, size, crystal structure and shape of the final product that allows the products for specific applications. Thirdly, materials with highly distinct, selective or multiple functions can be effectively developed via bio-inspired syntheses using biomolecules. Finally, availability of a range of synthetic and natural biomolecules makes it easy to find a biomolecule that can interact with, recognize, or direct the formation of an inorganic product [22, 23].

16.3.2 Bio-Inspired Synthesis of Nanocomposites

In general, the synthesis of high-performance nanocomposites involves high-energy input or expensive chemicals which may not be environmental-friendly. So, the use of biomolecules, microorganism or plant extract serves as a green and effective alternative to chemical and physical methods. Hazarika et al. [24], reported the biosynthesis of Fe_2O_3 nanoparticles loaded onto rice husk based bio-derived SiO_2 ($Fe_2O_3@SiO_2$) using the peel of *Musa balbisiana*. The photocatalytic activity of the biosynthesized $Fe_2O_3@SiO_2$ nanocomposite in degradation of methyl orange dye under visible light was reported.

Briefly, 2 g of powdered peel of *Musa balbisiana* was added to 20 ml of distilled water and boiled for 1 h. Then, 1 mM $FeCl_3 \cdot 6H_2O$ solution was added to the filtrate of the boiled peel and vigorously stirred. The Fe_2O_3 nanoparticles formed were mixed with a solution of SiO_2 and refluxed in 10 ml methanol for 3 h to form a photocatalytic bio-inspired nanocomposite. The SiO_2 nanoparticles were biosynthesized from rice husk, which is agricultural waste material [24].

The synthesis of a high-performance bio-inspired graphene–epoxy layered composite was reported by Ming et al. [25]. The fabrication process is briefly described. The graphene oxide (GO) sheets prepared by modified Hummers' method were dispersed into deionized water. The GO solution was vacuum-filtered into GO film and then subjected to 85% leavening ($N_2H_4 \cdot H_2O$) strategy treatment. The graphene foam (GF) obtained was impregnated with 5 wt% epoxy/acetone solution and the GF–epoxy preform was cured by hot-press to finally obtain the bio-inspired graphene–epoxy layered composites as shown in Fig. 16.6. The bio-inspired strategy enhances the graphene loading in the resultant composites and creates hierarchically layered composites with enhanced properties [25].

Parvathi et al. [26], biosynthesized graphene composite from sugar coated on river sand, characterized the bio-inspired products and applied it for pollution control. They reported that common sugar was employed as the source of carbon. The sugar solution (1 M) was added to sand (20 g), and the mixture was dried at ~95°C. The sugar-coated sand was heated in a muffle furnace under reducing atmosphere, and the resultant composite was treated with concentrated sulphuric acid (10 ml), dried at 120°C and named graphene sand composite [26]. Similarly, spherical Fe_3O_4/bacterial cellulose nanocomposites were biosynthesized from *Gluconacetobacter xylinum* by agitation fermentation technique for heavy metal remediation [27].

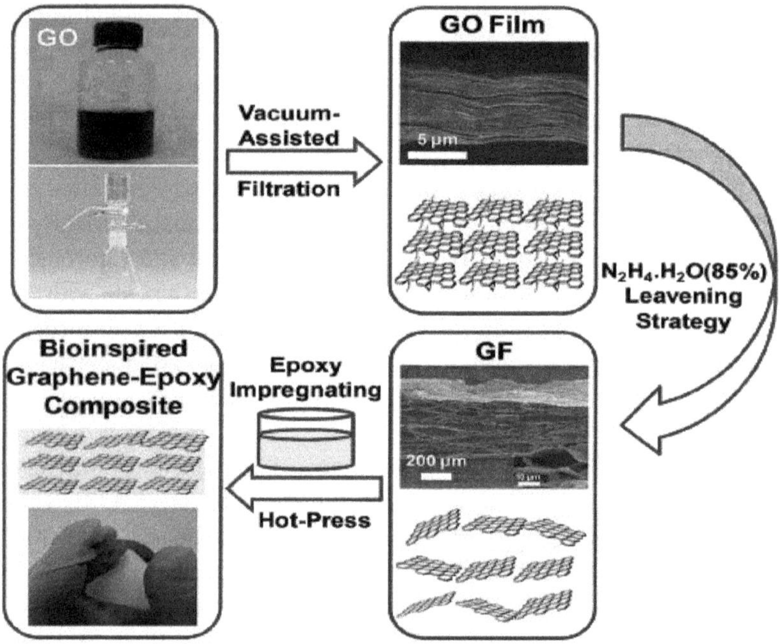

Figure 16.6 Schematic illustration of the preparation process of bio-inspired graphene–epoxy layered composite. Reproduced from RSC Advances [25]. Copyright © 2015, Royal Society of Chemistry.

The illustration of the biosynthesis of spherical Fe_3O_4/BC nanocomposites is shown in Fig. 16.7. The authors reported that during the initial fermentation, the pH of culture medium was adjusted to 6.5 and resulted to the aggregation of Fe_3O_4 nanoparticles. However, the nanoparticles were dispersed uniformly between the BC nanofibrils after 36 h when the medium pH reduced to 4. The embedded Fe_3O_4 nanoparticles formed hydrogen bonds with the hydroxyl groups of BC nanofibrils [24].

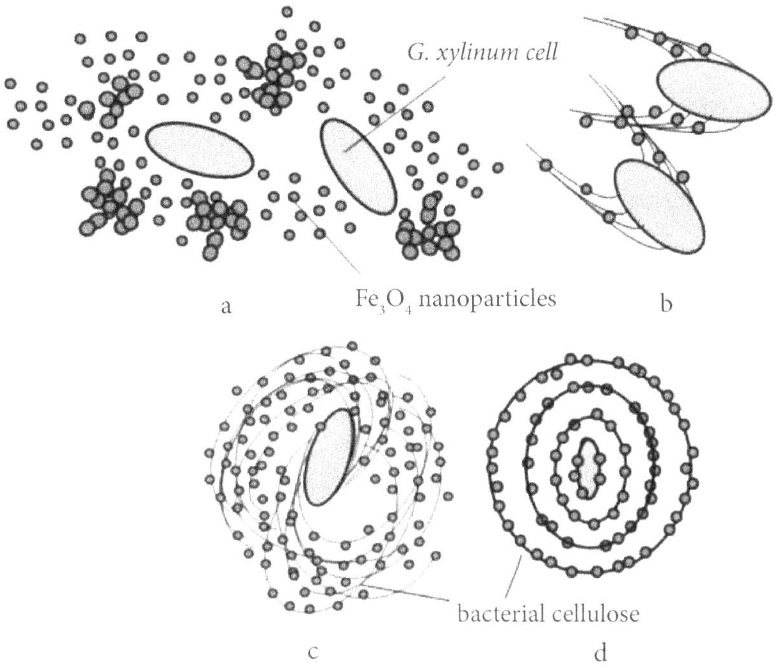

Figure 16.7 Schematic illustration of biosynthesized spherical Fe_3O_4/BC nanocomposites for heavy metal remediation. (a) Dispersing of Fe_3O_4 nanoparticles in the medium; (b, c) Fe_3O_4 nanoparticles embedded in BC; (d) spherical.

The authors reported that the spherical BC is opaque while the spherical Fe_3O_4/BC is black due to well-dispersed Fe_3O_4 between BC nanofibrils as shown in Fig. 16.8. In this study, pH controlling embedding method was applied as a new approach to biosynthesize high-performance nanocomposites. Magnetic responsive nanocomposites of hydroxyapatite/$CuFe_2O_4$ and collagen were prepared via a continuous flow system for pollution control. Briefly, synthetic Hap containing $CuFe_2O_4$ particles (2:1) was dispersed into a solution of telopeptide-free collagen molecules and subjected to microwave irradiation for 10 min to obtain protein-mediated bio-inspired nanocomposites. The protein-mediated bio-inspired nanocomposite exhibited high photocatalytic and adsorptive activities towards a range of selected dyes and heavy metal ions.

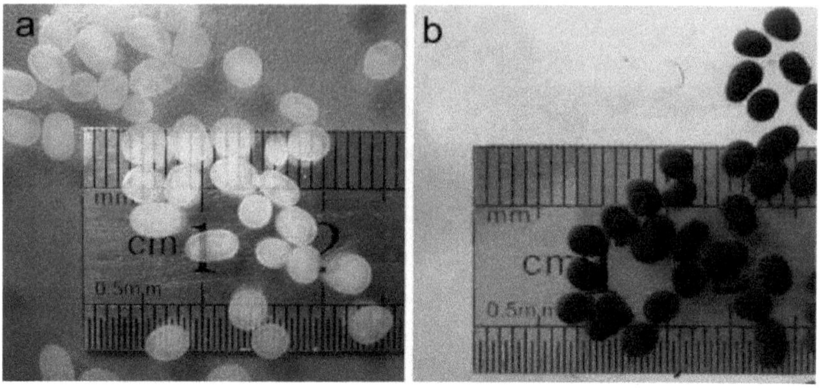

Figure 16.8 Photographic illustration of the opaque spherical BC and black spherical Fe_3O_4/BC nanocomposites. Reproduced with permission from Elsevier from Zhu et al. [27].

16.4 Adsorption Process Involving Biosynthesized and Bio-Inspired Nanocomposites

Biosynthesized and bio-inspired nanocomposites have emerged as one of the promising adsorbents in water treatment. In this perspective, biosynthesized nanocomposites were employed to remove heavy metals, dyes, etc. from wastewater under varying sorption reaction conditions. There are various factors that contribute in determining the whole efficacy of nanocomposites during the adsorption process of contaminants as described below. These factors include nature of the adsorbent and the adsorbate, surface area of the adsorbent, solution pH and reaction medium temperature, etc. [14, 16, 28]. For instance, porous and finely powdered adsorbents adsorb more as compared to the rigid non-porous materials. Also, the extent of adsorption is dependent upon the surface area of the adsorbent, viz. the smaller the surface area of the adsorbent, the lower is the extent of adsorption. Furthermore, the adsorption of contaminants on the adsorbents could be governed by multiple mechanisms as shown in Fig. 16.9.

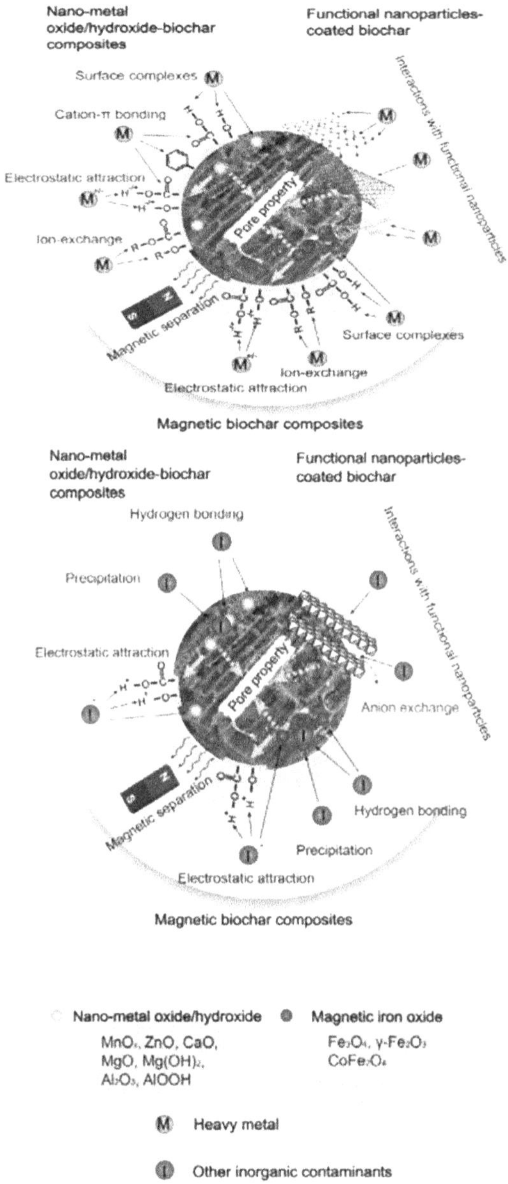

Figure 16.9 Schematic illustration of the general adsorption mechanisms by nanocomposite for the removal of various contaminants. Reproduced with permission from Elsevier from Tan et al. [29].

16.4.1 Effect of Initial Concentration of Pollutant and Adsorbent Dose

Increasing initial pollutant concentration provides a driving force to overcome mass transfer resistances of pollutant between the medium and adsorbent phases, hence increasing adsorption [14, 17]. According to Zhu et al. [27], when the concentrations of Cr^{3+} and Mn^{2+} were less than 60 mg/mL at a fixed adsorbent dosage, the quantities adsorbed on biosynthesized spherical Fe_3O_4/BC nanocomposites spheres increased, meanwhile the adsorption capacity decreased from 43% to 25% for Cr^{3+} and 46% to 33% for Mn^{2+} when the adsorbate concentration is greater than 60 mg/mL. The observation may be attributed to the fact that the total available adsorption sites are limited for a constant nanocomposite dosage, hence leading to a decrease in the percentage removal of the adsorbate [29] as shown in Fig. 16.10.

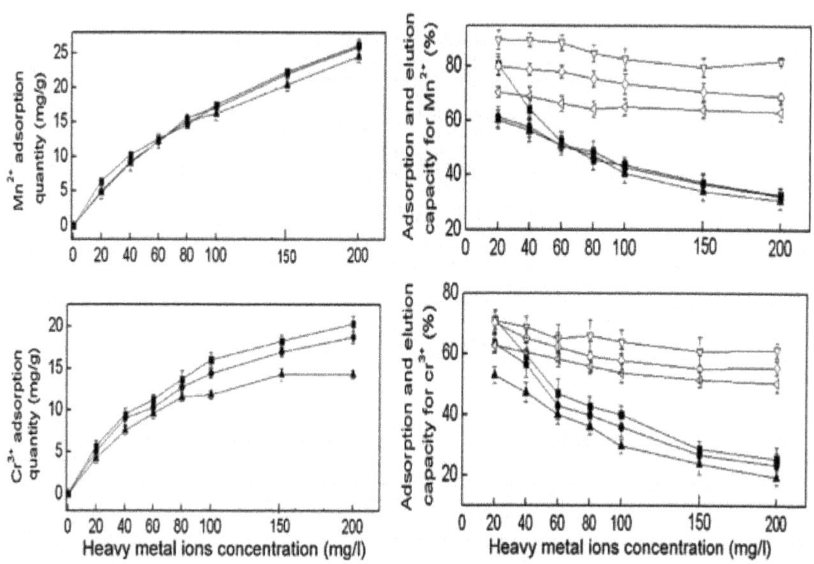

Figure 16.10 The adsorption capabilities of the biosynthesized spherical Fe_3O_4/BC nanocomposites for heavy metal ion removal. Reproduced with permission from Elsevier from Zhu et al. [27].

Chavez-Guajardo et al. [28] reported that when interacting 2 mg of PPY/g-Fe_2O_3 nanocomposite (2 mg) with 50 mg/L of

Cr(VI) solution (10 mL), about 0.41 mg Cr(VI) was adsorbed corresponding to 82% removal efficiency. Meanwhile, the same amount of nanocomposite only adsorbed 0.52 mg of Cr(VI) that is 52.2% removal efficiency when the initial Cr(VI) concentration increased to 100 mg as shown in Fig. 16.11 [28]. Thus, various nanocomposites exhibit different sorption mechanisms depending on the nature of the nanocomposite surface chemistry and porosity. Also, an optimum nanocomposite dose is needed to maximize the interactions between the pollutant ions and the available sorption sites on the nanocomposites.

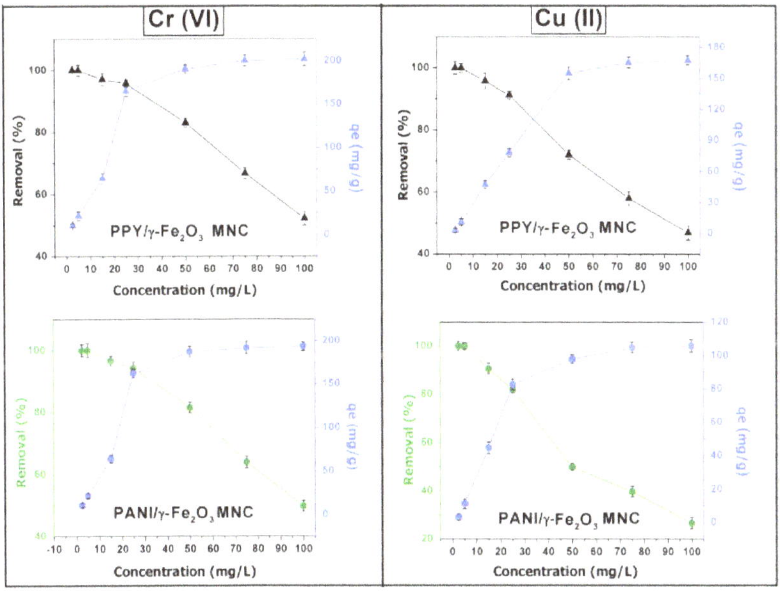

Figure 16.11 The effect of the initial Cu(II) and Cr(VI) concentration on the removal percent and quantity adsorbed by maghemite-based magnetic nanocomposites. Reproduced with permission from Elsevier from Chávez-Guajardo et al. [28].

16.4.2 Effect of Solution pH

The solution pH affects the functional groups of the nanocomposites and the chemistry of the adsorbates. The first approach to estimate the adsorption capacity of adsorbents towards pollutants is the determination of the pH point zero charge (pHpzc) and stability of the adsorbates at varying solution pH. The total surface charge on the adsorbents becomes negative when the pH of the solution is above the pHpzc thus limiting the approach of the negatively charged pollutant ions (electrostatic repulsion) [17, 30].

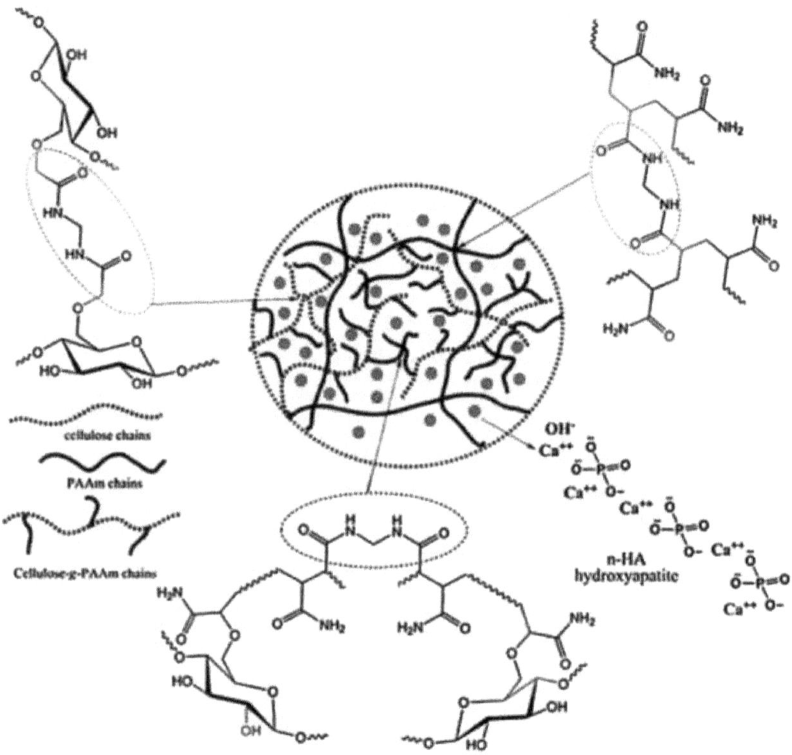

Figure 16.12 Hydroxyapatite mediated biosynthesized nanocomposite for efficient remediation of anthraquinone dyes. Reproduced with permission from Elsevier from Oladipo et al. (2014).

Pan et al. [31], reported that the adsorption capacities of mercapto-functionalized nanocomposites for Hg^{2+} increased with increasing solution pH until it reached an optimum state at pH between 4 and 6 [31]. At pH 2.03–2.72, the solution pH is lower than the pHpzc of the mercapto-functionalized nanocomposite indicating the presence of cations such as Hg^{2+}, $HgOH^{+}$ and $HgCl^{+}$, thus, resulting in low adsorption capacities due to electrostatic repulsion. Oladipo et al. [30], reported hydroxyapatite biosynthesized nanocomposite (EBH) for remediation of anthraquinone dyes. The EBH shown in Fig. 16.12 is a functional nanocomposite with various functional groups such as hydroxyl and amine groups which are majorly affected by the solution pH. At acidic domain, the amino and hydroxyl groups of the EBH are protonated and to maintain neutrality in the reaction medium, the negatively charged anthraquinone was electrostatically adsorbed by the EBH as shown schematically in Fig. 16.13.

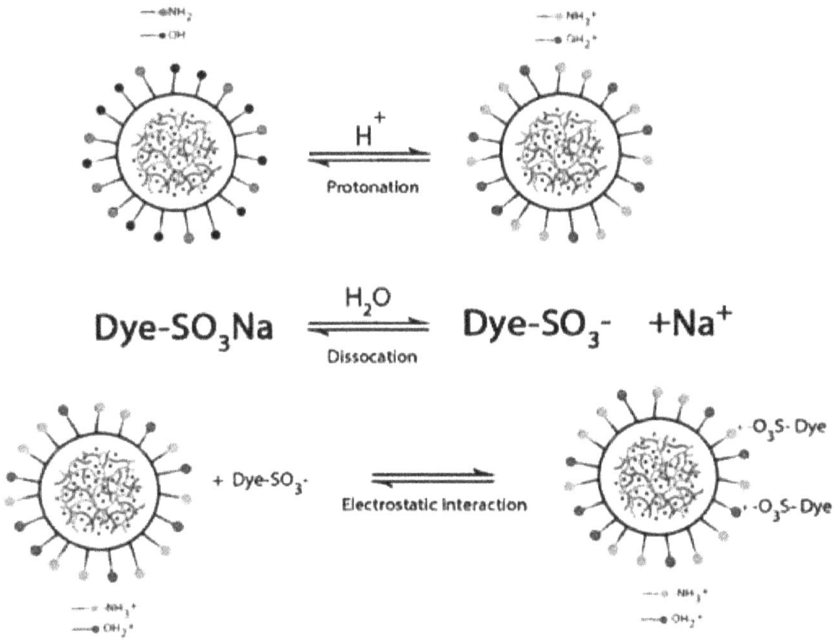

Figure 16.13 Mechanism of adsorption of reactive blue 2 dye by biosynthesized nanocomposite in acidic domain. Reproduced with permission from Elsevier from Oladipo et al., [30].

16.4.3 Effect of Temperature and Interaction Time

It has been reported that adsorption capacity of pollutant increases with increasing temperature, indicating that adsorption mechanism is an endothermic process. Also, the interaction time between the adsorbate and adsorbent is an important parameter determining the feasibility of a given adsorbent to reach the optimum capacity for capturing the pollutants that need to be separated [29]. Oladipo et al. [30], studied the adsorption of reactive blue 2 dye (RB2) by biosynthesized nanocomposite at varying temperature (25–75°C) and obtained a result is shown in Fig. 16.14. They reported that the adsorption of RB2 increases with increasing temperature. It is concluded that the number of active sites on the EBH increases as the temperature increases and resulting in increased adsorption capacity, indicating the adsorption mechanism is endothermic in nature.

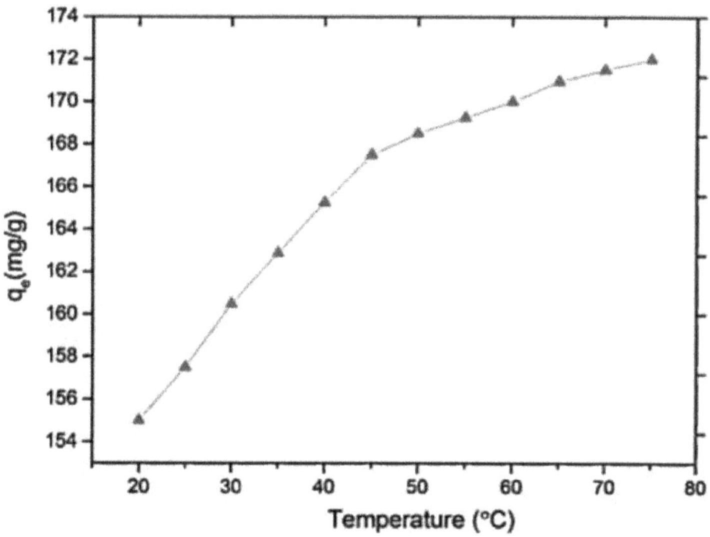

Figure 16.14 Effect of solution temperature on RB2 adsorption by biosynthesized. Reproduced with permission from Elsevier from Oladipo et al. [30].

Chávez-Guajardo et al. [28], reported the adsorption capacity of maghemite-based magnetic nanocomposites for removal of Cu(II) and Cr(VI) as a function of the interaction time. They reported that the adsorption capacity increases with the increase in the interaction time and type of the nanocomposite involved. For instance, Cr(VI) adsorbed faster on PPY/γ-Fe_2O_3 than on the PANI/γ-Fe_2O_3 maghemite-based magnetic nanocomposites. 50% of Cr(VI) present in the solution was adsorbed within the first 5 min by the PPY/γ-Fe_2O_3 nanocomposite whereas 49% of Cr(VI) was captured after 35 min by the PANI/γ-Fe_2O_3 nanocomposite, and similar observation was noted for Cu(II) adsorption as shown in Fig. 16.15.

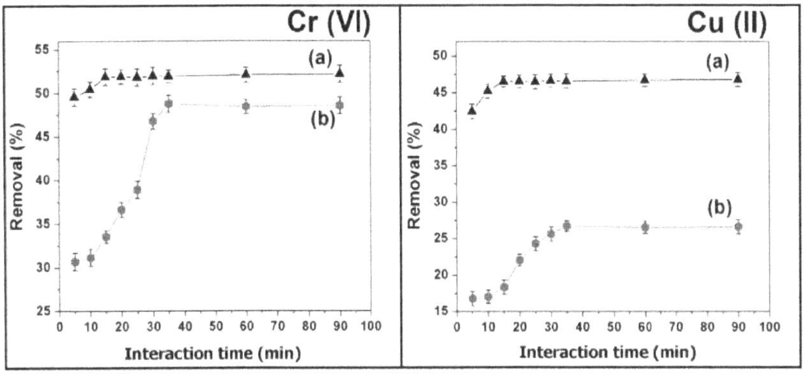

Figure 16.15 Effect of interaction time on the removal efficiency of Cu(II) and Cr(VI) by the (a) PPY/γ-Fe_2O_3 (b) and PANI/γ-Fe_2O_3 MNCs. Reproduced with permission from Elsevier from Chávez-Guajardo et al. [28].

16.5 Pollution Control via Advanced Oxidation Process Using Bio-Inspired Nanocomposites

AOPs have received increasing attention in the development of effective wastewater treatment technologies in recent years. Efficient AOPs can be employed to achieve complete mineralization of organic pollutants. Zhu et al. [32] reported chitosan-based nanocomposite catalyst for decolourization of Congo red (CR) dye under visible light irradiation. The effects various parameters including catalyst loading, pH of the solution, and reaction kinetics were studied.

A representative UV/vis spectra of CR as a function of reaction time are shown in Fig. 16.16 to clarify the changes of structural and molecular characteristics of CR as a result of photocatalytic degradation by chitosan-based nanocomposite. During the photodegradation, the absorbance values diminish due to the fragmentation of the azo bonds [32]. The authors concluded that not only decolorization but also degradation of CR took place on the nanocomposite catalyst under visible light irradiation.

Hazarika et al. [24] reported the photocatalytic degradation of methyl red using biosynthesized $Fe_2O_3@SiO_2$ nanocomposite under the visible light. The biosynthesized nanocomposite was found to be highly active, efficient and could be recycled without significant loss of photocatalytic activity. The absorption peak of the methyl red at 523 nm tested at a different time interval in the presence of the biosynthesized nanocomposite is shown in Fig. 16.17 and can be seen that the absorption peak at 415 nm increased steadily and the photocatalytic degradation reaction of the MR belongs to the pseudo first order reaction mechanism.

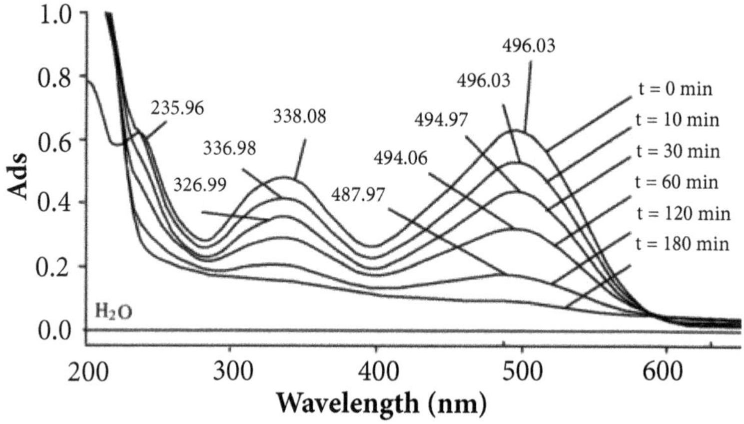

Figure 16.16 The UV–vis spectral changes of Congo red dye with reaction time. Reproduced with permission from Elsevier from Zhu et al. [32].

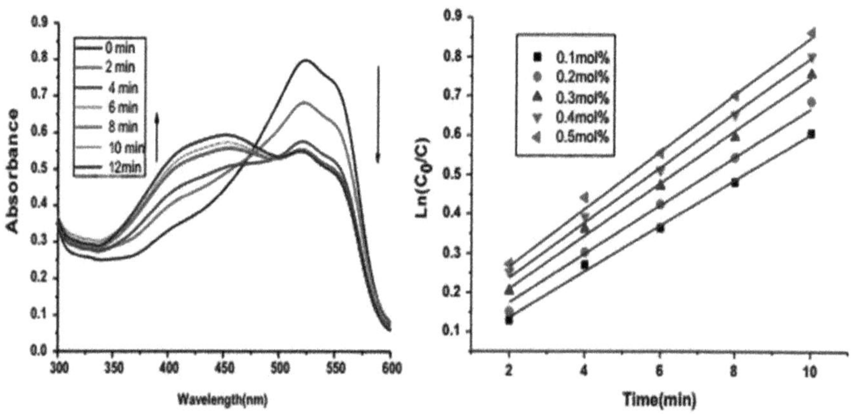

Figure 16.17 The absorption spectra of methyl red tested at different time in the presence of biosynthesized nanocomposite and kinetic result. Reproduced with permission from Elsevier from Hazarika et al. [24].

The mechanism of AOPs ultimately leads to the generation of highly reactive hydroxyl radicals, required for oxidative degradation of pollutant. Hydroxyl radicals mostly prefer the electrons from the surrounding electron rich organic compounds may lead to the complete degradation of organic pollutants. In the author's recent studies, bio-inspired collagen-mediated hydroxyapatite/$ZnFe_2O_4$ nanocomposite was fabricated and utilized for phenol degradation under sunlight irradiation.

In this study, approximately 78% phenol degradation at pH 7.0 was obtained even in the presence of co-existing heavy metal ions as shown in Fig. 16.18. The rate of degradation and efficiency of the bio-inspired collagen-mediated nanocomposite outperformed reported catalysts and adsorbents for phenol degradation.

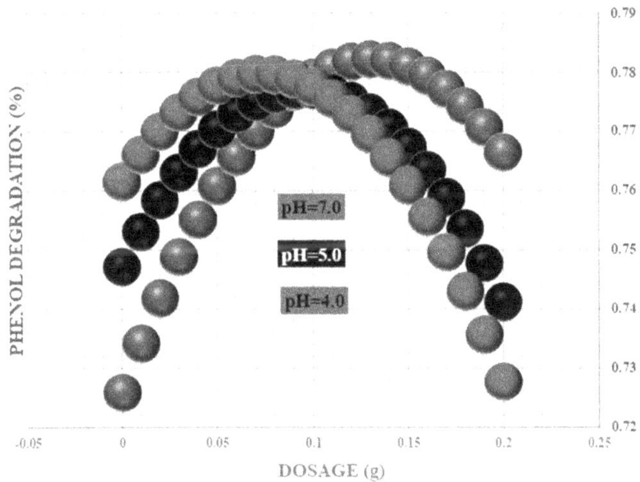

Figure 16.18 Degradation of phenol using bio-inspired collagen mediated hydroxyapatite/$ZnFe_2O_4$ nanocomposite under sunlight irradiation.

16.6 Summary and Future Perspectives

Based on our studies and the comprehensive literature survey, it is concluded that pollutant-laden wastewater is indeed a disconcerting problem. Naturally, hierarchical structure materials are obtained by biomineralization of macromolecules that act as templates for the growth and/or nucleation of the inorganic phases. Inspired by the development of biominerals in living organisms, novel hybrid composite materials have been formulated and developed by biomimetic routes. Thus, growing interest in using various engineered and synthetic polymer-based hybrid architectures as templates for bio-inspired synthesis has intensified.

The bio-inspired and biosynthesized nanocomposites have exhibited excellent potential in the adsorption and complete mineralization of different types of inorganic and organic pollutants. The bio-inspired nanocomposites have been reported to exhibit higher adsorption capacity, faster adsorption rate, and highly selective adsorption as compared to the nanocomposites synthesized via other conventional routes. The high-performance bio-inspired nanocomposites have a great potential for applications in various domains; however, most of the results are restricted to the laboratory scale. To expand the industrial applications of the bio-inspired and biosynthesized nanocomposites, several recommendations are suggested for the future work as follows.

Firstly, magnetically responsive bio-inspired nanocomposite should be prepared and effective separation technique must be utilized to separate the nanocomposite from the solution after adsorption process since the traditional methods of separation are both time-consuming and uneconomic. Secondly, sustainable reutilization of the spent biosynthesized nanocomposites must be taken into account. Hence, mechanical stable and recyclable bio-inspired nanocomposites should be prepared for use and reuse cycles. Finally, the incorporation of photocatalytic sensitive components within the matrix of bio-inspired nanocomposite will not only adsorb pollutions but also degrade pollutants into harmless products. In summary, various biomolecules mediated bio-inspired nanocomposite should develop for the treatment of air and soil pollutions.

Acknowledgements

The author is thankful to Polymeric Materials Research Laboratory of Chemistry Department (Eastern Mediterranean University) for providing access to the analytical equipment. Also, the inputs of Prof. Mustafa Gazi greatly helped improve this chapter, for which the author is grateful.

References

1. Oladipo, A. A., Abureesh, M. A., and Gazi, M. (2016). Bifunctional composite from spent "Cyprus coffee" for tetracycline removal and phenol degradation: Solar-Fenton process and artificial neural network, *Int. J. Biol. Macromol.*, **90**, pp. 89–99.
2. Ngwabebhoh, F. A., Gazi, M., and Oladipo, A. A. (2016). Adsorptive removal of multi-azo dye from aqueous phase using a semi-IPN superabsorbent chitosan-starch hydrogel, *Chem. Eng. Res. Des.*, **112**, pp. 274–288.
3. Dehghani, M. H., Niasar, Z. S., Mehrnia, M. R., Shayeghi, M., Al-Ghouti, M. A., Heibati, B., McKay, G., and Yetilmezsoy, K. (2016). Optimizing the removal of organophosphorus pesticide malathion from water using multi-walled carbon nanotubes, *Chem. Eng. J.*, doi: http://dx.doi.org/10.1016/j.cej.2016.10.057.
4. Oladipo, A. A., and Gazi, M. (2015). Two-stage batch sorber design and optimization of biosorption conditions by Taguchi methodology for the removal of acid red 25 onto magnetic biomass, *Korean J. Chem. Eng.*, **32**, pp. 1864–1878.
5. Gomes, R. F., de Azevedo, A. C. N., Pereira, A. G. B., Muniz, E. C., Fajardo, A. R., Rodrigues, F. H. A. (2015). Fast dye removal from water by starch-based nanocomposites, *J. Colloid Interface Sci.*, 454, pp. 200–209.
6. Jiang, L., Huang, T., Feng, S., and Wang, J. (2016). Zirconium(IV) functionalized magnetic nanocomposites for extraction of organophosphorus pesticides from environmental water samples, *J. Chromatogr. A*, **1456**, pp. 49–57.
7. Shen, B., Tian, L., Li, F., Zhang, X., Xu, H., and Singh, S. (2017). Elemental mercury removal by the modified bio-char from waste tea, *Fuel*, **187**, pp. 189–196.
8. Kumari, A. R., and Sobha, K. (2016). Removal of lead by adsorption with the renewable biopolymer composite of feather (*Dromaius novaehollandiae*) and chitosan (*Agaricus bisporus*), *Environ. Technol. Innov.*, **6**, pp. 11–26.
9. Scullion, J. (2006). Remediating polluted soils, *Naturwissenschaften*, **93**, pp. 51–65.
10. Hashim, M. A., Mukhopadhyay, S., Sahu, J. N., and Sengupta, B. (2011). Remediation technologies for heavy metal contaminated groundwater, *J. Environ. Manag.*, **92**, 2355–2388.
11. Fu, F., and Wang, Q. (2011). Removal of heavy metal ions from wastewaters: A review, *J. Environ. Manag.*, **92**, pp. 407–418.

12. Oladipo, A. A., and Gazi, M. (2016). High boron removal by functionalized magnesium ferrite nanopowders, *Environ. Chem. Lett.*, **14**, pp. 373–379.

13. Abas, S. N. A., Ismail, M. H. S., Kamal, M. L., and Izhar, S. (2013). Adsorption process of heavy metals by low-cost adsorbent: A review, *World Appl. Sci. J.*, **28**, pp. 1518–1530.

14. Oladipo, A. A., and Gazi, M. (2015). Nickel removal from aqueous solutions by alginate-based composite beads: Central composite design and artificial neural network modeling, *J. Water Proc. Eng.*, **8**, pp. e81–e91.

15. Coonery, D. O. (1999). *Adsorption Design for Wastewater Treatment*, Lewis Publishers, USA, p. 182.

16. Oladipo, A. A., Gazi, M., and Yilmaz, E. (2015). Single and binary adsorption of azo and anthraquinone dyes by chitosan-based hydrogel: Selectivity factor and Box-Behnken process design, *Chem. Eng. Res. Des.*, **104**, pp. 264–279.

17. Oladipo, A. A., and Gazi, M. (2016). Hydroxyl-enhanced magnetic chitosan microbeads for boron adsorption: Parameter optimization and selectivity in saline water, *React. Funct. Polym.*, **109**, pp. 23–32.

18. Mu, B., and Wang, A. (2016) Adsorption of dyes onto palygorskite and its composites: A review, *J. Environ. Chem. Eng.* **4**, pp. 1274–1294.

19. Antonopoulou, M., Evgenidou, E., Lambropoulou, D., and Konstantinou, I. (2014). A review on advanced oxidation processes for the removal of taste and odor compounds from aqueous media, *Water Res.*, **53**, pp. 215–234.

20. Asghar, A., Abdul Raman, A. A., and Daud, W. M. A. W. (2015). Advanced oxidation processes for in-situ production of hydrogen peroxide/hydroxyl radical for textile wastewater treatment: A review, *J. Cleaner Prod.*, 87, pp. 826–838.

21. George, A., and Ravindran, S. (2010). Protein templates in hard tissue engineering, *Nano Today*, **5**, pp. 254–266.

22. Liu, X., and Mallapragada, S. K. (2011). Bioinspired Synthesis of Organic/Inorganic Nanocomposite Materials Mediated by Biomolecules, in *Biomimetics*, Lilyana Pramatarova (ed.), InTech, DOI: 10.5772/18411.

23. Duan, J., Gong, S., Gao, Y., Xie, X., Jiang, L., and Cheng, Q. (2016). Bioinspired ternary artificial nacre nanocomposites based on reduced graphene oxide and nanofibrillar cellulose, *ACS Appl. Mater. Interfaces*, **8**, pp. 10545–10550.

24. Hazarika, M., Saikia, I., Das, J., Tamuly, C., and Das, M. R. (2016). Biosynthesis of Fe_2O_3@SiO_2 nanoparticles and its photocatalytic activity, *Mater. Lett.*, **164**, pp. 480–483.

25. Ming, P. Zhang, Y., Bao, J., Liu, G., Li, Z., Jiang, L., and Cheng, Q. (2015). Bioinspired highly electrically conductive graphene-epoxy layered composites. *RSC Adv.*, **5**, pp. 22283–22288.

26. Parvathi, P., Umadevi, M., and Bhaviya Raj, R. (2015). Improved waste water treatment by bio-synthesized graphene sand composite. *J. Environ. Manag.*, **162**, pp. 299–305.

27. Zhu, H., Jia, S., Wan, T., Jia, Y., Yang, H., Li, J., Yan, L., and Zhong, C. (2011). Biosynthesis of spherical Fe_3O_4/bacterial cellulose nanocomposites as adsorbents for heavy metal ions, *Carbohydr. Polym.*, 86, pp. 1558–1564.

28. Chavez-Guajardo, A. E., Medina-Llamas, J. C., Maqueira, L., Andrade, C. A. S., Alves, K. G. B., and Celso de Melo, P. (2015). Efficient removal of Cr (VI) and Cu (II) ions from aqueous media by use of polypyrrole/maghemite and polyaniline/maghemite magnetic nanocomposites, *Chem. Eng. J.*, **281**, pp. 826–836.

29. Tan, X. F., Liu, Y. G., Gu, Y. L., Xu, Y. Zeng, G. M., Hu, X. J., Liu, S. B., Wang, X., Liu, S. M., and Li, J. (2016). Biochar-based nano-composites for the decontamination of wastewater: A review, *Bioresour. Technol.*, **212**, pp. 318–333.

30. Oladipo, A. A., Gazi, M., and Saber-Samandari, S. (2014). Adsorption of anthraquinone dye onto eco-friendly semi-IPN biocomposite hydrogel: Equilibrium isotherms, kinetic studies and optimization, *J. Taiwan Inst. Chem. Eng.*, **45**, pp 653–664.

31. Pan, S., Zhang, Y., Shen, H., and Hu, M. (2012). An intensive study on the magnetic effect of mercapto-functionalized nano-magnetic Fe_3O_4 polymers and their adsorption mechanism for the removal of Hg(II) from aqueous solution, *Chem. Eng. J.*, **210**, pp. 564–574.

32. Zhu, H., Jiang, R., Xiao, L., Chang, Y., Guan, Y., Li, X., Zeng, G., Hazard, J. (2009). Photocatalytic decolorization and degradation of Congo Red on innovative crosslinked chitosan/nano-CdS composite catalyst under visible light irradiation, *J. Hazard. Mater.*, **169**, 933–940.

Part 5

Green and Sustainable Future: Nanocomposites

Chapter 17

Nanoscience and Its Role in the Solar Collectors' Future

Ahmed Kadhim Hussein,[a] Hussein A. Mohammed,[b] Kolsi Lioua,[c,d] Dong Li,[e] Rasoul Nikbakhti,[f] and B. Mallikarjuna[g]

[a]*College of Engineering, Mechanical Engineering Department, Babylon University, Babylon City, Hilla, Iraq*
[b]*WA School of Mines-Minerals, Energy & Chemical Engineering, Curtin University, 6102 Bentley, Western Australia, Australia*
[c]*College of Engineering, Mechanical Engineering Department, Haïl University, Haïl City, Saudi Arabia*
[d]*Unité de recherche de Métrologie et des Systèmes Énergétiques, College of Engineering of Monastir, Energy Engineering Department, University of Monastir, Tunisia*
[e]*School of Architecture and Civil Engineering, Northeast Petroleum University, Fazhan Lu Street, Daqing 163318, China*
[f]*Faculty of Engineering, Ferdowsi University of Mashhad, Mashhad, Iran*
[g]*Department of Mathematics, BMS College of Engineering, Bangalore 560019, Karnataka, India*

ahmedkadhim7474@gmail.com

This chapter gives a comprehensive overview of the recent advances related to the application of the nanotechnology in various kinds of the solar collectors. The papers reviewed include theoretical, numerical, and experimental up-to-date works related

Nanocomposites for Pollution Control
Edited by Chaudhery Mustansar Hussain and Ajay Kumar Mishra

ISBN 978-981-4774-45-1 (Hardcover), 978-1-315-14368-2 (eBook)
www.panstanford.com

with the nanotechnology applications in the flat plate, direct absorption, parabolic trough, wavy, heat pipe and other kinds of the solar collectors. A lot of literature is reviewed in this chapter and summarized carefully in Tables 17.1–17.7 to present a panoramic overview of the role of the nanotechnology in improving the various types of the solar collectors. It was found that the use of the nanofluid in the solar collector field can play a crucial role in increasing the efficiency of these devises. This chapter can be considered an important link between nanotechnology and all available kinds of the solar collectors. Further research is required to study the effect of nanotechnology to enhance the solar collector industry over the next several coming years.

Nomenclature

Symbol	Description	Unit
A	Surface area of solar collector [Eq. 17.2], [Eq. 17.5] and [Eq. 17.10]	m^2
A_c	Reduction in the size of collector's area [Eq. 17.1]	m^2
C_p	Specific heat of the working fluid [Eq. 17.1]	Kj/kg °C
C_{pro}	Water production cost [Eq. 17.4]	
c_w	Specific heat of water [Eq. 17.7]	Kj/kg °C
c_p	Specific heat of nanoparticle [Eq. 17.7]	Kj/kg °C
F_R	Collector heat removal factor [Eq. 17.2] and [Eq. 17.6]	
f	Plant availability [Eq. 17.4]	
G_t	Solar radiation on solar collector [Eq. 17.5]	W/m^2
h	Local heat transfer coefficient [Eq. 17.2]	$W/m^2 \cdot K$
I_b	Global solar radiation [Eq. 17.10]	W/m^2
I_T	Incident radiation or total radiation [Eq. 17.1] and [Eq. 17.6]	W/m^2
I	Intensity of solar radiation [Eq. 17.2]	W/m^2
$\dot{m}$	Mass flow rate of the working fluid [Eq. 17.1]	L/s
m_w	Mass flow rate of water [Eq. 17.7]	kg/s
m_p	Mass flow rate of nanoparticle [Eq. 17.7]	kg/s
Nu_{av}	Average Nusselt number [Eq. 17.3]	—

Nomenclature		
Symbol	**Description**	**Unit**
n	Number of the panels in the collecting system [Eq. 17.9]	
Pr	Prandtl number [Eq. 17.3]	
P_{net}	System collecting power [Eq. 17.9]	W
Q_{usfl}	Actual useful energy gain [Eq. 17.2]	W
Q_u	Rate of useful energy gained [Eq. 17.5] and [Eq. 17.10]	W
Q_{cpc}	Solar energy radiating on one CPC plate [Eq. 17.9]	W
Re	Reynolds number [Eq. 17.3] and [Eq. 17.8]	
SAR	Specific absorption rate [Eq. 17.7]	KW/g
TCO	Total cost of ownership [Eq. 17.4]	
T_{in}	Fluid inlet temperature [Eq. 17.1]	°C
T_{out}	Fluid outlet temperature [Eq. 17.1]	°C
T_a	Ambient temperature [Eq. 17.2] and [Eq. 17.6]	°C
T_m	Fluid mean temperatures [Eq. 17.2]	°C
U_L	Overall loss coefficient of solar collector [Eq. 17.6]	W/m^2 K
Greek Symbols		
α	Absorptivity [Eq. 17.6]	
κ	Absorption rate of the absorber [Eq. 17.2]	
τ	Transmissivity [Eq. 17.6]	
ϕ	Nanoparticles volume fraction [Eq. 17.8]	
η	Efficiency of the collector [Eq. 17.1] and [Eq. 17.8]	
η_i	Efficiency of the collector [Eq. 17.6] and [Eq. 17.10]	
η_{net}	System collecting efficiency [Eq. 17.9]	
λ	Rate of transmission of the solar collector cover [Eq. 17.2]	
ΔT_n	Temperature rise at the same time interval for nanofluid [Eq. 17.7]	°C
ΔT_w	Temperature rise at the same time interval for water [Eq. 17.7]	°C
Δt	Time interval [Eq. 17.7]	s

Table 17.1 Summary of investigations of nanofluid in a flat-plate solar collector

Model	Reference	Year	Nanofluid type	Results and remarks
Experimental	Otanicar and Golden [53]	2009	General	The nanofluid based solar collector had a lower embodied energy (about 9%) and approximately (3%) higher levels of pollution offsets than a conventional collector
Experimental	Natarajan and Sathish [54]	2009	MWCNT-water	Nanofluids were more effective than the conventional fluids and if were used as a heat transport medium, it increased the efficiency of the traditional solar water heater
Experimental	Polvongsri and Kiatsiriroat [55]	2011	Silver-water	Nanofluid improved the thermal performance of the collector especially at high inlet temperature
Experimental	Yousefi et al. [56]	2012	Al_2O_3-water	Using 0.2 wt% of nanofluid increased the efficiency of the collector in comparison with water by about 28.3%
Experimental	Yousefi et al. [57]	2012	MWCNT-water	Using 0.2 wt% of MWCNT nanofluid without surfactant decreased the collector efficiency, while with surfactant increased it

Model	Reference	Year	Nanofluid type	Results and remarks
Experimental	Yousefi et al. [58]	2012	MWCNT-H_2O	When pH of nanofluid was varied, a clear effect on the efficiency of the solar collector was observed
Numerical	Tora and Moustafa [59]	2013	Al_2O_3-water	Thermal conductively of the nanofluid and the collector efficiency had better values with higher particle sizes
Experimental	Jamal-Abad et al. [60]	2013	Cu–water	Efficiency of the collector at 0.05 wt% was approximately 24% more than that of the pure base fluids
Experimental	Faizal et al. [61]	2013	MWCNT	Collector's size can be reduced up to 37% of its original size when applying MWCNT nanofluid
Experimental	Gangadevi et al. [62]	2013	Al_2O_3-water	Efficiency of the collector was increased about 30% by using Al_2O_3-water nanofluid
Experimental	Chaji et al. [63]	2013	TiO_2-water	Collector efficiency was improved between 2.6 and 7% by using nanofluid
Theoretical	Tiwari et al. [64]	2013	Al_2O_3-water	Nanofluid increased the thermal efficiency in comparison with water by about 31.64%

(*Continued*)

Table 17.1 (*Continued*)

Model	Reference	Year	Nanofluid type	Results and remarks
Numerical	Faizal et al. [65]	2013	CuO-water SiO_2-water TiO_2-water Al_2O_3-water	25.6%, 21.6%, 22.1% and 21.5% solar collector area reduction were achieved for using CuO, SiO_2, TiO_2 and Al_2O_3 nanofluid
Experimental	Said et al. [66]	2013	AL_2O_3-water AL_2O_3-ethylene glycol/water mixture	Pressure drop and pumping power of the nanofluid flows were very close to that of the base liquid for low volume concentration
Theoretical	Alim et al. [67]	2013	Al_2O_3-water CuO-water SiO_2-water TiO_2 –water	CuO/water nanofluid was reduced the entropy generation by 4.34% and enhanced the heat transfer coefficient by 22.15%
Experimental	Colangelo et al. [68]	2013	Al_2O_3-water	Convective heat transfer coefficient was increased up to 25% at a concentration of 3% volume
Numerical	Shankar and Manivannan [69]	2013	CuO-water	About (10.88%) of an improved efficiency was observed at (0.025%) volume fractions
Numerical	Nasrin and Alim [70]	2014	Ag-water Cu-water CuO-water Al_2O_3-water	Cu-water nanofluid was more effective in order to promote heat loss system through the riser pipe of a flat-plate solar collector

Model	Reference	Year	Nanofluid type	Results and remarks
Experimental	Polvongsri and Kiatsiriroat [71]	2014	Silver-water	The overall heat loss coefficient of the solar collector with nanofluid was reduced and more solar heat gain was obtained
Numerical	Ekramian et al. [72]	2014	MWCNT-water CuO-water Al_2O_3-water	Thermal efficiency of CuO-water nanofluid were greater than other nanofluids
Theoretical	Mahian et al. [73]	2014	Al_2O_3-water	Entropy generation decreased with increasing the nanofluid concentration
Numerical	Nasrin and Alim [74]	2014	Alumina-water	When Reynolds number increased, the percentage of collector efficiency was enhanced by using nanofluid
Numerical	Kabeel and El-Said [75]	2014	Cu-water	Volume fractions of nanoparticles had a significant effect on increasing the fresh water production and decreasing its cost
Theoretical	Mahian et al. [76]	2014	Cu-water Al_2O_3-water TiO_2-water SiO_2-water	Efficiency decreased with increasing volume fraction of nanoparticles and SiO_2/water nanofluid gave the highest efficiency of the collector

(Continued)

Table 17.1 (*Continued*)

Model	Reference	Year	Nanofluid type	Results and remarks
Experimental	Moghadam et al. [77]	2014	CuO-water	Nanofluid with mass flow rate of 1 kg/min increased the collector efficiency by about 21.8%
Experimental	Zamzamian et al. [78]	2014	Cu-synthesized/EG	Efficiency of the collector was improved by increasing the nanofluid concentration
Experimental	He et al. [79]	2014	Cu-H_2O	Efficiency of the collector was enhanced by 23.83% by using nanofluid
Theoretical	Said et al. [80]	2014	SWCNT-water TiO_2-water Al_2O_3-water SiO_2-water	SWCNT nanofluid reduced the entropy generation by 4.34% and enhanced the heat transfer coefficient by 15.33%
Theoretical	Mahian et al. [81]	2014	Boehmite alumina-water and ethylene glycol	(1) Outlet temperature increased with an increase in volume fraction of nanoparticles (2) Entropy generation rate due to heat transfer was reduced for all shapes of nanoparticles
Experimental	Roy et al. [82]	2015	Silver-water	The maximum efficiency of the solar collector was found to be near 70% for 0.04% particle volume concentration at 6 L/min

Table 17.2 Summary of investigations of nanofluid in the direct absorption solar collector

Model	Reference	Year	Nanofluid Type	Results and remarks
Theoretical	Tyagi et al. [83]	2009	Aluminum-water	Nanofluid increased the absorption of incident radiation by more than nine times over that of pure water
Experimental and numerical	Otanicar et al. [84]	2010	Carbon nanotubes-water Graphite-water Silver-water	Efficiency improvements of up to 5% in solar thermal collectors by utilizing nanofluids as an absorption mechanism
Experimental	Taylor et al. [85]	2011	Graphite-water Aluminum-water Sliver-water Copper-water	Nanofluids could be used to absorb sunlight with a negligible amount of viscosity and/or density increase
Experimental	Poinern et al. [86]	2012	Carbon nanospheres (CNS)	Photo-thermal response of both nanofluids and films composed of CNS were investigated under 1000 W/m^2 solar irradiation
Theoretical	Saidur et al. [87]	2012	Aluminum-water	1. Al-water nanofluid was a good option for improving the performance of the collector

(*Continued*)

Table 17.2 (*Continued*)

Model	Reference	Year	Nanofluid Type	Results and remarks
				2. Collector efficiency increased slightly with an increase in the particle size
Numerical	Moradi et al. [88]	2013	Glycol-based and water-based nanofluid	When the concentration of nanoparticles increased, the efficiency of the collector increased up to a certain limit and then decreased
Experimental	Kundan and Sharma [89]	2013	CuO-water	Efficiency of the solar collector was increased by 4–6% compared to conventional water-based solar collector
Experimental	Verma and Kundan [90]	2013	Al_2O_3-water	Collector efficiency was increased about 3–5% when nanofluid was used as compared to a simple water
Numerical	Ladjevardi et al. [91]	2013	Graphite-water	Nanofluid capable to absorb more than 50% of incident irradiation energy
Theoretical	Hector and Singh [92]	2013	Graphene-Therminol VP-1 Aluminum-Therminol VP-1	Reducing DARS diameter was recommended to achieve higher mean nanofluid outlet temperatures

Model	Reference	Year	Nanofluid Type	Results and remarks
Experimental	Zhidong et al. [93]	2014	Magnetic nanofluids	Both magnetic field and magnetic nanofluids enhanced the heat transfer efficiency of the collector
Experimental and Theoretical	Lee et al. [94]	2014	MWCNT-water	DASC concept can further improved the efficiency of conventional flat-plate solar collectors
Experimental and Numerical	Luo et al. [95]	2014	TiO_2-Texatherm oil Al_2O_3-Texatherm oil Ag-Texatherm oil Cu-Texatherm oil SiO_2-Texatherm oil	Nanofluids improved outlet temperature by 30–100 K and efficiency by 2–25% than the base fluid
Experimental	Filho et al. [96]	2014	Silver-de-ionized water	Stored thermal energy increased by 52%, 93% and 144% for silver particle concentration of 1.62, 3.25 and 6.5 ppm, respectively at the peak temperature
Numerical	Parvin et al. [97]	2014	Cu-water Ag-water	Collector efficiency enhanced about two times with increasing Reynolds number and solid volume fraction

(Continued)

Table 17.2 (*Continued*)

Model	Reference	Year	Nanofluid Type	Results and remarks
Experimental	Hordy et al. [98]	2014	MWCNTs-water MWCNTs-ethylene glycol MWCNTs-propylene glycol MWCNTs-Therminol VP-1	MWCNTs were close to 100% solar energy absorption, even at low concentrations and small collection volumes
Experimental	Karami et al. [99]	2014	CNT-water	CNT-water nanofluid was very recommended for increasing the overall efficiency of direct absorption solar collectors
Experimental and theoretical	Zhang et al. [100]	2014	Ni/C-ionic liquid Ni-ionic liquid Cu-ionic liquid	Optical absorption property of the ionic liquid was greatly enhanced by adding a low volume fraction of nanoparticles in it
Experimental	Sadique and Verma [101]	2014	Graphite-water Carbon nanotube-water Silver-water	Nanofluids could be used to absorb sunlight with a negligible amount of viscosity and/or density increase
Numerical	Moradi et al. [102]	2015	Carbon-nanohorn	Heat losses due to conduction, convection and radiation at the boundaries were computed

Table 17.3 Summary of investigations of nanofluid in parabolic trough solar collector (PTSC)

Model	Reference	Year	Nanofluid Type	Results and remarks
Numerical	Khullar and Tyagi [103]	2010	Al-water	Thermal and optical efficiencies and outlet temperatures of the collector was improved with using nanofluid
Numerical	Khullar et al. [104]	2012	Aluminum-Therminol VP-1	Nanofluid collector had about 5–10% higher efficiency as compared to conventional parabolic solar collector
Numerical	De Risi et al. [105]	2013	Ni-CuO	Maximum thermal efficiency was about 62.5%, for a nanofluid outlet temperature of 650°C and a nanoparticles volume concentration of 0.3%
Experimental	Sunil et al. [106]	2014	SiO_2-H_2O based nanofluid	SiO_2-H_2O nanofluid had comparatively higher efficiency at higher volume flow rates
Numerical	Ghasemi and Ahangar [107]	2014	Cu-water nanofluid	1. Thermal efficiency of the collector decreased with increase receiver length. 2. By increasing volume fraction of nanoparticles, the performance of collector was enhanced

(Continued)

Table 17.3 (*Continued*)

Model	Reference	Year	Nanofluid Type	Results and remarks
Numerical	Sokhansefat et al. [108]	2014	Al_2O_3-Synthetic oil	Heat transfer coefficient of the working fluid in an absorber tube was enhanced with presence of nanoparticles
Experimental	Kasaeian et al. [109]	2015	MWCNT-Mineral oil	Global efficiency was enhanced about 4, 5% at 0.2% and 5–7%, at 0.3% when MWCNT/mineral oil nanofluid was used

Table 17.4 Summary of investigations of nanofluid in wavy solar collector

Model	Reference	Year	Nanofluid Type	Results and remarks
Numerical	Nasrin and Alim [110]	2013	Ag-water CuO-water	Better performance of heat transfer inside the collector was found by using the highest solid volume fraction of Ag-water nanofluid
Experimental	Alaeian et al. [111]	2014	MWCNT–Oil	Heat transfer coefficient was increased up to 56% compared to base fluid at highest concentration and lowest dimensionless curvature ratio

Table 17.5 Summary of investigations of nanofluid in heat pipe solar collector

Model	Reference	Year	Nanofluid Type	Results and remarks
Experimental	Lu et al. [112]	2011	CuO-water	Nanofluid significantly enhanced he thermal performance of the evaporator
Experimental	Moorthy et al. [113]	2012	TiO_2-water	The efficiency was enhanced by 16.7% by using nanofluid compared to pure water
Experimental	Chougule et al. [114]	2012	Carbon nanotube (CNT)	Nanofluid collector gave a better performance in all test conditions
Experimental	Senthil Kumar et al. [115]	2012	CNT-water	The nanofluid working fluid collector gave the better performance in all the operating conditions
Experimental	Chougule et al. [116]	2013	Carbon nanotube (CNT)	They obtained the optimal value of CNT nanofluid concentration for better performance of the heat pipe solar collector
Experimental	Liu et al. [117]	2013	CuO-water	Both the air outlet temperature and collector efficiency using nanofluid are higher than that using water

(Continued)

Table 17.5 (*Continued*)

Model	Reference	Year	Nanofluid Type	Results and remarks
Experimental	Saravanan and Karunakaran [118]	2014	TiO_2-DI water	Thermal efficiency of the collector was increased by using nanofluid
Experimental	Aruna et al. [119]	2014	TiO_2-DI water	Nano fluid gave a better performance when compared to propanol

Table 17.6 Summary of investigations of nanofluid in another solar collectors

Model	Reference	Year	Nanofluid/ Collector type	Results and remarks
Experimental	Li et al. [120]	2011	Al_2O_3-water ZnO-water MgO-water Tubular solar collector	ZnO nanofluid with 0.2% vol. concentration was a good option in solar energy utilization
Experimental	Taylor et al. [121]	2011	Aluminum-Therminol VP-1 Copper-Therminol VP-1 Graphite-Therminol VP-1 Silver-Therminol VP-1 Power tower solar collectors	Efficiency improvement on the order of 5–10% was possible with a nanofluid receiver
Experimental	Gangadevi et al. [122]	2013	Al_2O_3–water PVT solar collector	Both the electrical and thermal efficiencies of a hybrid solar system increased considerably by using the nanofluid

Model	Reference	Year	Nanofluid/ Collector type	Results and remarks
Experimental	Paul et al. [123]	2013	Ionic liquid-Al_2O_3 Concentrating Solar Power (CSP) system	Heat transfer performance was enhanced by 15% with using the nanofluid
Numerical	Hewakuruppu et al. [125]	2014	Not specified	A nanofluid with a short wavelength, optical depth of 3 with no scattering, and a long wavelength with optical depth of zero was required for optimum performance
Numerical	Rahman et al. [126]	2014	Cu-water Al_2O_3-water TiO_2-water Triangular solar collector with corrugated bottom wall	Cu-water nanofluid performed a better performance than other used nanofluids
Experimental	Goudarzi et al. [127]	2014	CuO-H_2O Cylindrical solar collector with receiver helical pipe	Maximum thermal efficiency was increased by about 25.6% by using 0.1 wt% nanofluid in 0.0083 kg/s mass flow rate of fluid
Numerical	Rahman et al. [128]	2014	CNT-water Quarter circular solar collector	Both solid volume fraction and tilt angle played vital roles for augmentation of heat transfer in solar collector

(*Continued*)

Table 17.6 (*Continued*)

Model	Reference	Year	Nanofluid/ Collector type	Results and remarks
Experimental	Goudarzi et al. [129]	2015	CuO-water Al_2O_3-water Cylindrical solar collector	Efficiency of the collector was increased when there was more differences between pH of nanofluid and pH of iso-electric point

Table 17.7 Summary of review investigations of nanofluid in solar collector

Reference	Number of reviewed papers	Year	Applications	Results and remarks
Sruthi [130]	5	2012	1. Solar water heater 2. Solar cell	Various advantages of using nanomaterials in solar water heater were presented
Khanafer and Vafai [131]	104	2013	1. Thermal energy storage systems 2. Photovoltaic systems 3. Direct absorption solar collectors 4. Solar desalination	For optimum performance of a solar collector, the solar radiation should be absorbed within a small wave length range (0.25 mm $< \lambda <$ 2.5 mm)
Javadi et al. [132]	134	2013	Direct absorption solar collector	Further research must be focused on the two-phase analysis of nanofluids
Mahian et al. [133]	93	2013	1. Solar collectors and solar water heaters	1. Different volume fractions should be tested to find the optimum volume fraction

Reference	Number of reviewed papers	Year	Applications	Results and remarks
			2. Photovoltaic/ thermal systems 3. Solar ponds 4. Solar cells 5. Thermal energy storage 6. Solar still	2. Using of nanofluids in collectors led to a reduction in CO_2 emissions, annual electricity and fuel savings
Al-Shamani et al. [134]	100	2014	1. Cooling solar collector 2. Photovoltaic/ thermal (PV/T) collector systems	Nanofluids could be used to cool photovoltaic/thermal (PV/T) collector systems
Chaudhari and Walke [135]	20	2014	1. Solar collector 2. Solar still 3. Thermal energy storage 5. Solar cell	Nanofluid minimized temperature difference between absorber and the heat transfer fluid
Nerella et al. [136]	15	2014	Solar collector	The improvements of solar collector efficiency by using nanofluids with base fluid were highly comparable with the solar collector efficiency when only base fluid was used
Kasaeian et al. [137]	127	2015	1. Solar collector 2. Photovoltaic systems 3. Energy storage system 4. Solar thermoelectrics devices 5. Solar cells	Development of the particle production and decreasing in costs was essential for the nanofluid research

17.1 Introduction

Solar energy (or radiant light and heat from the sun) is currently one of the most important sources of clean, free, inexhaustible and renewable energy with minimal environmental impact. The power from the sun intercepted by the earth is approximately 1.8×10^{11} MW [1]. About 30% of the solar power actually reaches the earth and at every 20 min, the sun produces enough power to supply the earth with its needs for an entire year [2]. The solar energy can be defined as the energy which comes from the sun and can be converted to electricity and heat. It has produced energy for billions of years, so the utilization of the solar energy and the technologies of its materials has received much attention especially in the last 10 years [3–4]. For example, some studies have indicated that about 1000 times from the global energy requirements can be achieved by using the solar energy; however, only 0.02% of this energy is currently utilized [5]. The main reasons of this huge attention in the solar energy applications are due to the growing demand of energy, limited availability of fossil fuels and environmental problems associated with them such as carbon dioxide emissions. Energy experts predict that the world would need 30 TW of energy resources by the year 2050 to maintain economic growth. Many scientists believe that the sun is the only candidate that can offer a fully developed solution for the energy crisis. Moreover, the rapid increase in the human population can be considered as an additional serious problem, since the global population has increased by nearly 2 billion with a major contribution from developing countries [6]. Furthermore, it is proved that the consumption rate of fossil fuels by humans is much faster than they are replaced by geologic processes. In fact, the sun radiates every day, enormous amount of energy and the hourly solar flux incident on the earth's surface is greater than all of human consumption of energy in a year [7]. In spite of this huge amount of available solar energy, approximately 80% of energy used worldwide still predominantly comes from fossil fuels such as coal, petroleum and natural gas [8].

17.2 What Is the Nanofluid?

Nanofluid or suspensions of nanoparticles in liquids is defined as a smart mixture consists from a normal fluid such as (water, light and Diesel oils, ethylene glycol, propylene glycol, terpineol, ethanol, decane, acetone, toluene, sodium chloride, glycerol and molten salts) with a very small amount of solid metallic or metallic oxide nanoparticles (1–100 nm) [where (nm = 1 billionth of a meter)] or nanotubes which was firstly suggested by Choi [9] in 1995 at the Argonne National Laboratory, U.S.A. For comparison, 10 nm is about 1000 times smaller than the diameter of a human hair. Nanofluids have significantly improved the fluid thermophysical properties, such as thermal conductivity, viscosity, and convective heat transfer coefficients, compared with conventional fluids. The nanofluid was considered as the new generation of advanced heat transfer fluids or a two-phase system which used for various engineering and industrial applications due to its excellent performance. Some of these applications including nuclear reactors, transportation industry, cooling of transformer oil, electrical energy, mechanical, magnetic, cooling of microchips, improving diesel generator efficiency, solar absorption, microelectronics, paints and coatings, biomedical fields and many another areas where the heat removal is involved [10]. Nanofluids are called also as a super-coolant fluids due to their high ability to absorb heat more than any traditional fluids, so they can reduce the size of system and increase its efficiency. It is well known that metals have higher thermal conductivities than those of fluids. For example, the thermal conductivity of copper at room temperature is about 700 times greater than that of water and about 3000 times greater than that of engine oil [11]. Various types of nanoparticles are utilized in the preparation process of the nanofluid. Examples of these types include Cu, Ag, Au, Fe, Zn Al_2O_3, CuO, TiO_2, SiO_2, SiC, ZnO, MgO, Fe_3O_4 and TiC. Also, carbon nanotubes (CNTs) are used nowadays widely due to their extremely high thermal conductivity in the longitudinal direction and their large aspect ratios. The first decade of nanofluid research primarily focused on measuring the thermo-physical properties of these fluids such as the thermal conductivity, density, viscosity and heat transfer coefficient [12]. Nanofluid have a good properties of radiation absorption and it has a high thermal conductivity. For example, the thermal conductivity at the room temperature of individual multi-walled carbon nanotubes (MWCNTs) were found to have values greater than 3000 W/m·K [13]. Moreover, Assael et al. [14] indicated that about 1% volumetric fraction of MWCNT was enhanced the thermal conductivity of water by about 40%. In order to prepare nanofluids by dispersing nanoparticles in a base fluid, a proper mixing and stabilization of the particles is required. It is necessary to mention that the traditional fluids have many serious problems related with their heat transfer performance. Some of these problems can be summarized as follows [15]:

(1) The particles settle rapidly, forming a layer on the surface and reducing the heat transfer capacity of the fluid.
(2) When the circulation rate of the fluid is increased, sedimentation is reduced, but the erosion of the heat transfer device and pipe lines increased rapidly.
(3) The pressure drop in the fluid increases significantly.
(4) The large size of the particle leads to clog the flow channels, especially if the cooling channels are narrow.

Also, it is highly recommended not to add large solid particles in the base fluids (more than 100 nm) due to the following main drawbacks [7]:

(1) Mixtures become unstable and hence, sedimentation occurs.
(2) Existence of large solid particles require a large pumping power and this increases the cost.
(3) Large quantities of solid particles erode the channel walls and increase the pressure drop.
(4) They lead to clogging of pumps and valves used in the overall system.

Therefore, nanofluid can be used efficiently to solve these drawbacks, since it has many advantages such as the following:

(1) It increases the effective thermal conductivity of the suspension and as a result enhances the heat transfer characteristics. Since, high thermal conductivity of nanofluids and Brownian motion of nanoparticles increase the heat transfer performances.
(2) It has a very small size, so it fluidizes easily inside the base fluids and can be moves faster inside solid blocks such as the porous media.
(3) It has a large surface area (more than 100 m^2/g) to volume ratio, dimension-dependent physical properties and lower kinetic energy. In fact, the large surface area increases the heat transfer rate between the base fluid and solid particles.

(4) The nanofluid reduces the problem of rapid settling of micro or millimeter sized particles when they used in the conventional collectors.
(5) The properties like viscosity, specific heat, thermal conductivity and density may be varied easily by changing particle concentrations to be suitable with different industrial applications [16].
(6) The pumping power required for the equivalent heat transfer is less than that compared to pure liquids [17].
(7) Nanofluid can pass through pumps and pipes without adverse effects such as clogging and fouling. Therefore, it is very useful in micochannel applications.
(8) The heat transfer increases as a result of increase in the heat transfer surface area between the particles and fluids.
(9) In contrast to conventional heat transfer fluids, nanofluids are not transparent to solar radiant energy, but they absorb and scatter significantly the solar irradiance passing through them.
(10) The high stability of nanofluid make them to stay in the liquid phase for months or even years and its stability can be increased by the Brownian motion.
(11) Nanoparticles dispersed faster in liquids, so it reduce the friction and wear occur in the pipelines and pumps.
(12) Nanfluid has a high thermal capacity, since the small volume of nanoparticles make them easily to store a large quantity of heat. This of course will reduce the energy losses and increasing the efficiency of the system [18].
(13) Nanoparticles increase significantly the mixing fluctuation and turbulence of the fluid [19].
(14) The dispersion of nanoparticles flattens the transverse temperature gradient of the fluid.
(15) Nanoparticles in fluids have an extremely low momentum, which greatly reduces abrasive wear if it exists.

However, the science which deals with the nanofluids is called the nanoscience or nanotechnology and it provides a new area of research to deal with these new types of fluids. Nanoscience term is a mixture of two words. The first one is nano, coming from the Greek word "nanos" (or Latin "nanus"), which means the "Dwarf", while the second word is "Science" which means knowledge [20]. Nanoscience can be defined as the study of phenomena and manipulation of materials at atomic, molecular, and macromolecular scales, where properties differ significantly from those at a larger scale. In fact, the term "Nanotechnology" was first defined by Norio Taniguchi of the Tokyo Science University in 1974. This technology has the potential to dramatically redefine the methods used for developing lighter, stronger and high-performance structures and processes with clear and non-traditional properties. For comprehensive details about the applications and challenges of nanofluids, see the review by Saidur et al. [21].

17.3 Solar Collector

The solar collector is one of the most important components of a solar energy and water heating systems which can be defined as a low, medium, or high temperature green heat exchanger device which converts the energy in sunlight or incident solar radiation either to the thermal energy in solar thermal applications, or to an electrical energy directly in PV (photovoltaic) applications. Therefore, the main job of the solar collector is that it collects the solar energy and transfers it to a fluid passing in contact with it. The ideal solar collector absorbs the concentrated solar radiation and converts it to a heat and then transfer this heat to the collector fluid. Therefore, higher the heat transfer to fluid, means higher outlet temperature and higher the collector efficiency in the power cycle [22]. So, the major challenge is how we can improve this device to increase its efficiency to convert the solar energy into a thermal or electrical energy. The solar collectors are classified to non-tracking and tracking collectors. The non-tracking collectors are kept at rest and also known as fixed or stationary collectors, while the tracking collectors are designed to track the movement of the sun, so that the received solar radiations always fall perpendicular to them. Solar collectors can be used for a variety of residential and small commercial applications such as water heating systems in homes, solar space heating, solar desalination, solar drying devices, electricity production and small solar power plants. For solar thermal applications, the solar irradiation is absorbed by a solar collector as a heat and then transferred to its working fluid (air, water or oil). The heat carried by the working fluid can be used to either provide domestic hot water/heating, or to charge a thermal energy storage tank where the heat can be used later at night or cloudy days. For photovoltaic applications, a PV module not only converts solar irradiation directly into the electrical energy, but it also produces plenty of waste heat, which can be recovered for thermal use by attaching PV board with recuperating tubes filled with carrier fluids [23]. In fact, classical solar thermal collectors have a metal sheet as an absorber, designed such that water has the minimum temperature in each transversal section, in order to collect as much as possible from the solar thermal energy. The performance of the solar collector depends upon the properties of the working fluid which are used to maximize the solar energy absorption in the solar collector. Examples of solar thermal collectors are solar water heaters, solar cookers and solar ponds. Many researchers are presented a literature review papers about the solar collectors such as Kalogirou [24], Jaisankar et al. [25], Alkilani et al. [26], Hossain et al. [27], D'Antoni and Saro [28], Mathur et al. [29], Tian and Zhao [23], Shukla et al. [30], Devanarayanan and Murugavel [31] and very recently by Hussein [32].

17.3.1 Solar Collector Types

(1) Conventional flat-plate solar collector: It has a simple construction as shown in Fig. 17.1 [33], which consists of an absorber [a metal plate (Copper or Aluminum) coated with a black chrome layer] to collect the sun's heat energy which is then transferred to a fluid running in tubes attached to the underside of the collector. Insulation (mineral wool) surrounds the plate to prevent the heat loss due to the conduction. A glass plate is placed above the absorber plate to prevent convective heat losses. This collector is used for many technological applications such as hot water system, air heating and solar cooker. In this type, the fluid absorbs heat energy through a surface absorber (fluid flow through pipes). The absorber is a metal sheet designed such that the water has the minimum temperature in each transversal section, in order to collect as much as possible from the solar thermal energy. Flat-plate solar collectors are used broadly to increase the working fluid temperature in the range of 30 to 100°C above the ambient temperature. The performance of this collector depends on geographic factors, geometry and orientation of the collector. However, this type suffers from several defects which can be summarized as follows:

(a) The heat losses due to the conduction, natural convection and radiation from the collector surface due to the temperature difference between the plate and ambient air are considerably large which reduces its efficiency.
(b) The working life of the collector is somewhat short due to the pipe corrosion by the water [34].
(c) The water may freeze in cold days; therefore, a pump is required to maintain the forced circulation of the water. This leads to increase the cost and the space required of the collector.
(d) The incident flux density in this collector is limited.
(e) The limited quantity of the heat transferred by the fluid.
(f) The outlet temperatures of this collector are low.
(g) The using of this type is limited in sunny and warm climates, while its performance is greatly diminished during cold, cloudy and windy days.

Figure 17.1 Photograph of the flat-plate solar collector [33].

(2) **Uncovered or unglazed solar collector**: It is one of various types of the flat-plate solar collector. This type is usually the cheaper option, but still offers an effective solar thermal energy in applications such as water preheating for domestic

or industrial use, heating of swimming pools, space heating and air heating for industrial or agricultural applications [35].

(3) **Direct absorption solar collector (DASC)**: This type was first proposed in 1975 by Minardi and Chuang [36] and used to enhance the efficiency of the flat-plate collector by making the fluid to directly absorb the solar radiation. It is also called as a volumetric solar collector and has the ability to offer an unlimited source of renewable energy with minimal environmental impact. This type has some advantages compared to the conventional one. Besides larger solar absorption area and actual installation surface area ratio, it is capable to avoid surface heat losses due to the excessive temperature on the surface absorption collector. However, the main disadvantage of this device is that its efficiency is limited by the absorption properties of the working fluid, which is very poor for typical fluids used in solar collectors [37]. In this configuration, the hottest part of the system is the operating fluid and this allows to have a more efficient conversion. There are many kinds of the direct absorption solar collector such as volume trap solar collectors, black liquid collectors and small particle collectors.

(4) **Concentrating solar collector**: This type is used in power generation plants, which heating the water with a high mass flow rate. There are many types of concentrating solar collector such as Praboloid Dish, Parabolic Trough and Heileostat. This type is more efficient than the direct absorption solar collector, but it has a high installation cost and needs a modern tracking system.

(5) **Parabolic trough solar collector (PTSC)**: This type of the solar collector is one of the linear concentrating solar collectors which are appropriate for working in the range of 150–400°C. It is straight in one dimension (z-direction) and curved as a parabola in the other two dimensions (x–y plane) as shown in Fig. 17.2 [38]. It uses mirrored surfaces of a linear parabolic reflector to focus solar radiation into an evacuated tubular receiver placed along the focal line of the parabola. The receiver mainly includes an inner absorber tube surrounded by an outer glass cover and supported brackets [39, 40]. This type can be used to drive machinery

or to generate electricity. The largest solar power plant using a parabolic trough technology is called SEGS plants which is located in the California, United States, and produces about 354 MW of electricity. Two important factors must be carefully considered in the construction of this collector which are the accuracy of the parabolic shape and the torsional resistance of the collector. Moreover, the stresses and deflections experienced by the receiver and the reflector must remain below specified levels under various thermal, wind and gravitational loads as indicated recently by Coccia et al. [41]. For further details about the parabolic trough solar collector, see the comprehensive overview by Garcýa et al. [38].

Figure 17.2 Photograph of the parabolic trough solar collector [38].

(6) **Evacuated tube collector (ETC)**: This type depends on the concept of the heat pipe which transfers heat more efficiently than solid conductors as shown in Fig. 17.3 [42]. It uses a series of borosilicate glass tubes with each containing a finned heat pipes (typically made from copper). The fins are coated with a black chrome coating and pipes contain a liquid which undergoes an evaporating-condensing cycle of the pure working fluid as it is heated and cooled. The heat pipes are connected to a manifold which facilitates the heat exchange from the liquid to a heat transfer fluid flowing

through the manifold. The framework of the collector is made from stainless steel and rock wool is used for insulation [43]. A structure of this type is shown in Fig. 17.4 as explained by Wei et al. [34], while their components are illustrated in Fig. 17.5. This type of the solar collectors has a low thermal losses, low cost and high efficiency than the conventional collectors. In fact, heat pipes that working under gravity with the condenser above the evaporator does not require an external power or a capillary action to return the working fluid from the condenser to the evaporator. This type of the heat pipe is known as a thermosyphon or wickless gravity assisted heat pipe. Common thermosyphon is a tubular construction and can be easily integrated into a flat-plate or evacuated tubular solar heating systems. Figure 17.6 illustrates an example of an evacuated tube collector which is called the solar heat pipe vacuum collectors as given by Bourdoukan et al. [44]. The evacuated tube collector has a two types of arrangements. The first one is called H-type and the other is T-type collector. It is useful to mention that T-type collector collect slightly more radiation compared to H-type collector on an annual basis. For more details about the evacuated tube collector, see the review by Shah et al. [45].

Figure 17.3 Photograph of the heat pipe solar collector [42].

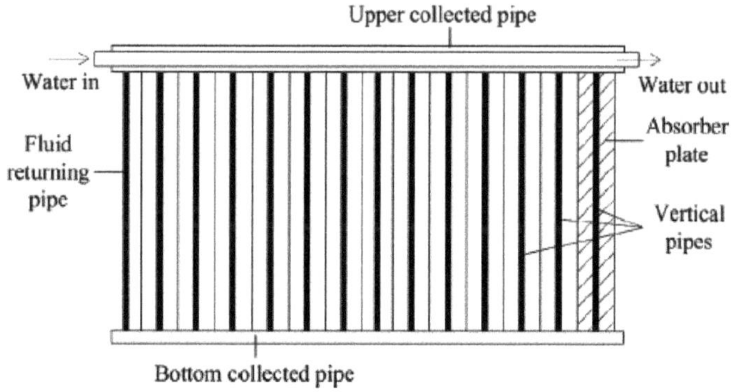

Figure 17.4 Structure of solar heat collector with integrated heat pipe [34].

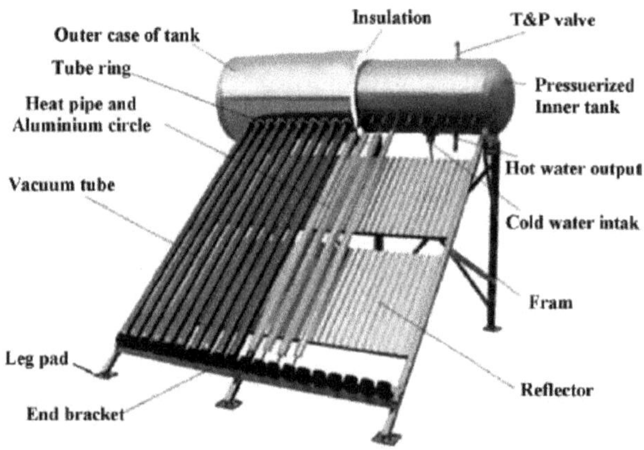

Figure 17.5 Components of evacuated tube solar water heater collectors [27].

Figure 17.6 Photograph of the solar heat pipe vacuum collectors [44].

(7) **Cylindrical solar collector**: It is a type of tubular solar collectors which has a copper coil in the shape of a helical pipe (Fig. 17.7) instead of absorbent coating in the cylindrical center and it has a high thermal efficiency than a flat-plate type. The major advantage of this type is that it is not necessary to direct it to the sun because of its circular shape, while the flat-plate collector should always be directed to face the sun with a certain tilted angle to get the best efficiency [46].

Figure 17.7 Photograph of the cylindrical solar collector [46].

(8) **Hybrid photovoltaic/thermal solar collector (PV/T)**: This type can be considered as a thermal and electrical energy system which combines between a photovoltaic (PV) cell [which converts the electromagnetic radiation (photons) into electricity] with a solar thermal collector, which captures the remaining energy and removes waste heat from the PV module as shown in Fig. 17.8 Shan et al. [47]. This combination of both electricity and heat make this collector more efficient than the conventional one [48]. The (PV/T) collector consists from a PV module and an attached pipe for flowing working fluid as shown in Fig. 17.9 [49]. This collector can be utilized for simultaneous production of the electricity and low temperature hot water for domestic purposes. So, the two main goals of the (PV/T) solar collector are summarized as follows [50]:

(a) Thermal energy can be produced and used for heating purposes.
(b) It cools the solar cell and decreases its temperature and as a result, the electrical conversion efficiency is increased.

For more details about the (PV/T) solar collector, see the comprehensive review by Ibrahim et al. [51].

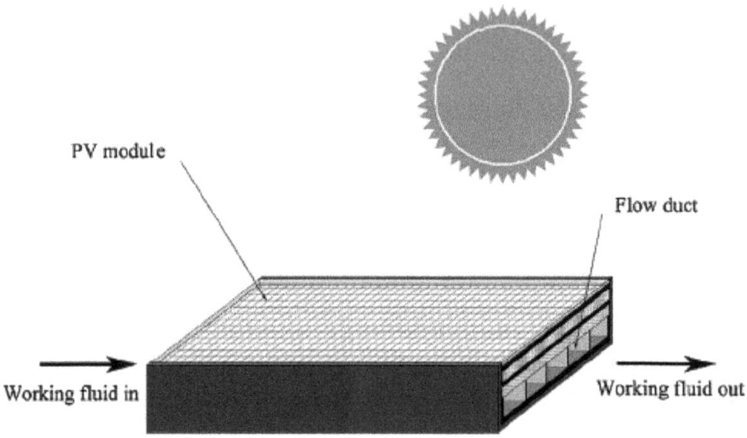

Figure 17.8 The brief structure of a typical hybrid PVT collector [47].

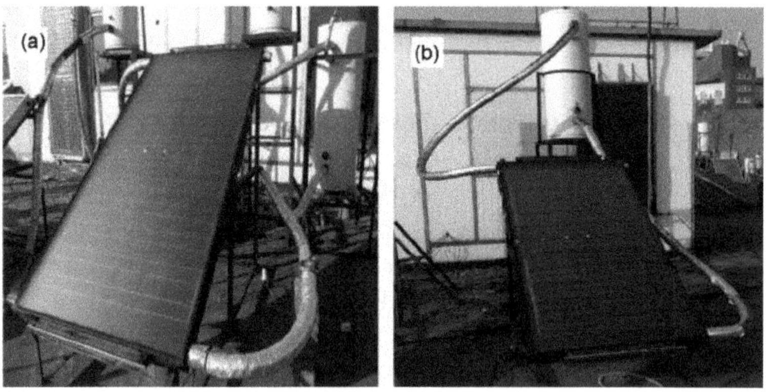

Figure 17.9 Photograph of PVT solar collector [49].

(9) **Transpired solar collector (TSC)**: This type is a highly efficient collector system particularly used for preheating the fresh air and was widely used in Canada and USA. It uses the solar energy to heat the absorber surface, which transmits

the thermal energy to the ambient air. The absorber surface is generally a metallic sheet (usually made from steel or aluminum), which can be integrated to the building façade (Fig. 17.10). The contact surface between the metal skin and air is increased by drawing air through the multiple small perforations into the cavity between the skin and facade. Finally, the heated air is drawn into the building to provide space heating. During the summer season, the warm air in the cavity can be released using a by-pass damper to avoid over heating inside the building or can be used for water heating in order to maximize the utilization of TSC. For additional information about the transpired solar collector, see the review paper by Shukla et al. [52].

Figure 17.10 Photograph of transpired solar collector [52].

(10) Nanofluid-based solar collector: This new generation of solar collectors depends on the concept of nanoscience where nanoparticles in a liquid medium can scatter and absorb the solar radiation. It benefits from the efficiency improvements that arise from using a nanofluid as a working fluid in the collector. This collector have a layer of a nanofluid which lies on the top of the collector to directly absorbs the sun's radiation. Therefore, this nanofluid layer eliminates the need for the absorber plate and tubes which are found in the conventional solar collector as mentioned by Otanicar and Golden [53]. Figure 17.11 shows schematics and materials for both conventional and nano collectors. Therefore, the main difference between the conventional and nanofluid-based collectors lies in the mode of the working fluid heating. In the conventional collector, the sunlight is absorbed by a surface and then transferred to the fluid. However, in the nanofluid collector, the sunlight is directly absorbed by the working fluid through the radiative heat transfer.

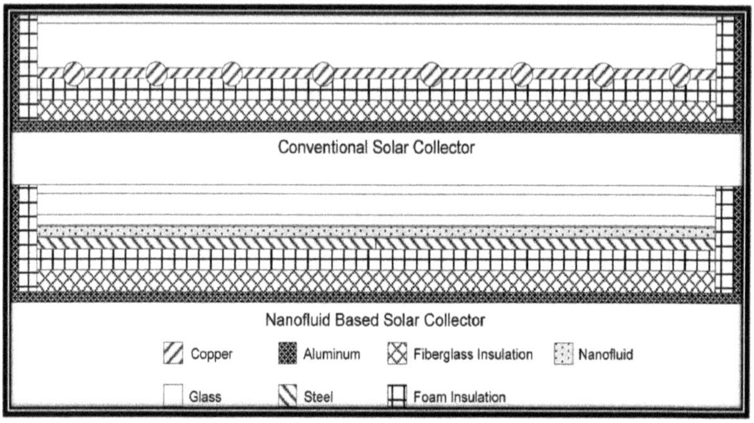

Figure 17.11 Schematics and materials for both conventional and nano collectors [53].

17.4 Benefits of Using Nanofluid in Solar Collectors

Nanofluids have many advantages as compared to classical fluids which make them very efficient in solar collectors. Some of these advantages can be summarized as follows [1]:

(1) The easy change in the shape, material, size and solid volume fraction of nanoparticles make them very effective to maximize the absorption of the solar energy.

(2) The solid nanoparticles increase the surface area and the heat capacity of the fluid due to their very small particle size.

(3) Nanofluid can be optically selective, it shows high absorption in the solar range and low emittance in the infrared range. Therefore, the optical characteristics (light absorption and emission behavior) of a base liquid such as the extinction coefficient are enhanced by using the nanofluid.

(4) Solid nanoparticles increase dramatically the thermal conductivity of the base fluid and as a result increase the

efficiency of solar collectors. Since, the thermal conductivity increases with increasing the concentration and the temperature of the nanofluid.

(5) Nanofluid has a high absorption coefficient in comparison with the base fluid. Also, it has a good stability under moderate temperature gradients which makes it as a one of excellent absorbing fluids.

(6) Nanofluids improve highly the radiative properties of the basic fluids and leading to increase the efficiency of solar collectors.

(7) Nanofluid can be used to reduce the surface temperature by enhancing the fluid properties instead of pumping water at higher flow rate which is considered not advantageous as the overall efficiency of the solar collector is lowered.

(8) Nanofluid has the ability to avoid the sedimentation, fouling, clogging of pumps and pipes due to its extremely small size.

(9) Nanofluid was used efficiently to reduce the required heat transfer area of tubes and heat exchangers which were utilized in the parabolic trough collectors and as a result reduces the total cost of these collectors.

(10) Nanofluids enhance the scattering and absorption of the incident radiation when passing through it. Also, it is found that the optical properties of the pure water are significantly improved by adding nanoparticles in it.

(11) The using of nanofluid in the solar collector leads to increase the desired output temperature of the system which is required to increase the collector efficiency. While, in the conventional solar collector, the output temperature increasing requires an increase in the heat transfer area and this leads to increase the size and the cost of the collector.

(12) The nanofluid can be used efficiently to reduce the size and cost of the solar collector system. This is due to the elimination of the complex manufacturing processes encountered in creating surface-absorbing plates.

(13) The nanofluid increases the collector efficiency due to its high density, low specific heat of nanoparticles and high convective heat transfer coefficient.

(14) The nanofluid can be used effectively to reduce the convection and emissive heat loss encountered in the conventional collector.

17.5 Applications of Nanofluid in the Flat-Plate Solar Collector

Otanicar and Golden [53] performed a comparative environmental and economic analysis of conventional and nanofluid solar hot water systems. They concluded that the nanofluid based solar collector had a slightly longer payback period, but at the end of its useful life had the same economic savings as a conventional solar collector. Also, the results showed that the nanofluid based solar collector had a lower embodied energy (about 9%) and approximately (3%) higher levels of pollution offsets than a conventional collector. Natarajan and Sathish [54] investigated experimentally the role of nanofluid (MWCNT-water) in the solar water heater. Thermal conductivities had been measured by the transient hot-wire method. The results proved that nanofluids were more effective than the conventional fluids and if were they used as a heat transport medium, it increased the efficiency of the traditional solar water heater. Polvongsri and Kiatsiriroat [55] investigated experimentally the performance of a flat-plate solar collector when silver-water nanofluid was used as a working fluid. The silver nanoparticle size was 20 nm with concentrations at 1000 and 10000 ppm, respectively. The experiments were undertaken with three identical flat-plate solar collectors each had an area of 0.15 × 1.0 m^2. They concluded that the nanofluid improved the thermal performance of the collector compared with water, especially at high inlet temperature. Yousefi et al. [56] investigated experimentally the effect of using the Al_2O_3-water nanofluid as an absorbing medium in a flat-plate solar collector (Fig. 17.12). The effect of mass flow rate, nanoparticles mass fraction and the presence of surfactant on the efficiency of the collector was studied. The weight fraction of nanoparticles were taken as 0.2% and 0.4%, while the particles dimension was considered as 15 nm. The results showed that using the 0.2 wt% of nanofluid increased the efficiency of the collector in comparison with water by about 28.3% as shown in Fig. 17.13.

Figure 17.12 The experimental flat-plate solar collector [56].

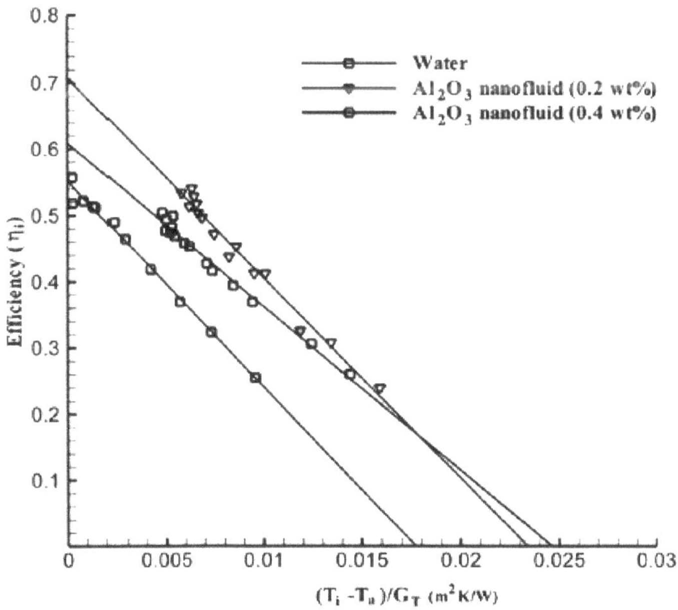

Figure 17.13 The effect of mass fraction of Al_2O_3-water nanofluid on the efficiency of a flat-plate solar collector [56].

Yousefi et al. [57] investigated experimentally the effect of using the MWCNT-water nanofluid as an absorbing medium in a flat-plate solar collector. The effect of Triton X-100 as a surfactant on the stability of the nanofluid was studied. The weight fraction of CNTs were taken as 0.2% and 0.4%, while the mass flow rates of nanofluid were varied from 0.0167 to 0.05 kg/s. Results showed that by increasing the weight fraction from 0.2% to 0.4%, a substantial increase in the collector efficiency was observed. They concluded that using the 0.2 wt% MWCNT nanofluid without surfactant decreased the collector efficiency, while with surfactant increased it as shown in Fig. 17.14. Yousefi et al. [58] investigated experimentally the effect of pH variation of MWCNT-H_2O nanofluid on the efficiency of a flat-plate solar collector. The experiments were carried out using 0.2 wt.% MWCNT with various pH values (3.5, 6.5 and 9.5) and Triton X-100 as an additive. Results showed that by increasing or decreasing the pH values with respect to the pH of iso-electric point, more enhancement in the efficiency of the solar collector was observed. Tora and Moustafa [59] simulated numerically the heat transfer performance of an Al_2O_3-water based nanofluid for the flat-plate solar collector. It was found that the thermal conductively of the nanofluid and the collector efficiency had better values with higher particle sizes. Also, the collector using an alumina-water nanofluid had higher efficiency than that using the water only. Jamal-Abad et al. [60] examined experimentally the effect of Cu-water nanofluid on the performance of a flat-plate solar collector. The results showed that the collector efficiency was higher when the concentration of nanoparticles was raised. They concluded that the efficiency of the collector at 0.05 wt% was approximately 24% more than that of the pure base fluids. Faizal et al. [61] analyzed the potential of the size reduction of the flat-plate solar collector by using MWCNT as an absorbing medium. The analysis was based on different mass flow rate, nanoparticles mass fraction and presence of surfactant in the fluid. For the same output temperature, it was observed that the collector's size can be reduced up to 37% of its original size when applying MWCNT nanofluid and as a result decreased the overall cost of the system. They computed the reduction in the size of collector's area from the following equation:

$$A_c = \frac{\dot{m}C_p(T_{out} - T_{in})}{I_T\eta} \tag{17.1}$$

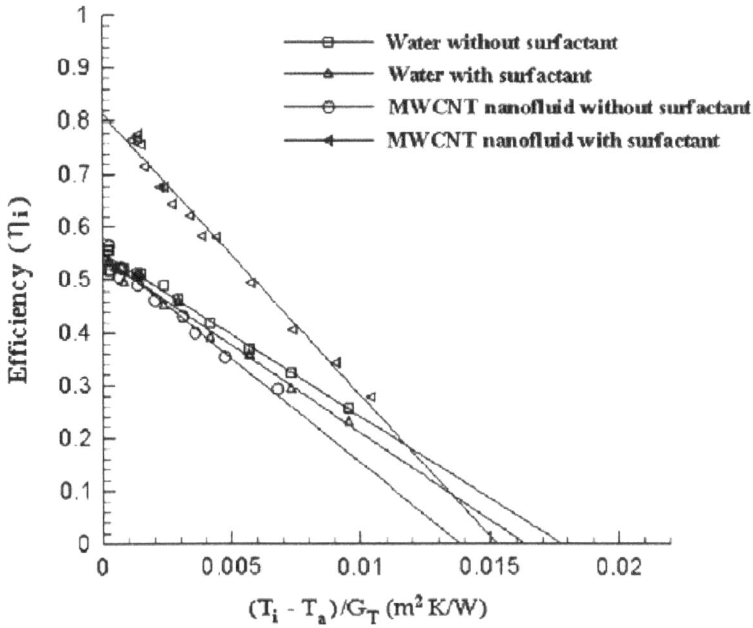

Figure 17.14 The efficiency of the flat-plate solar collector containing 0.2 wt% MWCNT nanofluid and pure water for both with and without surfactant cases [57].

Gangadevi et al. [62] analyzed experimentally the performance of a thermal solar flat-plate collector by using Al_2O_3-water nanofluid and used it to enhance the efficiency and decrease the electricity consumption of a solar assisted water heater. It was found that the efficiency of the solar collector was increased about 30% by using Al_2O_3-water nanofluid than that the water. Chaji et al. [63] fabricated and tested a small flat-plate solar collector filled with TiO_2-water nanofluid (Fig. 17.15). Three flow rates (36, 72 and 108 lit/m^{2}h) and four particles concentration ratios (0, 0.1, 0.2 and 0.3% wt) were investigated. It was found that the collector efficiency was improved between 2.6 and 7% by using the nanofluid compared to the base fluid. Tiwari et al. [64] analyzed theoretically the thermal performance of the flat-plate solar collector for water heating by using Al_2O_3/water nanofluid as an absorbing medium with different particle volume concentrations (0.5–2%). The effect of mass flow rate and particle volume fraction on the collector efficiency was investigated. The results showed that using 1.5% (optimum) particle volume fraction of nanofluid increased the thermal efficiency in comparison with water by about 31.64%. Faizal et al. [65] investigated numerically the energy, economic and environmental aspects of using metal oxides nanofluid in a flat-plate solar collector. It was estimated that 25.6%, 21.6%, 22.1% and 21.5% solar collector area reduction were achieved for using CuO, SiO_2, TiO_2 and Al_2O_3 water based nanofluid, respectively as shown in Fig. 17.16. Moreover, the average value of 220 MJ embodied energy was saved for each collector and around 170 kg less CO_2 emissions was reduced for the nanofluid solar collector compared to a conventional one. Said et al. [66] investigated experimentally the thermo-physical properties of ethylene glycol/water mixture and water based alumina nanofluids and its effect on a flat-plate solar collector. It was found that Al_2O_3-water nanofluid was more preferable against sedimentation and aggregation than Al_2O_3-ethylene glycol/water mixture nanofluid. They concluded that the pressure drop and pumping power of the nanofluid flows were very close to that of the base liquid for low volume concentration. Alim et al. [67] theoretically analyzed the entropy generation, heat transfer and pressure drop of (Al_2O_3, CuO, SiO_2 and TiO_2) nanoparticles suspended in water inside a flat-plate solar collector. These nanofluids had different nanoparticle volume fractions and volume flow rates in the range of 1–4% and –4 L/min, respectively. They concluded that, the CuO/water nanofluid was reduced the entropy generation by 4.34% and enhanced the heat transfer coefficient by 22.15% compared to the water as an absorbing fluid. Colangelo et al. [68] investigated experimentally the performance of the flat panel solar collectors [Fig. 17.17] using Al_2O_3-water nanofluid in order to avoid the effect of nanoparticles sedimentation. It was found that, the convective heat transfer coefficient was increased up to 25% at a concentration of 3% volume. The results showed that the main sedimentation parameter was the flow velocity and to better control it, a standard flat panel was modified by changing the cross section of the lower and top header of the panel in order to keep the fluid axial velocity constant. Shankar and Manivannan [69] developed a numerical model of the solar water heating system by using CuO-water nanofluid. The theoretical results showed that about (10.88%) of an improved efficiency was observed at (0.025%) volume fractions of the nanofluids. While the experimental results by using the water as a working fluid in the flat-plate collector type was found to be with a maximum temperature of about (68.2°C). Nasrin and Alim [70] investigated numerically the heat transfer by a water based nanofluid containing double nanoparticles (alumina and copper) and various nanofluids inside the riser pipe of a flat-plate solar collector (Fig. 17.18). The results showed that the better performance of heat loss through the riser pipe of the flat-plate solar collector was found by using the double nanoparticles (alumina and copper) than single nanoparticle (only alumina). They concluded that Cu-water nanofluid was more effective in order to promote heat loss system through the riser pipe of a flat-plate solar collector. They stated that the actual useful energy gain in a solar collector was computed from the following equation:

$$Q_{\mathrm{usfl}} = F_{\mathrm{R}} A(I(\lambda\kappa) - h(T_{\mathrm{m}} - T_{\mathrm{a}})) \tag{17.2}$$

Figure 17.15 The experimental flat-plate solar collector [63].

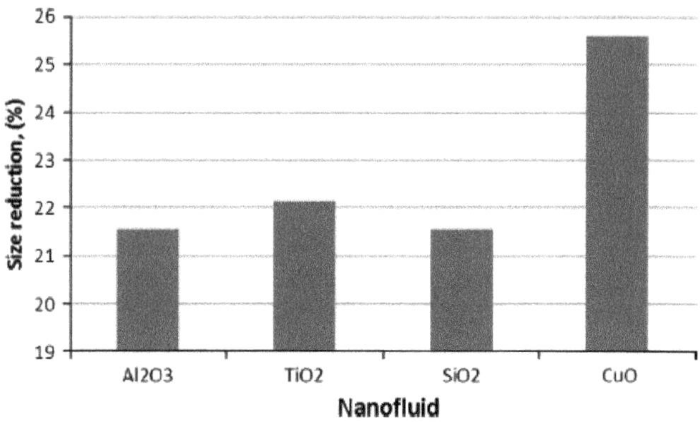

Figure 17.16 Percentage of size reduction for solar collector by applying different nanofluids [65].

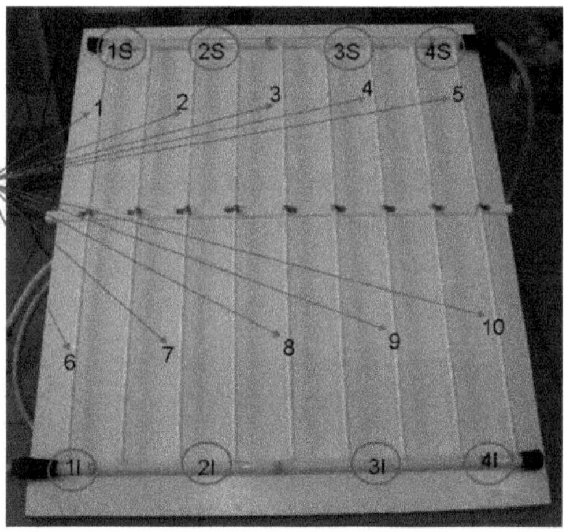

Figure 17.17 Flat panel solar thermal collector [68].

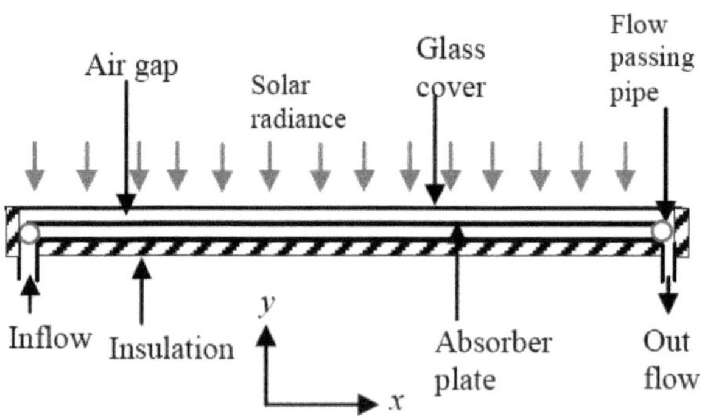

Figure 17.18 Schematic of flat-plate solar collector [70].

Polvongsri and Kiatsiriroat [71] investigated experimentally the water solar collector performance with a silver-water nanofluid. In their study, a 20 nm silver particles mixed with water at concentrations of 1000 and 10000 ppm were undertaken in a three small identical closed-loop flat-plate solar collectors, each with an area of (0.15 m × 1.0 m). It was found that the overall heat loss coefficient of the solar collector with nanofluid was reduced and more solar heat gain was obtained, especially with a high inlet temperature of the nanofluid. The results indicated also, that when the flow rate was different from the standard value, the solar thermal characteristics were also improved by using the nanofluid. Ekramian et al. [72] studied numerically the heat transfer performance of MWCNT-water, Al_2O_3-water and CuO-water nanofluids with mass percents of 1, 2 and 3 wt.% in a flat-plate solar collector. Effects of the inlet temperature, nanoparticle mass percent and mass flow rate on the heat transfer coefficients and thermal efficiency of pure water and nanofluids were studied. The results showed that the thermal efficiency of CuO-water nanofluid were greater than other nanofluids as shown in Fig. 17.19. Mahian et al. [73] performed an analytical study on the entropy generation and the heat transfer due to Al_2O_3/water nanofluid flow in a flat-plate solar collector as shown in Fig. 17.20. Four different particle sizes, including 25, 50, 75 and 100 nm and volume concentrations up to 4% were considered. Effects of the tube roughness, nanoparticle size and different thermophysical models on the Nusselt number, heat transfer coefficient, outlet temperature of the collector, entropy generation and Bejan number were investigated. It was found that, the entropy generation was decreased with increasing the nanofluid concentration. Nasrin and Alim [74] investigated numerically the forced convection by a water-alumina nanofluid through the horizontal flat-plate solar collector. Comprehensive average Nusselt number, average temperature, mean velocity, percentage of collector efficiency, mid-height temperature for both nanofluid and base fluid through the collector pipe were presented. The results showed that, when the Reynolds number increased, the percentage of collector efficiency was enhanced by using the nanofluid as shown in Fig. 17.21. They suggested a semi-empirical correlation to compute the average Nusselt number through the riser pipe of the collector which was given by

$$Nu_{av} = (0.9223 + 0.2327Pr)(Re)^{0.2120}, \tag{17.3}$$

when $4.2 \le Pr \le 10.2$ and $200 \le Re \le 1700$

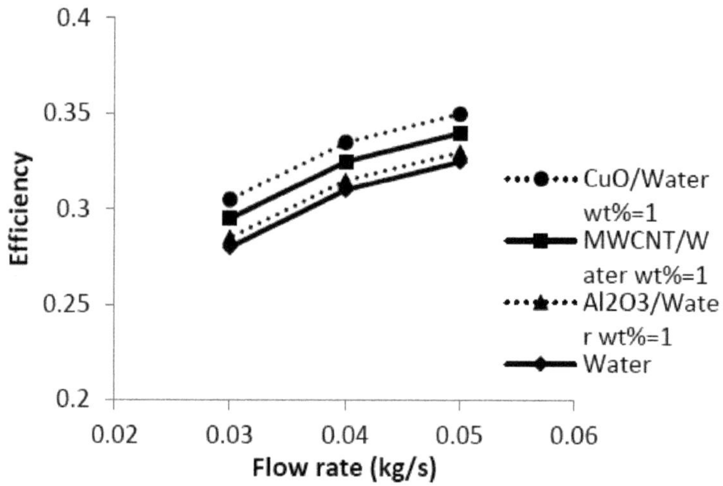

Figure 17.19 Comparison between thermal efficiencies of various nanofluids under different mass flow rates [72].

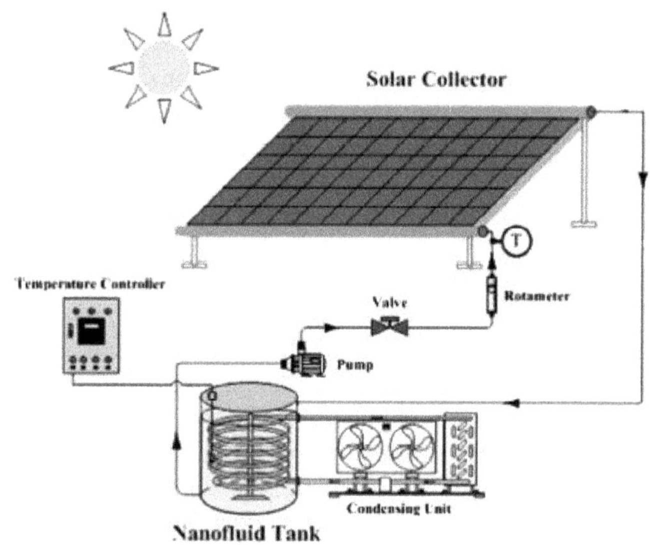

Figure 17.20 A schematic of the nanofluid cycle in the flat-plate solar collector [73].

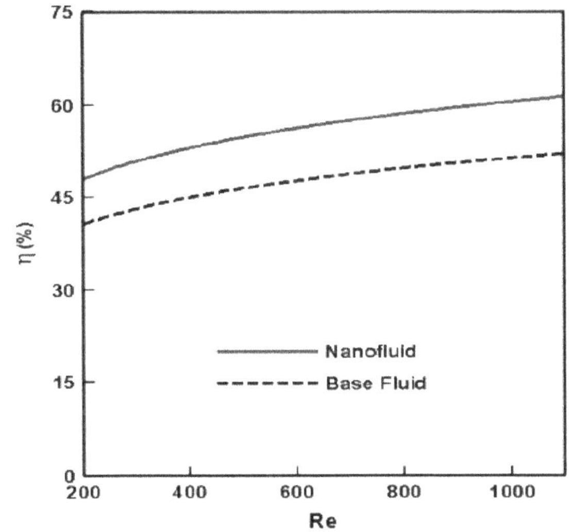

Figure 17.21 The variation of percentage of collector efficiency as a function of Reynolds numbers [74].

Kabeel and El-Said [75] investigated numerically the applicability of flashing desalination technique for small scale needs using a novel integrated system coupled with Cu-water nanofluid in a flat-plate solar collector. The thermal properties of the working fluid in the collector was improved by using different concentrations of Cu nanoparticles to determine the best volume fraction which gave a higher fresh water productivity. The results showed that the solar water heater collecting area was considered as a significant factor for reducing the water production cost. They concluded that, the volume fractions of nanoparticles had a significant effect on increasing the fresh water production and decreasing its cost. They computed the water production cost (C_{pro}) from the following relation assuming 365 working days:

$$C_{\mathrm{pro}} = \frac{\mathrm{TCO}}{f \times \mathrm{CA} \times 365} \tag{17.4}$$

Mahian et al. [76] performed a theoretical study to evaluate the performance of a minichannel-based flat-plate solar collector using four different nanofluids including Cu/water, Al_2O_3/water,

TiO_2/water, and SiO_2/water. The results were presented for volume fractions up to 4% and nanoparticle size of 25 nm where the inner diameter of the risers of the collector was assumed to be 2 mm. They concluded that the efficiency was decreased with increasing the volume fraction of nanoparticles and SiO_2/water nanofluid gave the highest efficiency of the collector as shown in Fig. 17.22. It was found also that the entropy generation was reduced by adding nanoparticles. They estimated the efficiency of the collector from the following equation:

$$\text{Efficiency (\%)} = \frac{Q_u}{AG_t} \times 100 \tag{17.5}$$

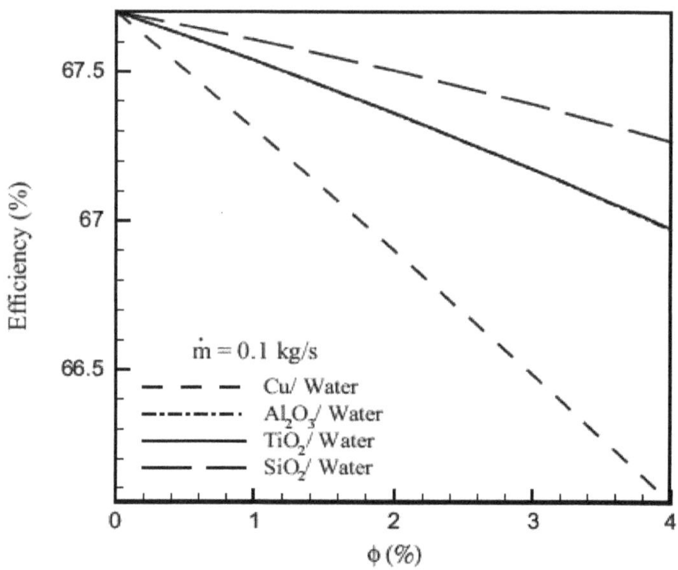

Figure 17.22 Variations of efficiency with volume fraction for different nanofluids [76].

Moghadam et al. [77] investigated experimentally the effect of CuO-water nanofluid as an absorbing medium on the performance and the efficiency of a flat-plate solar collector as shown in Fig. 17.23. The volume fraction of nanoparticles was taken as 0.4%, the mean particle dimension was kept constant at 40 nm, while the working fluid mass flow rate was varied from 1 to 3 kg/min. The experimental results explained that

Figure 17.23 The experimental flat-plate solar collector [77].

the nanofluid with mass flow rate of (1 kg/min) increased the collector efficiency by about 21.8%. Zamzamian et al. [78] performed an experimental study to investigate the effect of Cu-synthesized/ ethylene glycol nanofluid on the efficiency of a flat-plate solar collector (Fig. 17.24). The weight fractions of the tested nanoparticles with an average diameter of 10 nm, were taken as 0.2% and 0.3% of the nanofluid. The experiments were performed at different volume flow rates of the nanofluid ranging from 0.016 to 0.050 kg/s. It was found that by increasing the nanofluid concentration, the efficiency of the collector was improved. He et al. [79] investigated experimentally the effect of Cu-H_2O nanofluid on the efficiency of the flat-plate solar collector (Fig. 17.25). The experimental results showed that the efficiency of the collector was enhanced by 23.83% by using the nanofluid (25 nm, 0.1 wt%) as an absorbing medium. It was found also, that the efficiency of the collector was decreased with the nanoparticle size increasing. Said et al. [80] analyzed theoretically the entropy generation, heat transfer enhancement and the pressure drop for a flat-plate solar collector operated with four different nanoparticles suspended in the water (single-wall carbon nanotubes (SWCNTs), TiO_2, Al_2O_3 and SiO_2). It was observed that the SWCNTs nanofluid reduced the entropy

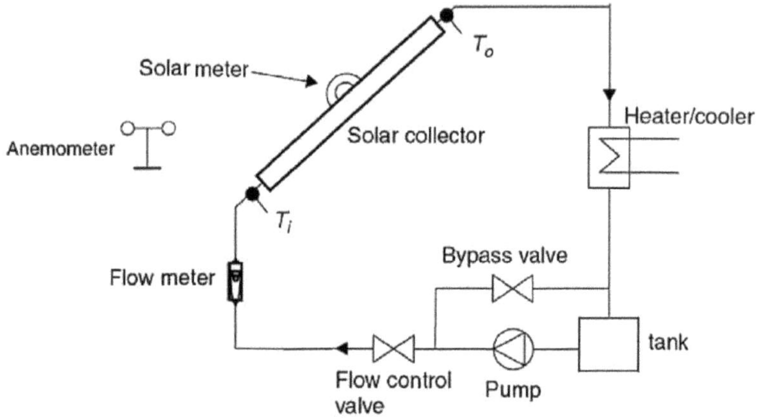

Figure 17.24 Schematic of closed loop test system for flat-plate solar collector [78].

Figure 17.25 The experimental flat-plate solar collector [79].

generation by 4.34% as shown in Fig. 17.26 and enhanced the heat transfer coefficient by about 15.33%. Mahian et al. [81] examined theoretically the effects of different shapes of boehmite alumina nanoparticles suspended in a mixture of water and ethylene glycol together with the tube materials on the performance of a flat-plate minichannel-based solar collector. Two different materials were considered for the solar collectors including copper and steel. It was found that the outlet temperature increased with an increase in the volume fraction of nanoparticles. They concluded that with an increase in the volume fraction of nanoparticles, the entropy generation rate due to the heat transfer was reduced for all shapes of nanoparticles. Roy et al. [82] investigated experimentally the heat transfer characteristics of silver/water nanofluid in a flat-plate solar collector (Fig. 17.27 and Fig. 17.28). The solar radiation heat flux varied between 800 W/m^2 and 1000 W/m^2, respectively, while the particle concentration varied between 0.01%, 0.03%, and 0.04%. The fluid Reynolds number varied from 5000 to 25000. The influence of radiation heat flux, mass flow rate of nanofluid, solar collector inlet temperature and volume concentration of the particle on the convective heat transfer coefficient and the collector efficiency were studied. They concluded that, the efficiency of the flat-plate collector was increased with increasing particle concentration and flow rate. The maximum efficiency of the system was found to be near 70% for 0.04% particle volume concentration at 6 L/min. They computed the efficiency of the flat-plate collector from the following equation:

$$\eta_{\mathrm{i}} = F_{\mathrm{R}}(\tau\alpha) - F_{\mathrm{R}}U_{\mathrm{L}}\frac{T_{\mathrm{fi}} - T_{\mathrm{a}}}{I_{\mathrm{T}}} \tag{17.6}$$

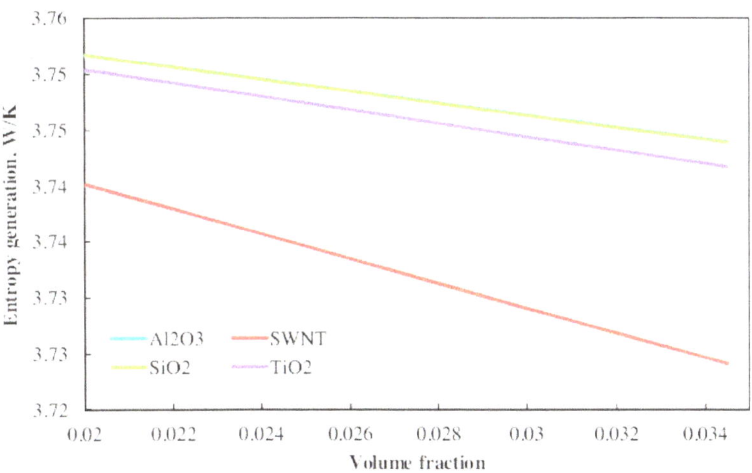

Figure 17.26 The variation of the entropy generation with volume fraction in the flat-plate solar collector [80].

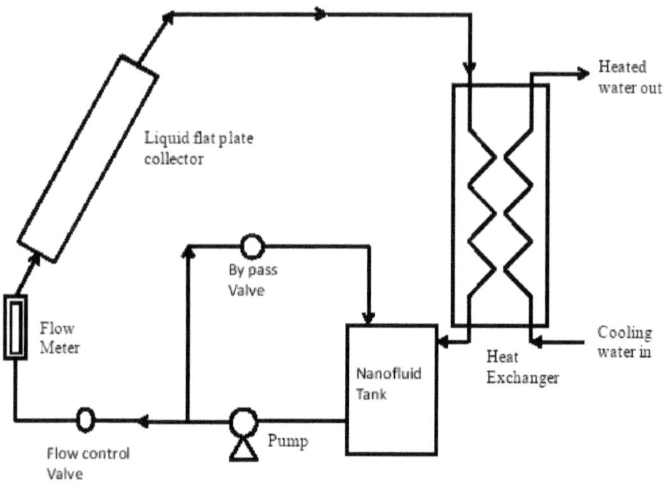

Figure 17.27 Schematic of nanofluid-based liquid flat-plate solar collector components used in [82].

Figure 17.28 Photographic view of the test section [82].

17.6 Applications of Nanofluid in the Direct Absorption Solar Collector

Tyagi et al. [83] investigated theoretically the possibility of using water and aluminum nanofluid as an absorbing medium for a low-temperature (<100°C) direct absorption solar collector. The results showed that the nanofluid increased the absorption of the incident radiation by more than nine times over that of the pure water. Moreover, the efficiency of the collector was found to be up to 10% higher than that of a flat-plate collector under similar operating conditions. Otanicar et al. [84] investigated both numerically and experimentally the performance of nanofluid-based direct absorption solar collector. Three different groups of nanofluids, with water as a base fluid, were considered which are graphite (30 nm diameter), carbon nanotube (6–20 nm diameter) and silver (20 and 40 nm diameters). They demonstrated an efficiency improvements of up to 5% in solar thermal collectors by utilizing nanofluids as an absorption mechanism. Taylor et al. [85] examined experimentally the effectiveness of various nanofluids in direct absorption solar collectors by testing their absorption of the solar spectrum. They observed that for materials used in their study, over 95% of incoming sunlight could be absorbed (in a nanofluid thickness ≥10 cm) with extremely low nanoparticle volume fractions less than 1×10^{-5}, or 10 parts per million. They concluded, that nanofluids could be used to absorb sunlight with a negligible amount of viscosity and/or density increase. Poinern et al. [86] investigated experimentally the photo-thermal response of nanoparticles of functionalized carbon nanospheres (CNS) for potential application in direct solar absorption collectors. The synthesized CNS were examined and characterized using field-emission scanning electron microscopy, transmission electron microscopy, X-ray diffraction spectroscopy, Raman spectroscopy, thermal gravimetric analysis and ultraviolet-visible analysis. The photo-thermal response of both nanofluids and films composed of CNS were investigated under 1000 W/m^2 solar irradiation. Saidur et al. [87] investigated theoretically the effect of aluminum-water nanofluid on the performance of the direct absorption solar collector. They concluded that the volume fraction of just 1.0% gave a satisfactory improvement to the solar absorption and as a result, the Al-water nanofluid was a good option for improving the performance of the direct absorption solar collector. Also, it was found that, the collector efficiency was increased slightly with an increase in the particle size as shown in Fig. 17.29. Moradi et al. [88] performed a CFD modeling of the direct absorption solar collector with nano-fluid used in civil

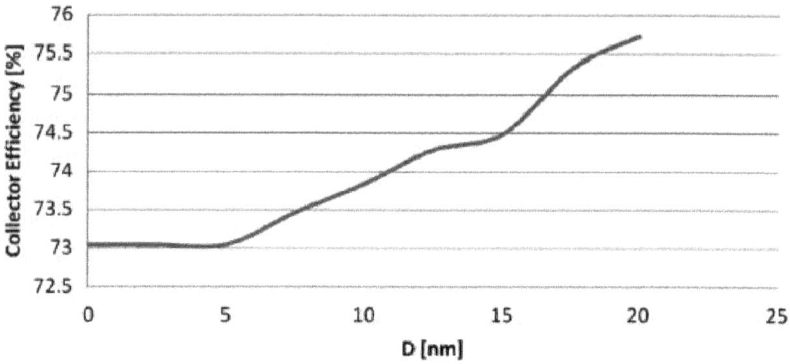

Figure 17.29 Collector efficiency as a function of the particle size (D) [87].

applications. Recent measurements of the optical properties of nano-fluids with different concentrations were used for the radiation heat transfer and the fluid dynamic modeling. They concluded as shown in Fig. 17.30 that the increasing concentration of nanoparticles initially increased the efficiency of the solar collector but, beyond a certain value of concentration, a further increase in the nanoparticle concentration decreased the efficiency due to the high surface temperature. Kundan and Sharma [89] performed an experimental study to improve the efficiency of the direct absorption solar collector by using a CuO-water based nanofluid in it. They concluded that the efficiency of the solar collector was increased by 4–6% compared to the conventional water-based solar collector. They mentioned that one of the main reasons of getting a high efficiency was the very small particle size, which enhanced the absorption capacity of nanofluids and improved the solar collector efficiency. Verma and Kundan [90] investigated experimentally the effect of Al_2O_3-H_2O based nanofluids as an absorbing medium in a direct absorption solar collector (Fig. 17.31). The volume fractions of Al_2O_3 nanoparticles used were 0.005% and 0.05%, respectively. Efficiency of the collector was calculated for different mass flow rates (60, 80 and 100 ml/h) of the nanofluid. It was found that the collector efficiency was increased about 3–5% when the nanofluid was used as compared to a simple water. This behavior was illustrated in Fig. 17.32. They concluded also, that the collector efficiency depended on the size, shape and the volume fraction of nanoparticles. Ladjevardi et al. [91] investigated

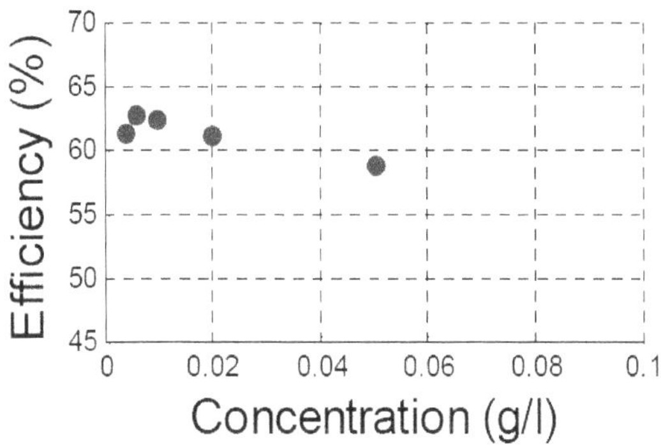

Figure 17.30 The effect of nanoparticle concentration on the efficiency of solar collector [88].

Figure 17.31 Direct absorption solar collector [90].

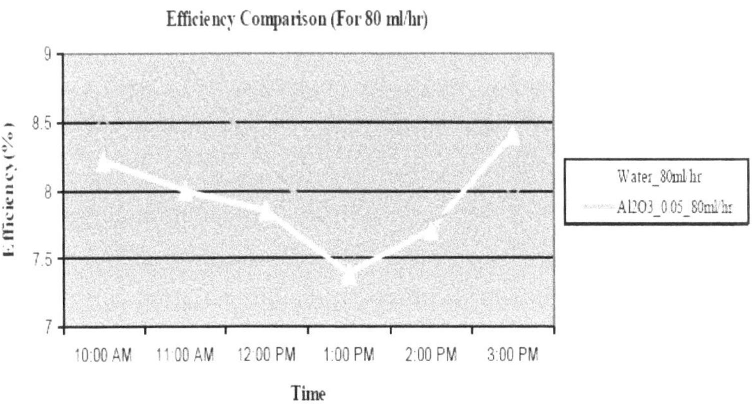

Figure 17.32 Variation of efficiency of nanofluid and water for (80 ml/h) at different instants of time [90].

numerically the effects of using graphite/water nanofluid in the improvement of solar radiation absorption efficiency in a volumetric solar collector (Fig. 17.33) to understand the appropriate values of nanoparticles volume fractions and diameters which provided a better efficiency and lowest cost. It was found that, by using nanofluid with a volume fraction around 0.000025%, it would be possible to absorb more than 50% of the incident irradiation energy, while pure water solar collector absorbed around 27% of the incident irradiation energy under the same conditions. Hector and Singh [92] investigated theoretically the development of a nano-heat transfer fluid carrying direct absorbing receiver system (DARS) for concentrating solar collectors. Graphene and aluminum nanosphere-based suspensions in Therminol VP-1 were simulated to identify the optimum thermo-geometric configuration of DARS. It was found that reducing DARS diameter was recommended to achieve higher mean nanofluid outlet temperatures. Zhidong et al. [93] investigated the thermal performance of a simulated direct absorbing solar collector with the application of magnetic nanofluid as a heat transfer media. It was found that the collector efficiency by using magnetic nanofluid was greater than that by using pure ethylene glycol. The higher efficiency could be obtained at a lower particle volume fraction. It was indicated that the use of both magnetic field and magnetic nanofluids enhanced the heat transfer efficiency of the collector. Lee et al. [94] experimentally measured, the extinction coefficient of water-based nanofluids containing MWCNTs. With the obtained extinction coefficient, the efficiency of a flat-plate direct-absorption solar collector (DASC) was theoretically estimated. The results showed that the DASC concept can further improved the efficiency of the conventional flat-plate type solar collectors. Luo et al. [95] studied both numerically and experimentally the performance of a nanofluid direct absorption solar collector. Nanoparticles such as TiO_2, Al_2O_3, Ag, Cu and SiO_2, as well as graphite and carbon nanotubes were added directly into Texatherm oil to prepare a stable suspension colloids. They concluded that, nanofluids improved the outlet temperature by 30–100 K and the efficiency by 2–25% than the base fluid. It was found also that nanofluids, even of low-content, had a good absorption of the solar radiation. Moreover, it was indicated that the photo-thermal efficiency was decreased with the increase of the incident radiation as shown in Fig. 17.34. Filho et al. [96] investigated experimentally the photo-thermal conversion characteristics of silver-de-ionized water nanofluids. The results showed that silver nanoparticles had an excellent photo-thermal conversion capability even under very low concentrations. Also, it was found that the stored thermal energy was increased by 52%, 93% and 144% for silver particle concentration of 1.62, 3.25 and 6.5 ppm, respectively at the peak temperature. They characterized the photo-thermal conversion efficiency of nanoparticles by the specific absorption rate (SAR), which described the particle's capability in the absorbing energy per unit mass and was given by

$$\mathrm{SAR} = \frac{(m_{\mathrm{w}}c_{\mathrm{w}} + m_{\mathrm{p}}c_{\mathrm{p}})\Delta T_{\mathrm{n}} - m_{\mathrm{w}}c_{\mathrm{w}}\Delta T_{\mathrm{w}}}{1000 m_{\mathrm{p}}\Delta t}. \tag{17.7}$$

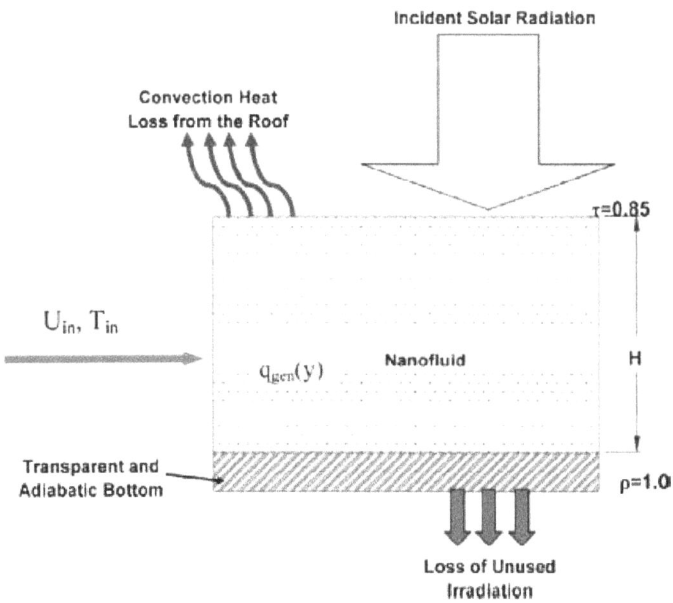

Figure 17.33 The volumetric solar collector [91].

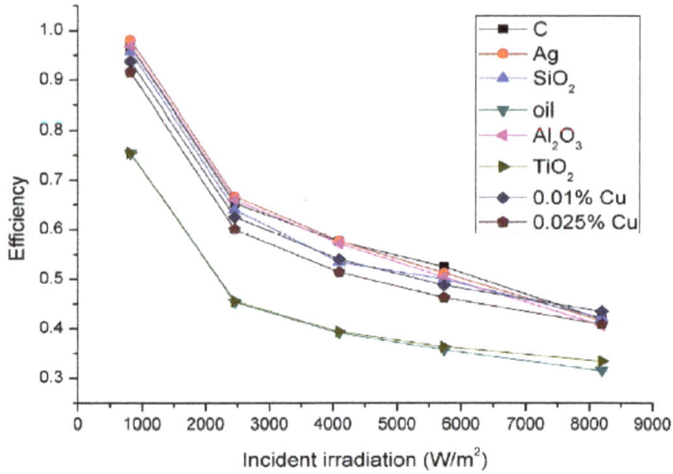

Figure 17.34 Photo-thermal efficiency against the incident radiation intensity of the direct absorption solar collector [95].

Parvin et al. [97] investigated numerically the heat transfer performance and the entropy generation of forced convection through a direct absorption solar collector with Cu-water and Ag-water nanofluids (Fig. 17.35). The effects of solid volume fraction of nanoparticles and Reynolds number on the mean Nusselt number, mean entropy generation, Bejan number and the collector efficiency were studied. The results showed that the collector efficiency enhanced about two times with increasing Reynolds number and solid volume fraction. They suggested a correlation to compute the collector efficiency which was given by

$$\eta = (2.488 + 0.327\phi)(\mathrm{Re})^{0.4684}, \quad (17.8)$$

when $0\% \le \phi \le 3\%$ and $200 \le \mathrm{Re} \le 1000$.

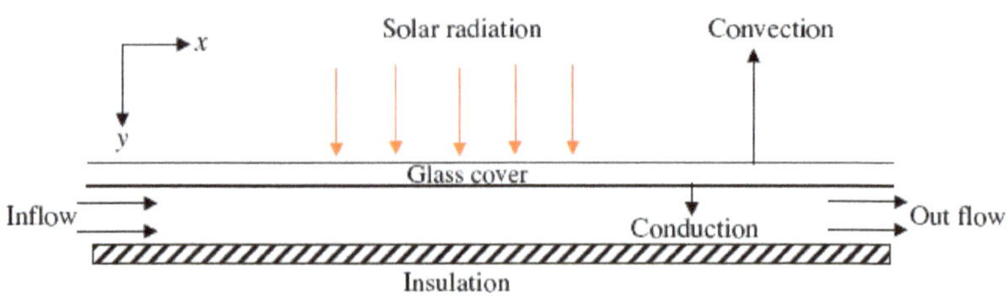

Figure 17.35 Schematic of direct absorption solar collector [97].

Hordy et al. [98] quantitatively examined both the long-term and high-temperature stability of MWCNTs nanofluids for use in the direct absorption solar collector. The optical properties of four base fluids (water, ethylene glycol, propylene glycol and Therminol VP-1 [mixture of biphenyl and diphenyl oxide]) were characterized with a range of concentrations of corresponding nanofluids. Optical characterization of nanofluids demonstrated that MWCNTs were close to 100% solar energy absorption, even at low concentrations and small collection volumes which made them as an ideal candidate in the direct absorption solar collectors. Karami et al. [99] examined experimentally the dispersion stability, optical properties and the thermal conductivity of CNTs suspension in water as a nanofluid for application in low-temperature direct absorption solar collector. They demonstrated that the thermal conductivity improvements was reached to 32% by adding only 150 ppm of CNTs to water as an absorbing medium. They concluded that this kind of nanofluids was highly recommended for increasing the overall efficiency of the direct absorption solar collectors. Zhang et al. [100] investigated both experimentally and theoretically the radiative properties of ionic liquid-based nanofluids for medium-to-high-temperature direct absorption solar collectors. Three different types of nanoparticles with an average sizes of 40 nm were considered (carbon-coated Ni (Ni/C), Ni and Cu). It was found that the optical absorption property of the ionic liquid was greatly enhanced by adding a low volume fraction of nanoparticles in it. They concluded that the excellent radiative properties of the ionic liquid-based nanofluids made them a good option to be used as an absorber for direct absorption solar collectors. Sadique and Verma [101] performed an experimental study on the effect of the nanofluid on the performance of a direct solar thermal collector. Three different groups of nanofluids with water were considered [Graphite sphere-based 30 nm diameter, carbon nanotube-based 6–20 nm diameter and silver sphere-based 20 and 40 nm diameters]. They concluded that nanofluids could be used to absorb sunlight with a negligible amount of viscosity and/or density increase. Very recently, Moradi et al. [102] investigated numerically the utilization of carbon-nanohorn based nanofluids for a direct absorption solar collector used in the civil applications. In their work, a three-dimensional model of the absorption phenomena in nanofluids within a cylindrical tube was coupled with a CFD analysis of the flow and temperature fields. They computed also the heat losses due to the conduction, convection and radiation at the boundaries.

17.7 Applications of Nanofluid in the Parabolic Trough Solar Collector

Khullar and Tyagi [103] analyzed numerically the heat transfer and fluid flow characteristics of the linear parabolic solar collectors. They added aluminum nanoparticles into the base fluid (water) to improve its absorption characteristics. The effect of various parameters such as concentration ratio, receiver length, fluid velocity and volume fraction of nanoparticles were studied. They concluded that nanofluid based collector was better than the conventional collector under similar working conditions. Khullar et al. [104] analyzed numerically the idea of harvesting solar radiant energy by using nanofluid-based concentrating parabolic solar collectors. It was observed that nanofluid collector had about 5–10% higher efficiency as compared to the conventional parabolic solar collector. They concluded also that nanofluid collector had the potential to harness the solar radiant energy more efficiently than a conventional one. De Risi et al. [105] investigated numerically the modeling and optimization of transparent parabolic trough solar collector working with gas-based nanofluid. It was found, that the transparent receivers combined with nanofluids were able to directly adsorb solar radiation due to the very high total surface of nanoparticles. They concluded that the maximum thermal efficiency was about 62.5%, for a nanofluid outlet temperature of 650°C and a nanoparticles volume concentration of 0.3%. Sunil et al. [106] performed an experimental study to investigate the performance of a parabolic solar collector using SiO_2-H_2O based nanofluid (Fig. 17.36). Volumetric concentration of 0.01% and 0.05% were used in the experiment to prepare the nanofluid. Different volume flow rates employed in the experiment which varied as 20, 40 and 60 L/h, respectively. Magnetic stirrer with a hot plate system was used to mix the nanoparticles in water

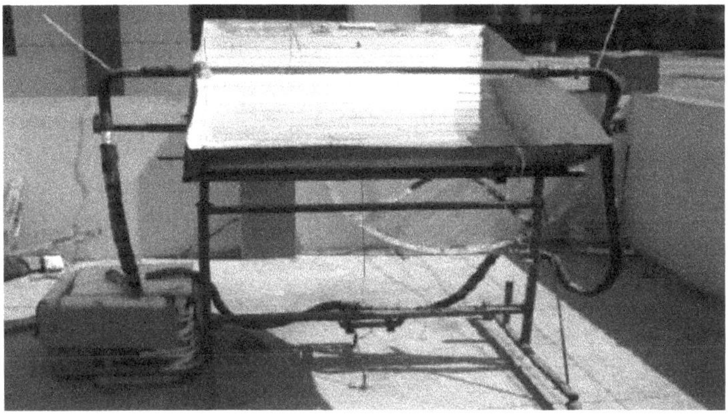

Figure 17.36 SCHEMATIC of parabolic solar collector [106].

before sonication. It was found that SiO_2-H_2O based nanofluid had comparatively higher efficiency at higher volume flow rates. Ghasemi and Ahangar [107] studied numerically the effect of Cu-water nanofluid, as a heat transfer fluid, on the performance of a solar parabolic trough collector. The temperature field, thermal efficiency, mean-outlet temperatures were evaluated and compared with the conventional parabolic collectors and nanofluid based collectors. Also, the effect of various parameters such as fluid velocity, volume fraction of nanoparticles, concentration ratio and the receiver length were investigated. It was found that the thermal efficiency of the collector was decreased with the increase in the receiver length as shown in Fig. 17.37. The results explained also that, by increasing the volume fraction of nanoparticles, the performance of the collector was enhanced. Sokhansefat et al. [108] investigated numerically the heat transfer enhancement in a parabolic trough collector tube by using Al_2O_3/synthetic oil nanofluid with a non-uniform heat flux. The effect of Al_2O_3 particle concentration in the synthetic oil on the rate of heat transfer from the absorber tube was also investigated. It was found that the heat transfer coefficient of the working fluid in an absorber tube was enhanced with the presence of nanoparticles. The results explained also that the thermal conductivity was remarkably improved by using a nanofluid as shown in Fig. 17.38. Very recently, Kasaeian et al. [109] investigated experimentally the performance of a solar parabolic trough collector (Fig. 17.39). The multi-wall carbon nanotube

(MWCNT)/oil based nanofluids with 0.2% and 0.3% were prepared as a working fluid. The results showed that, the global efficiency of a parabolic collector was enhanced about 4–5% at 0.2% and 5–7% at 0.3% when MWCNT/mineral oil nanofluid was used instead of pure oil.

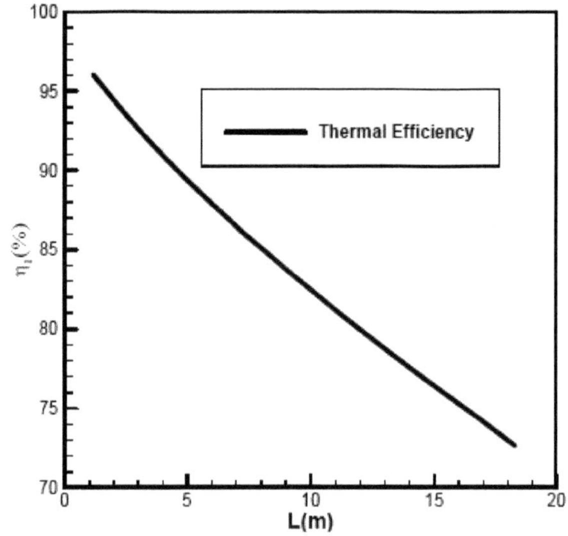

Figure 17.37 Thermal efficiency variation with receiver length [107].

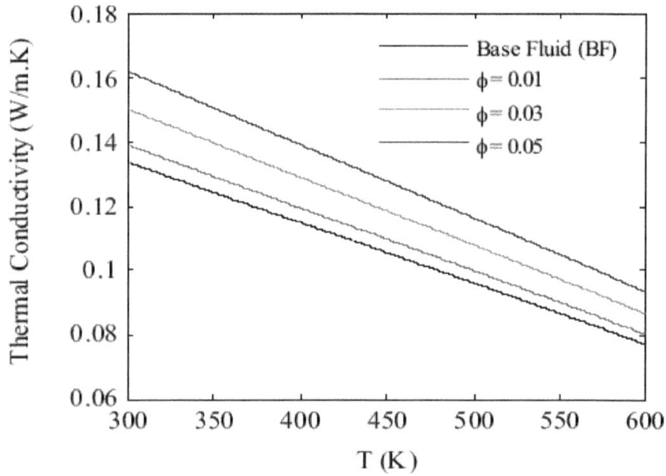

Figure 17.38 The variation of thermal conductivity with the operational temperature [108].

Figure 17.39 Schematic of solar parabolic trough collector [109].

17.8 Applications of Nanofluid in the Wavy Solar Collector

Nasrin and Alim [110] investigated numerically the performance of Ag-water and CuO–water nanofluids on the heat transfer in a solar collector with a flat-plate cover and a sinusoidal wavy absorber (Fig. 17.40). The behavior of both nanofluids related to performance such as temperature and velocity distributions, radiative and convective heat transfers, mean temperature and velocity of the nanofluid was investigated systematically. The results showed that the better performance of the heat transfer inside the collector was found by using the highest solid volume fraction of Ag-water nanofluid. Alaeian et al. [111] investigated experimentally the indirect absorption of the solar energy by MWCNT-oil nanofluid with mass fraction concentration of 0.1%, 0.2% and 0.4% in wavy tubes for use in solar collectors. Different configurations of U-bend wavy copper tubes were investigated based on the dimensionless curvature radius and the number of sequences of U-bends in the laminar flow regime. It was observed that, the heat transfer coefficient was increased up to 56% compared to the base fluid at the highest concentration and the lowest dimensionless curvature ratio.

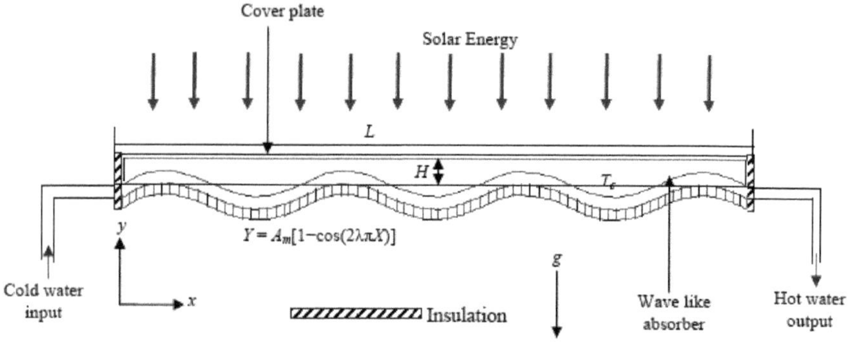

Figure 17.40 Schematic of wavy solar collector [110].

17.9 Applications of Nanofluid in the Heat Pipe Solar Collector

Lu et al. [112] investigated experimentally the thermal performance of an open thermosyphon using both deionized water and CuO-water nanofluid for high-temperature evacuated tubular solar collectors (Fig. 17.41). Experimental results showed that the nanofluid significantly enhanced the thermal performance of the evaporator and evaporating heat transfer coefficients were increased by about 30% compared with those of the deionized water. It was found also, that the wall temperatures of the open thermosyphon using nanofluids decreased after substituting CuO nanoparticles into the water. Moorthy et al. [113] investigated experimentally the efficiency of the evacuated tube solar collector using water-based titanium oxide (TiO_2) nanofluid for conversion of the solar thermal energy. They concluded that the efficiency of the collector using TiO_2 nanofluid with 0.3% concentration was about 73%, compared to that using pure water which was about 58%. Chougule et al. [114] examined experimentally the performance of two identical flat-plate collectors. In each collector, three identical wickless copper heat pipes were used. The working fluid used in one collector was a pure water and in the another a pure water with carbon nanotubes (CNTs). Both collectors were tested using the solar tracking system. They concluded that the nanofluid collector gave a better performance in the all test conditions. Senthil Kumar et al. [115] performed an

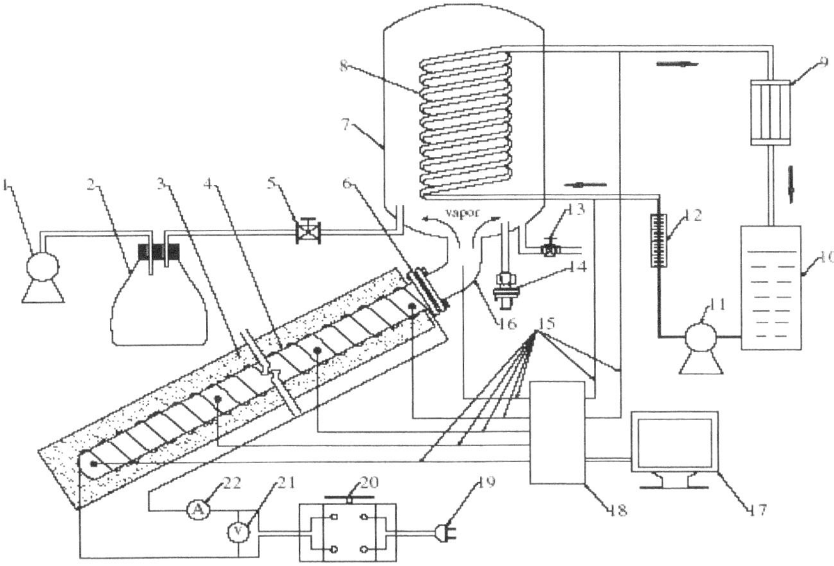

Figure 17.41 Schematic of experimental apparatus. (1) Vacuum pump (2) regulators box (3) thermal insulator (4) evaporator tube (5) vacuum valve (6) flange plate (7) condenser box (8) condensing coil (9) heat exchanger (10) water tank (11) pump (12) rotameter (13) water valve (14) relief valve (15) thermocouples (16) elbow tube (17) computer (18) data acquisition system (19) DC power supply (20) transformer (21) voltmeter and (22) ammeter [112].

experimental analysis of nano fluid-charged wickless solar heat pipe collector by using the solar tracking system. In their work, two identical experimental set up of flat-plate collectors with same dimensions, using heat pipes were fabricated. In each set up, three identical wickless copper heat pipes were used which having a length of (620 mm) and outer diameter of (18 mm). The working fluid used in one set up was the pure water and in the another was pure water with CNT nanoparticles. They concluded that the nanofluid collector gave the better performance in all the operating conditions. Chougule et al. [116] investigated experimentally the thermal performance of a solar wickless heat pipe collector at the outdoor test condition for pure water and carbon nanotubes (CNT) with various concentration (0.15%, 0.45%, 0.60% and 1% by volume) and various tilt angles (20°, 32°, 40°, 50° and 60°). They obtained the optimal value of

CNT nanofluid concentration for better performance of the heat pipe solar collector. Liu et al. [117] designed experimentally an evacuated tubular solar air collector integrated with simplified CPC (compound parabolic concentrator) and an open thermosyphon (Fig. 17.42) using CuO-water nanofluid. The thermal performance of an open thermosyphon using nanofluid for this collector was evaluated. Experimental results showed that both the air outlet temperature and the collector efficiency using the nanofluid are higher than that using water. They computed the system collecting efficiency by using the following equation:

$$\eta_{\text{net}} = \frac{P_{\text{net}}}{nQ_{\text{cpc}}} \tag{17.9}$$

Figure 17.42 Photograph of the evacuated tubular solar air collector [117].

Saravanan and Karunakaran [118] investigated experimentally the performance of V-type absorber plate solar collector with a heat pipe by using different working fluids such as nanofluid (TiO_2 + DI water), methanol, ethanol and DI water. The results showed that there was a significant increase in the thermal efficiency of the collector by using the nanofluid. They concluded that the solar collector using V-type absorber plate with the heat pipe was recommended for domestic solar water heater applications due to its low weight, low cost and the long life.

Aruna et al. [119] investigated experimentally the performance of the heat pipe solar collector by using two different working fluids (TiO_2 + DI water nanofluid and propanol) with a constant concentration of nanoparticle and size as 80 ml/lit and 40 nm, respectively. It was found that the nanofluid gave a better performance when compared to the propanol.

17.10 Applications of Nanofluid in the Other Solar Collectors

Li et al. [120] investigated experimentally the heat transfer performance of the tubular solar collector containing Al_2O_3, ZnO and MgO nanoparticles with distilled water as a base fluid. The experimental results showed that the heat transfer efficiencies of all types of nanofluids were increased in comparison to a distilled water. They concluded that according to the low viscosity and the excellent heat transfer performance, the ZnO nanofluid with 0.2% vol. concentration was a good option in the solar energy utilization. Taylor et al. [121] investigated experimentally the applicability of using nanofluids in high flux solar collectors. In their experiments, they used the aluminum, copper, graphite and silver nanoparticles together with the Therminol VP-1 as a base fluid. Figure 17.43 shows images of the reflective dish and the receiver with instrumentation used in their experiments. It was found that the efficiency improvement was on the order of 5–10% by using a nanofluid receiver. They showed also, that graphite nanofluids with volume fractions on the order of 0.001% or less were suitable for 10–100 MW power plants. Gangadevi et al. [122] investigated experimentally the performance of the hybrid solar system (PVT) which consisted from a flat-plate solar collector attached to a solar photovoltaic cell by using Al_2O_3-water nanofluid. The results showed that both the electrical and thermal efficiencies of a hybrid solar system was increased considerably by using the nanofluid as a working fluid. Paul et al. [123] and Paul [124] investigated experimentally the thermal performance of ionic liquid-Al_2O_3 nanofluid as a heat transfer fluid in the solar collector. Viscosity, heat capacity and the thermal conductivity were measured and compared with existing theoretical models for liquid-solid suspensions.

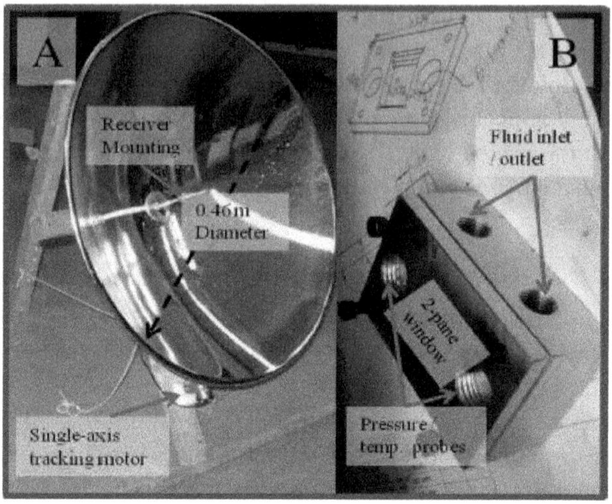

Figure 17.43 Images of the reflective dish and receiver with instrumentation used by [121].

It was found that the heat transfer performance was enhanced by about 15% by using ionic liquid-Al_2O_3 nanofluid. Hewakuruppu et al. [125] investigated numerically the optical properties of nanofluid-based direct absorption solar collector components (Fig. 17.44), when they were optimized for selective absorption. It was found that a nanofluid with a short wavelength, optical depth of 3 with no scattering, and a long wavelength with optical depth of zero were considered for the optimum nanofluid. A cover of the solar collector which perfectly transmitted the short wavelength radiation and perfectly reflected long wavelength radiation was also required. Rahman et al. [126] studied numerically the heat transfer augmentation in a triangular solar collector with a corrugated bottom wall (Fig. 17.45) by utilizing water based nanofluids. They considered three types of nanoparticles (Cu, Al_2O_3 and TiO_2). It was found that Cu-water nanofluid performed a better performance from the heat transfer point of view than other used nanofluids. Also, they concluded that the heat transfer was increased up to 24.28% from the heated surface as the volume fraction was increased from 0% to 10%. Goudarzi et al. [127] investigated experimentally the effect of CuO-H_2O nanofluid and distilled water on the efficiency of a cylindrical solar collector with receiver helical pipe (Fig. 17.46). In their experiments, the mass flow rate of fluid was changed from 0.0083 to 0.033 kg/s while the weight fraction of nanoparticles were varied as 0.1%, 0.2% and 0.4%, respectively. Results showed that the collector efficiency was enhanced by using nanofluid compared with the pure water. They observed also, that the maximum thermal efficiency was increased by about 25.6% by using 0.1 wt% nanofluid in 0.0083 kg/s mass flow rate of the fluid. They evaluated the performance of the collector by computing the instantaneous efficiency which was given by

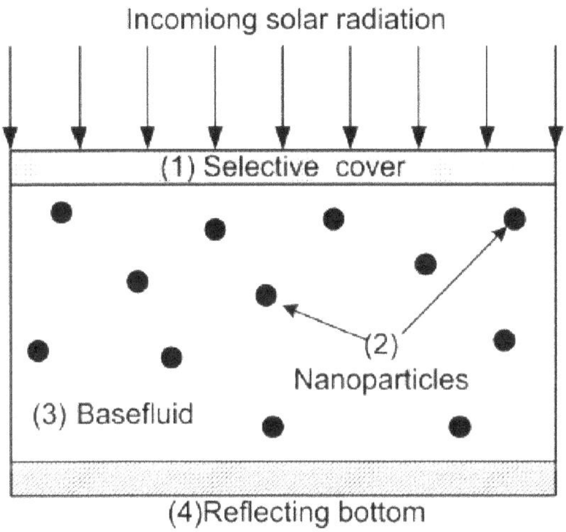

Figure 17.44 Schematic of nanofluid-based direct absorption solar collector components used in [125].

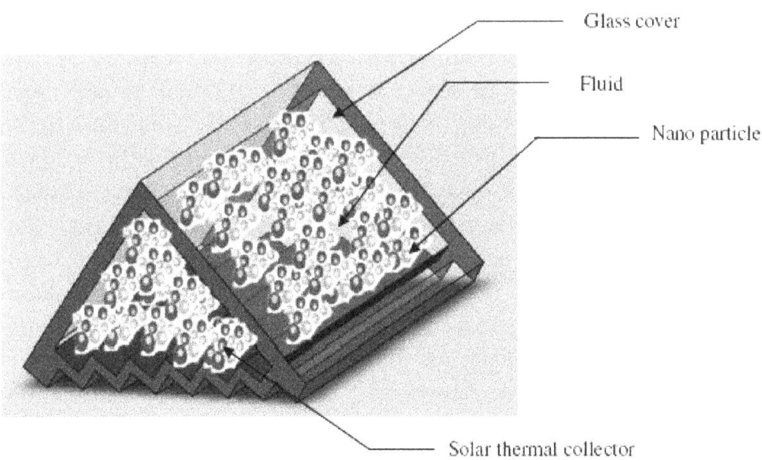

Figure 17.45 3D view of a triangular solar collector [126].

$$\eta_i = \frac{Q_u}{AI_b}. \tag{17.10}$$

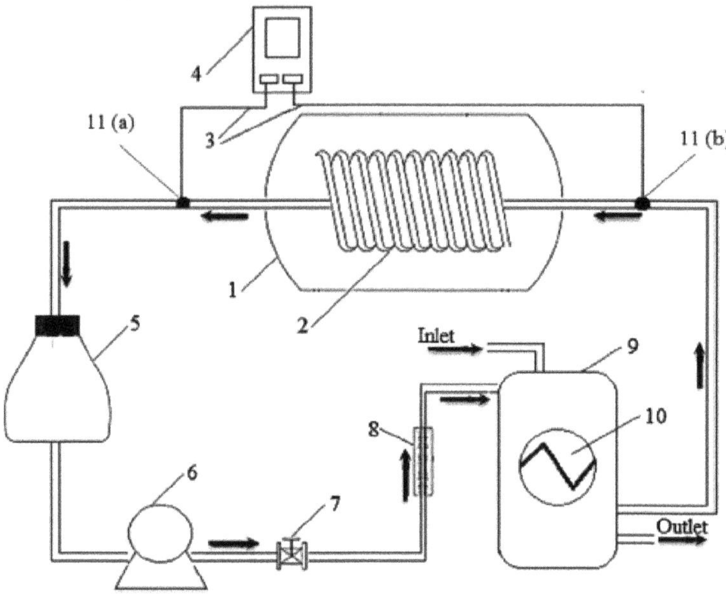

Figure 17.46 Schematic of the cylindrical solar collector (1) cylindrical glass (2) copper coil (3) thermocouple wires (4) data logger (5) water supply and drainer (6) pump (7) line valve (8) rotameter (9) reservoir tank (10) heat exchanger (11a and 11b) thermocouples [127].

Rahman et al. [128] studied numerically the effect of solid volume fraction and tilt angle in a quarter circular solar thermal collectors filled with CNT-water nanofluid. A wide range of solid volume fraction (0 to 0.12) and tilt angle (0 to 60°) was investigated for Rayleigh number ($Ra = 10^5–10^8$) and with various dimensionless times. It was found that both solid volume fraction and tilt angle played vital roles for the augmentation of heat transfer in the collector. Also, they concluded that the effect of inclination angle was more severe than the effect of solid volume fraction. Very recently, Goudarzi et al. [129] investigated experimentally the effect of pH variation of two nanofluids ($CuO-H_2O$ and $Al_2O_3-H_2O$) on the efficiency of a cylindrical solar collector. The collector consisted of a cylindrical glass tube with a helical pipe as a solar energy receiver as shown in Fig. 17.47. Their experiments were performed using 0.1 wt.%

CuO and 0.2 wt.% Al_2O_3 with various pH values. The results showed that the thermal efficiency of the solar collector was increased when there was more differences between pH of nanofluid and pH of iso-electric point. This increasing in the efficiency was estimated, respectively by about 52% for CuO nanofluid and 64.5% for Al_2O_3 nanofluid.

Figure 17.47 Cylindrical solar collector and the experimental setup [129].

17.11 Review Papers Related to Application of Nanotechnology in a Solar Collector

Sruthi [130] presented a review about the application of solar panels in water heater system and the use of nanotechnology in solar water heater (SWH). They presented various advantages of using nanomaterials in the solar water heater such as

(1) Easy and safe to use.
(2) Wide array of commercial applications due to its wireless capabilities.
(3) Energy conserving and save money out of electricity.
(4) Excellent thermal insulation in a thin layer system which does not increase the weight.

(5) No moisture infiltration, which gives the insulation the benefits of corrosion resistance.
(6) Easy installation and the high efficiency.

Khanafer and Vafai [131] presented an overview of the use of nanomaterials in solar energy and desalination sectors. They reviewed the most advances of the nanotechnology in thermal energy storage systems, photovoltaic systems, direct absorption solar collectors and solar desalination. They mentioned that, for an optimum performance of a solar collector, the solar radiation should be absorbed within a small wave length range (0.25 mm < l < 2.5 mm) and converted directly to a heat inside the working fluid to minimize the heat losses and the effect of both fouling and pumping cost. Javadi et al. [132] presented an overview of studies related with the performance of solar collector, especially the direct absorption solar collector by using nanofluid as a working fluid. They concluded that further research must focus on the two-phase analysis of nanofluids in order to find more accurate relationships between properties of nanoparticles and the nanofluid. Moreover, they indicated that there was a lack of study on the effect of nanofluid's optical properties such as transmittance and extinction coefficient on the performance of the solar collector. Mahian et al. [133] presented a review of the applications of nanofluids in solar thermal engineering. The papers related with the applications of nanofluids in solar collectors, solar water heaters, photovoltaic/thermal systems, solar ponds and solar thermoelectric cells were reviewed. They indicated that the experimental and numerical studies for solar collectors showed that in some cases, the efficiency could increased remarkably by using nanofluids. Moreover, the challenges of using nanofluids in solar energy devices were discussed. It was suggested that the nanofluids in different volume fractions should be tested to find the optimum volume fraction. They presented various suggestions such as

(1) It was worth to carry out an experimental work on the effect of particle size on the collector efficiency.
(2) From the economic and environmental point of view, the using of nanofluids in collectors led to a reduction in CO_2 emissions, annual electricity and fuel savings.

Al-Shamani et al. [134] presented an overview of the effects of nanofluids on the performance of cooling solar collectors from the considerations of efficiency and environmental benefits. Also, they presented an overview of the research, performance and development of photovoltaic/thermal (PV/T) collector systems. An efficiency fact was introduced in their paper to provide a general understanding for designers and researchers. They presented various conclusions such as

(1) The thermal conductivity enhancement of nanofluids depended on the volume fraction, size, type of nanoparticles and the base fluid.
(2) Suspended nanoparticles remarkably increased the forced convection heat transfer performance of the base fluid.
(3) The sheet and tube collector was highly efficient and less expensive in practical application of the water-based PV/T, such as building-integrated systems.
(4) Nanofluids could be used to cool photovoltaic/thermal (PV/T) collector systems.

Chaudhari and Walke [135] reviewed some studies related with the application of nanofluids in solar thermal engineering systems such as solar collectors, solar cells, thermal energy storage and solar stills. They suggested that, it was important to carry out an experimental work on the effect of particle size on the collector efficiency. They concluded that the volumetric absorption of nanofluid in solar collectors reduced the thermal resistance at interfaces and minimized temperature difference between the absorber and the heat transfer fluid and as a result the efficiency was increased. Nerella et al. [136] explained in their review paper that the efficiency of the solar collectors was limited by the absorption properties of the working fluid. The improvements of the efficiency in solar collectors by utilizing nanofluids with base fluid were highly comparable with the efficiency of solar collector when only base fluid was used. They indicated that the experimental and numerical results demonstrated an initial rapid increase in the solar collectors efficiency with volume fraction, followed by a leveling off in the efficiency as the volume fraction further increased. Very recently, Kasaeian et al. [137] reviewed the applications of nanofluids on different types of solar collectors, photovoltaic systems, solar cells, energy storage system and solar thermoelectrics devices. They concluded that the volumetric absorption of nanofluid in solar collectors reduced the thermal resistance at interfaces and minimized the temperature difference between the absorber and the heat transfer fluid and as a result the efficiency was increased. They suggested that the development of the particle production and decreasing in costs were essential for the nanofluid research.

17.12 Challenges and Difficulties

The application of the nanofluid in the solar collectors suffers from many problems which can be summarized as follows:

(1) The nanofluid requires a long time in order to be stable with base fluids.
(2) The specific heat of the nanofluid is low in comparison with the base fluid. Many studies [138] states that the optimum heat transfer process requires that the working fluid have a high specific heat in order to exchange more heat.
(3) The toxicity of the nanofluid is high; so one needs to be careful during its preparation.
(4) The preparation and testing of the nanofluid is highly costly.
(5) The boiling characteristics of the nanofluid are low. Therefore, when the concentration of nanoparticles increases, it leads to increase the surface temperature of nanofluid and make a severe overheating.
(6) The high viscosity of the nanofluid leads to the increase of the pressure drop and the required power for pumping is increased also.
(7) The presence of nanoparticles in the nanofluid may leads to a corrosion and erosion of solar collector for a long time.
(8) The utilizing of nanofluids suffers from some technical problems such as the nanoparticles sedimentation.

17.13 Summary Remarks and Outlook

The present chapter gives a comprehensive overview about the recent advances related with the application of the nanotechnology in various kinds of the solar collectors. The results presented in this study provide a very useful source of references for enhancing the solar collector performance by using the nanofluid technology. Some important conclusions are summarized below:

(1) The future research must be directed towards inventing efficient energy transport methods of nanofluid in solar collectors such as the enhancement of the heat transfer rate by studying the effect of particle shape on the thermal conductivity of nanofluid.
(2) The future research must be directed towards inventing a non-toxic and low-cost nanoparticle to reduce further the cost of nanofluid based solar collector and to meet quickly with the market needs.
(3) More research works are needed to study the effect of nanotechnology on both heat pipe and PVT solar collectors, since the number of published papers related to these types are still limited compared with the corresponding papers related with the another types of the solar collector.
(4) It is very useful to invent a new design of solar collectors to capture and minimize the sedimentation phenomena in the solar collector pipes. Also, they have a high operating temperature and a high heat storage capacity.
(5) More efforts are needed to study the reliability of using nanofluids in solar collectors from both environmental and economical point of view.
(6) Nanoparticles must be dispersed uniformly in the base fluid to enhance the solar-weighted absorption and increase the efficiency of the solar collector.
(7) Volume fraction of nanoparticles must be chosen accurately to enhance the performance of nanofluid collector. This is because the high volume fractions of nanoparticles increase the viscous force of nanofluids and reduce the heat transfer. While, the low quantity makes the nanofluid not able to absorb all of the incident solar radiation.
(8) It is recommended to use carbon nanohorns (CNHs) as a nanoparticles to improve the optical properties of the direct solar collectors. This is due to their large surface area and large number of cavities.
(9) Further research must be directed towards various significant challenges in the field of nanotechnology and its

application in the solar collector such as: Brownian motion of particles, particle migration, changing thermophysical properties with temperature, tendency of nanoparticles to agglomeration, changing nanofluid properties by using additives and the stability of nanofluids.

(10) More efforts are needed to study the benefits of using nanofluids in order to develop the performance of the transpired solar collector since no paper exists up to date to consider this problem.

(11) More research is needed to study the effect of nanoparticles sedimentation on the performance of various types of the solar collectors.

Acknowledgements

The first author would like to express about his deepest gratitude to his wife and Mrs. Topsy N. Smalley from the United States of America for their kind assistance in completing this huge work.

References

1. Sukhatme S., Nayak K. *Solar Energy Principles of Thermal Collection and Storage*. Tata McGraw Hill Education Private Limited 2009.
2. Duffie J., Beckman W. *Solar Engineering of Thermal Processes*. John Wiley and Sons, New York 1991.
3. Kandasamy R., Muhaimin I., Rosmila, A. The performance evaluation of unsteady MHD non-Darcy nanofluid flow over a porous wedge due to renewable (solar) energy. *Renewable Energy*, 2014; 64: 1–9.
4. Shan F., Tang F., Cao L., Fang G. Comparative simulation analyses on dynamic performances of photovoltaic-thermal solar collectors with different configurations. *Energy Conversion and Management* 2014; 87: 778–786.
5. Xia X., Xia J., Virkar A. Evaluation of potential for developing renewable sources of energy to facilitate development in developing countries. *Proceedings of the Asia-Pacific Power and Energy Engineering Conference 2010* Chengdu, China: pp. 1–3.
6. Allamraju K. Materials used for renewable energy resources. *Advanced Materials Manufacturing and Characterization*, 2013; 3: 243–248.

7. Hussein A. Applications of nanotechnology in renewable energies: A comprehensive overview and understanding. *Renewable and Sustainable Energy Reviews*, 2015; 42: 460–476.
8. Thirugnanasambandam M., Iniyan S., Goic, R. A review of solar thermal technologies. *Renewable and Sustainable Energy Reviews*, 2010; 14: 312–322.
9. Choi S. Enhancing thermal conductivity of fluids with nanoparticles. In: *Developments and Applications of Non-Newtonian Flows, ASME FED*, vol. 231/MD-66 1995: pp. 99–105.
10. Rao N., Gahane L., Ranganayakulu, S. Synthesis, applications and challenges of nanofluids: Review. *IOSR Journal of Applied Physics*, 2014: 21–28.
11. Bejan A., Karaus A. *Heat Transfer Handbook*. John Wiley and Sons, 2003.
12. Faiz F., Zahir E. A comparative study of nanofluids for tuneable filter operation. *International Journal of Engineering Research*, 2014; 3: 9–12.
13. Hone J. Carbon nanotubes: Thermal properties. In *Dekker Encyclopedia of Nanoscience and Nanotechnology*, 2004.
14. Assael M., Chen C., Metaxa, N., Wakeham, W. Thermal conductivity of suspensions of carbon nanotubes in water. *International Journal of Thermophysics*, 2004; 25: 971–985.
15. Yadav D., Agrawal G., Bhargava, R. The onset of convection in a binary nanofluid saturated porous layer. *International Journal of Theoretical and Applied Multiscale Mechanics*, 2012; 2: 198–224.
16. Ravisankar R., Venkatachalapathy, V., Alagumurthy N., Thamizhmaran K. A review on oxide and metallic form of nanoparticle in heat transfer. *International Journal of Engineering Science and Technology*, 2014; 6: 63–68.
17. Adil A., Gupta, S., Ghosh, P. Numerical prediction of heat transfer characteristics of nanofluids in a minichannel flow. *Journal of Energy*, 2014; Article ID 307520: 1–7.
18. Chieruzzi M., Cerritelli G., Miliozzi A., Kenny J. Effect of nanoparticles on heat capacity of nanofluids based on molten salts as PCM for thermal energy storage. *Nanoscale Research Letters*, 2013; 8: 448–456.
19. Gupta H., Agrawal G., Mathur J. An overview of nanofluids: A new media towards green environment. *International Journal of Environmental Sciences*, 2012; 3: 433–440.

20. Rashid F., Dawood K., Hashim, A. Maximizing of solar absorption by (TiO_2-water) nanofluid with glass mixture. *International Journal of Research in Engineering & Technology*, 2014; 2: 87–90.
21. Saidur R., Leong K., Mohammad H. A review on applications and challenges of nanofluids. *Renewable and Sustainable Energy Reviews*, 2011; 15: 1646–1668.
22. Nanofluids in solar collectors, from wikipedia, the free encyclopedia.
23. Tian Y., Zhao C. A review of solar collectors and thermal energy storage in solar thermal applications. *Applied Energy*, 2013; 104: 538–553.
24. Kalogirou S. Solar thermal collectors and applications. *Progress in Energy and Combustion Science*, 2004; 30: 231–295.
25. Jaisankar S., Ananth J., Thulasi S., Jayasuthakar S., Sheeba K. A comprehensive review on solar water heaters. *Renewable and Sustainable Energy Reviews*, 2011; 15: 3045–3050.
26. Alkilani M., Sopian K., Alghoul M., Sohif M., Ruslan M. Review of solar air collectors with thermal storage units. *Renewable and Sustainable Energy Reviews*, 2011; 15: 1476–1490.
27. Hossain M., Saidur R., Fayaz H., Rahim N., Islam M., Ahamed J., Rahman M. Review on solar water heater collector and thermal energy performance of circulating pipe. *Renewable and Sustainable Energy Reviews*, 2011; 15: 3801–3812.
28. D'Antoni M., Saro O. Massive solar-thermal collectors: A critical literature review. *Renewable and Sustainable Energy Reviews*, 2012; 16: 3666–3679.
29. Mathur A., Agrawal G., Chandel M. Recent developments in the field of solar water heater using flat plate collector: A review. *International Journal of Advanced Engineering Technology*, 2012; 3: 68–70.
30. Shukla R., Sumathy K., Erickson P., Gong J. Recent advances in the solar water heating systems: A review. *Renewable and Sustainable Energy Reviews*, 2013; 19: 173–190.
31. Devanarayanan K., Murugavel K. Integrated collector storage solar water heater with compound parabolic concentrator: Development and progress. *Renewable and Sustainable Energy Reviews*, 2014; 39: 51–64.
32. Hussein, A. K. Applications of nanotechnology to improve the performance of solar collectors—Recent advances and overview. *Renewable and Sustainable Energy Reviews*, 2016, 62: 767–792.

33. Kong W., Perers, B., Fan J., Furbo S., Bava, F. A new Laplace transformation method for dynamic testing of solar collectors. *Renewable Energy*, 2015; 75: 448–458.

34. Wei L., Yuan D., Tang D., Wu B. A study on a flat-plate type of solar heat collector with an integrated heat pipe. *Solar Energy*, 2013; 97: 19–25.

35. Soltau H. Testing the thermal performance of uncovered solar collectors. *Solar Energy*, 1992; 49: 263–272.

36. Minardi J., Chuang H. Performance of a black liquid flat-plate solar collector. *Solar Energy*, 1975; 27: 179–183.

37. Yu W., Xie H. A review on nanofluids: Preparation, stability mechanisms and applications. *Journal of Nanomaterials*, 2012; Article ID 435873: 1–17.

38. Garcýa A., Zarza E., Valenzuela L., Perez M. Parabolic-trough solar collectors and their applications. *Renewable and Sustainable Energy Reviews*, 2010; 14: 1695–1721.

39. Wikipedia, the free encyclopedia

40. Cheng Z., He Y., Qiu Y. A detailed non-uniform thermal model of a parabolic trough solar receiver with two halves and two inactive ends. *Renewable Energy*, 2015; 74: 139–147.

41. Coccia G., Di Nicola, G., Sotte, M. Design, manufacture and test of a prototype for a parabolic trough collector for industrial process heat. *Renewable Energy*, 2015; 74: 727–736.

42. Du B., Hu E., Kolhe M. An experimental platform for heat pipe solar collector testing. *Renewable and Sustainable Energy Reviews*, 2013; 17: 119–125.

43. Greening B., Azapagic, A. Domestic solar thermal water heating: A sustainable option for the UK? *Renewable Energy*, 2014; 63: 23–36.

44. Bourdoukan P., Wurtz E., Joubert P., Sperandio M. Potential of solar heat pipe vacuum collectors in the desiccant cooling process: Modelling and experimental results. *Solar Energy*, 2008; 82: 1209–1219.

45. Shah V., Bhatt N. Review on evacuated glass tube based solar liquid heaters. *International Journal of Engineering Development and Research*, 2014; 2: 2727–2733.

46. Sadhishkumar S., Balusamy T. Performance improvement in solar water heating systems: A review. *Renewable and Sustainable Energy Reviews*, 2014; 37: 191–198.

47. Shan F., Tang F., Cao L., Fang G. Dynamic characteristics modeling of a hybrid photovoltaic–thermal solar collector with active cooling in buildings. *Energy and Buildings*, 2014; 78: 215–221.

48. Pathak M., Sanders P., Pearce J. Optimizing limited solar roof access by exergy analysis of solar thermal, photovoltaic, and hybrid photovoltaic thermal systems. *Applied Energy*, 2014; 120: 115–124.

49. Ji J., Guo C., Sun W., He W., Wang Y., Li G. Experimental investigation of tri-functional photovoltaic/thermal solar collector. *Energy Conversion and Management*, 2014; 88: 650–656.

50. Tagliafico L., Scarpa F., De Rosa M. Dynamic thermal models and CFD analysis for flat-plate thermal solar collectors: A review. *Renewable and Sustainable Energy Reviews*, 2014; 30: 526–537.

51. Ibrahim A., Othman M., Ruslan M., Mat S., Sopian K. Recent advances in flat plate photovoltaic/thermal (PV/T) solar collectors. *Renewable and Sustainable Energy Reviews*, 2011; 15: 352–365.

52. Shukla A., Nkwetta D., Cho Y., Stevenson V., Jones P. A state of art review on the performance of transpired solar collector. *Renewable and Sustainable Energy Reviews*, 2012; 16: 3975–3985.

53. Otanicar T., Golden J. Comparative environmental and economic analysis of conventional and nanofluid solar hot water technologies. *Environmental Science and Technology*, 2009; 43: 6082–6087.

54. Natarajan E., Sathish R. Role of nanofluids in solar water heater. *International Journal of Advanced Manufacturing Technology*, 2009; DOI 10.1007/s00170-008-1876-8.

55. Polvongsri S., Kiatsiriroat, T. Enhancement of flat-plate solar collector thermal performance with silver nanofluid. *The Second TSME International Conference on Mechanical Engineering*, Krabi, Thailand 2011: pp. 1–7.

56. Yousefi T., Veysi F., Shojaeizadeh E., Zinadini S. An experimental investigation on the effect of Al_2O_3-H_2O nanofluid on the efficiency of flat-plate solar collectors. *Renewable Energy*, 2012; 39: 293–298.

57. Yousefi T., Veysi F., Shojaeizadeh E., Zinadini S. An experimental investigation on the effect of MWCNT-H_2O nanofluid on the efficiency of flat-plate solar collectors. *Experimental Thermal and Fluid Science*, 2012; 39: 207–212.

58. Yousefi T., Shojaeizadeh E., Veysi F., Zinadini S. An experimental investigation on the effect of pH variation of MWCNT–H_2O nanofluid on the efficiency of a flat-plate solar collector. *Solar Energy*, 2012; 86: 771–779.

59. Tora E., Moustafa T. Numerical simulation of an Al_2O_3-H_2O nanofluid as a heat transfer agent for a flat-plate solar collector. *International Journal of Scientific and Engineering Research*, 2013; 4: 762–773.

60. Jamal-Abad M., Zamzamian A., Imani, E., Mansouri, M. Experimental study of the performance of a flat-plate collector using Cu–water nanofluid. *Journal of Thermophysics and Heat Transfer*, 2013; 27: 756–760.

61. Faizal M., Saidur R., Mekhilef, S. Potential of size reduction of flat-plate solar collectors when applying MWCNT nanofluid. *4th International Conference on Energy and Environment* (ICEE 2013) 2013: pp. 1–4.

62. Gangadevi R., Senthilraja S., Imam, S. Efficiency analysis of flat plate solar collector using Al_2O_3-water nanofluid. *Methods Enriching Power and Energy Developments (MEPED'13)* 2013: pp. 1–4.

63. Chaji H., Ajabshirchi, Y., Esmaeilzadeh E., Heris S., Hedayatizadeh M., Kahani, M. Experimental study on thermal efficiency of flat plate solar collector using TiO_2/water nanofluid. *Modern Applied Science*, 2013; 7: 60–69.

64. Tiwari A., Ghosh P., Sarkar J. Solar water heating using nanofluids: A comprehensive overview and environmental impact analysis. *International Journal of Emerging Technology and Advanced Engineering*, 2013; 3: 221–224.

65. Faizal M., Saidur R., Mekhilef, S., Alim, M. Energy, economic and environmental analysis of metal oxides nanofluid for flat-plate solar collector. *Energy Conversion and Management*, 2013; 76: 162–168.

66. Said Z., Sajid M., Alim M., Saidur R., Rahim N. Experimental investigation of the thermophysical properties of AL_2O_3-nanofluid and its effect on a flat plate solar collector. *International Communications in Heat and Mass Transfer*, 2013; 48: 99–107.

67. Alim M., Abdin Z., Saidur R., Hepbasli A., Khairul M., Rahim N. Analyses of entropy generation and pressure drop for a conventional flat plate solar collector using different types of metal oxide nanofluids. *Energy and Buildings*, 2013; 66: 289–296.

68. Colangelo G., Favale E., De Risi A., Laforgia D. A new solution for reduced sedimentation flat panel solar thermal collector using nanofluids. *Applied Energy*, 2013; 111: 80–93.

69. Shankar S., Manivannan A. Performance evaluation of solar water heater using nanofluid. *International Journal of Engineering Research and Applications*, 2013; 3: 793–798.

70. Nasrin R., Alim M. Finite element simulation of forced convection in a flat plate solar collector: Influence of nanofluid with double nanoparticles. *Journal of Applied Fluid Mechanics*, 2014; 7: 543–556.

71. Polvongsri S., Kiatsiriroat T. Performance analysis of flat-plate solar collector having silver nanofluid as a working fluid. *Heat Transfer Engineering*, 2014; 35: 1183–1191.

72. Ekramian E., Etemad S., Haghshenasfard, M. Numerical investigations of heat transfer performance of nanofluids in a flat plate solar collector. *International Journal of Theoretical and Applied Nanotechnology*, 2014; 2: 30–39.

73. Mahian O., Kianifar A., Sahin A., Wongwises S. Entropy generation during Al_2O_3/water nanofluid flow in a solar collector: Effects of tube roughness, nanoparticle size, and different thermophysical models. *International Journal of Heat and Mass Transfer*, 2014; 78: 64–75.

74. Nasrin R., Alim M. Semi-empirical relation for forced convective analysis through a solar collector. *Solar Energy*, 2014; 105: 455–467.

75. Kabeel A., El-Said E. Applicability of flashing desalination technique for small scale needs using a novel integrated system coupled with nanofluid-based solar collector. *Desalination*, 2014; 333: 10–22.

76. Mahian O., Kianifar A., Sahin A., Wongwises S. Performance analysis of a minichannel-based solar collector using different nanofluids. *Energy Conversion and Management*, 2014; 88: 129–138.

77. Moghadam A., Farzane-Gord M., Sajadi, M., Hoseyn-Zadeh, M. Effects of CuO/water nanofluid on the efficiency of a flat-plate solar collector. *Experimental Thermal and Fluid Science*, 2014; 58: 9–14.

78. Zamzamian A., KeyanpourRad M., KianiNeyestani M., Jamal-Abad M. An experimental study on the effect of Cu-synthesized/EG nanofluid on the efficiency of flat-plate solar collectors. *Renewable Energy*, 2014; 71: 658–664.

79. He Q., Zeng S., Wang S. Experimental investigation on the efficiency of flat-plate solar collectors with nanofluid. *Applied Thermal Engineering*, 2014; 88: 165–171.

80. Said Z., Saidur R., Rahim N., Alim M. Analyses of exergy efficiency and pumping power for a conventional flat plate solar collector using SWCNTs based nanofluid. *Energy and Buildings*, 2014; 78: 1–9.

81. Mahian O., Kianifar A., Heris S., Wongwises S. First and second laws analysis of a minichannel-based solar collector using boehmite alumina nanofluids: Effects of nanoparticle shape and tube materials. *International Journal of Heat and Mass Transfer*, 2014; 78: 1166–1176.

82. Roy S., Asirvatham L., Kunhappan D., Cephas, E., Wongwises, S. Heat transfer performance of silver/water nanofluid in a solar flat-plate collector. *Journal of Thermal Engineering*, 2015; 1: 104–112.

83. Tyagi H., Phelan P., Prasher R. Predicted efficiency of a low-temperature nanofluid based direct absorption solar collector. *Journal of Solar Energy Engineering*, 2009; 131: 041004.

84. Otanicar T., Phelan P., Prasher, R., Rosengarten, G., Taylor, R. Nanofluid-based direct absorption solar collector. *Journal of Renewable and Sustainable Energy*, 2010; 2: 1–13.

85. Taylor R., Phelan, P., Otanicar, T., Adrian, R., Prasher, R. Nanofluid optical property characterization: Towards efficient direct absorption solar collectors. *Nanoscale Research Letters*, 2011; 6: 225–235.

86. Poinern G., Brundavanam S., Shah M., Laava I., Fawcett D. Photothermal response of CVD synthesized carbon (nano)spheres/aqueous nanofluids for potential application in direct solar absorption collectors: A preliminary investigation. *Nanotechnology, Science and Applications*, 2012; 5: 49–59.

87. Saidur R., Meng T., Said Z., Hasanuzzaman M., Kamyar A. Evaluation of the effect of nanofluid-based absorbers on direct solar collector. *International Journal of Heat and Mass Transfer*, 2012; 55: 5899–5907.

88. Moradi A., Sani E., Simonetti M., Francini, F., Chiavazzo, E., Asinari, P. CFD modeling of solar collector with nano-fluid direct absorption for civil application. *The 3rd International Conference on Microgeneration and Related Technologies*, Napoli, Italy 2013: pp. 1–10.

89. Kundan L., Sharma, P. Performance evaluation of a nanofluid (CuO-H_2O) based low flux solar collector. *International Journal of Engineering Research*, 2013; 2: 108–112.

90. Verma V., Kundan L. Thermal performance evaluation of a direct absorption flat plate solar collector (DASC) using Al_2O_3-H_2O based nanofluids. *IOSR Journal of Mechanical and Civil Engineering*, 2013; 6: 29–35.

91. Ladjevardi S., Asnaghi A., Izadkhast P., Kashani, A. Applicability of graphite nanofluids in direct solar energy absorption, *Solar Energy*, 2013; 94: 327–334.

92. Hector A., Singh H. Development of a nano-heat transfer fluid carrying direct absorbing receiver for concentrating solar collectors. *International Journal of Low-Carbon Technologies*, 2013; 1–6.

93. Zhidong P., Innocent N., Minghui, Z., Yanmin, W., Huining, H., Zhiyuan, L. Thermal performance of simulated direct absorbing solar collector with magnetic nanofluid. *Journal of The Chinese Ceramic Society*, 2014; 42: 522–527.

94. Lee S., Kim H., Kim, K., Jang, S. Extinction coefficient of water-based multi-walled carbon nanotube nanofluids for application in direct-absorption solar collectors. *Micro and Nano Letters*, 2014; 9: 635–638.

95. Luo Z., Wang C., Wei, W., Xiao, G., Ni, M. Performance improvement of a nanofluid solar collector based on direct absorption collection (DAC) concepts. *International Journal of Heat and Mass Transfer*, 2014; 75: 262–271.

96. Filho E., Mendoza O., Beicker C., Menezes, A., Wen, D. Experimental investigation of a silver nanoparticle-based direct absorption solar thermal system. *Energy Conversion and Management*, 2014; 84: 261–267.

97. Parvin S., Nasrin R., Alim M. Heat transfer and entropy generation through nanofluid filled direct absorption solar collector. *International Journal of Heat and Mass Transfer*, 2014; 71: 386–395.

98. Hordy N., Rabilloud, D., Meunier J., Coulombe, S. High temperature and long-term stability of carbon nanotube nanofluids for direct absorption solar thermal collectors. *Solar Energy*, 2014; 105: 82–90.

99. Karami M., Akhavan Bahabadi M., Delfani S., Ghozatloo A. A new application of carbon nanotubes nanofluid as working fluid of low-temperature direct absorption solar collector. *Solar Energy Materials and Solar Cells*, 2014; 121: 114–118.

100. Zhang L., Liu J., He G., Ye Z., Fang X., Zhang Z. Radiative properties of ionic liquid-based nanofluids for medium-to-high-temperature direct absorption solar collectors. *Solar Energy Materials and Solar Cells*, 2014; 130: 521–528.

101. Sadique M., Verma A. Nano fluid-based receivers for increasing efficiency of solar panels. *International Journal of Advanced Mechanical Engineering*, 2014; 4: 77–82.

102. Moradi A., Sani E., Simonetti M., Francini, F., Chiavazzo, E., Asinari, P. Carbon-nanohorn based nanofluids for a direct absorption solar collector for civil application. *Journal of Nanoscience and Nanotechnology*, 2015; 15: 3488–3495.

103. Khullar V., Tyagi H. Application of nanofluids as the working fluid in concentrating parabolic solar collectors. *Proceedings of the 37th*

National and 4th International Conference on Fluid Mechanics and Fluid Power, Chennai, India 2010: pp. 1–9.

104. Khullar V., Tyagi H., Phelan P., Otanicar T., Singh H., Taylor, R. Solar energy harvesting using nanofluids-based concentrating solar collector. *ASME Journal of Nanotechnology in Engineering and Medicine*, 2012; 3: 031003 (doi: http://dx.doi.org/10.1115/1.4007387).

105. De Risi A., Milanese M., Laforgia D. Modelling and optimization of transparent parabolic trough collector based on gas-phase nanofluids. *Renewable Energy*, 2013; 58: 134–139.

106. Sunil K., Kundan L., Sumeet, S. Performance evaluation of a nanofluid based parabolic solar collector—an experimental study. *Proceedings of Twelfth IRF International Conference*, Chennai, India 2014: pp. 29–35.

107. Ghasemi S., Ahangar G. Numerical analysis of performance of solar parabolic trough collector with Cu-water nanofluid. *International Journal of Nano Dimension*, 2014; 5: 233–240.

108. Sokhansefat T., Kasaeian, A., Kowsary F. Heat transfer enhancement in parabolic trough collector tube using Al_2O_3/synthetic oil nanofluid. *Renewable and Sustainable Energy Reviews*, 2014; 33: 636–644.

109. Kasaeian A., Daviran, S., Azarian, R., Rashidi, A. Performance evaluation and nanofluid using capability study of a solar parabolic trough collector. *Energy Conversion and Management*, 2015; 89: 368–375.

110. Nasrin R., Alim M. Performance of nanofluids on heat transfer in a wavy solar collector. *International Journal of Engineering, Science and Technology*, 2013; 5: 58–77.

111. Alaeian M., Sedaghat, A., Bahabadi, M. Heat transfer enhancement of MWCNT/HT-Oil nanofluid in U-bend wavy tubes for use in solar collectors. *Journal of Energy and Power Sources*, 2014; 1: 134–140.

112. Lu L., Liu Z., Xiao H. Thermal performance of an open thermosyphon using nanofluids for high-temperature evacuated tubular solar collectors Part 1: Indoor experiment. *Solar Energy*, 2011; 85: 379–387.

113. Moorthy M., Chui L., Sharma K., Anuar S. Performance evaluation of evacuated tube solar collector using water-based titanium oxide (TiO_2) nanofluid. *Journal of Mechanical Engineering and Sciences*, 2012; 3: 301–310.

114. Chougule S., Pise, A., Madane, A. Performance of nanofluid-charged solar water heater by solar tracking system. *International Conference*

on Advances in Engineering, Science and Management (ICAESM), Nagapattinam, Tamil Nadu 2012: pp. 247–253.

115. Senthil Kumar R., Manimaran R., Ramadoss K., Shankar N. Experimental analysis of nano fluid-charged solar water heater by solar tracking system. *Archives of Applied Science Research*, 2012; 4: 2582–2590.
116. Chougule S., Sahu, S., Pise, A. Thermal performance of two phase thermosyphon on flat-plate solar collectors using nanofluid. *Journal of Solar Energy Engineering*, 2013; 136: 1–5.
117. Liu Z., Hu R., Lu L., Zhao F., Xiao H. Thermal performance of an open thermosyphon using nanofluid for evacuated tubular high temperature air solar collector. *Energy Conversion and Management*, 2013; 73: 135–143.
118. Saravanan M., Karunakaran N. Experimental analysis of heat pipe with V-trough solar collector. *International Journal of Research in Advent Technology*, 2014: 13–17.
119. Aruna V., Channakaiah D., Murali, G. A study on a flat plate type of solar water heater with an thermosyphon using different working fluid. *Singaporean Journal of Scientific Research*, 2014; 6: 132–135.
120. Li Y., Xie H., Yu W., Li J. Investigation on heat transfer performances of nanofluids in solar collector. *Materials Science Forum*, 2011; 694: 33–36.
121. Taylor R., Phelan P., Otanicar, T., Walker, C., Nguyen, M., Trimble, S., Prasher, R. Applicability of nanofluids in high flux solar collectors. *Journal of Renewable and Sustainable Energy*, 2011; 3: 1–15.
122. Gangadevi R., Agarwal S., Roy S. A novel hybrid solar system using nanofluid. *International Journal of Engineering Research and Technology*, 2013; 6: 747–752.
123. Paul T., Morshed A., Fox E., Visser A., Bridges N., Khan J. Enhanced thermal performance of ionic liquid-Al_2O_3 nanofluid as heat transfer fluid for solar collector. *ASME—Seventh International Conference on Energy Sustainability*, Minneapolis, Minnesota, USA 2013: pp. 1–7.
124. Paul T. Investigation of thermal performance of nanoparticle enhanced ionic liquids (NEILs) for solar collector applications. PhD dissertation, University of South Carolina, U.S.A. 2014: pp. 1–172.
125. Hewakuruppu Y., Taylor R., Tyagi H., Otanicar, T. Limits of nanofluid based direct absorption solar collection. *Proceedings of the 52nd Annual Conference, Australian Solar Energy Society* (Australian Solar Council) 2014: pp. 1–10.

126. Rahman M., Mojumder S., Saha, S., Mekhilef, S., Saidur, R. Augmentation of natural convection heat transfer in triangular shape solar collector by utilizing water based nanofluids having a corrugated bottom wall. *International Communications in Heat and Mass Transfer*, 2014; 50: 117–127.

127. Goudarzi K., Shojaeizadeh E., Nejati F. An experimental investigation on the simultaneous effect of CuO-H_2O nanofluid and receiver helical pipe on the thermal efficiency of a cylindrical solar collector. *Applied Thermal Engineering*, 2014; 73: 1236–1243.

128. Rahman M., Mojumder S., Saha S., Mekhilef S., Saidur R. Effect of solid volume fraction and tilt angle in a quarter circular solar thermal collectors filled with CNT-water nanofluid. *International Communications in Heat and Mass Transfer*, 2014; 57: 79–90.

129. Goudarzi K., Nejati F., Shojaeizadeh E., Asadi Yousef-abad, S. Experimental study on the effect of pH variation of nanofluids on the thermal efficiency of a solar collector with helical tube. *Experimental Thermal and Fluid Science*, 2015; 60: 20–27.

130. Sruthi B. Nanotechnology in solar water heater. *National Conference on Developing Scenario in Applied Sciences and Communicative English*, Kumaraguru College of Technology, Coimbatore 2012: pp. 27–29.

131. Khanafer K., Vafai K. Applications of nanomaterials in solar energy and desalination sectors. *Advances in Heat Transfer*, 2013; 45: 303–329.

132. Javadi F., Saidur R., Kamalisarvestani, M. Investigating performance improvement of solar collectors by using nanofluids. *Renewable and Sustainable Energy Reviews*, 2013; 28: 232–245.

133. Mahian O., Kianifar A., Kalogirou S., Pop I., Wongwises S. A review of the applications of nanofluids in solar energy. *International Journal of Heat and Mass Transfer*, 2013; 57: 582–594.

134. Al-Shamani A., Yazdi M., Alghoul M., Abed A., Ruslan M., Mat S., Sopian K. Nanofluids for improved efficiency in cooling solar collectors—A review. *Renewable and Sustainable Energy Reviews*, 2014; 38: 348–367.

135. Chaudhari K., Walke P. Applications of nanofluid in solar energy—a review. *International Journal of Engineering Research and Technology*, 2014; 3: 460–463.

136. Nerella S., Sudheer N., Bhramara, P. Enhancement of heat transfer by nanofluids in solar collectors. *International Journal of Innovations in Engineering and Technology*, 2014; 3: 115–120.

137. Kasaeian A., Eshghi A., Sameti M. A review on the applications of nanofluids in solar energy systems. *Renewable and Sustainable Energy Reviews*, 2015; 43: 584–598.

138. Ravi kumar J., Goud P. Nanofluids: A promising future. *Journal of Chemical and Pharmaceutical Sciences*, 2014: 57–61.

Chapter 18

Green and Sustainable Future Nanocomposites

Vaneet Kumar,[a] **Saruchi**[b] **and Ajay Kumar Mishra**[c]

[a]*CT Group of Institutions, Shahpur, Jalandhar, Jalandhar 144011, Punjab, India*
[b]*Sardar Swaran Singh National Institute of Renewable Energy (SSS-NIRE), Jalandhar-Kapurthala Road, Wadala Kalan, Kapurthala 144601, Punjab, India*
[c]*Nanotechnology and Water Sustainability Research Unit, College of Science Engineering & Technology, University of South Africa, Florida Campus, Johannesburg, South Africa*

vaneet2106@gmail.com

18.1 Introduction

Nanoscience is one of the most important scientific and industrial breakthroughs of the 21st century. In scientific limits from electronics to medicine in advanced manufacturing, to cosmetics, nanotechnology has the potential to dramatically change lifestyles, jobs, and whole economies with applications that cross. The combination of green chemistry techniques with nanotechnology science applications has thus become a key component of the nanotechnology future. The use of natural ingredients to synthesize nanomaterials and design environmentally friendly synthetic processes has been extensively explored. While many of these, also called 'green nanotechnologies', are now finding

Nanocomposites for Pollution Control
Edited by Chaudhery Mustansar Hussain and Ajay Kumar Mishra

ISBN 978-981-4774-45-1 (Hardcover), 978-1-315-14368-2 (eBook)
www.panstanford.com

their way from the laboratory to commercial application, green nanotechnology still faces significant challenges.

In the past few decades, nanomaterials have demonstrated advanced performance in numerous applications, including medicine, energy, agriculture, industries and advanced manufacturing.

Green nanotechnology, the combination of nanotechnology science and the principles/practices of green chemistry, may hold the key to building an environmentally sustainable society in the 21st century. Green chemistry is a set of principles or rather a chemical philosophy that encourages the design of products and processes that reduce or eliminate the use and generation of hazardous substance.

Recently green nanotechnology science practices often involve the use of natural sources, non-hazardous solvents, and energy-efficient processes in the preparation of nanomaterials.

In recent times, attention has been drawn to the use of bio-reinforced composites in automobile, construction, packaging, agriculture and medical applications due to increased concern for environmental sustainability. Green polymer nanocomposites always demonstrate unique properties of combining the advantages of natural fillers and organic polymers. Plant fibres are also found suitable to reinforce polymers. These types of polymer have relatively high tensile strength, low cost of acquisition, low density and produce low CO_2 emission. These types of nanocomposites are biodegradable and are annually renewable as compared to other fibrous materials. On the other hand, organic type's polymers are sometime desirable because they are either recyclable or biodegradable without causing environmental hazards.

It has been reported [1, 2] that the addition of nanoparticles to base polymers confers improved properties that make them usable in the fields of automobiles, agriculture, construction, textiles and medicine. Properties of nanocomposites have been shown to improve substantially are mechanical properties like strength, elastic modulus and dimensional stability, thermomechanical and permeability properties like gases, water and hydrocarbons. Others are thermal stability and heat distortion temperature, flame retardancy and smoke emissions, chemical resistance, surface appearance, physical weight and electrical as well as magnetic conductivity. Nanocomposite is a class of composites in which the measurements of the reinforcing material are in the order of nanometres. Because of this nanometre size characteristic, nanocomposites possess superior properties than the conventional composites due to maximizing the interfacial adhesion.

In the past, major interest has been shown in the use of synthetic materials such as aliphatic polyesters, aliphatic-aromatic polyesters, polyvinyl alcohols, polyesteramides, polystyrene, nanoclays, glass/carbon fibres and carbon nanotubes, etc., for the synthesis of nanocomposites. The use of these types of materials, however, at present has great challenges [3, 4]. These include shortage of the organic compounds due to decreasing oil and gas resources day by day and increasing oil and gas prices [5]. Other major effects include environmental challenges for their degradation or incineration and global warming, uneconomical costs, cross contaminations in their recycling and toxicity risks of the customer [6–8]. These types of concerns always gave birth to the quest for materials that can overcome these challenges and maintain the required properties for the different applications [9, 10].

The use of polymer nanocomposites from renewable sources has advantages over synthetic sources, particularly as a solution to the environmental problems generated by plastic waste materials [11]. They always offer alternatives to maintaining sustainable development of economic and ecologically attractive technology [12]. Green nanocomposites are today widely researched because of the need for innovations in the development of new materials from biodegradable polymers, preservation of fossil-based raw materials, complete biological degradability and reduction in the volume of CO_2 release into the environment. Applications of agricultural resources (wastes and products) for the production of green materials are some of the important reasons why green nanocomposites have attracted tremendous research interests worldwide. The use of these green nanocomposites is expected to improve manufacturing speed and recycling with enhanced environmental compatibility.

Materials are said to be 'green material' when they are purely biodegradable and renewable in nature. The major attractions about green nanocomposites are that they are environmentally friendly, fully biodegradable and sustainable in every way. At the end of the service life of nanocomposites, they can be easily disposed of or composted without harming the environment. The challenge of green nanocomposites involves basically the challenge of obtaining 'green' polymers matrix for the production of the green nanocomposites. Polymer is said to be green when it possesses environmentally favourable properties such as renewability and degradability in nature. Biodegradation implies degradation of a polymer in natural environment that includes changes in the chemical-based structure, loss of mechanical and structural properties and changing into other compounds that are favourable to the environment. Polymers which are obtained from natural sources (e.g. starch, lignin, cellulose acetate, poly-lactic acid (PLA), polyhydroxylalkanoates (PHA), polyhydroxylbutyrate (PHB)) and some synthetic sources (e.g. aliphatic and aromatic polyesters, polyvinyl alcohol, modified polyolefins) that are degradable in nature are classified as biopolymers [13]. However, sometimes those which are obtained from synthetic sources are not renewable in nature and therefore do not conform wholly to the concept of renewability and degradability.

The green nanocomposite concept for sustainable future represents a stimulating route for creating new and innovative materials, in the area of natural polymers. Materials with a large assortment of properties have been realized, and even more are due to be realized. The green nanocomposites materials gained by mixing natural polymers and sheets of crystalline solid layered, offer a vast variety of property profiles. They are even capable of competing equally on price and performance, with synthetic polymeric materials in packaging. In spite of the enormous possibilities existing for packaging in bio-based nanocomposite materials, the future scenario is difficult to predict.

According to the Indian National Environment Policy 2006, the dominant theme of this policy is that while conservation of environmental resources is essential to secure livelihoods and well-being for all, the most secure basis for conservation is to ensure that people dependent on particular resources obtain enhanced livelihoods from the fact of conservation than from biodegradation of the resources. This type of policy stresses the two-way linkages between poverty and environmental quality. Figure 18.1 shows the application of green nanocomposites in various fields.

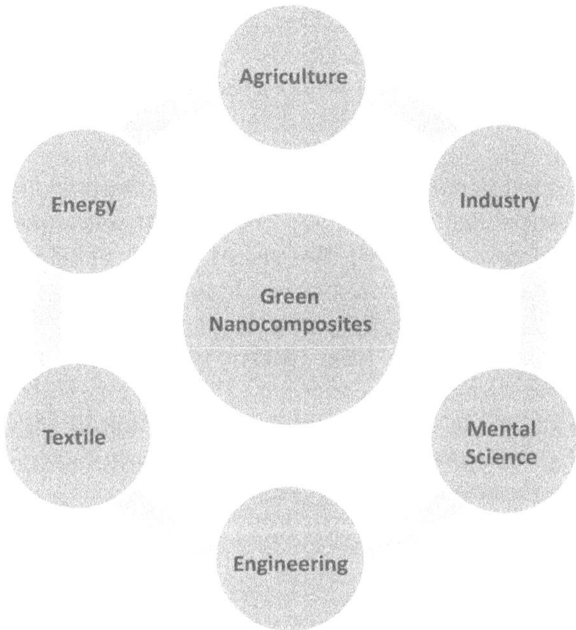

Figure 18.1 Application of green nanocomposites in various fields.

A number of natural and other biodegradable polymers have been used in green nanocomposites and some of these are discussed as follows:

18.2 Nanocomposites Based on Cellulose

Cellulose-fibre-reinforced polymer green nanocomposites have received much attention because of their low density, nonabrasive, combustible, nontoxic, low cost, and biodegradable properties. A number of research works have been done for the production of nanoscale green cellulose fibres and their application in nanocomposites materials have gained increasing attention due to their high tensile strength and stiffness combined with low molecular weight, biodegradability and renewability in nature. The major reason for using green cellulose nanofibres in composite materials is that one can potentially exploit the high stiffness of the cellulose crystal for reinforcement. This can be performed by breaking down the hierarchical structure of the plant into individualized nanofibres of high crystallinity, with a reduction of amorphous parts [14, 15].

The cellulosic plastic with 80 wt% pure cellulose acetate and 20 wt% triethyl citrate plasticizer was used as the polymer matrix for the synthesis of green nanocomposites. Mechanical properties of the nanocomposites were determined and correlated with observations from X-ray diffraction and transmission electron microscopy (TEM). Results show that Cellulosic plastic-based nanocomposites containing 5 and 10 wt% organoclay (organically modified phyllosilicate, derived from a naturally occurring clay mineral) have better exfoliated and intercalated structure as compared to those of 15 wt% organoclay. Tensile strength and modulus of cellulosic plastic reinforced with 10 wt% organoclay get better by 75 and 180%, respectively. Thermal stability of the cellulosic plastic also enhanced.

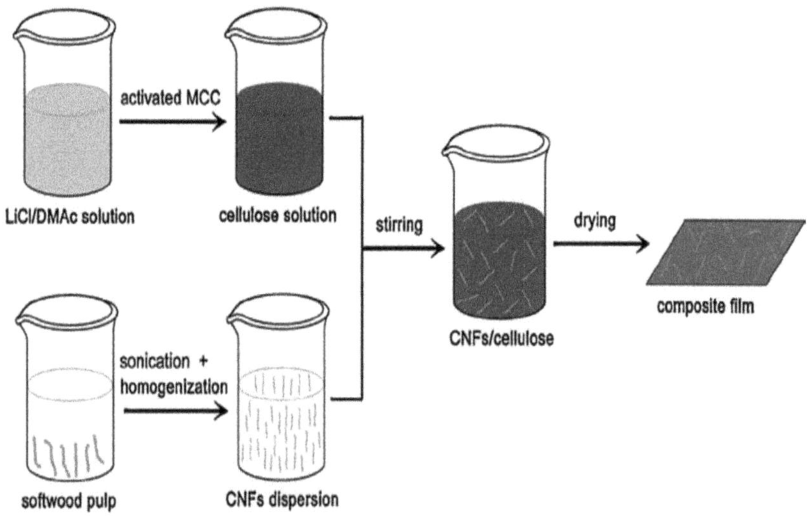

Figure 18.2 Procedures for manufacturing CNFs reinforced all-cellulose green nanocomposites.

In a recent study, a cellulose-based nanocomposite material has been investigated as a flexible humidity and temperature sensor [16]. Cellulose was obtained from cotton pulp by acid hydrolysis using a solution of $LiCl_2$ and N,N-dimethylacetamide. Nanoscaled polypyrrole can be used as the second component of the nanocomposite. Nanocomposites were generated by polymerization induced adsorption process. Nanoscale polypyrrole were coated on the surface of the cellulose and its influence on the cellulose membrane was identified by atomic force microscopy (AFM). Figure 18.2 indicates the general procedures for manufacturing CNFs reinforced all-cellulose green nanocomposites.

18.3 Nanocomposites Based on Plant Oil

Plant oils constitute a rich source for many different polymers and polymer forerunner and they are being considered for the production of 'greener' nanocomposites. The wide range of possible combinations of plant oils, chemical modifications, polymerization methods, nature of the fillers and fibres used as reinforcement materials allows tailoring the composite properties to fit the requirements of structural as well as functional materials. Therefore, a wide range of macro, micro- and nanoscale-sized particles and fibres have been projected as reinforcements/fillers, including organic and inorganic ones, natural or synthetic, in order to give satisfactory answers to specific requirements. Role of plant oil-based products may seem modest in some cases like partial replacement of synthetic materials. There is a clear trend to enhance the percentage of 'green'-based raw materials in the formulations of commodities as well as specialty green polymers nanocomposites for high added value applications.

Soy-based polyurethane can be used as a matrix for the synthesis of green nanocomposites [17]. The polyurethane modified with halloysite nanotubes (HNT) nanoparticles was used to synthesize E-glass reinforced composites using low charge vacuum assisted resin transfer moulding process and characterized for vital compressive strength, flexural strength, flexural modulus and interlaminar shear strength (ILSS). Results indicate that interlaminar shear strength increased with increase in percentage of HNT used. Eventual compressive strength, flexural modulus and strength did not show appreciable changes; 0.8 wt% HNT gave the best combination of properties with 6%, 28%, and 82% increase in flexural strength, flexural modulus and ILSS, respectively.

On the other hand, the attempts to prepare natural polymeric-matrix-based green nanocomposites are quite limited in the literature. Most of the research studies reported concentrate on the renewable polymeric matrices only [18]. These are the works generally on the preparation of clay nanocomposites based on renewable plant oils in the presence of different synthetic intercalants such as alkyl tallow quaternary ammonium salts [19] and (4-vinylbenzyl)triethylammoniu m chloride which might be co-cross-linked with the matrix. Figure 18.3 indicates the common procedure for the synthesis of plant oil nanocomposites.

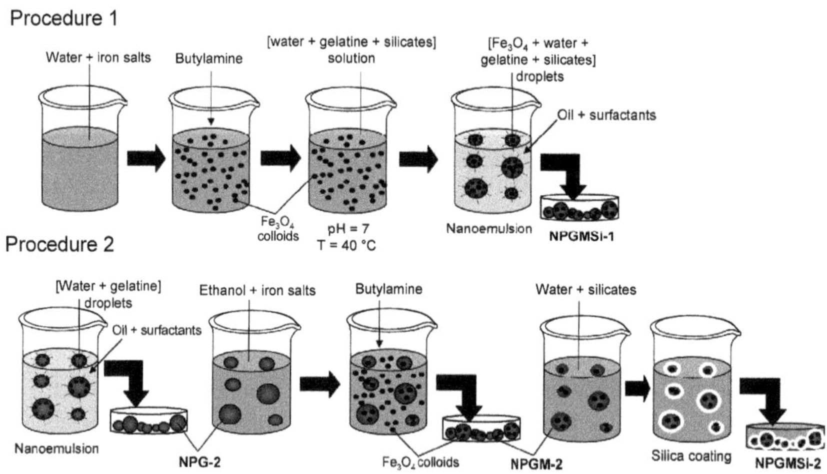

Figure 18.3 Common method for synthesis of plant oil-based nanocomposites.

18.4 Nanocomposites Based on Thermoplastic Starch

Thermoplastics are polymers that can flow when heated beyond a melting or vitrification temperature condition. They undergo plastic deformation, meaning viscous flow with often-complex rheology due to their huge molar mass, embarrassments, interactions and chain brushwood. Starch is a natural polymer with complex levels of structure that impose upon thermoplastic deformation.

Thermoplastic starch (TPS) has attracted much attention due to its thermoplastic-like processability with temperature and shear, though the structures being disrupted are more complicated as compared to those of synthetic thermoplastics. However, Thermoplastic starch is no dissimilar to any other polymer with respect to linear, branched structures, molar mass, glass transition temperature, plasticizer modification, crystallinity and melting temperature. Starch is a stereo-regular polymer with chirality, chains directionality, branching and hydrogen bonding with high density.

Earlier research of thermoplastic starch-based composites focused on the use of plasticized starch as matrix for green nanocomposites. De Carvalho et al. [18] first reported the use of thermoplastic starch for the synthesis of composites by dissolve intercalation in double screw extruder. The composites were synthesized with usual cornstarch plasticized with glycerin and reinforced with hydrated kaolin. The study indicated significant increase in the tensile strength from 5 to 7.5 MPa for the composite having 50 phr clay compositions. The modulus of elasticity increased from 120 to 290 MPa whereas, tensile strain at break decreased from 30% to 14%. The highest value for the modulus of elasticity found and the tensile strength corresponded to the maximum quantity of clay that was incorporated in the matrix.

Pandey and Singh [20] reported the sequence of addition of plasticizers to determine the effect of plasticizers on the mechanical and structural properties of resulting green nanocomposites using solution method. Thermal stability, mechanical properties and water absorption studies were carried out to examine the properties of material, whereas FT-IR spectroscopy was used to study the microdomain structure of green nanocomposites. It was found that the sequence of addition of components like starch/plasticizer, glycerol/clay had a significant consequence on the nature of nanocomposites formed and consequently the properties were modified with no well-established sequence to find out appropriate process method for the green nanocomposites. The filler dispersion was observed to become highly heterogeneous and more brittle when starch was plasticized before filling with clay. The modulus of nanocomposites increased significantly for all compositions relative to the unfilled matrix, irrespective of their method of preparation.

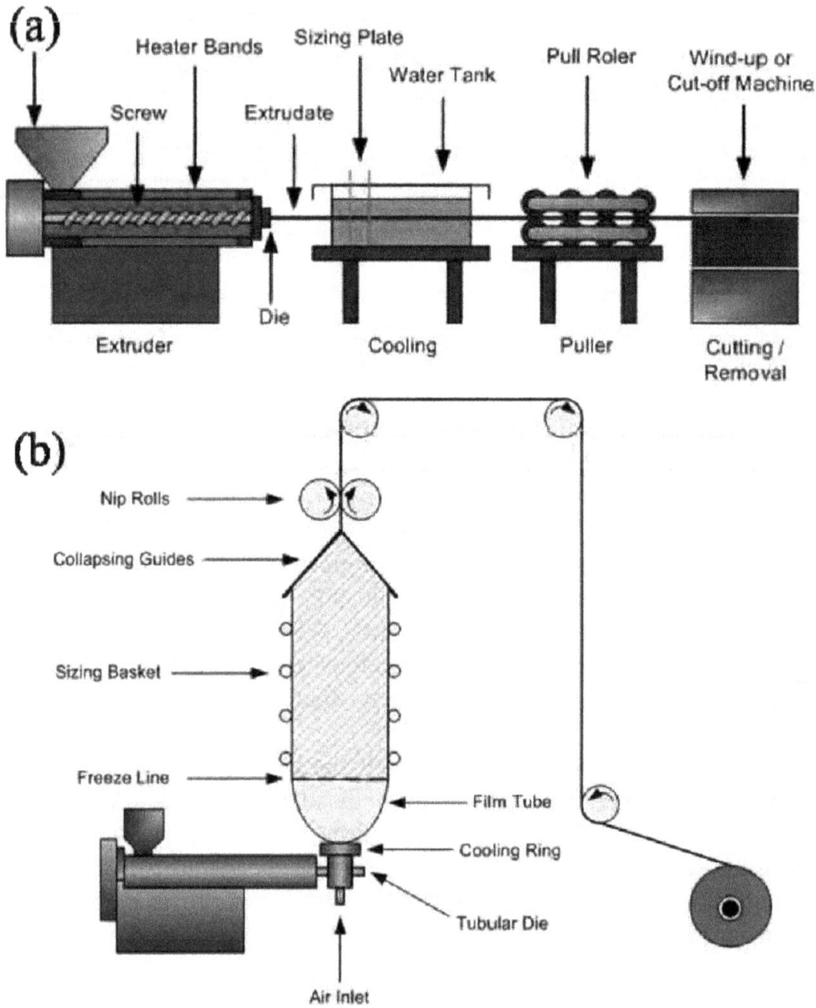

Figure 18.4 Schematic for processing of a thermoplastic polymer and its nanocomposite (a) Extrusion and (b) film blowing.

Extrusion is the constantly shaping of a fluid polymer and its nanocomposites through the orifice of a suitable tool (die) (Fig. 18.4a), followed by solidifying it into extrudate of steady cross section. The feed material is regularly thermoplastics powder or pellets format. In this process, the feed material is heated first to a fluid state via a screw extruder, followed by pumping into the die, which is finally solidified by cooling method. Extrusion products are often subdivided into small groups which include filaments of circular cross-section, profiles of irregular cross section, axis-symmetric tubes and pipes, and flat products such as films as well as sheets. Another method used for the processing of thermoplastics into tubular product several times of its initial diameter that can be use directly or made into film, is known as film blowing (Fig. 18.4b) [21]. Nanocomposites films generated by the film blowing process are widely used for agricultural, construction, and industrial applications, including covers for silage, greenhouses, chemical/solar ponds, flat cars, etc., or for a mixture of packing applications, such as wrapping, can lining, fabricated bags such as garbage.

18.5 Nanocomposites Based on Poly Lactic Acid

Poly(lactic acid) (PLA)-based green nanocomposites, based on 5 wt% of an organically modified montmorillonite (CLO), unmodified sepiolite (SEP) and organically modified zirconium phosphonate (ZrP), which were obtained by melt blending.

In broad, commercial PLA grades are copolymers of poly (L-lactic acid) and poly(D,L-lactic acid), which are generated from L-lactides and D,L-lactides, respectively. The ratio of L-enantiomers to D,L-enantiomers is known to influence the properties of PLA [22, 23], i.e. whether the materials are semicrystalline or amorphous in nature. There is increasing interest in using PLA for disposable degradable plastic articles; however, there are many properties such as flexural and gas barrier properties, high melt viscosity and melt strength/'elasticity' during processing, that are often not good enough for some end-use applications, such as blow moulding [24, 25]. In the order to improve the physical properties of PLA, especially in terms of thermomechanical stability, addition of different fillers (nanoparticles) in PLA was explored [26–28].

Studies on the thermal, mechanical and morphological properties of PLA-based composites have been presented by Lee et al. [29]. Green nanocomposites were synthesized by melt compounding and injection moulding. Thermal degradation, thermal transition, morphological, and mechanical properties of the green nanocomposites were evaluated. Tensile modulus of the nanocomposites increased from 63.5% to 170% contrary to the earlier studies where no improvement was recorded. More recently, green nanocomposites of PLA with a compatibilizer and cellulose fibrils have been developed by Qu, et al. [30]. Bleached wood pulp was used as the fibre and commercial grade PLA as the matrix. A chemo-mechanical method was used to prepare cellulose nanofibrils dispersed uniformly in an organic solvent. Poly ethylene glycol (PEG) was added to the matrix as a compatibilizer to improve the interfacial bonding/adhesion between the matrix and the fibre. The composites were obtained by solvent casting methods using N,N-dimethylacetamide (DMAC) and characterized PLA reinforced with cellulose nanofibrils resulted in no improvement in tensile strength (30 MPa compared with pure PLA) and percent elongation (2.5% compared with pure PLA) of the composites. The authors attributed this to the poor interfacial bonding between cellulose nanofibrils and the PLA matrix. Addition of PEG to the blend of PLA resulted in significant improvement in tensile strength (28.2%) and percentage elongation (25%) compared to pure PLA. The authors posited that PEG covers the surface of the cellulose nanofibrils and act not only as a plasticizer for PLA to improve its elongation, but also as a compatibilizer between the hydrophobic PLA and the hydrophilic cellulose nanofibrils. It was also evident that PEG also prevents the aggregation of the nanofibrils, so that the cellulose nanofibrils disperse in the PLA matrix homogeneously to form a network structure. It was also noted by the authors that the optimum composition of cellulose nanofibrils to obtain the best properties was 3% above which tensile strength and percentage elongation decreased. Figure 18.5 indicates the common method for the preparation of green nanocomposites based on PLA.

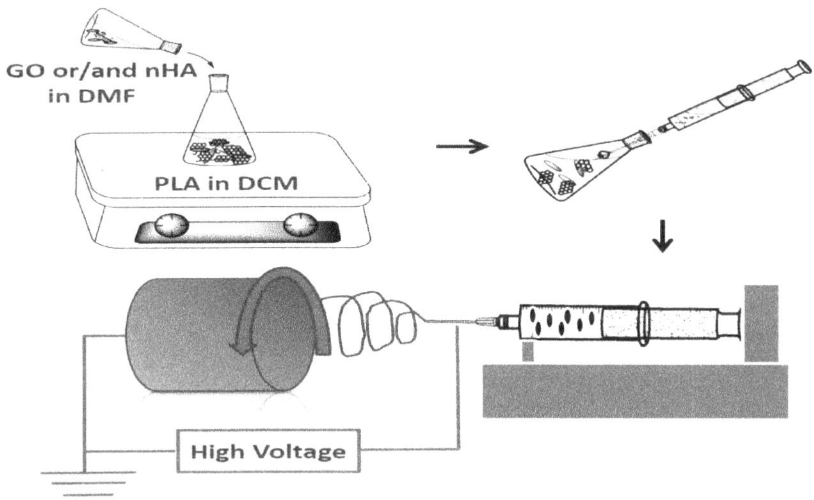

Figure 18.5 Common methods for the preparation of green nanocomposites based on PLA.

18.6 Nanocomposites Based on Biopolymers

Biopolymers used in green nanocomposites synthesized are polyhydroxyl butyrate (PHB)—a natural occurring polyester generated by numerous bacteria in nature, gelatin—a biopolymer found from thermal denaturation of collagen isolated from skin of animal and bones with dilute acid and chitosan—a natural polymer found in exoskeletons of crustaceans and insects and in the cell wall of fungi and micro organisms [31–34]. Mechanical and water vapour barrier properties of chitosan-based green nanocomposites were improved by addition of cellulose nanofibre, whereas chitosan filled with hydroxyapatite, tricalcium phosphate, carbon fibre and montmorillonite will not demonstrate substantial improvement in mechanical properties.

Table 18.1 indicates the list of natural fibres that have been extracted and their properties. These fibres have been applied as reinforcement for composites and nanocomposites.

Table 18.1 Natural fibres and their properties

Fibre source	Treatment	% cellulose content	Tensile strength (MPa)	Elastic modulus (GPa)	% Elongation	Reference
Banana	NS	NS	779	32	2	[37]
Cotton	Alkaline boiling modified with carbomethyl cellulose	NS	NS	82	13.57	[38]
Bark of cotton stalks	NaOH treatment	79	377	18.7	3.0	[39]
Sisal	NS	NS	600–700	38	2.3	[40]
Flax	NS	NS	800–1500	60–80	1.2–1.6	[40]
Hemp	NS	NS	550–900	70	1.6	[40]
Jute	NS	NS	400–800	10–30	1.8	[40]
Coir (coconut husk)	NS	NS	220	6	1.5–2.5	[40]
Velvet Leaf	Alkaline treatment	69	325–500	18–38	1.6–2.6	[41]
Switch grass leaves	Simple alkaline treatment	61.2	715	31	2.2	[41]
Switch grass stems	Simple alkaline treatment	68.2	351	9.1	6.8	[42]
Stinging nettle	NS	NS	1594	87	2.11	[43]
Kenaf	Alkaline treatment	NS	130	11	1.3	[44]
Ramie	Untreated	68.6–76.2	560	24.5	2.5	[45]
Soybean straw	Alkaline treatment	85	351	12	3.9	[46]
Flax	NS	NS	1339	54	3.27	[47]

Fibre source	Treatment	% cellulose content	Tensile strength (MPa)	Elastic modulus (GPa)	% Elongation	Reference
Wheat straw	Mechanical Processing	47.55	58.7	3.7	NS	[48]
Wheat straw	Microbial retting	NS	139.9	4.8	NS	[49]
Wheat straw	Chemical processing	63.14	146	7.9	NS	[50]
Hop stem	Alkaline treatment	84	53.3	20	3.3	[50]
Cornstalks	Alkaline treatment	52	286	16.5	2.2	[51]

NS: Not specified.

18.7 Green Fillers in Nanocomposites

The challenge of green nanocomposites expands to the quest of reinforcements that are green, biodegradable and renewable in nature. Cellulose is the most abundant renewable fibre in nature because cellulose gives plant to their structural rigidity. Moreover, they can be generated from agricultural waste which makes it more economically viable as compared to any other source of fibres currently in use. Cellulose fibre has presents an outstanding balance of excellent mechanical properties, low density, safer handling and working conditions compared to synthetic fibres [35]. Efforts, in the last few decades have been focused on the exploitation of natural fibres as load bearing constituents in nanocomposites materials.

The use of natural fibre in the synthesis of nanocomposites in the early 1990s was largely found in the use of steam exploded hemp fibre as reinforcement in polypropylene-based composites [36]. Hemp fibres were initially purified by steam explosion in a flash hydrolysis laboratory pilot unit and compounded directly with polypropylene. Surface treatment was given with propylene maleic anhydride co-polymer. Tensile and microstructural properties were studied with a SEM and Instron tensile testing machine, respectively. The steam exploded fibre was had a cellulose content of 55%. Slight enhance in strength was observed in the treated fibre reinforced composite at a higher filler content, strength of untreated fibre reinforced composites decreased in because of poor adhesion between fibre and matrix. The percentage elongation generally decreased with increase in fibre content, while tensile modulus increased with percentage of fibre in the polymer.

18.8 Challenges and Prospects in the Field of Green Nanocomposites

In this context, the real challenge is to incorporate nanomaterials in green products with ELSI (Ethical, Legal, and Societal Issues) to satisfy modern consumers. This ultimately open new research sectors which can establish relationship among green economy, nano-products and their impact on health and safety of workers.

18.9 Challenges and Prospects

The compatibility of the hydrophobic polymer matrix and hydrophilic fibres in making of nanocomposites is of greatest challenge, because it leads to non-uniform dispersion of fibres within the matrix and have poor mechanical properties. The affinity and adhesion between thermoplastic matrices and fibres in composites production can be improved by using different chemical coupling agents like carboxylated polyethylene (CAPE), Maleic anhydride polypropylene (MAPP), titanium derived mixture (TDM), etc. Before preparing green polymer nanocomposite, a better understanding of the molecular structure, interfacial interaction with the matrix and the fibres is very important. A vast knowledge of the relationship between the structure and property of the polymer matrix and nanocomposite would be a major breakthrough in this particular area of research. Another challenge is the quest for true green polymers, which can be used as a matrix material and have good mechanical properties. Starch biopolymer is highly brittle, have poor water resistance and inferior tensile properties. Therefore, to have biopolymer which is compatible with nanoparticle and get a product with strong mechanical strength, water resistance, thermostability and lower toxicity is a great challenge. Plasticization is the common problem while synthesizing green nanocomposite. The biopolymer like chitosan, cellulose, gelatin, and plant-based oils are inadequate sources and it involves costly and tiresome production process. Nanotechnology require to separate filler particles into the right shape and layer structure. Very thin and very wide nanoparticles are required to achieve Maximum properties of the nanoparticle. Achieving nanoparticle of this size is an enormous task and it requires sophisticated instrument like high-pressure homogenizers and inline dispersers. Though current technology are very promising for achieving these nanosize particles produce wide size ranges but it also leads to inconsistencies. The use of current technology also makes the process costly.

Tensile properties of the nanoparticle are effective by the orientation of the particle and it is very difficult to take the orientation of nanosize fibres into consideration. As the small size of the nanoparticle, there is a possibility of re-aggregation. This occurred at the time of synthesis and its prevention must be taken into account when synthesizing the nanocomposites. There are lots of applications of nanocomposites due to its chemical resistance, extraordinary barrier properties and surface appearance. It can be used as mirror housings on various vehicle types, door handles, roofs, door panels, headrests and engine covers, timing belt covers, in beer and carbonated drinks bottles, paperboard in dairy and juice industry. Nanocomposites also have prospect in aerospace applications because of their light weight.

18.10 Concluding Remarks and Future Perspectives

The aim of this overview was to discuss the current status of research on green nanocomposites and sustainable future. At present, this is an thrilling and quickly developing area. Humankind is on the threshold of large-scale technological application of green nanocomposites, which are considered as the only alternative to synthetic, petroleum-based polymers. Green nanocomposites are abundant, renewable resources that can provide a basis for sustainable and economic development. Furthermore, green nanocomposites are biocompatible, biodegradable, and environmentally friendly. Their use has a wide range of environmental benefits, including reduced reliance on fossil carbon, decreased greenhouse gas emissions, and biodegradation to harmless products by the action of different micro-organisms. It has been proved by many researchers that green nanocomposites will allow us to avoid problems caused by synthetic polymers.

There is a wide variety of sources of biological as well as chemical origin to make green nanocomposites. Green nanocomposites are highly versatile, having a broad range of physical properties suitable for miscellaneous applications. Nevertheless, green nanocomposites themselves have the restricted potential to replace petroleum-based polymers because their properties are not good enough for material fabrication. Green nanocomposites are generally not competitive with polymers in mechanical strength. Nanosized particles are introduced to improve the properties and functionalities of green nanocomposites. The advantages of nanoparticles reveal themselves only if they are appropriately distributed in a polymer matrix. Because of a very large contacting area, there are excess interactions and linkages between the mixed components, influencing the mobility and relaxation behaviour of macromolecules that is reflected in the first place in the mechanical and thermal properties of materials. The effect of homogeneously distributed nanoparticles peaks at the concentration of 3–5 wt%. This is a real challenge to gain the homogeneous dispersion of nanosized additive in the green polymer matrix. The major obstruction consists of the strong aggregation and pronounced tendency to the agglomeration of nanoparticles owing to their very high surface and surface energy. If they are inadequately dispersed and not de-aggregated in the course of preparation or agglomerate with time because of inadequate stabilization, one has a common composite material through the filler. The latter modifies the properties at the concentration of only some tens of percentage. The materials, which are composed of biopolymers and fillers, do not demonstrate proper properties inherent to green nanocomposites. They should be classified as bio-composites. Green nanocomposites are still a loosely defined family of nanomaterials. Nowadays, this is a rapidly developing area in the making. One may find a sharply increased number of research publications devoted to composites on the basis of green polymers with nanosized additives during the past few decades in which they are denoted in different manners. A universally accepted term is found to be absent. Material of this kind is called nanocomposites, nanobiocomposites, bio-composites, green composites, bio-hybrids, bio-based plastics, and bioplastics. In our opinion, the term 'green nanocomposite for sustainable future' is best suited. It always points out the similarity with green nanocomposites, which are well known at present, and takes account of the difference in the polymer origin. A wide range of available green polymers and inorganic nanoparticles offers a means of developing a diversity of green nanocomposites with different structure, properties, functionalities, and applications. This is achieved by mixing them in various combinations. Green polymers form the matrix, determining the shape, structural organization, and main functionalities of green nanocomposites. Dispersed nanoparticles always modify the matrix. Which allow the tuning of the structure, properties, and, as a result, functionality. The nanoparticles can be also added for introducing a special functionality that is not provided by green polymers themselves. A wide variety of both green polymers and nanosized particles makes, in principle, possible materials for any desired application. As of now, only certain ones of them are put to fabricate the green nanocomposites. They await wider applications an inestimable advantage of bio-nanocomposites over green nanocomposites based on synthetic made polymers is the biocompatibility. This makes them suitable for wide range of applications in medicine, agriculture and industries. Nowadays, this is an area of great activity in which the green nanocomposites have dominant significance. They are also used to prepare biomaterials like scaffolds and implants, drug-delivery systems, diagnostics, and biomedical devices. The biocompatibility makes them also suitable for cosmetics and biotechnology. In principle, green nanocomposites will substitute the current materials on the basis of petroleum-based polymers that are in contact with the living body.

There is a worldwide realization of the loss that a rapidly rising indiscriminate industrial processes can do to the ecosystem and environmental balance of the planet. There is general agreement that future technology development will require concepts such as biological sustainability, minimum need of energy and renewable raw materials that will probably be set at international level. Current R&D needs emphasize the development of high-value and safer products. Advanced green nanocomposites materials are being seen to have benefits compared with traditional materials in many of these areas.

References

1. Hay, J. N., and Shaw, S. J. (2010). Nanocomposites-properties and applications, *A Rev. Nanocomposites*, **2001**, pp. 1–6.
2. Njuguna, J., Pielichowski, K., and Desai, S. (2008). Nanofiller-reinforced polymer nanocomposites, *Polym. Adv. Technol.*, **19**, pp. 947–959.
3. Leja, K., and Lewandowicz, G. (2010). Polymer biodegradation and biodegradable polymers—a review, *Polish J. Environ. Stud.*, **19**, pp. 255–266.
4. Sugg, R. M., Pary, J. K., Uchino, K. (2006). Argatroban TPA stroke study: Study design and results in the first treated cohort, *Arch Neurol.*, **63**, pp. 1057–1062.
5. Pandey, J. K., Kumar, A., Pratheep, M., Manjusri, M., Amar, K., Drzal, L. T., Singh, R. P. (2005). Recent advances in biodegradable nanocomposites, *J. Nanosci. Nanotechnol.*, **5**(4), pp. 497–526.
6. Jamshidian, M., Tehrany, E. A., Imran, M., Jacquot, M., and Desobry, S. (2010). Poly-lactic acid: Production, applications, nanocomposites, and release studies, *Compr. Rev. Food Sci. Food Saf.*, **9**, pp. 552–571.
7. Amass, W., Amass, A., Tighe, B. (1998). A review of biodegradable polymers: Uses, current developments in the synthesis and characterization of biodegradable polyesters, blends of biodegradable polymers and recent advances in biodegradation studies, *Polym. Int.*, **47**, pp. 89–144.
8. Chandra, R., and Rustgi, R. (1998). Biodegradable polymers, *Prog. Polym. Sci.*, **23**, pp. 1273–1335.

9. Mohanty, A. K., Misra, M., and Hinrichsen, G. (2000). Biofibres, biodegradable polymers and biocomposites: An overview, *Macrmol. Mater. Eng.*, **276/277**, pp. 1–24.

10. Siracusa, V., Rocculi, P., Romani, S., and Rosa, M. D. (2008). Biodegradable polymers for food packaging: A review, *Trends Food Sci. Technol.*, **19**, pp. 634–643.

11. Pandey, J. K., Chu, W. S., Lee, C. S., and Ahn, S. H. (2007). Preparation characterization and performance evaluation of nanocomposites from natural fiber reinforced biodegradable polymer matrix for automotive applications. *Presented at the International Symposium on Polymers and the Environment: Emerging Technology and Science, BioEnvironmental Polymer Society (BEPS)*, Vancouver, WA, USA, pp. 17–20.

12. Sinha, S. R., and Bousmina, M. Biodegradable polymer/layered silicate nanocomposites. In *Polymer Nanocomposites* (Mai, Y., Yu, Z., eds.), Woodhead Publishing and Maney Publishing: Cambridge, England, pp. 57–129.

13. John, M. J., and Thomas, S. (2008). Biofibres and biocomposites, *Carbohyd. Polym.*, **71**, pp. 343–364.

14. Siro, I., and Plackett, D. (2010). Microfibrillated cellulose and new nanocomposite materials: A review, *Cellulose*, **17**(3), pp. 459–494.

15. Eichhorn, S. J., Dufresne, A., and Aranguren, M. (2010). Review: Current international research into cellulose nanofibres and nanocomposites, *J. Mater. Sci.*, **45**(1), pp. 1–33.

16. Guan, J., and Hanna, M. A. (2006). Selected morphological and functional properties of extruded acetylated starch-cellulose foams, *Bioresource Technol.*, **97**, pp. 1716–1726.

17. Tate, J. S., Akinola, A. T., Kabakov, D. (2010). Bio-based nanocomposites: An alternative to traditional composites, *J. Technol. Stud.*, **1**, pp. 25–32.

18. Carvalho, A. J. F., Curvelo, A. A. S., Agnelli, J. A. M. A. (2001). First insight on composites of thermoplastic starch and kaolin, *Carbohyd. Polym.*, **45**, pp. 189–194.

19. Pandey, J. K., and Singh, R. P. (2005). Green nanocomposites from renewable resources: Effect of plasticizer on the structure and material properties of clay-filled starch, *Starch/Stärke*, **57**, pp. 8–15.

20. Haider, S., Khan, Y., Almasry, W. A., and Haider, A. *Thermoplastic Nanocomposites and Their Processing Techniques*, In Tech, ISBN 978-953-51-0310-3, 146 pages.

21. Sinha, Ray, S., and Okamoto, M. (2003). Biodegradable polylactide and its nanocomposites: Opening a new dimension for plastics and composites, *Macromol. Rapid Commun.*, **24**, 815–840.

22. Feijoo, J. L., Cabedo, L., Giménez, E., Lagaron, J. M., and Saura, J. J. (2005). Development of amorphous PLA-montmorillonite nanocomposites, *J. Mater. Sci.*, **40**, 1785–1788.

23. Singh, R. P., Pandey, J. K., Rutot, D., Degée, Ph., and Dubois, Ph. (2003). Biodegradation of poly(#-caprolactone)/starch blends and composites in composting and culture environments: The effect of compatibilization on the inherent biodegradability of the host polymer, *Carbohydr. Res.*, **338**, 1759–1769.

24. Di, Y., Iannace, S., Di, Maio, E., and Nicolais, L. (2005). Poly(lactic acid)/organoclay nanocomposites: Thermal, rheological properties and foam processing, *J. Polym. Sci. Part B: Polym. Phys.*, **43**, 689–698.

25. Pollet, E., Paul, M.-A., and Dubois, P. (2003). New aliphatic polyester layered-silicate nanocomposites. In *Biodegradable Polymers and Plastics* (Chiellini, E., Solaro, R., eds.), vol. 1., Kluwer, New York.

26. Utracki, L. A. (2004). Basic elements of polymeric nanocomposites technology. In *Clay-containing Polymeric Nanocomposites* (Utracki, L. A., ed.), vol. 1. Rapra, Shropshire.

27. Pluta, M., Galeski, A., Alexandre, M., Paul, M.-A., and Dubois, P. (2002). Polylactide/montmorillonite nanocomposites and microcomposites prepared by melt blending: Structure and some physical properties, *J. Appl. Polym. Sci.*, **86**, pp. 1497–1506.

28. Lee, S. Kang, I., Doh, G., Yoon, H., Park, B., and Wu, Q. (2008). Thermal and mechanical properties of wood flour/talc-filled polylactic acid composites: effect of filler content and coupling treatment, *J. Thermoplast. Compos. Mater.*, **21**, pp. 209–223.

29. Qu, P., Gao, Y., Wu, G., and Zhang, L. (2010). Nanocomposites of poly (lactic acid) reinforced with cellulose nanofibrils, *Bio. Resources,* **5**, pp. 1811–1823.

30. Maiti, P., Batt, C. A., Giannelis, E. P. (2003). Renewable plastics: Synthesis and properties of PHB nanocomposites, *Polym. Mater. Sci. Eng.*, **88**, pp. 58–59.

31. Zheng, J. P., Li, P., Ma, Y. L., and Yao, K. D. (2002). Gelatine/montmorillonite hybrid nanocomposite. I. Preparation and properties, *J. Appl. Polym. Sci.*, **86**, pp. 1189–1194.

32. Takegawa, A., Murakami, M., Kaneko, Y., and Kadokawa, J. (2010). Preparation of chitin/cellulose composite gels and films with ionic liquids, *Carbohyd. Polym.*, **79**, 85–90.

33. Nunes, M. R. S., Silva, R. C., Silva, J. G., Tonholo, J., and Ribeiro, A. S. (2009). Preparation and morphological characterization of chitosan/clay nanocomposites. In *Proceedings of the 11th International Conference on Advanced Materials*, Rio de jenero, Brazil, 20–25 September 2009; pp. 20–25.

34. Svagan, A. (2008). Bio-inspired cellulose Nanocomposites and foams based on starch matrix. PhD thesis, Department of Fiber and Polymer Technology, KTH Chemical Science and Engineering, SE-100 44, Stockholm, Sweden.

35. Lee, J. H., Park, T. G., Park, H. S., Lee, D. S., Lee, Y. K., Yoon, S. C., and Nam, J. D. (2002). Thermal and mechanical characteristics of poly(L-lactic acid) nanocomposite scaffold, *Biomaterials*, **24**, pp. 2773–2778.

36. Pothan, L. A., Thomas, S. (2003). Polarity parameters and dynamic mechanical behavior of chemically modified banana fiber reinforced polyester composites, *Compos. Sci. Technol.*, **63**, pp. 1231–1240.

37. Zemljic, L. F., Stenius, P., Stana-kleinschek, J., and Ribitsch, V. (2006). Characterization of cotton fibers modified by carboxymethyl cellulose, *Lenzinger Berichte*, **85**, pp. 68–76.

38. Reddy, N., and Yang, Y. (2009). Properties and potential application of natural cellulose fibers from the bark of cotton stalks, *Bioresource Technol.*, **100**, pp. 3563–3569.

39. Wambua, P., Ivens, J., and Verpoest, I. (2003). Natural fibers: Can they replace glass in fiber reinforced plastics, *Compos. Sci. Technol.*, **63**, pp. 1259–1264.

40. Reddy, N., and Yang, Y. (2008). Characterizing natural cellulose fibers from velvet leaf (*Abutilon theophrasti*) stems, *Bioresource Technol.*, **99**, pp. 2449–2454.

41. Reddy, N., and Yang, Y. (2007). Natural Cellulose fibers from switchgrass with tensile properties similar to cotton and linen, *Biotechnol. Bioeng.*, **97**, pp. 1021–1027.

42. Bodros, E., and Baley, C. (2008). Study of the tensile properties of stinging nettle fibers (*Urtica dioica*), *Mater. Lett.*, **62**, pp. 2143–2145.

43. Batra, S. K. (1998). Other long vegetable fibers. In *Handbook of Fiber Science and Technology* (Lewi, N. M., Pearce, E. M., eds.), Marcel Dekker Fiber Chemistry: New York, NY, USA, vol. 4, p. 727.

44. Goda, K., Sreekala, M. S., Gomes, A., Kaji, T., and Ohgi, J. (2006). Improvement of plant based natural fibers for toughening green composites—Effect of load application during mercerization of ramie fibers, *Compos. Part A Appl. Sci. Manuf.*, **37**, pp. 2213–2220.

45. Reddy, N., and Yang, Y. (2009). Natural cellulose fibers from soybean straw, *Bioresource Biotechnol.*, **100**, pp. 3593–3598.

46. Baley, C. (2002). Analysis of the flax fiber tensile behavior and analysis of the tensile stiffness increase, *Compos. Part A Appl. Sci. Manuf.*, **33**, pp. 939–948.

47. Sain, M., and Panthapulakkal, S. (2006). Bioprocess preparation of wheat straw fibers and their characterization, *Ind. Crops Prod.*, **23**, pp. 1–8.

48. Panthapulakka, S., Zereshkian, A., and Sain, M. (2006). Preparation and characterization of wheat straw for reinforcing application in injection molded thermoplastic composites, *Bioresource Biotechnol.*, **97**, pp. 265–272.

49. Reddy, N., and Yang, Y. (2009). Properties of natural cellulose fibers from hop stems, *Carbohyd. Polym.*, **77**, pp. 898–902.

50. Reddy, N., and Yang, Y. (2005). Structure and properties of high quality natural cellulose fibers from corn stalks, *Polymer*, **46**, pp. 5494–5500.

Chapter 19

Concluding Notes

Chaudhery Mustansar Hussain and Ajay Kumar Mishra

[a]*Department of Chemistry and Environmental Science, NJIT, University Heights, Newark, New Jersey 07102, USA*
[b]*Nanotechnology and Water sustainability Research Unit, College of Science, Engineering and Technology, University of South Africa, Florida Campus, Johannesburg, South Africa*

chaudhery.m.hussain@njit.edu

Nanocomposites for pollution control techniques and devices are getting impetus globally. The exceptional properties of nanocomposites and their merging with current pollution devices show abundant prospects to modernize pollution control discipline. Even though many nanocomposites emphasized in this book are still in the laboratory research phase, some have made their way to pilot testing or even commercialization. Magnetic-based nanocomposites and multifunctional nanocomposites show most promise in full-scale application in the near future based on their stages in research and development, commercial availability and cost involved, and compatibility with the existing devices and proved excellent synergistic materials when compared to conventional ones. However, these nanocomposites have few

Nanocomposites for Pollution Control
Edited by Chaudhery Mustansar Hussain and Ajay Kumar Mishra

ISBN 978-981-4774-45-1 (Hardcover), 978-1-315-14368-2 (eBook)
www.panstanford.com

commercial products available, but still they have not been applied in large-scale pollution control techniques. Additionally, ZnO-based nanocomposites have demonstrated potential application in photocatalytic conversion of organic pollutants. ZnO is believed to be a better starting UV material to form nanocomposites with semiconductor photocatalysts. One of the virtues of ZnO incorporation into the nanocomposites is served to promote the separation of charge carriers and transport of photogenerated electrons and holes. The first part of the book comprises those chapters which have gathered all those innovations and dimensions provided by nanocomposites to the pollution control discipline.

Carbon-based nanomaterials are one of the most interesting materials with unusual properties and are expected to make a large impact on the environment in future. A number of synthesis route, treating agents and experimental conditions were used for the synthesis of CNT-based nanomaterials. These materials have been attracted significant attention due to their unique electronic, optical, thermal, mechanical, and chemical properties. Carbon nanomaterial-based nanocomposites for pollution control have found successful use in the detection of a wide range of environmental pollutants. Their utilization is very important for capturing heavy metal ions and organic contaminants. Application of carbon nanomaterial-based nanocomposites in pollution control is highly interdisciplinary, which dictates combined efforts of the materials science, chemistry, engineering, and environmental science communities. Their inclusion is very attractive for environmental pollutant control. The second part of the book attempts to address the special kind of nanocomposites in a sustainable environment and green technologies perspectives.

Despite of the recent advances in nanocomposites technology, the applications of nanocomposite membranes for environmental pollutants removal are still in infancy. Currently many encouraging laboratory-based studies have been performed, but the large-scale production of nanocomposite membranes and their industrial scale implementation have been rarely reported. The fabrication of dual-layer hollow fibre membranes with different TiO_2 loadings in the outer layer by a single step co-extrusion technique is one of those advancements. The addition of TiO_2 nanoparticles into the membrane outer layers has improved the membrane properties by increasing their hydrophilicity, pore size and permeability and greatly promotes the photocatalytic degradation of pollutant. The use of dual-layer hollow fiber (DLHF) membranes enables an efficient photocatalytic degradation of NP without additional separation steps for the catalyst particles. In terms of commercialization, as the exploding benefits and potential of nanocomposite membranes can be retrofitted into the existing technologies without significant changes, it is believed that this state-of-the-art innovation will take relatively short time from laboratory exploration to achieving operational maturity. It should be noted that the complexity and economic measures related to the technological scale up should be taken into consideration from as early as in the research process until the product maturity period. The current approaches adopted for the synthesis and fabrication of nanocomposite membranes or their devices may not be price competitive in comparison with existing membranes, but looking at the broader sense, keeping the overall membrane separation and treatment investment and processing costs low enough through the improvement in energy and material efficiency could also prove to be a commercial success. The third part of the book is dedicated to these challenges and issues in recent scientific and technological advances in nanocomposite-enabled membranes which can undoubtedly provide more efficient, cost-effective and sustainable solutions for pollutant removal.

Recently, there has been increasing interest in the development of bio-inspired or biosynthesized composite materials for pollution control. The unique properties of bio-inspired nanoscale materials have given rise to immense research activity directed towards the applications of bio-inspired nanocomposites for pollution control. The properties of bio-inspired materials may differ from materials prepared by other traditional methods. The potential range of compositions, shapes, and sizes of biosynthesized materials translates into a broad domain of new and existing nanocomposite applications. Although the use of bio-inspired nanocomposites has extended to almost all fields, Chapter 16 considers biosynthesized inorganic semiconductors being utilized with other biological and polymer components for driving pollution control process. This chapter (Chapter 16) also provides important updates on advances in the fabrication of novel bio-inspired functional nanocomposites for pollution control application.

At present, mankind is at the threshold of large-scale technological application of green nanocomposites, which are considered as the only alternative to synthetic, petroleum-based polymers. Green nanocomposites are abundant, renewable resources that can provide a basis for sustainable and economic development. Furthermore, green nanocomposites are bio-compatible, biodegradable, and environmentally friendly. Their use has a wide range of environmental benefits, including reduced reliance on fossil carbon, decreased greenhouse gas emissions, and biodegradation to harmless products by the action of different micro-organisms. Green nanocomposites are highly versatile, having a broad range of physical properties suitable for miscellaneous applications. A wide range of available green polymers and inorganic nanoparticles offers a means of developing a diversity of green nanocomposites with different structure, properties, functionalities, and applications. This is achieved by mixing them in various combinations. Green polymers form the matrix, determining the shape, structural organization, and main functionalities of green nanocomposites. Dispersed nanoparticles always modify the matrix, which allow tuning of the structure, properties, and, as a result, functionality. Part 5 of the book explored the green and sustainable future with nanocomposites compared with traditional materials.

Overall, an extensive use of nanocomposites can significantly increase their emissions into the environment, through air, groundwater and soil. Nanocomposites can enter the environment in the course of their lifecycle and can be toxic, which increases their risk. A crucial factor for the determination of a risk of exposure to nanocomposite is their stability and modifications forms once entered into the environment. Continued innovation and development of industrial nanoscience is dependent on maintaining public confidence. It is therefore critical that the regulation of the industry allows it to develop in a safe, sustainable and transparent manner. The study of the fate and transport of nanocomposite is largely concerned with determining how their properties and behaviour change over time, particularly after release into the environment. Currently, little is known about how nanocomposites kinetically behave in different environments, including the time-scale related with whether they remain relatively stable or change in ways that alter their anticipated impacts which will determine future challenges and prospects of nanocomposites pollution control application.

In conclusion, we could say that nanocomposites are making great strides in innovation for pollution control techniques in the lab and at the industrial level. Therefore, it is anticipated that within no time nanocomposites will be critical components for pollution control tools/devices not only at the lab scale but also in the industry leading to a safer and more protected world.

References

1. Hannah, W., and P. B. Thompson. 2008. Nanotechnology, risk and the environment: A review. *J. Environ. Monit.*, 10(3): 291–300.
2. Liang, M., and Guo L. H. 2009. Application of nanocomposite in pollution control and monitoring. *J. Nanosci. Nanotechnol.,* 9(4): 2283–2289.
3. Leitch, M. E., E. Casman, and G. V. Lowry. 2012 Nanotechnology patenting trends through an environmental lens: Analysis of materials and applications. *J. Nanoparticle Res. C7*, 1283, 14(12): 1–23.
4. Kharisov, B. I., O. V. Kharissova, and H. V. R. Dias. 2015. *Nanocomposite for Environmental Protection*: Wiley.
5. Hussain, C. M., and Kharisov, B., *Advanced Environmental Analysis: Applications of Nanomaterials*, Royal Society of Chemistry 2017.

Index

For Product Safety Concerns and Information please contact our EU
representative GPSR@taylorandfrancis.com
Taylor & Francis Verlag GmbH, Kaufingerstraße 24, 80331 München, Germany

www.ingramcontent.com/pod-product-compliance
Ingram Content Group UK Ltd.
Pitfield, Milton Keynes, MK11 3LW, UK
UKHW022200190926
475702UK00003B/128
* 9 7 8 9 8 1 4 7 7 4 4 5 1 *